Minoru Asada Hiroaki Kitano (Eds.)

RoboCup-98: Robot Soccer World Cup II

Springer

Series Editors

Jaime G. Carbonell, Carnegie Mellon University, Pittsburgh, PA, USA
Jörg Siekmann, University of Saarland, Saarbrücken, Germany

Volume Editors

Minoru Asada
Osaka University, Graduate School of Engineering
Department of Adaptive Machine Systems
Suita, Osaka 565-0871, Japan
E-mail: asada@ams.eng.osaka-u.ac.jp

Hiroaki Kitano
Sony Computer Science Laboratories, Inc.
3-14-13 Higasha-Gotanda, Shinagawa
Tokyo, 141-0022, Japan
E-mail: kitano@csl.sony.co.jp
and
ERATO Kitano Symbiotic Systems Project
Japan Science and Technology Corporation
Suite 6A, M31, 6-31-15 Jinguu-mae, Shibuya
Tokyo, 150-0001, Japan

Cataloging-in-Publication data applied for

Die Deutsche Bibliothek - CIP-Einheitsaufnahme

RoboCup <2, 1998, Paris>:
Robot Soccer World Cup II / RoboCup-98. Minoru Asada ; Hiroaki Kitano (ed.). - Berlin ; Heidelberg ; New York ; Barcelona ; Hong Kong ; London ; Milan ; Paris ; Singapore ; Tokyo : Springer, 1999
(Lecture notes in computer science ; Vol. 1604 : Lecture notes in artificial intelligence)
ISBN 3-540-66320-7

CR Subject Classification (1998): I.2, C.2.4, D.2.7, H.5, I.5.4, I.6, J.4

ISBN 3-540-66320-7 Springer-Verlag Berlin Heidelberg New York

Typesetting: Camera-ready by author
SPIN 10704850 06/3142 – 5 4 3 2 1 0 Printed on acid-free paper

Lecture Notes in Artificial Intelligence 1604

Subseries of Lecture Notes in Computer Science
Edited by J. G. Carbonell and J. Siekmann

Lecture Notes in Computer Science

Edited by G. Goos, J. Hartmanis and J. van Leeuwen

Springer
Berlin
Heidelberg
New York
Barcelona
Hong Kong
London
Milan
Paris
Singapore
Tokyo

Preface

RoboCup-98, the Second Robot World Cup Soccer Games and Conferences, was held on July 2-9, 1998 at La Cité des Sciences et de l'Industrie (La Cité) in Paris, the same city where the real world cup finals were being held at almost the same time. This book is the second official archival publication of the long range international research initiative, mainly focusing on the papers discussed at the Second International Workshop on RoboCup on July 2,3, and 9, 1998.

This book consists of three parts: overview and award papers, technical papers on focused topics, and team descriptions. The overview paper provides overall perspectives including the differences from the first RoboCup-97 and challenging issues. The award papers consist of scientific award ones and champion team description ones in all the competition leagues. In RoboCup-98, the Scientific Challenge Award was given to three research groups for their simultaneous development of fully automatic commentator systems for the RoboCup simulator league. Technical papers focused on topics related to infrastructure, basic skills of players, teamwork and evaluation, learning and evolution, and real robot perception, mechanism, and integration. The team description papers address the technical and strategic aspects of participating teams. They consist of four parts: three of them are official RoboCup leagues and the final one is Sony legged robot league which was held as an exhibition game and demonstration. RoboCup-99 will have it as an official league with more than three teams from RoboCup-98. Authors were requested to include an analysis of their team on the results of RoboCup-97. Also, for editing reasons, some papers on their team description are included in the corresponding technical papers. Therefore, team description parts do not involve all participating team descriptions. We hope this volume contributes to the progress of the field, and to the accomplishment of our dream some day.

March, 1999

Minoru Asada and Hiroaki Kitano

RoboCup-98 Organization

Organized by

Université Pierre et Marie Curie (Paris-6)

Centre National de la Recherche Scientifique (CNRS)

Supported by

La Cité des Sciences et de l'Industrie (La Cité)

Centre de Robotique Integrée d'Ile de France (CRIIF)

European Community (EC)

Nihon Keizai Shimbun, Inc.

RoboCup World Wide Sponsors:

Sony Corporation

NAMCO Limited

SUNX Limited

Official Supplier:

Molten Corporation (Balls for the middle-size league)

Table of Contents

Overview and Award Papers

Overview Paper

Scientific Challenge Award Papers

Champion Teams

Technical Papers

Infrastructure

Basic Skill

Teamwork and Evaluation

Learning and Evolution

Real Robot Perception, Mechanism, and Integration

Team Description

Simulation

Real Robot Small-Size League

Real Robot Middle-Size League

Legged robot league

Overview of RoboCup-98

M. Asada[1], M. Veloso[2], M. Tambe[3], I. Noda[4], H. Kitano[5], and G. K. Kraetzschmar[6]

[1] Adaptive Machine Systems, Osaka University, Suita, Osaka 565-0871, Japan
[2] Computer Science Department, Carnegie Mellon University, Pittsburgh, PA 15213, USA
[3] Information Sciences Institute, USC, Marina del Rey, CA 90292, USA
[4] Electrotechnical Laboratory, Tsukuba 305-8568, Japan
[5] Computer Science Lab, Sony Corp., Tokyo 141-0022, Japan
[6] Neural Information Processing, University of Ulm, Oberer Eselsberg, 89069 Ulm, Germany

Abstract. RoboCup is an increasingly successful attempt to promote the full integration of AI and robotics research. Following the astonishing success of the first RoboCup-97 at Nagoya [1], the Second Robot World Cup Soccer Games and Conferences, RoboCup-98, was held at Paris during July 2nd and 9th, 1998 at the partly same place and period of the real world cup. There are three kinds of leagues: the simulation league, the real robot small-size league, and the real robot middle-size league. The champion teams are CMUnited-98 (CMU, USA) for both the simulation and the real robot small-size leagues, and CS-Freiburg (Freiburg, Germany) for the real robot middle-size league. The Scientific Challenge Award was given to three research groups (Electrotechnical Laboratory (ETL), Sony Computer Science Laboratories, Inc., and German Research Center for Artificial Intelligence GmbH (DFKI)) for their simultaneous development of fully automatic commentator systems for RoboCup simulator league. Over 15,000 spectators and 120 international media covered the competition worldwide. RoboCup-99, the third Robot World Cup Soccer Games and Conferences, will be held at Stockholm in conjunction with the Sixteenth International Joint Conference on Artificial Intelligence (IJCAI-99) at the beginning of August, 1999.

1 Introduction

RoboCup-98, the Second Robot World Cup Soccer Games and Conferences, was held on July 2-9, 1998 at La Cite des Sciences et de l'Industrie (La Cite) in Paris (See Figure 1)[7] . It was organized by University of Paris-VI and CNRS, and it was sponsored by Sony Corporation, NAMCO Limited, and SUNX Limited. The official balls for the middle-size league were supplied by Molten Corporation. Over 15,000 people watched the games and over 120 international media (such as CNN, ABC, NHK, TV-Aich, etc,..) as well as prominent scientific magazines covered them. The Second RoboCup Workshop was held [2].

[7] the same city where the real world cup finals were held almost the same period

RoboCup-98 had three kinds of leagues: (1)the simulation league, (2) the real robot small-size league, (Please see Figure 2 for the real robot small-size league competition site where a match between J-Star (Japan) and CMUnited-97 (USA) is taking place.) and (3) the real robot middle-size league. (Please see

Fig. 1. RoboCup-98 in Paris

Figure 3 for the real robot middle-size league competition site where a semi-final between Osaka Trackies (Japan) and CS-Frieburg (Germany) is taking place.)

Although it was not an official RoboCup competition, Sony Legged Robot Competition and Demonstration have attracted many spectators, especially boys and girls, for its cute style and behaviors. Figure 4 shows a scene from their demonstrations. Three teams from Osaka University, CMU, and University of Paris-VI have show their exhibition games. In 1999, Sony Legged Robot league will be one of the RoboCup official competitions with more teams around the world [3]. Also, University of Aahus has built an exciting soccer stadium using Lego Mind Storm with many figures of supporters that could wave and give great cheers for the play.

Aside from the world championship award, the RoboCup Scientific Challenge Award is created as an equally or more prestigious award. This year, the Scientific Challenge Award was given to three research groups (Electrotechnical Laboratory (ETL), Sony Computer Science Laboratories, Inc., and German Research Center for Artificial Intelligence GmbH (DFKI)) for their simultaneous development of fully automatic commentator systems for RoboCup simulator league. Detailed information is given at *http://www.robocup.org/*. In this article, we review the challenge issues of each league and analyze the results of RoboCup-98. We compare the architectural differences between the leagues, and overview which research issues have been solved and how, and which have been left unsolved and remain as future issues.

Fig. 2. Real robot small-size league competition site

2 Leagues and Approaches

RoboCup-98 had three kinds of leagues:

1. **Simulation league:** Each team consists of eleven programs, each controlling separately each of eleven team members. The simulation is run using the Soccer Server developed by Noda[8] . Each player has distributed sensing capabilities (vision and auditory) and motion energy both of which are resource bounded. Communication is available between players and strict rules of the soccer game are enforced (e.g. off-sides). This league is mainly for researchers who may not have the resources for building real robots, but are highly interested in complex multiagent reasoning and learning issues.

[8] for the details, please visit RoboCup web site

Fig. 3. Real robot middle-size league competition site

2. **Small-size real robot league:** The field is of the size and color of a ping-pong table and up to five robots per team play a match with an orange golf ball. The robot size is limited to approximately 15cm^3. Typically robots are built by the participating teams and move at speeds of up to 2m/s. Global vision is allowed, offering the challenge of real-time vision-based tracking of five fast moving robots in each team and the ball.
3. **Middle-size real robot league:** The field size is of the size and color of three by three ping-pong tables, and up to five robots per team play a match with a Futsal-4 ball. The size of the base of the robot is limited to approximately 50cm diameter. Global vision is not allowed. Goals are colored and the field is surrounded by walls to allow for possible distributed localization through robot sensing.

Each league has its own architectural constraints, and therefore research issues are slightly different from each other. We have published proposal papers [4, 5] about research issues in RoboCup initiative. For the synthetic agent in the simulation league, the following issues are considered:

- Teamwork among agents, from low-level skills like passing the ball to a teammate, to higher level skills involving execution of team strategies.
- Agent modeling, from primitive skills like recognizing agents' intents to pass the ball, to complex plan recognition of high-level team strategies.
- Multi-agent learning, for on-line and off-line learning of simple soccer skills for passing and intercepting, as well as more complex strategy learning.

For the robotic agents in the real robot leagues, for both the small and middle-size ones, the following issues are considered:

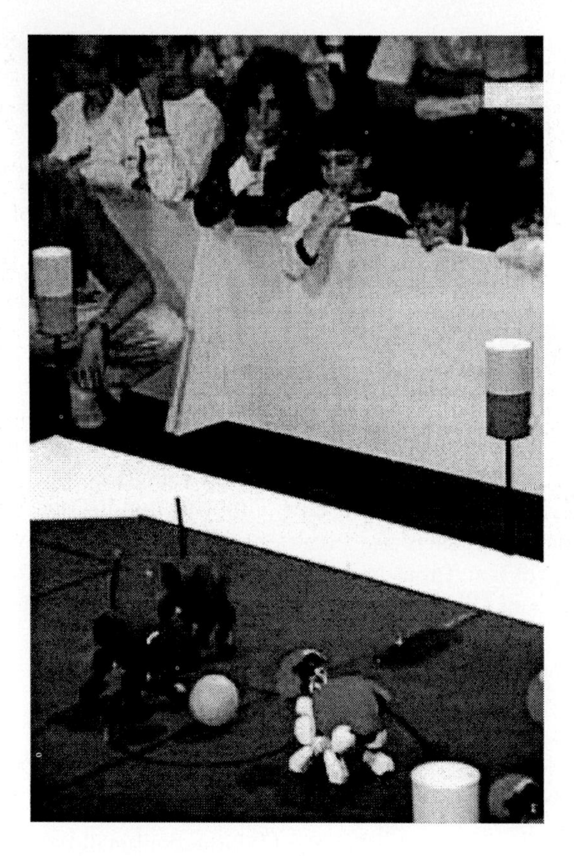

Fig. 4. Sony legged robot league competition site

- Efficient real-time global or distributed perception possibly from different sensing sources.
- Individual mechanical skills of the physical robots, in particular target aim and ball control.
- Strategic navigation and action to allow for robotic teamwork, by passing, receiving and intercepting the ball, and shooting at the goal.

More strategic issues are dealt in the simulation league and in the small-size real robot league while acquiring more primitive behaviors of each player is the main concern of the middle-size real robot league.

We held the first RoboCup competitions in August 1997, in Nagoya, in conjunction with IJCAI-97 [6]. There were 28, 4, and 5 participating teams in the simulation, small-size robot, and middle-size robot leagues, respectively.

The second RoboCup workshop and competitions took place in July 1998, in Paris [2] in conjunction with ICMAS-98 and AgentsWorld. The number of teams increased significantly from RoboCup-97 to 34, 11, and 16 participating teams in the simulation, small-size robot, and middle-size robot leagues respectively. More than twenty countries participated. Every team had its own features some of which have been exposed during their matches with different degrees of success.

3 RoboCup Architectural Approaches

There are two kinds of aspects in designing a robot team for RoboCup:

1. Physical structure of robots: actuators for mobility, kicking devices, perceptual (cameras, sonar, bumper sensor, laser range finder) and computational (CPUs, microprocessors) facilities.
2. Architectural structure of control software.

In the simulation league, both of the above issues are fixed, and therefore more strategic structure as a team has been considered. On the other hand, in the real robot leagues, individual teams have devised, built, and arranged their robots. Although the small league and the middle one have their own architectural constraints, there are variations of resource assignment and control structure of their robots. Table 1 shows the variations in architectural structure in terms of number of CPUs and cameras, and their arrangement.

Table 1. Variations in architectural structure

Type	CPU	Vision	issues	league
A	1	1 global	strategy	small-size
B	n	1 global	sharing of information	small-size
C	1	1 global + n local	sensor fusion; coordination	small-size
D	1+n	n local	multiple robots	middle-size
E	n	n local	sensor fusion; teamwork	middle-size & simulation

Communication between agents is possible in all of the leagues. The simulation league is the only that uses it except one team Uttori in the real robot middle-size league.

4 Simulation League

The simulation league continues to be the most popular part of the RoboCup leagues, with 34 teams participating in RoboCup-98, which is a slight increase over the number of participants at RoboCup-97. As with RoboCup-97, teams were divided into leagues. In the preliminary round, teams played within leagues in a round-robin fashion, and that was followed by a double-elimination round to determine the first three teams. Appendix A shows the results of all matches including preliminary and final ones.

Teams in the RoboCup simulation league are faced with three strategic research challenges: multi-agent learning, teamwork and agent modeling. All three are fundamental issues in multi-agent interactions. The learning challenge has been categorized into on-line and off-line learning both by individuals and by teams (i.e., collaborative learning). One example of off-line individual learning is learning to intercept the ball, while an example of on-line collaborative learning is to adaptively change player positions and formations based on experience in a game.

The RoboCup Teamwork Challenge addresses issues of real-time planning, re-planning, and execution of multi-agent teamwork in a dynamic adversarial environment. A team should generate a strategic plan, and execute it in a coordinated fashion, monitoring for contingencies and select appropriate remedial actions. The teamwork challenge interacts also with the third challenge in the RoboCup simulation league, that of agent modeling. Agent modeling refers to modeling and reasoning about other agent's goals, plans, knowledge, capabilities, or emotions. The RoboCup opponent modeling challenge calls for research on modeling a team of opponents in a dynamic, multi-agent domain. Such modeling can be done on-line to recognize a specific opponent's actions, as well as off-line for a review by an expert agent.

At least some researchers have taken these research challenges to heart, so that teams at RoboCup97 and RoboCup98 have addressed at least some of the above challenges. In particular, out of the three challenges outlined, researchers have attacked the challenge of on-line and off-line learning (at least by individual agents). Thus, in some teams, skills such as intercept, and passing are learned off-line. The two final teams, namely CMUnited simulation (USA) as winner of the first place and AT-Humboldt-98 (Germany) as runner-up, included an impressive combination of individual agent skills and strategic teamwork.

Research in teamwork has provided concepts such as exhibiting reusability of domain-independent teamwork skills (i.e., skills that can be transferred to domains beyond RoboCup), about roles and role reorganization in teamwork. RoboCup opponent modeling, in terms of tracking opponents' mental state, has however not received significant attention by researchers. There are however some novel commentator agents that have used statistical and geometric techniques to understand the spatial pattern of play.

5 Small-Size Real Robot League

The RoboCup-98 small-size real robot league provides a very interesting framework to investigate the full integration of action, perception, and high-level reasoning in a team of multiple agents. Therefore, three main aspects need to be addressed in the development of a small-size RoboCup team: (i) hardware of physical robots; (ii) efficient perception; and (iii) individual and team strategy.

Although all of the eleven RoboCup-98 teams included distinguishing features at some of these three levels, it showed crucial to have a *complete* team with robust hardware, perception, and strategy, in order to perform overall well.

This was certainly the case for the four top teams in the competition, namely CMUnited-98 (USA), Roboroos (Australia), 5DPO (Portugal), and Cambridge (UK), who classified in first, second, third, and fourth place respectively.

Fig. 5. Real robot small-size final match

Figure 5 shows a scene from the final match between CMUnited-98 and Queensland Roboroos (Ausuralia). Appendix B shows the results of all matches including preliminary and final ones. We overview now the characteristics of the RoboCup-98 teams and the research issues addressed.

Hardware: All of the eleven RoboCup-98 participating teams consisted of robots built by each participating group. The actual construction of robots within the strict size limitations offered a real challenge, but gave rise to a series of interesting physical and mechanical devices. Remarkably, the robots exhibited sensor-activated kicking devices (iXs and J-Star, Japan, Paris-6, France, and CMUnited-98, USA), sophisticated ball holding and shooting tools for the goalie robot (Cambridge, UK), and impressive compact and robust designs (Roboroos, Australia, and UVB, Belgium). All of the robots were autonomously controlled through radio communication by off-board computers.

Perception: Ten out of the eleven teams used a single camera overlooking the complete field. The ISpace (Japan) team included one robot with an onboard vision camera.

Global perception simplifies the sharing of information among multiple agents. However global perception presents at the same time a real challenging research opportunity for reliable and real-time detection of the multiple mobile objects – the ball, and five robots on each team. In fact, both detection of robot position and orientation and robot tracking need to be very effective. The frame rate of the vision processing algorithms clearly impacted the performance of the team. Frame rates reached 30 frames/sec as in the CMUnited-98 team.

In addition to the team color (blue or yellow), most of the teams used a second color to mark their own robots and provide orientation information, hence only about their own robots. Robot identification was achieved in general by greedy data association between frames. The 5DPO (Portugal) and the Paris-6 (France) teams had a robust vision processing algorithm that used patterns to discriminate among the robots and to find their orientation.

The environment in the small-size league is highly dynamic with robots and the ball moving at speeds between 1m/s and 2m/s. An interesting research issue consists of the prediction of the motion of the mobile objects to combine it with strategy. It was not clear which teams actually developed prediction algorithms. In the particular case of the CMUnited-98 team, prediction of the movement of the ball was successfully achieved and highly used for motion (e.g., ball interception) and strategic decisions (e.g., goaltender behavior and pass/shoot decisions).

Motion: In this RoboCup league, a foul should be called when robots push each other. This rule offers another interesting research problem, namely obstacle avoidance and path planning in a highly dynamic environment. The majority of the teams in RoboCup-98 successfully developed algorithms for such difficult obstacle avoidance and the semi final and final games showed smooth games that demonstrated impressive obstacle avoidance algorithms.

Strategy: Following up on several of the research solutions devised for RoboCup-97 both in simulation and in the small-size robots, at RoboCup-98, all of the small-size teams showed a role-based team structure. As expected, the goaltender played a very important role in each team. Similarly to the goaltender of CMUnited-97, the goaltender of most of the teams stayed parallel to the goal line and tried to stay aligned with or intercept the ball. The goaltender represented a very important and crucial role. To remark were the goaltenders of Roboroos, CMUnited-98, and Cambridge.

Apart for CMUnited-98 which had a single defender and three attackers, most of the other teams invested more heavily on defense, assigning two robots as defenders. In particular, defenders in the Belgium and in the Paris-8 teams occupied key positions in front of the goal making it difficult for other teams to path plan around them and to try to devise shots through the reduced open goal areas. Defending with polygonally-shaped robots proved to be hard, as the ball is not easily controlled at a fine grain. In fact a few goals for different teams were scored into their own goals due to small movements of the defenders or goaltender very close to the goal. It is clearly still an open research question how to control the ball more accurately.

Finally, it is interesting to note that one of the main features of the winning CMUnited-98 team is its ability to collaborate as a team. Attacking robots continuously evaluate (30 times per second) their actions, namely either to pass the ball to another attacking teammate or to shoot directly at the goal. A decision-theoretic algorithm is used to assign the heuristic and probabilistic based values to the different possible actions. The action with the maximum value is selected. Furthermore, in the CMUnited-98 team, a robot who was not the one actively pursuing the ball is not merely passive. Instead each attacking robot *anticipates* the needs of the team and it positions itself in the location that maximizes the probability of a successful pass. CMUnited-98 uses a multiple-objective optimization algorithm with constraints to determine this strategic positioning. The objective functions maximize repulsion from other robots and minimize attraction the ball and to the attacking goal.

6 Middle-Size Real Robot League

RoboCup-98 League Statistics and Results: The middle-size league this year had 18 entries, but the Deakin Black Knights (Deakin Univ., Andrew Price, Australia) had a fatal machine trouble, and the Iranian team could not attend the official games because of their late arrival due to the visa problem [9] . 16 teams were divided into four groups each of which consisted of four teams considering regional distribution, and preliminary games were took place in each group. Then, the best two teams from each group advanced to the final tournaments. Figure 6 shows a quarter final match between Osaka Trackies and NAIST.

Excitement both among participants and spectators reached new heights in the semi-finals, both of which were matches of Japanese against German teams (Appendix C shows the results of all matches including preliminary and final ones.). In the first semi-final, University of Freiburg won 3:0 against Osaka University. The second semi-final between Uttori United and University of Tübingen ended with a draw after regular time. Penalty shootouts did not produce a decision either, and a so-called technical challenge had to decide. In the technical challenge, a goal to shoot at is selected and the ball is placed in the middle of the field. A single robot is positioned on the field in the middle between the goal and the ball, heading towards the goal. The task is to find the ball, move it towards the goal, and finally shoot it into the goal. The time a robot takes to complete the task is taken as decision criterion. Tübingen won the technical challenge and proceeded to the finals. The finals itself were convincingly won 3:0 by University of Freiburg. This game also saw the nicest goal shot in the whole tournament, when the Freiburg robot took the ball from its "left hand" onto its "right hand" and scored.

Team Development Report: A very encouraging result from Paris is that all except two scheduled games could actually be played. Considering the large

[9] however, they played several exhibition games.

2. Many teams seem to have now available vision systems that work reasonably well, at least much better than what we saw in Nagoya. However, there are still many problems with the perceptual capabilities of the robots, especially detecting other agents, and vision will remain a central research topic in RoboCup.
3. A number of teams featured kicking mechanisms on their robots. A simple, yet powerful approach were pneumatic kickers. Other robots used a solenoid-based activation device. The kicking devices produced much higher ball accelerations than the robots could achieve by simply pushing the ball. One robot even scored a goal directly after kickoff. Overall, with kicking devices robots could move the ball significantly better, which is one of the research issues in the mid-size robot league.
4. Several teams attached passive devices such as shaped metal sheets or springs (nicknamed "fingers" or "hands") to their robots, thereby creating a concave surface for improved ball handling (moving, receiving, passing). With hands, robots could better move and turn with the ball, and often could retrieve the ball once it was stuck against the walls and bring the ball back into play although the use of such "hands" is still under discussion.

Despite of the architectural structure shown in Table 1, many teams used some kinds of radio communications to control their robots. However, frequency conflicts, noise produced by mobile phones and equipment used by film teams and the press often caused serious problems to communication. The less dependency on physical communication line is expected in future.

Research Results: One observation from the games in Paris is that creating a good goalie can dramatically improve overall team performance, but is somewhat simpler to build than a good field player. Several teams used omnidirectional vision systems that allowed their robots both to track their position in front of the goal as well as ball position [9, 8] since Osaka used it in the first RoboCup in Nagoya. USC's Ullanta used a fast B14 base as goalie, together with a rotating "hand" and a Cognachrome vision system; it did not allow a single goal. Probably the most successful goalie was the one by University of Tübingen, which did not allow a single goal, not even in penalty shootouts, until the final game and was the main reason why Tübingen made it to the finals.

Two Japanese teams, Uttori United[11] and Osaka University, demonstrated excellent ball handling capabilities. The Uttori robots feature a sophisticated omnidirectional drive system that allowed their robots to closely circle around the ball once they found it without visually loosing track of the ball (which happened often to other teams) until the robot's kicking paddle is heading towards the ball and the goal. Then the robot starts to move slowly towards the goal. The kicking device is designed such that the robot can push the ball across the floor without the ball starting to roll, thereby reducing the risk to loose the ball. Near the goal, the kicking paddle gives the ball a sufficiently strong kick to roll it away about half a meter. The robot then turns in order to head two fans towards the ball, activates the fans and blows the ball into the goal.

The new robots by Osaka University also exhibited very strong ball handling.

Fig. 6. A match from real robot middle-size

number of new teams, which were built within the nine months since Nagoya, this is a considerable achievement for most groups. Teams can use their technological base to investigate open problems, engineer new solutions, and conduct interesting experiments (see [7, 8, 9, 10]).

Technological State-of-the-Art Report: All participants agreed that the overall level of play improved dramatically since Nagoya. What are the major technological innovations that contributed to this improvement?

1. Many of the new teams used off-the-shelf platforms, such as Activmedia's Pioneer-1 and Pioneer-AT robots (used by six teams) or Nomadics' Scout robot (used by one team). These platforms are not perfect, therefore many teams substantially modified the robot and added additional equipment like vision systems, kicking devices, communication devices, and embedded PCs for onboard computation.

Once it found the ball it could move across the field in fast pace, guiding the ball closely to the base, all the way into the opponents goal. The main advantage over Uttori's approach is the higher speed they could achieve.

The winning strategy applied by Freiburg[12] was a combination of issues. The distinguishing feature of their robots was the use of a laser range finder, which provides fast and accurate range data, on each of their five Pioneer-1 robots. Freiburg applied their extensive work in laser-based self-localization to outperform teams using just vision systems. By matching the laser scan data against the known walls surrounding the field, they could not only determine their own position and orientation on the field, but also the position of the other robots. Via radio LAN the robots exchanged messages with a central server, which integrated all individual world models. By asking each of their robots about its own position, they could distinguish between teammates and opponents. The server in turn sent out a global, integrated world model to the robots, which was used to determine actions and to plan paths. The world model was precise enough to allow robots to choose and aim at the corner of the goal into which they would kick, or to give a pass to a teammate. However, team play would beverly suffer or be impossible in this case. As a result, their approach seems to be much more based on global positioning by LRF and centralized control (Type D in Table 1) although each player has its own CPU to detect a ball and to control its body than type E in Table 1 which is a typical architecture in the middle size league.

7 Future Issues

Simulation League

The major progress from RoboCup-97 to RoboCup-98 has been shown in the aspect of more dynamic and systematic teamworks. Especially, introduction of offside rule and improvement of individual plays force frexible team plays. However, the stage in RoboCup-98 is still in the preliminal level. For example, tactics to escape from off-side traps was still passive even in champion teams. In future RoboCup, such tactics will require recognition of intention of opponent players/teams. In this stage, opponent modeling and management of team strategies would become more important. Similarly, on-line learning will become more important, because team strategies should be changed during a match according to strategies of opponent teams.

While the research displayed in the RoboCup simulation league is encouraging, it is fair to remark that it has been difficult for researchers to extract general lessons learned and to communicate such lessons to a wider audience in multi-agents or AI. To facilitate such generalization, a new domain, *RoboCup rescue* is being designed. In RoboCup rescue, the focus will be on rescuing people stranded in a disaster area (where the disaster may be earthquake, fire, floods, or some combination of these events). This domain will not only emphasize the research issues of teamwork, agent modeling and learning, but in addition, raise novel issues in conflict resolution and negotiation. This domain will enable researchers to test the generality of their ideas and test their effectiveness in two separate domains.

Real Robot Small-Size League

The small-size RoboCup league provides a very rich framework for the development of multiagent real robotic systems. We look forward to understanding better several issues, including the limitations imposed by the size restrictions on on-board capabilities; the robustness of global perception and radio communication; and strategic teamwork. One of the main interesting open questions is the development of algorithms for on-line learning of the strategy of the opponent team and for the real-time adaptation of one's strategy in response. Finally, similarly to the simulation and middle-size leagues, we want to abstract from our experience algorithms that will be applicable beyond the robotic soccer domain.

Real Robot Middle-Size League

Despite the encouraging development of the middle-size league, we have to carefully review our current testbed and slowly adapt it to foster research in new directions and new areas. In most cases, this will require a slow evolution of rules.

The focus on colors to visibly distinguish objects exerts a strong bias for research in *color-based* vision methods. It is desirable to permit other approaches as well, such as using *edges*, *texture*, *shape*, *optical flow* etc., thereby widening the range of applicable vision research within RoboCup.

Another issue is the study of a better obstacle avoidance approaches. Currently, most robots except NAIST [10] and a few cannot reliably detect collisions with walls or other robots. Solving the charging problem using a rich set of onboard sensors is another major field of future research for RoboCup teams.

Finally, the use of communication in the different leagues is also an active research topic. Communication allows interesting research[11] in a variety of topics, including multi-robot sensor fusion and control. We want to explore limited communication environments and its relationship to agent autonomy, and learning of cooperative behavior.

8 Conclusion

As a grand challenge, RoboCup is definitely stimulating a wide variety of approaches, and has produced rapid advances in key technologies. With a growing number of participants RoboCup is set to continue this rapid expansion. With its three leagues, RoboCup researchers face an unique opportunity to learn and share solutions in three different agent architectural platforms.

RoboCup-99, the third Robot World Cup Soccer Games and Conferences, will be held at Stockholm in conjunction with the Sixteenth International Joint Conference on Artificial Intelligence (IJCAI-99) at the beginning of August,

1999. In addition to the existing leagues, a new league, Sony legged robot league will be introduced as an official RoboCup competitions with more teams than 1998 exhibition games and demonstrations.

Appendix A: Final resuluts for the simulation league

Preliminary Results

Group A

A1 Andhill,Japan A2 UBU, Sweden
A3 Windmillwanderers, Netherlands A4 Delphine, Germany
A5 UFSC, Brazil

VS	A1	A2	A3	A4	A5	W/L	Pnt	Rank
A1	-	22-0	0-1	21-0	11-0	3/1	9	2
A2	0-22	-	0-16	0-5	12-0	1/3	3	4
A3	1-0	16-0	-	9-0	11-0	4/0	12	1
A4	0-21	5-0	0-9	-	6-0	2/2	6	3
A5	0-11	0-12	0-11	0-6	-	0/4	0	5

Group B

B AT-Humboldt'97, Germany B2 Louvains, Belgium
B3 pippo, France B4 Hicpobs, Iran

VS	B1	B2	B3	B4	W/L	Pnt	Rank
B1	-	8-0	5-0	8-0	3/0	9	1
B2	0-8	-	1-0	0-3	1/2	3	3
B3	0-5	0-1	-	0-1	0/3	0	4
B4	0-8	3-0	1-0	-	2/1	6	2

Group C

C1 AT-Humboldt'98, Germany C2 TU-Cluj, Romania
C3 AIACS, Netherlands C4 ERIKA, Japan
C5 Darbotics, USA

VS	C1	C2	C3	C4	C5	W/L	Pnt	Rank
C1	-	23-0	6-0	17-0	13-0	4/0	12	1
C2	0-23	-	0-7	7-0	1-0	2/2	6	3
C3	0-6	7-0	-	17-0	16-0	3/1	9	2
C4	0-17	0-7	0-17	-	0-2	0/4	0	5
C5	0-13	0-1	0-16	2-0	-	1/3	3	4

Group D

D1 Gemini, Japan D2 Cosmoz, Germany
D3 Footux, France D4 Texas, USA

VS	D1	D2	D3	D4	W/L	Pnt	Rank
D1	-	6-0	21-0	11-0	3/0	9	1
D2	0-6	-	16-0	12-0	2/1	6	2
D3	0-21	0-16	-	2-0	1/2	3	3
D4	0-11	0-12	0-2	-	0/3	0	4

Group E

E1 PasoTeam, Italy E2 Ulm-Sparrow, Germany
E3 Miya2, Japan E4 DarwinUnited, USA

VS	E1	E2	E3	E4	W/L	Pnt	Rank
E1	-	6-2	0-1	5-0	2/1	6	2
E2	2-6	-	0-3	0-1	0/3	0	4
E3	1-0	3-0	-	0-0	2/0	7	1
E4	0-5	1-0	0-0	-	1/1	4	3

Group F

F1 CAT-Finland, Finland F2 MainzRollingBrains, Germany
F3 NIT-Stone, Japan F4 Dynamo98, Canada

VS	F1	F2	F3	F4	W/L	Pnt	Rank
F1	-	0-4	7-0	1-0	2/1	6	2
F2	4-0	-	10-1	0-0	2/0	7	1
F3	0-7	1-10	-	1-4	0/3	0	4
F4	0-1	0-0	4-1	-	1/1	4	3

Group G

G1 USCI, USA G2 Brainstorms, Germany
G3 LinkopingLizards, Sweden G4 kappa, Japan

VS	G1	G2	G3	G4	W/L	Pnt	Rank
G1	-	12-0	2-0	5-0	3/0	9	1
G2	0-12	-	0-4	0-4	0/3	0	4
G3	0-2	4-0	-	4-0	2/1	6	2
G4	0-5	4-0	0-4	-	1/2	3	3

Group H

H1 CMUnited, USA H2 UU, Netherlands
H3 TUM, Germany H4 Kasugabito-II, Japan

VS	H1	H2	H3	H4	W/L	Pnt	Rank
H1	-	22-0	2-0	5-0	3/0	9	1
H2	0-22	-	0-16	0-8	0/3	0	4
H3	0-2	16-0	-	9-1	2/1	6	2
H4	0-5	8-0	1-9	-	1/2	3	3

Appendix B: Final resuluts for the real robot small-size league

Preliminary Results

Group A

A1 Univ. Vrij Brussel (UVB), Belgique A2 Paris-6, France
A3 Univ. Western Australia (UWA), Australia

VS	A1	A2	A3	W/L	Pnt	Rank
A1	-	2-3	-	0/1	0	2
A2	3-2	-	-	1/0	3	1
A3	-	-	-	-	-	3

Group B

B1 5DPO/FEUP B2 iXs, Japan
B3 CMUnited98, USA

VS	B1	B2	B3	W/L	Pnt	Rank
B1	-	7-0	2-0	2/0	6	1
B2	0-7	-	2-16	0/2	0	3
B3	0-2	16-2	-	1/1	3	2

Group C

C1 Cambridge Univ., UK
C2 I-Space, Univ. Tokyo, Utsunomiya Univ,
C3 Roboroos, Univ. Queensland, Australia
and Shibaura Inst. Tech., Japan

VS	C1	C2	C3	W/L	Pnt	Rank
C1	-	6-2	5-4	2/0	6	1
C2	2-6	-	4-6	0/2	0	3
C3	4-5	16-2	-	1/1	3	2

Group D

D1 Paris-8, France D2 J-Star, Japan
D3 CMUnited97, CMU, USA

VS	D1	D2	D3	W/L	Pnt	Rank
D1	-	Cncld	forfeitO	-	-	3
D2	Cncld	-	3-2	1/0	3	1
D3	forfeitX	2-3	-	0/1	0	2

Appendix C: Final resuluts for the real robot middle-size league

Preliminary Results

Group A

A1 ISocRob, Portugal A2 Osaka Univ., Japan
A3 Ullanta Performance Robotics, USA A4 Ulm Sparrows, Germany

VS	A1	A2	A3	A4	W/L	Pnt	Rank
A1	-	0-0	0-0	1-1	0/0	3	2
A2	0-0	-	0-0	4-1	1/0	5	1
A3	0-0	0-0	-	0-0	0/0	3	3
A4	1-1	1-4	0-0	-	0/1	2	4

Group B

B1 CS-Freiburg, Germany B2 NAIST, Japan
B3 Dreamteam, ISI/USC, USA B4 Real Magicol, France-Columbia

VS	B1	B2	B3	B4	W/L	Pnt	Rank
B1	-	1-1	1-0	3-0	2/0	7	1
B2	1-1	-	1-1	2-0	1/0	5	2
B3	0-1	1-1	-	Cncl	0/1	4	3
B4	0-3	0-2	cncl	-	0/2	0	4

Group C

C1 GMD, Germany C4 LRP-Paris-6, France
C2 Uttori United, Utsunomiya Univ, Toyo Univ, and Riken, Japan
C3 Yale Univ., USA

VS	C1	C2	C3	C4	W/L	Pnt	Rank
C1	-	0-5	0-1	3-1	1/2	3	3
C2	5-0	-	1-0	2-0	3/0	9	1
C3	1-0	0-1	-	Cncl	1/1	6	2
C4	1-3	0-2	cncl	-	0/3	0	4

Group D

D1 Tubingen Univ., Germany D2 RMIT, Australia
D3 Munich Univ., Germany D4 RoboCup-Italy, Italy

VS	D1	D2	D3	D4	W/L	Pnt	Rank
D1	-	2-0	0-0	0-0	1/0	5	1
D2	0-2	-	1-2	0-0	0/2	1	4
D3	0-0	2-0	-	0-0 PK0-1	2/0	4	3
D4	0-0	0-0	0-0 PK1-0	-	1/0	5	2

References

1. Itsuki Noda, Shoji Suzuki, Hitoshi Matsubara, Minoru Asada, and Hiroaki Kitano. Overview of robocup-97. In Hiroaki Kitano, editor, *RoboCup-97: Robot Soccer World Cup I*, pages 20–41. Springer, Lecture Note in Artificail Intelligence 1395, 1998.
2. M. Asada, editor. *Proc. of the second RoboCup Workshop*. The RoboCup Federation, 1998.
3. Manuela Veloso, William Uther, Masahiro Fujita, Minoru Asada, and Hiroaki Kitano. Playing soccer with legged robots. In *Proc. of IROS'98*, pages 437–442, 1998.
4. Hiroaki Kitano, Milind Tambe, Peter Stone, Manuela Veloso, Silvia Coradeschi, Eiichi Osawa, Hitoshi Matsubara, Itsuki Noda, and Minoru Asada. The robocup synthetic agent challenge 97. In Hiroaki Kitano, editor, *RoboCup-97: Robot Soccer World Cup I*, pages 62–73. Springer, Lecture Note in Artificail Intelligence 1395, 1998.
5. Minoru Asada, Peter Stone, Hiroaki Kitano, Aalexis Drogoul, Dominique Duhaut, Manuela Veloso, Hajime Asama, and Shoji Suzuki. The robocup physical agent challenge: Goals and protocols for phase i. In Hiroaki Kitano, editor, *RoboCup-97: Robot Soccer World Cup I*, pages 41–61. Springer, Lecture Note in Artificail Intelligence 1395, 1998.
6. Hiroaki Kitano, editor. *RoboCup-97: Robot Soccer World Cup I*. Springer, Lecture Note in Artificail Intelligence 1395, 1998.
7. Wei Min Shen, Jafar Adibi, Rogelio Adobbati, Srini Lanksham, Hadi Moradi Benham Salemi, and Sheila Tejada. Integrated reactive soccer agents. In *RoboCup-98:Proceedings of the second RoboCup Workshop*, pages 251–264, 1998.
8. Andrew Price. Orsan: Omnidirectional radial signature analysis network. In *RoboCup-98:Proceedings of the second RoboCup Workshop*, pages 265–280, 1998.
9. Shóji Suzuki, Tatsunori Katoh, and Minoru Asada. An application of vision-based learning for a real robot in robocup learning of goal keeping behaviour for a mobile robot with omnidirectional vision and embeded servoing. In *RoboCup-98:Proceedings of the second RoboCup Workshop*, pages 281–290, 1998.
10. T. Nakamura, K. Terada, A. Shibata, J. Morimoto, H. Adachi, and H. Tadeka. The robocup-naist: A cheap multisensor-based mobile robot with on-line visual learning capability. In *RoboCup-98:Proceedings of the second RoboCup Workshop*, pages 291–304, 1998.
11. K.Yokota, K. Ozaki, N. Watanabe, A. Matsumoto, D. Koyama, T. Ishikawa, K. Kawabata, H. Kaetsu, and H. Asama. Cooperative team play based on communication. In *RoboCup-98:Proceedings of the second RoboCup Workshop*, pages 491–496, 1998.
12. Steffen Gutmann and Bernhard Nebel. The cs freiburg team. In *RoboCup-98:Proceedings of the second RoboCup Workshop*, pages 451–458, 1998.

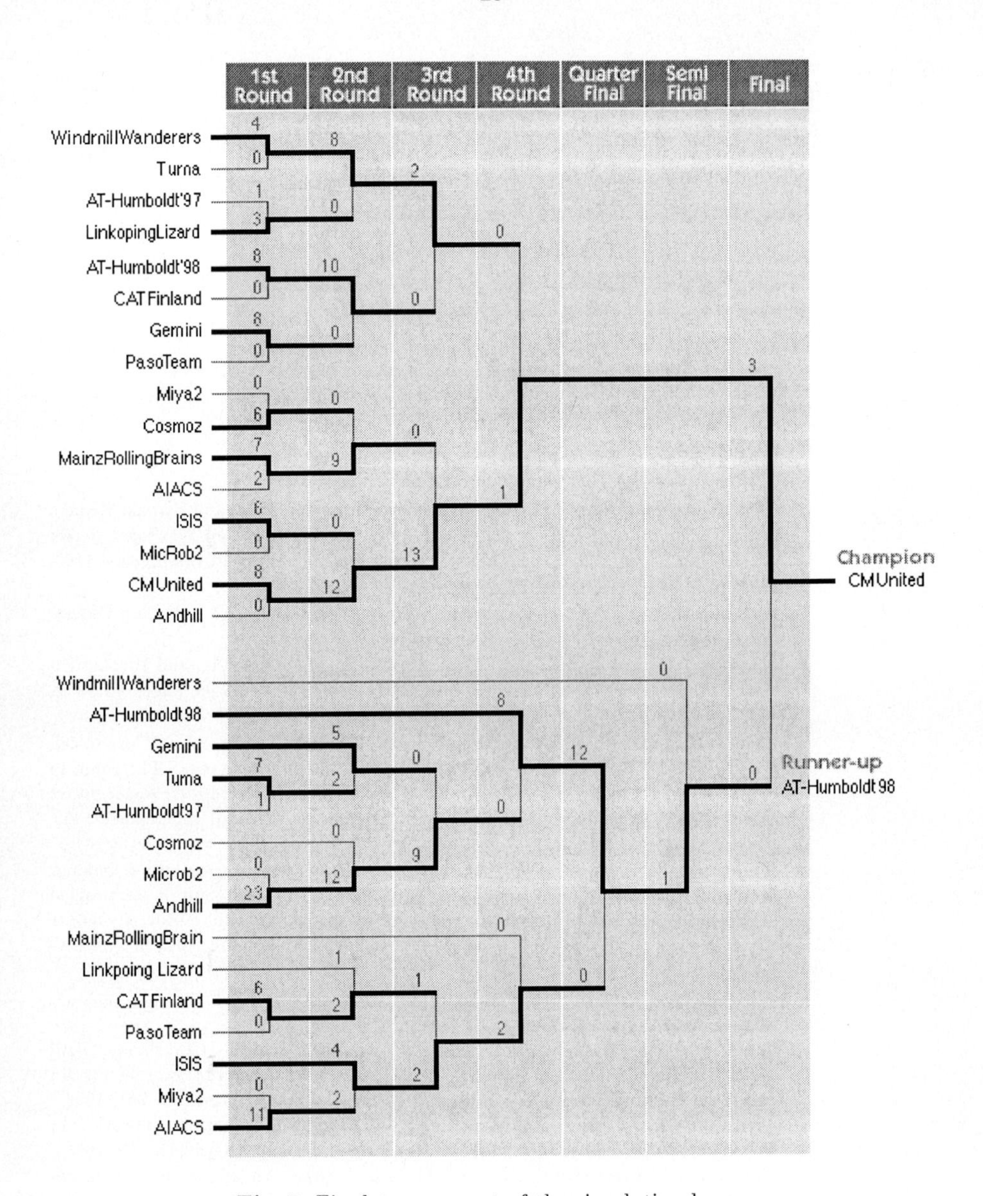

Fig. 7. Final tournament of the simulation league

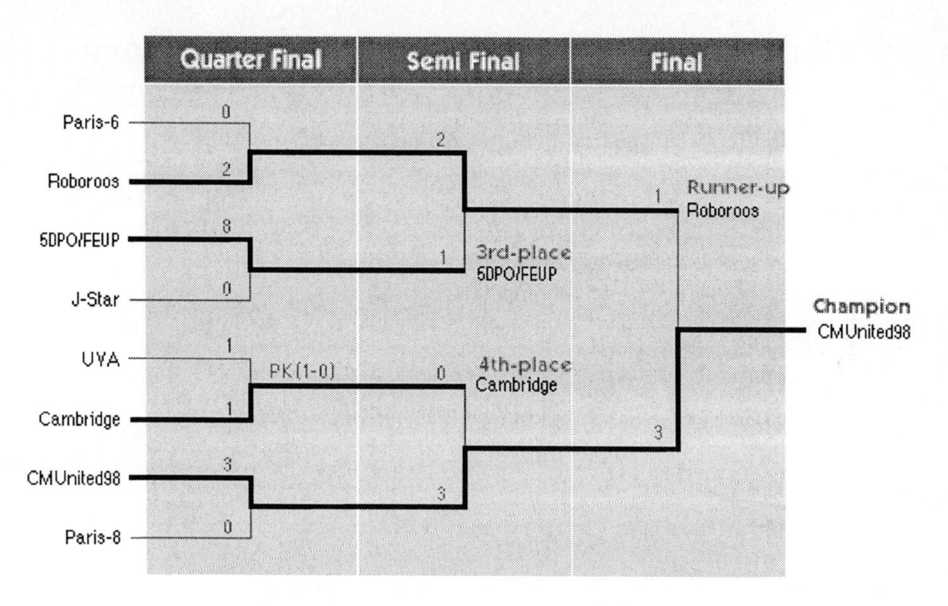

Fig. 8. Final tournament of the small-size league

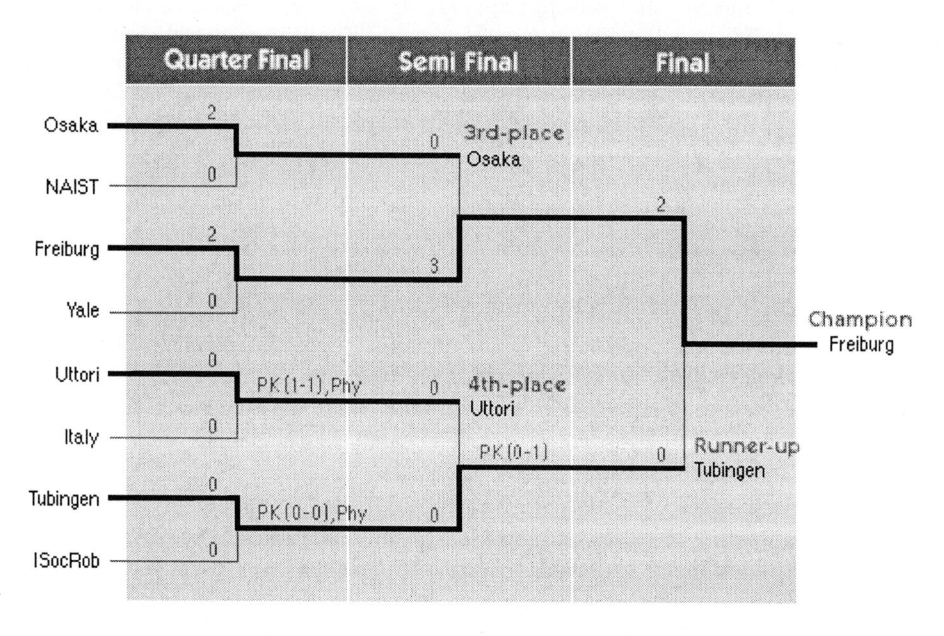

Fig. 9. Final tournament of the middle-size league

Character Design for Soccer Commentary

Kim Binsted[1] and Sean Luke[2]

[1] Sony Computer Science Lab
3-14-13 Higashigotanda
Shinagawa-ku, Tokyo 141 Japan
kimb@csl.sony.co.jp
http://www.csl.sony.co.jp

[2] Department of Computer Science
A. V. Williams Building
University of Maryland
College Park, MD 20742 USA
seanl@cs.umd.edu
http://www.cs.umd.edu/users/seanl

Abstract. In this paper we present early work on an animated talking head commentary system called **Byrne**. The goal of this project is to develop a system which can take the output from the RoboCup soccer simulator, and generate appropriate affective speech and facial expressions, based on the character's personality, emotional state, and the state of play. Here we describe a system which takes pre-analysed simulator output as input, and which generates text marked-up for use by a speech generator and a face animation system. We make heavy use of inter-system standards, so that future versions of Byrne will be able to take advantage of advances in the technologies that it incorporates.

1 Introduction

Many natural systems have behaviour complex enough that people will tend to ascribe personalities to them, and use those personalities as flawed but powerfully predictive tools. For example, we might summarize a dog's behavioral tendencies as "eager to please" or "yappy and spoiled", and use this assigned personality to predict its future behaviour.

Designed characters — such as characters in films or novels — exploit this tendency, expressing their personality so as to manipulate the observer's expectations to the designer's ends. For example, the villain in a novel might sneer and speak in a menacing voice, cueing us to expect villainous behaviour. This expectation might be reinforced by more static characteristics, such as a strong Transylvanian accent or a cheekbone scar. Consistency between expression and action, and also between modalities of expression, contributes to a character's believability. Believability, in turn, contributes to the expected predictive value of the character's perceived personality.

We are interested in the relationship between consistency and believability, and between expression and perceived personality. To explore these issues, we are developing a talking head system which can generate entertaining, believable commentary on RoboCup simulator league games [11], complete with facial expressions and affective speech, in (close to) real time.

A goal of this research is to develop an architecture which allows artists (whose technological skills may vary) to design expressive, believable characters (in the first instance, talking heads). The emphasis is on the expression, not the content — we assume the pre-linguistic raw content is generated by another system, specialized to the application. In the context of a particular character, our system:

- generates appropriate natural language text to express that content,
- transforms that text into natural affective speech, and
- controls a face animation, so that appropriate lip movements and facial expressions are generated.

In this early stage of the project, we are mostly interested in the speech and facial animation components of the system.

The key issue here is *appropriateness.* How do we ensure that the behaviour of the system is appropriate for the designed character in the given situation? For example, when reporting a scored goal, the language and facial expressions used might depend strongly on which team the character supports. Moreover, how do we ensure that the designed character is appropriate for the application? A character which might be perfect for soccer commentary might not be right for, say, a medical advisory system.

This begs the question: is a talking head appropriate for soccer commentary at all? After all, the heads of human sports commentators are rarely seen on screen during the main action of the game. In fact, our attraction to this domain is due more to its usefulness for our research (please see Section 3.1) than because soccer 'needs' talking heads in any way. Nonetheless, we do believe that an expressive entertaining talking head commentary would add to the fun of watching (or in the case of of a video game, playing) soccer. This has yet to be seen, of course.

2 Related work

2.1 Believable characters

Recently there has been a great deal of interest in the design and implementation of characters with personality. For example, the Virtual Theatre Project at Stanford [8] is working on a number of animated virtual actors for directed improvisation, basing their efforts on Keith Johnstone's theories of improvisational theatre [9]. They make use of character animations developed as part of the IMPROV project [6], which can take fairly high-level movement directions and carry them out in a natural, expressive manner. Related work on agent action selection and animation has been done as part of the ALIVE [3] and OZ projects [12] [16].

Although these projects have similar goals and assumptions to ours, our approach differs from theirs on several points. First, our focus on talking heads (rather than fully embodied agents in virtual environments) leads to a stronger emphasis on language-related behaviours. Also, we do not attempt to have the personality of the character control content selection, or any other action of the agent, for that matter. Although this sharp distinction between content and expression might negatively affect the consistency of the character, a clear separation between content and expression allows the character to be portable across content generators. For example, you could have essentially the same character (an aggressive older Scottish man, for example) commentate your soccer games and read your maps. In the first case, the content generating application is the RoboCup soccer simulator, and in the second case it is an in-car navigation system — but the character remains the same.

We also do not make a great effort to make the emotion component of our system cognitively plausible. Designed characters, such as characters in novels or films, are generally both simpler and more exaggerated than natural personalities, such as those we perceive in each other. The goal is not to develop a psychologically realistic personality, but to generate a consistent and entertaining character for some application. For this reason, what psychology there is in Byrne is more folk psychology than modern cognitive science.

2.2 Game analysis and commentary

There are at least two existing systems which generate analysis and commentary for the RoboCup simulation league: MIKE [20] and Rocco [1].

Rocco is a system for analysing simulation league games and generating multimedia presentations of games. Its output is a combination of spoken natural language utterances and a 3-D visualization of the game itself. The generated language has appropriate verbosity, floridity, specificity, formality and bias for the game context. It uses a text-to-speech system to synthesize the spoken utterances.

MIKE is a system developed at ETL which, given raw simulation data as input, analyses the state and events of the game, chooses relevant comments to make, and generates natural language commentary in real time and in a range of languages. It also uses a text-to-speech synthesizer to generate spoken output.

Our approach differs from the above in that we do not, at present, do any game analysis within the Byrne system itself; instead, we assume pre-analysed game information as input. This is because our emphasis is on expression, rather than utterance content. To our knowledge, neither MIKE nor Rocco attempt to generate affective speech, and neither use face animation.

3 Enabling technologies

In this section we outline some of the existing technologies that Byrne uses, and discuss some issues related to inter-system standards.

3.1 The RoboCup simulation league

The simulation league of RoboCup [11] [14] features teams of autonomous players in a simulated environment. This is an interesting domain for several reasons. There is no direct interaction with a user, which simplifies things a great deal — no natural language understanding is necessary, for example. More importantly, having direct access to the state of the simulation simplifies perception. If our system had to commentate a real (non-simulated) game, vision and event-recognition issues arise, which we'd rather avoid.

However, there are some problems with using the output of the RoboCup simulator directly:

- The output of the simulator is at a much lower level of description than that typically used by a football commentator.
- The simulator has no sense of which actions and states of play are relevant or interesting enough to be worthy of comment.

For this reason, some kind of game-analysis system is a necessary intermediary between the soccer simulator and Byrne.

3.2 Mark-up languages

SGML-based [7] mark-up languages are a useful tool for for controlling the presentation of information at a system-independent level. Here we make use of three different mark-up languages: one to indicate the linguistic structure of the text, one to determine how the text is to be spoken, and one to control the facial animation.

Global Document Annotation Global Document Annotation (GDA) is an SGML-based standard for indicating part of speech (POS), syntactic, semantic and pragmatic structure in text [13]. Although our main interest in this work is emotional expression, this must be underlaid by linguistically appropriate intonation and facial gestures. Moreover, the quality of speech synthesis is generally improved by the inclusion of simple phrase structure and part of speech information, which can be encoded in GDA.

Sable SABLE [17] is a SGML-based text-to-speech mark-up system developed by an international consortium of speech researchers. It is based on several earlier attempts to develop a standard, namely SSML [21], STML [18] and JSML [10]. The goal is to have a system- and theory-independent standard for text mark-up,

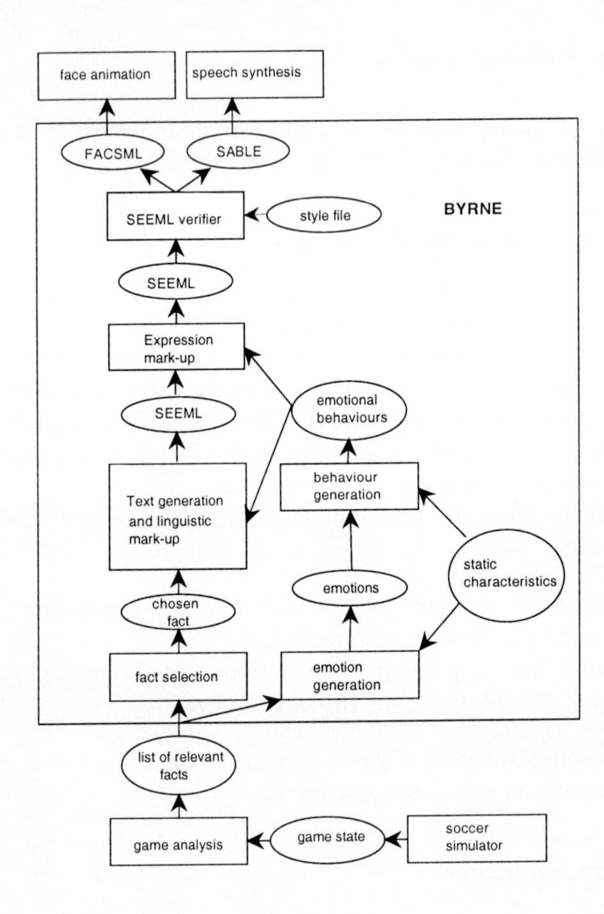

Fig. 1. The Byrne system architecture

so that non-experts in speech synthesis can mark up text in an intuitive manner which produces reasonable output from a variety of systems.

Although SABLE is in its early stages and does not yet have the richness required for the generation of affective speech (such as that described in [4]), it is a useful standard.

FACS FACS [5] stands for the Facial Action Coding System, and is a set of all the Action Units (AUs) which can be performed by the human face. It is often used as a way of coding the articulation of a facial animation [15]. For example, AU9 is known as the "nose wrinkler", and is based on the facial muscles Levator Labii Superioris and Alaeque Nasi. There are 46 AUs.

Although FACS is not an SGML-based mark-up language, it can be trivially converted into one by treating each AU as empty SGML element. Here we refer to such a mark-up as FACSML.

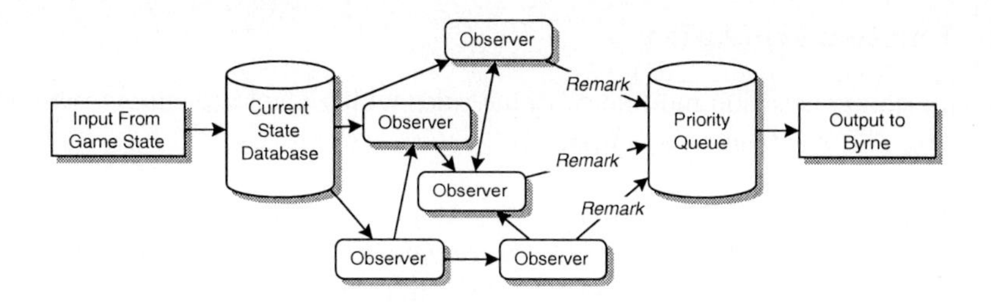

Fig. 2. The game analysis system.

4 Character architecture

Byrne is a system for expressive, entertaining commentary on a RoboCup Simulation League soccer game. Here we describe the Byrne character architecture (see Figure 1).

4.1 Input

Byrne can use any modular game analysis system as its input module (MIKE, for example). For RoboCup98, however, Byrne used a simple but effective play-by-play game analysis system designed for the competition (see Figure 2).

Byrne's input module listens for game state information reported from the RoboCup soccer monitor, and correspondingly updates a database of current state of play. The input module filters this information through *observers*, special objects which make abstract conclusions about the game, everything from ball ownership to noticing goals and keeping score to managing statistics. These conclusions take two forms. First, observers make many rudimentary *observations*, acknowledgement of changes in observed features which may or may not be very important. Many observers depend on both the state database and other observers' observations in order to make their own. Second, observers make *remarks*, occasional facts which could be commented on by Byrne.

Byrne's input module produces remarks much faster than Byrne is capable of saying them. To compensate for this, the input module feeds its remarks into a priority queue. Each remark has a *birthday* (the time when it was entered into the queue), a *deadline* (a time beyond which it is "old news"), and a *priority*. When Byrne requests a new fact to say, the queue returns one using a simple priority-scheduling algorithm. First, any fact in the queue whose deadline has past is deleted from the queue. Then the queue picks the fact **F** with the highest priority, and secondarily (to break ties) the earliest birthday. Finally, every fact with a birthday earlier than **F** is deleted from the queue. **F** is removed from the queue and returned to Byrne to comment on.

4.2 Emotion generation

The emotion generation module contains rules which generate simple *emotional structures*. These structures consist of:

- **a type**, e.g. *happiness, sadness*, etc. At present we are using Ekman's six basic emotions (*fear, anger, sadness, happiness, disgust* and *surprise*) [5], as research has shown that these are clearly and unambiguously expressible. We also include *interest* [1], as it is important in our domain of sports reporting, and is relatively easy to express in speech. In the future, however, we plan to have a hierarchy of emotions available, to allow for a richer range of resultant behaviours.
- **an intensity**, scored from 1 to 10. An emotion with intensity less than one is considered inactive, and is deleted from the emotion pool. Note that apparently opposite emotions (such as *happiness* and *sadness*) are not considered to be true opposites in Byrne. That is, a *sadness* structure is not merely a *happiness* structure with an intensity of -10. This is because *happiness* and *sadness* are not opposites in the sense that they cannot coexist, but only in that the behaviours they tend to inspire often conflict — that is, it is hard to express joy and sadness at the same time. Thus, this apparent conflict is resolved in the emotional behaviour rules, rather than in the emotional structures themselves. Moreover,emotion structures of the same type, but with different causes and/or targets, can coexist in the emotion pool.
- **a target** [optional]. Some emotions, such as *anger* and *interest*, are usually directed at some person or object. This is the **target** of that emotion.
- **a cause**. This is the fact about the world (in this case, the soccer game) which caused the emotion to come into being.
- **a decay function**. This is an equation describing how the intensity of the emotion decays over time, where time is given in seconds. Feelings which do not decay (e.g. a permanent dislike of a particular player) are not considered to be emotions for our purposes, and belong among the static characteristics of the character.

So, if a character is very sad about Team A having just scored, the relevant emotional structure might be:

(type:sadness, intensity:10, target:nil, cause:(scored team:a time:125) decay:1/t)

The intensity of this emotion would go down each second for ten seconds, then when it goes below one on the eleventh second, the emotional structure would be deleted from the emotion pool.

An emotion structure generation rule consists of a set of preconditions, which are to be filled by matching them on the currently true facts about the world

[1] For our purposes, "interest" is that emotion which at high intensity is called "excitement" and low intensity is called "boredom".

and about the character, and currently active emotion structures, the emotional structures to be added to the emotion pool, and the emotional structures to be removed. For example:

Preconditions:
(supports team: ?team)
(scores team: ?team)
Emotional structures to add:
(type: happiness intensity: 8 target: nil cause: (scores team: ?team) decay: 1/t)
Emotional structures to delete:
none

This rule indicates that, if the team that the commentator supports scores, a happiness structure should be added to the emotion pool.

There are only two ways for an emotion structure to be removed from the emotion pool: it can be explicitly removed by an emotion structure update rule, or its intensity can decay to below one, in which case it is automatically removed. In future, it might be necessary to develop a more sophisticated emotion maintenance system; however, since we have no ambitions to cognitive plausibility, we will only add such complications as necessary. We expect that this very simple emotional maintenance method will suffice for most of the situations a soccer commentator is likely to face.

Both emotion generation and behaviour generation are influenced by the **static characteristics** of the commentator character. This is a set of static facts about the character, such as his nationality, the team he supports, and so on. It is used to inform emotion and behaviour generation, allowing a character to react in accordance with his preferences and biases. For example, if a character supports the team which is winning, his emotional state is likely to be quite different that if he supports the losing team.

These emotion-generation rules can be arbitrarily complex, to take into account the context of both the character's static characteristics and the state of the world (in the case of soccer, the game).

4.3 Emotional behaviours

Emotion structures and static characteristics are preconditions to the activation of high-level emotion-expressing behaviours. These in turn decompose into lower-level behaviours. The lowest level behaviours specify how the text output by the text generation system is to be marked up.

Emotionally-motivated behaviours are organized in a hierarchy of mutually inconsistent groups. If two or more activated behaviours are inconsistent, the one with the highest activation level is performed. This will usually result in the strongest emotion being expressed; however, a behaviour which is motivated by several different emotions might win out over a behaviour motivated by one strong emotion.

It is entirely possible for mixed emotional expressions to be generated, as long as they are not inconsistent. For example, a happy and excited character might express excitement by raising the pitch of his voice and happiness by smiling. However, it is less likely that a character will have a way to express, say, happiness and sadness in a consistent manner.

4.4 Text generation

Character has a role to play in natural language generation. For example, a character from a particular country or region might use the dialect of that area, or a child character might use simpler vocabulary and grammar than an adult. The current emotional state of the character would also have an effect: an excited, angry character might use stronger language and shorter sentences than a calm happy one, for example. Loyall's work in the OZ project [12] discusses some of these issues.

Despite these possibilities for affective generation, text generation is an area in which Byrne does almost nothing of interest at present. It is done very simply through a set of templates. Each template has a set of preconditions which constrain the game situations they can be used to describe. If more than one template matches the chosen content, then the selection is based on how often and how recently the templates have been used.

Byrne's text generation module does not generate plain text, but rather text marked up with SEEML. Although the speech synthesis system we use can generate reasonable speech from plain text, it is helpful to retain some phrase structure and part of speech (POS) information from the natural language generation process to help the speech synthesis system to generate appropriate prosody.

Moreover, linguistic information embedded in the text also helps determine appropriate interruption points, should a more important fact need to be expressed. Here we assume that Byrne should finish the phrase it is currently uttering before it interrupts and starts a new utterance. This is a very simplistic approach, and may not be adequate.

Finally, linguistically-motivated facial gestures and speech intonation are now hard-coded into the templates. If the natural language generation system were more sophisticated, then a post-generation gesture and intonation system might be necessary, but with simple template generation this is the most effective method.

4.5 Expressive mark-up

SEEML[2] (the speech, expression and emotion mark-up language) is really just a slightly supplemented superset of three different mark-up systems, namely FACSML, SABLE and GDA. GDA is used to inform linguistically motivated expressive behaviours, and also to aid the speech synthesizer in generating appropriate prosody. FACSML is used to add facial behaviours, and SABLE is used to control the speech synthesis.

[2] SEEML is pronounced "seemly".

The expression mark-up module adds emotionally motivated mark-up to the already marked-up text from the text generation system. Conflicts are resolved in a simple (perhaps simplistic) manner. The combination rules are:

- Unless identical, tags are assumed to be independent. Any potential practical conflicts are left to be resolved by the speech synthesizer and/or facial animation systems.
- If two identical tags are assigned to a piece of text, the one with the smaller scope is assumed to be redundant, and removed.
- If two otherwise identical tags call for a change in some parameter, it is assumed that that change is additive.

4.6 SEEML verifier and style file

The SEEML verifier interprets SEEML tags in the context of a style file, adds time markers and lip synching information, and sends appropriate FACS to the facial animation system and SABLE (supplemented with phrase structure and POS tags) to the speech synthesis system.

Although sophisticated lip-synchronization algorithms have been developed (e.g. in [22]), they are not necessary for our purposes. Instead, we use a simple 'cartoon style' lip animation, which only shows the more obvious phoneme-viseme matches, as described in [15].

The style file contains speech and animation system specific interpretation rules. For example, it would determine the FACS which are to be used to indicate a **smile** for this particular face model, and the sound file to be used for a **hiccup**.

5 Implementation

In the first implementation, Byrne uses Franks and Takeuchi's facial animation system [19] and the Festival speech system [2]. We hope that the standardized mark-up of the output will allow the use of other systems as well.

The facial animation system is based on Waters' [22] polygon face model, and was implemented by Franks in C++ to run on a Silicon Graphics machine. It has a wide range of expressions, although the lip movement does not allow for sophisticated lip synching.

Festival is a concatenative speech synthesis system. It can generate speech in a number of different languages and voices, and the resulting speech is reasonably natural-sounding, although the user's control of the speech synthesis is quite coarse. Its other important feature for our purposes is its support of the Sable speech markup system.

Byrne itself is implemented in C++.

6 Future work

Although the emotional behaviours outlined above are very simple, this architecture allows for quite complex behaviours to be implemented. Maintaining consistency and believability in these behaviours is a significant problem. We plan to develop a set of tools for character designers to help them in in this task.

Also, the current interruption mechanism is too simplistic, and sometimes results in clumsy commentary. We intend to devise a more sophisticated mechanism.

Although the character described throughout is a play-by-play commentator character, we hope to develop other characters for the soccer simulator domain, such as a coach or a soccer fan. We would also like to develop a colour commentator [3] to work with the play-by-play, which would necessitate introducing a turn-taking mechanism into the system.

7 Conclusion

In this paper we have motivated and described an architecture for a soccer commentator system, Byrne. Byrne generates emotional, expressive commentary of a RoboCup simulator league soccer game. It makes heavy use of inter-system standards, so that the system can take advantage of advances in speech synthesis, facial animation and natural language generation.

References

1. Elisabeth Andre, Gerd Herzog, and Thomas Rist. Generating multimedia presentations for robocup soccer games. Technical report, DFKI GmbH, German Research Center for Artificial Intelligence, D-66123 Saarbrucken, Germany, 1998.
2. Alan W. Black, Paul Taylor, and Richard Caley. *The Festival Speech Sythesis System.* CSTR, University of Edinburgh, 1.2 edition, September 1997.
3. Bruce Blumberg and Tinsley Galyean. Multi-level control for animated autonomous agents: Do the right thing... oh, not that... In Robert Trappl and Paolo Petta, editors, *Creating Personalities for Synthetic Actors*, pages 74–82. Springer-Verlag Lecture Notes in Artificial Intelligence, 1997.
4. Janet Cahn. Generating expression in sythesized speech. Master's thesis, Massachusetts Institute of Technology Media Laboratory, Boston, May 1989.
5. Paul Ekman and Erika L. Rosenberg, editors. *What the face reveals: Basic and applied studies of spontaneous expression using the facial action coding system.* Oxford University Press, 1997.
6. Athomas Goldberg. IMPROV: A system for real-time animation of behavior-based interactive synthetic actors. In Robert Trappl and Paolo Petta, editors, *Creating Personalities for Synthetic Actors*, pages 58–73. Springer-Verlag Lecture Notes in Artificial Intelligence, 1997.
7. Charles Goldfarb. *The SGML Handbook.* Clarendon Press, 1991.
8. Barbara Hayes-Roth, Robert van Gent, and Daniel Huber. Acting in character. In Robert Trappl and Paolo Petta, editors, *Creating Personalities for Synthetic Actors*, pages 92–112. Springer-Verlag Lecture Notes in Artificial Intelligence, 1997.
9. Keith Johnstone. *Impro.* Routledge Theatre Arts Books, 1992.

[3] A colour commentator provides background details on teams and players, such as statistics or amusing anecdotes.

10. Java speech markup language specification [0.5 beta]. Technical report, Sun Microsystems, 1997.
11. Hiroaki Kitano. Robocup. In Hiroaki Kitano, editor, *Proceedings of the IJCAI workshop on Entertainment and AI/ALife*, 1995.
12. A. Bryan Loyall. Some requirements and approaches for natural language in a believable agent. In Robert Trappl and Paolo Petta, editors, *Creating Personalities for Synthetic Actors*, pages 113–119. Springer-Verlag Lecture Notes in Artificial Intelligence, 1997.
13. Katashi Nagao and Koiti Hasida. Automatic text summarization based on the global document annotation. Technical report, Sony Computer Science Laboratory, 1998.
14. Itsuki Noda. *Soccer Server Manual Rev. 2.00*, May 1997.
15. Frederic I Parke and Keith Waters. *Computer Facial Animation.* A K Peters Ltd, Wellesley, MA, 1996.
16. W. Scott Neal Reilly. *Believable social and emotional agents.* PhD thesis, School of Computer Science, Carnegie Mellon University, May 1996.
17. Draft specification for sable version 0.1. Technical report, The Sable Consortium, 1998.
18. R. Sproat, Paul Taylor, and Amy Isard. A markup language for text-to-speech synthesis. In *Proceedings of EUROSPEECH*, Rhodes, Greece, 1997.
19. Akikazu Takeuchi and Steven Franks. A rapid face construction lab. Technical Report SCSL-TR-92-010, Sony Computer Science Laboratory, Tokyo, Japan, May 1992.
20. Kumiko Tanaka-Ishii, Itsuki Noda, Ian Frank, Hideyuki Nakashima, Koiti Hasida, and Hitoshi Matsubara. MIKE: An automatic commentary system for soccer. Technical Report TR-97-29, Electrotechnical Laboratory, Machine Inference Group, Tsukuba, Japan, 1997.
21. Paul Taylor and Amy Isard. SSML: A speech synthesis markup language. *Speech communication*, 1997.
22. Keith Waters and Thomas M. Levergood. DECFace: An automatic lip-synchronization algorithm for sythetic faces. Technical Report CRL 93/4, Digital Cambridge Research Laboratory, September 1993.

Automatic Soccer Commentary and RoboCup

Hitoshi Matsubara Ian Frank Kumiko Tanaka-Ishii
Itsuki Noda Hideyuki Nakashima Kôiti Hasida

Complex Games Lab, Language Learning Lab
Electrotechnical Laboratory (ETL)
Umezono 1-1-4, Tsukuba
Ibaraki, Japan 305

Abstract. This paper suggests that automated soccer commentary has a key role to play within the overall RoboCup initiative. Firstly, we identify soccer commentary as allowing and requiring investigation of a wide variety of research topics, many of which could not be addressed by the simple development of teams for the RoboCup leagues themselves. Secondly, we highlight a key task of soccer commentary: the expert analysis of a game. We suggest that this expert analysis task has the potential to make a significant impact on RoboCup challenges such as learning, teamwork, and opponent modeling. We illustrate our arguments by discussing the progress on soccer commentary systems to date, in particular reviewing our own system, MIKE.

1 Introduction

In real-life, commentary adds so much to the coverage of football that no TV company would contemplate screening a game without it. Indeed, sometimes the commentary itself is just as memorable as the scenes it describes (for example, the well-known "They think it's all over... It is now!" as Geoff Hurst scored in the dying seconds of extra time in the 1966 World Cup final, and the less well-known but equally noteworthy "17 minutes gone and already no goals!" [Jones 96]). At the first RoboCup tournament in Japan, however, the quantity of matches made it impractical to recruit human volunteers to describe all the games. Since RoboCup also does not (yet) generate crowd noise comparable to that of real soccer, it was therefore common to hear the comment that some games seemed 'flat'. It was to fill this gap, and to provide extra atmosphere, interest and context for RoboCup games that we originally started to develop our automatic commentary system, MIKE (Multi-agent Interactions Knowledgeably Explained). This system, which produces text or spoken commentary from the output of the Soccer Server [Noda *et al* 98], was first used to commentate public games at the Japan Open in April 1998, and will be used again in Paris this year to provide commentary for the second RoboCup contest.

We believe, however, that automated soccer commentary has a far greater role to play within the overall RoboCup initiative than simply adding atmosphere. In this paper we clarify this role by suggesting that soccer commentary allows for the investigation of a number of research topics that are outside the scope of the original RoboCup challenges. Also, within the framework of RoboCup itself, we identify how a key task of soccer commentary — the expert analysis of games — has the potential to make a significant impact on existing RoboCup challenges such as learning, teamwork, and opponent modeling. We demonstrate these points by examining the work to date on automatic commentary systems. We focus on our own system MIKE, but also review systems produced by others. We hope that our summary of expert analysis techniques, in particular, will be of general interest to a wide variety of researchers involved in RoboCup. We ourselves identify a number of specific tasks for which such techniques should be indispensable.

2 Why Soccer Commentary?

One of the strengths of RoboCup is its ability to attract researchers from many different domains. For example, [Kitano *et al* 97a] discusses how a large number of technologies need to be integrated to produce successful RoboCup teams. In the same vein, we see soccer commentary as providing challenges of its own. Let us illustrate this by examining the research topics involved in generating convincing game descriptions:

- **Natural language generation**. The generation of text or speech is the most obvious aspect of soccer commentary. The time-critical nature of the domain forces real-time decisions about what to utter, when the current commentary would be better interrupted by a new description, and how to maintain overall coherency.
- **Understanding of multi-agent systems**. To describe soccer it is necessary to understand how the multiple players interact. This understanding can be pursued on numerous levels, such as the recognition of low-level tactics or of high-level strategy, the following of a focus point (such as the ball), the analysis of the territories established by the players, or by a general analysis of the nature of teamwork. Note that the attribution of player *intentions* is also an issue here.
- **Machine vision**. For the robotic leagues (as for real-life soccer commentary), vision and image understanding is critical.
- **Presentation techniques**. Commentary is not necessarily restricted to natural language, but may also include the generation of replays to highlight interesting events, the incorporation of the visual display of statistics or graphs, and the superimposition of graphics on the screen to highlight interesting (or even upcoming) events.
- **Re-use of stored knowledge and experiences**. Many of the comments made by real-life announcers add context to a game by reviewing the past performances of a team and its players, or (in tournaments or league promotion/demotion situations) analyzing the consequences of the possible game results for each team. Also, past experience is often brought to bear to identify situations (goals, comebacks, sendings-off) similar to the current one,

and may even be used to make predictions ("Shearer scored from this position last weekend...").

- **Rules of communication**. Depending on what the audience is assumed to know the appropriate commentary will change. For example, the English expression "Back to square one" comes from the early days of radio football commentary when listeners followed the game using a template of a football pitch divided into numbered regions: the square numbered one was the center spot. The flexibility to commentate for different audiences allows investigation of issues such as Grice's cooperative principles of conversation [Grice 75] (*e.g.*, the maxim of quantity, that your contribution should be no more and no less informative than is required). Rules of communication become especially significant if a commentary *team* is being modeled, for example with an announcer following the ball-by-ball action and an expert providing higher-level analysis.
- **Incorporation of emotion**. Soccer commentary offers plentiful opportunities for studying the expression of emotion in language. Meaning (and emotion) can be expressed in speech through prosody (the phrasing and the tones used for speech), and even via changes in vocabulary. Further, the effects of the natural bias of a commentator (in favor of teams from one country, of certain styles of play, or in support of the underdogs) can also be studied.
- **Focusing the complex**. Humans watching a game of soccer can view it as a single process. However, at any one time, the individual players involved will give rise to many possible focus points, both for the game itself and for a commentary. This interaction between a medium-sized number of adaptive agents meets the criteria of a complex, adaptive system described, for example, by [Casti 97]. As yet, there are no good mathematical theories for understanding the overall behavior of such systems.

Note that a significant proportion of the topics listed here are not addressed by simply developing teams for the RoboCup leagues themselves. Thus, we view the challenge of automated soccer commentary as extending the scope of the RoboCup goals. In fact, we additionally believe that research on commentary systems also has the potential to make a significant impact on the original goals of the RoboCup project. For instance, the 1997 RoboCup Synthetic Agent Challenge [Kitano *et al* 97b] presents three specific challenges for RoboCup research:

- **Learning**. The learning of individual agents and teams.
- **Teamwork**. Multi-agent team planning and plan execution in service of teamwork.
- **Opponent Modeling**. On-line tracking of opponents' aims, on-line strategy recognition, and off-line review.

We make the observation that the automatic production of *expert analysis* of soccer play can directly facilitate each of these challenges. This is illustrated in Figure 1. The expert analysis module in this figure is an automatic system

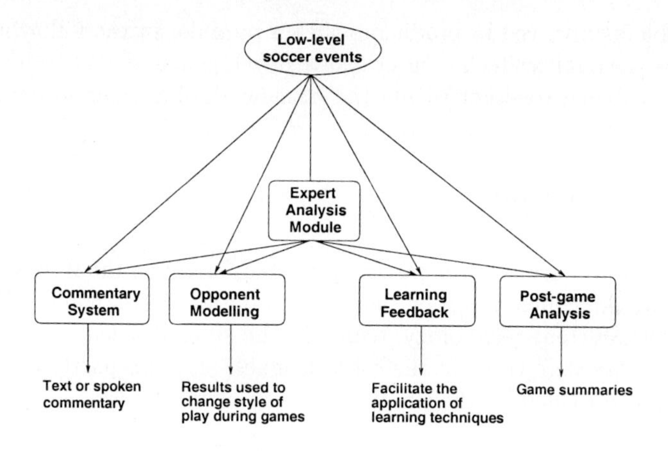

Fig. 1. Possible uses of an expert analysis module

that analyzes low-level events in a soccer game to produce an assessment of each team's playing style, tactics, strengths and weaknesses. We envisage the results of such analysis then being used for tasks such as:

- Informing a commentary system. This is our main focus in the current paper. Convincing commentary is probably not achievable without an expert analysis module.
- Opponent Modeling. This addresses one of the main RoboCup challenges — how a team should modify their play to deal with a particular opponent. Clearly, analyzing and understanding the opponents' play is a critical step in this process.
- Learning feedback. Another of the RoboCup challenges is learning. The 'credit-assignment problem' poses a serious barrier to automated learning techniques, however, since the number of goals scored by a team is only a rough indicator of how well it actually plays. An effective expert analysis module will make the use of learning techniques more efficient.
- Match analysis. An expert analysis module will also enable further interesting applications, such as post-game (or half-time) analysis. The ability to select highlights from a game and to demonstrate the strengths and weaknesses of each team (suggesting how a game was won and lost, and allowing the easy creation of game digests) will add to the interest in RoboCup competitions.

This list is only a broad characterization of the possible uses for expert analysis. However, we believe it demonstrates that an expert analysis module — in addition to being an integral part of a commentary system — directly enables work on the RoboCup learning, opponent modeling, and teamwork challenges described above. In our own work on the MIKE system, we have addressed some of the problems involved in producing such a module. In the following sections, we describe the analysis techniques we have developed, and also outline how their output is combined together within the framework of a coherent commentary.

3 MIKE — A Commentator System For Soccer Server

Although commentary systems can be envisaged for any of the RoboCup leagues, the MIKE system concentrates on just the simulation tournament. This is partly because, with current technology, teams in the simulator league produce more 'soccer-like' play than those in the robot leagues, and also partly because of the level of detail provided by the simulator.

3.1 MIKE's Input and Output

Games in RoboCup's simulator league are conducted using the Soccer Server [Noda *et al* 98]. In this simulation, the soccer field and all objects on it are 2-dimensional so that, unlike in real football, the ball cannot be kicked in the air and the players cannot make use of skills such as heading or volleying. The Soccer Server provides a real-time game log of a very high quality, sending detailed information on the positions of the players and the ball to a monitoring program every 100ms. Specifically, this information consists of: player location and orientation (for all players), ball location, and the game score and play modes (such as throw-ins, goal kicks, *etc*).

From the Soccer Server input, MIKE creates a commentary that can consist of any combination of the possible repertoire of remarks shown in Figure 2. Currently, this output is produced with the simple mechanism of template matching, converting the system's internal language into appropriate expressions in either

- **Explanation of complex events**. Formation changes, position change, and advanced plays.
- **Evaluation of team play**. Average formations, formations at a certain moment, player locations, indication of active or problematic players, winning passwork patterns, wasteful movements.
- **Suggestions for improving play**. Loose defense areas, better locations for inactive players, and 'should-have' comments about failed passes.
- **Predictions**. Prediction of passes and shots at goal. Also, prediction of game result by comparing team performance metrics against statistics compiled from a database of played matches.
- **Set pieces**. Goal kicks, throw ins, kick offs, corner kicks, and free kicks.
- **Passwork**. Basic tracking of the ball-by-ball play.

Fig. 2. MIKE's commentary repertoire

English, Japanese or French. To reduce repetition, this matching process is non-deterministic, and several templates are available for each decision. An example of MIKE's English language commentary might be "Interception by the Yellow-Team,... Yellow10 shoots!... Red4,... Yellow11's shot!... The Yellow-Team's 7th goal!! 7 to 0! Another goal by Yellow11!" MIKE also uses off-the-shelf text-to-speech packages to produce spoken commentary.

3.2 Summary of MIKE's Overall Structure

MIKE's design is described in detail in [Tanaka-Ishii *et al* 98b]. Here, we therefore give just a brief overview by presenting Figure 3. In this figure, the rectangles represent data and the ovals represent processes, which run concurrently, carrying out the following tasks:

- **Communicator**. Receives log data from the Soccer Server every 100ms and writes it into shared memory.
- **Soccer Analyzer Modules**. There are six Soccer Analyzers, of which three analyze basic events (shown in the figure as the 'Basic', 'Techniques', and 'Shoot' processes), and three carry out more high-level analysis (shown as the 'Bigram', 'Voronoi', and 'Statistic' processes). These six processes analyze the information posted to the shared memory by the Communicator, communicate with each other via the shared memory, and also post *propositions* to the proposition pool. Propositions are MIKE's internal represen-

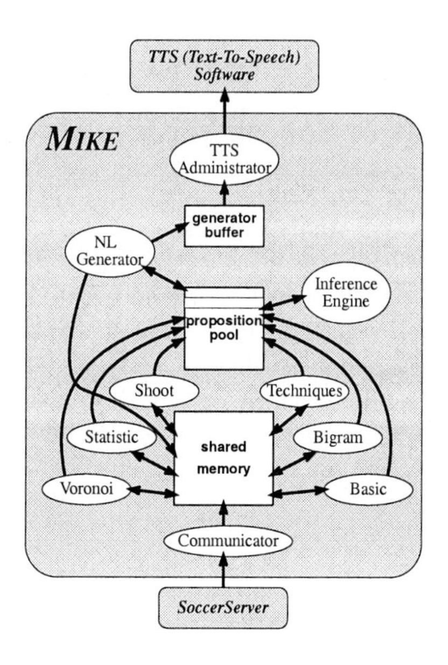

Fig. 3. MIKE's structure

tation of commentary fragments, and consist of a tag accompanied by some attributes. For example, a kick by player No.5 is represented as `(Kick 5)`, where `Kick` is the tag and `5` is the attribute. In total, MIKE has about 80 different kinds of propositions. Every proposition posted to the pool is also given an *importance* value by the Analyzer that generated it. The importance of a proposition decreases with time until it reaches zero, when it is deleted from the proposition pool without having been uttered.

- **Inference Engine**. A collection of over 50 forward chaining inference rules that identify relations between propositions in the pool. Successful firing of a rule may add new propositions (logical consequences), change the importance values of existing propositions (logical subsumption or identification of state change), or identify relations between two or more propositions (second-order relations).
- **Natural Language Generator**. Selects the proposition from the proposition pool that best fits the current state of the game (considering both the situation on the field, the importance values of the available propositions, and the commentary currently being made). Translates the proposition into natural language, using pattern-matching.
- **Text-To-Speech Administrator**. Synchronizes the Natural Language Generator's output with a text-to-speech software program.

The primary research goals in developing MIKE have been to investigate the automatic analysis of multi-agent systems, and to study the task of natural language generation in a fast-moving, real-time domain. For a detailed discussion of the natural language issues involved (such as interruption control), readers are referred to [Tanaka-Ishii *et al* 98a]. In this paper, we will concentrate on the central theme identified in §2: the generation of expert soccer analysis.

4 Expert Analysis of Soccer Within MIKE

In MIKE, the workings of the multi-agent system that is soccer are tracked and interpreted by six Soccer Analyzers. Three of these Analyzers are very low-level, 'event-based' modules that identify individual incidents on the field. These are the 'Basic' module, which identifies simple events such as ordinary kicks, the 'Techniques' module, which identifies passes, dribbles, interceptions, one-two passes, and through passes (all defined in terms of patterns of successive kicks), and the 'Shoot' module, which details shots and goals. The remaining three Analyzers are more high-level 'state-based' modules that keep track of and interpret accumulated events on the field of play. It is these that provide the expert analysis of a game.

4.1 Simple Statistics

Even basic statistics can be very useful for interpreting a game and assessing the performance of teams. MIKE's Statistics Analyzer is responsible for maintaining the following figures:

- Each player's average location and its variance. For example, Figure 4 shows the average positions of the players in the first half of the RoboCup'97 World Cup Final held in Nagoya, Japan (players numbered from 1 to 11 play left to right, and those numbered 12 to 22 play right to left). Such analysis is used by MIKE to make simple guesses about the roles each player is taking.

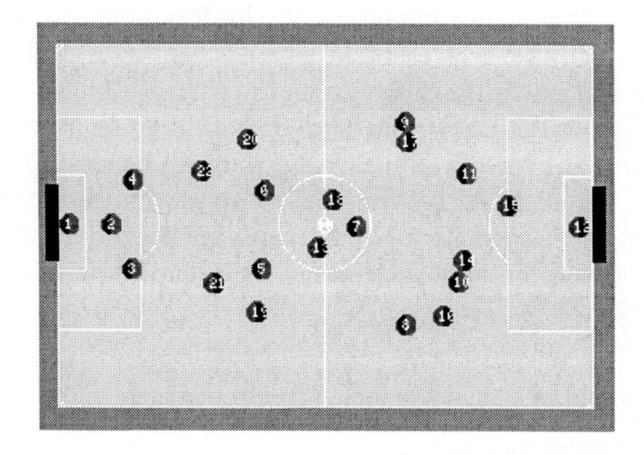

Fig. 4. Average positions of players in the RoboCup'97 final

- The average position of *all* of a team's players taken as a whole (and its variance), expressed as a distance (positive or negative) past the centerline. This figure indicates whether a team favors defense or attack. (Note that the average position of a team may also vary towards the left or right wing.)
- Average position of ball. Again, expressed as a distance past the centerline.
- The average duration of each team's possession spells, and each team's overall percentage share of the possession.
- Number of free kicks and corners taken by each team.
- Average pass distance (for each team).
- Average length of ball-play chains (for each team).

Much information on passing (such as the number of successful passes, the number of successful steals, or the length of pass sequences) is absent from this list. This is because, for MIKE, such information is the responsibility of the Bigrams Analyzer.

4.2 Analysis Based on Ball-play Chains (Bigrams)

In MIKE, ball-play chains are regarded as first-order Markov processes. Matrices representing these processes are automatically maintained by MIKE and used to describe the activity of each player. These 24×24 matrices (22 players and 2 goals) are referred to as *pass bigrams*, and record the *numbers* of each possible ball transition. As an example, consider Table 1, which shows the pass distribution and reception bigrams produced by MIKE for part of a real game (again, the first half of the RoboCup'97 World Cup Final). Only those transitions occurring more than once, plus all goals, are included in a bigram. In our examples, the left column shows the champion team, Humboldt, whose players are numbered from 1 to 11, and the right column shows Andhill, whose players are numbered 12 to 22. The letter 'G' denotes the scoring of a goal. Transitions involving giving the ball to an opponent are marked with a '+' sign.

MIKE can use these bigrams to interpret why Humboldt outperform Andhill. Of course, Humboldt scored seven goals, whereas Andhill scored none, but more than this, MIKE can assess the difference in the passing abilities of the teams by comparing the number of transitions marked with a '+' sign. Also, playmakers can be identified. For example, player No.7 (who can be seen to be a midfielder from analysis of Figure 4), successfully passes to players No.11 and No.10 (forwards), and even scores a goal himself. The main goalscorer, Player No.11, scored four times even though Figure 4 suggests he could have been marked by players No.14 and No.15. On the other hand, Andhill did not score any goals despite

Table 1. RoboCup'97 final: ball-play transitions represented as bigrams

Pass Distribution					
Humboldt			Andhill		
from	to	freq	from	to	freq
2	4	2	+12	10	4
3	5	2	+12	11	3
5	8	2	+13	5	4
+5	19	2	+13	6	4
6	7	3	+13	7	3
+6	18	2	13	14	2
7	8	2	+14	6	2
7	10	2	+14	7	2
7	11	3	+14	10	5
+7	18	2	+15	7	2
+7	20	2	+15	11	2
7	G	1	15	14	2
8	10	2	16	14	3
+8	12	2	+17	7	2
+8	14	3	19	13	3
8	G	1	+20	4	2
10	8	2	+21	5	3
+10	12	5			
+10	15	3			
+10	16	3			
10	G	1			
+11	12	3			
+11	15	3			
11	G	4			

Pass Reception					
Humboldt			Andhill		
to	from	freq	to	from	freq
4	2	2	+12	10	5
+4	20	2	+12	11	3
+5	13	4	+12	8	2
+5	21	3	13	19	3
5	3	2	14	13	2
+6	13	4	14	15	2
+6	14	2	14	16	3
+7	13	3	+14	8	3
+7	14	2	+15	10	3
+7	15	2	+15	11	3
+7	17	2	+16	10	3
7	6	3	+18	6	2
8	10	2	+18	7	2
8	5	2	+19	5	2
8	7	2	+20	7	2
+10	12	4			
+10	14	5			
10	7	2			
10	8	2			
+11	12	3			
+11	15	2			
p11	7	3			

the ball reaching player No.21, which Figure 4 identifies as a forward. Players No.12, 13, 14, 15, 20, and 21 often give the ball to the opponents, so they are either severely marked, or just very bad at passing.

To help interpret the information in a bigram, MIKE introduces the notion of a *winning passwork pattern*. This is defined as any chain of three players from the same team A, B, C, such that both the transition from A to B and the transition from B to C appear in the bigram, and C scores at least one goal. In the example of Table 1, Humboldt has six winning passwork patterns and Andhill have none, as they failed to score any goals. An example winning passwork pattern is from Player No.6 to No.7 to No.11. MIKE uses winning passwork patterns to predict passwork and shots on goal during a game.

4.3 Analysis Based on Voronoi Diagrams

To analyze the area of influence of each player, MIKE divides the field of play into *Voronoi regions*. These regions are technically defined in terms of a set of n points (or *sites*) in a plane. Given such a set of sites, the Voronoi region associated with site p is the locus of all points in the plane closer to p than to any of the other $n-1$ sites. A graphical rendering of the borders of the Voronoi regions for a given set of sites is known as the Voronoi diagram for that set. In MIKE, such diagrams are calculated for each team by using the team's players as sites. To do this efficiently, the technique of [Oishi & Sugihara 95] is used, which has a cost of $O(n)$ on average, and $O(n^2)$ at worst. This technique allows MIKE to calculate complete diagrams for each team every simulation cycle (100ms).

Using Voronoi diagrams, MIKE can determine the defensive areas covered by players and also assess overall positioning. As an example, Figure 5 shows the Voronoi diagram of a moment from the first half of the RoboCup'97 final. Humboldt's players are represented by a '+' sign, Andhill's players by a '$\diamond$', and the ball by a '⊟'. In this figure, the positions of Humboldt's No.7 and No.10 have also been marked. We have already seen how MIKE can use bigrams to identify Humboldt's No.7 as a playmaker. Now we see that at the instant depicted by the Voronoi diagram, No.7 is close to the meeting point of a number of the opponent's Voronoi regions, and is thus as far as possible from the neighboring Andhill players. Also, No.10 has an excellent chance to shoot at goal: there are no opponents near him and he has the ball. Further, right in front of the left goal, Humboldt has created a triangular Voronoi region. As the average shape of Voronoi regions is hexagonal, this area demonstrates Humboldt's tight defense.

Note that [Taki *et al* 96] have proposed a method of calculating players' areas of influence that takes into account each player's current speed and orientation. This model generates defensive borders in the form of higher-ordered curves. In MIKE, however, the speed and orientation of a player are taken into consideration via a simple yet very effective short-cut: before calculating the Voronoi regions for a team's players, the location of the sites is modified by displacing each player in the direction they are currently facing, for a distance proportional to their current speed.

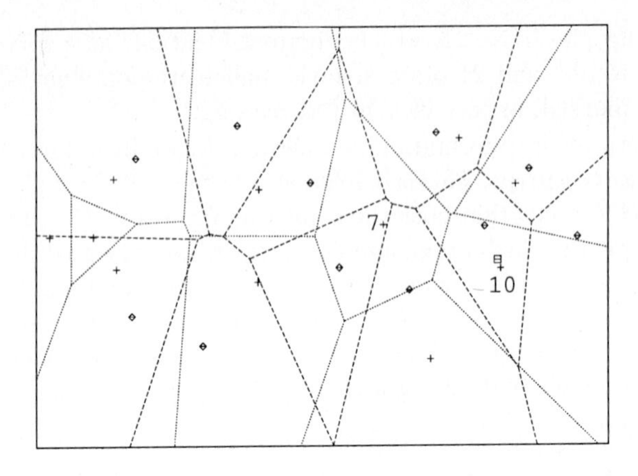

Fig. 5. Voronoi diagram (RoboCup'97 final)

4.4 Re-use of Stored Game Logs

To demonstrate the general applicability of MIKE's analyses, we processed the logs of the 26 games from the 1998 Japan Open tournament (the tournament conducted with the version of Soccer Server most similar to the one that will be used for the 1998 RoboCup in Paris). To identify the statistics most useful for highlighting differences between teams, we collected figures separately for the winners and losers of each half of each game (looking at halfs of games reduces the number of draws, and therefore also reduces the amount of data that is discarded). The results, shown in Table 2, demonstrate that, with the exception of the number of steals and the number of corners, many of the statistics have

Table 2. Average statistics, non-drawn games from the Japan Open 1998

	Winners	Losers
Goals	5.3	0.8
Shots	16.1	12.9
Average distance of a team's players past centerline (m)	-3.91	-10.84
Average number of players in ball-play chains	2.1	1.8
Total successful passes	16.45	10.36
Total number of dribbles	1.7	0.6
Total successful steals	17.6	22.9
Total winning passwork patterns	0.31	0
Number of corners taken by each team	0.23	0.13

good predictive value for identifying the stronger team. In the case of corners, the low values are explained by the nature of collisions in the simulator; rather than calculating deflections, the Soccer Server simply *stops* moving objects when their paths cross.

Base values such as those in Table 2 are employed by MIKE for predicting which team is likely to win a game. This is a simple example of how the re-use of stored knowledge, as mentioned in §2, can be applied to analyze games in real-time.

5 Related Systems

To date, there has been only limited research on automatically analyzing soccer. In terms of expert analysis techniques, the main drive for progress has been in commercial systems to assist professional coaches. Academic research on soccer commentary has also been carried out, however, focusing on the multimedia presentation of visual data and the incremental recognition of events.

5.1 SECOND LOOK

A comprehensive commercial package for analyzing football games is SECOND LOOK, a system that enables users to produce detailed statistical analyses. SECOND LOOK has been developed by Zvi Friedman, an international soccer coach, and Jon Kotas, a computer programmer, who founded a company in 1993 to market their product. On their Web pages [Sof] they claim to provide services for the American Major League Soccer (MLS), the European soccer finals in England, and to have provided the Women's Olympics Soccer Team with analysis of their opponents at the 1996 Olympics Games. A 1997 job advert found on the Internet also suggests that Manchester United may also now be making use of their system (`http://www.umist.ac.uk/INDEX_ONE/softsport.html`).

SECOND LOOK requires a user to input a game by sequentially recording the *position* on the pitch of each kick of the ball, and also the player in possession. This restriction to individual kicks facilitates the process of data entry, since the positions of the remaining players (who don't have the ball) do not have to be specified. The user-entered data is processed to automatically generate graphics, charts and detailed pitch diagrams that analyze the game. In the previous section we have already described the expert analysis capabilities of MIKE. Now we can extend the list of possible soccer analyses by summarizing the further game interpretations suggested by SECOND LOOK.

- **Pass bigrams**. Like MIKE, SECOND LOOK maintains pass bigrams, but adds the option of displaying these bigrams for individual players rather than for a whole team. This emphasis on individual players is repeated in many of SECOND LOOK's analyses. For example, the number of different players that each team member manages to pass to is compared. Also, each player's passing success rate and ball possession time is maintained separately (and compared to a database giving statistics collected over many matches for defenders, midfielders, attackers and goalkeepers).

- **Shots on goal**. How many shots does each player make? SECOND LOOK also analyses what proportion of a team's shots come from defenders, attackers, or midfielders, and where a team's shots come from on the field. In the latter case, a graphical display can be particularly revealing, indicating for example whether a team favors a particular side of the field, or close-range or long distance strikes (an example of this analysis is shown in Figure 6). Note that identifying a 'shot' is not a straightforward task for automated systems, since it involves reasoning about players intentions. In MIKE's bigrams, for instance, shots are not recorded unless they produce goals or go to the opposing keeper.
- **Impact passes**. SECOND LOOK introduces this useful concept, defined as all those passes made by a team in a series that results in a shot on goal. A graphic displaying all the chains of impact passes can reveal information such as 'all plays started from the right-hand side of the Mexican attack'. For real soccer, such pass chains are also known to often display directional changes. A shot or a goal is also counted as an impact pass.
- **Completed passes**. A completed pass is a pass between two players of the same team. Completed passes can be displayed on a pitch diagram for individual players or for an entire team. Analyzing these diagrams can lead to analysis such as which side of the pitch a team is favoring: 'the Mexican

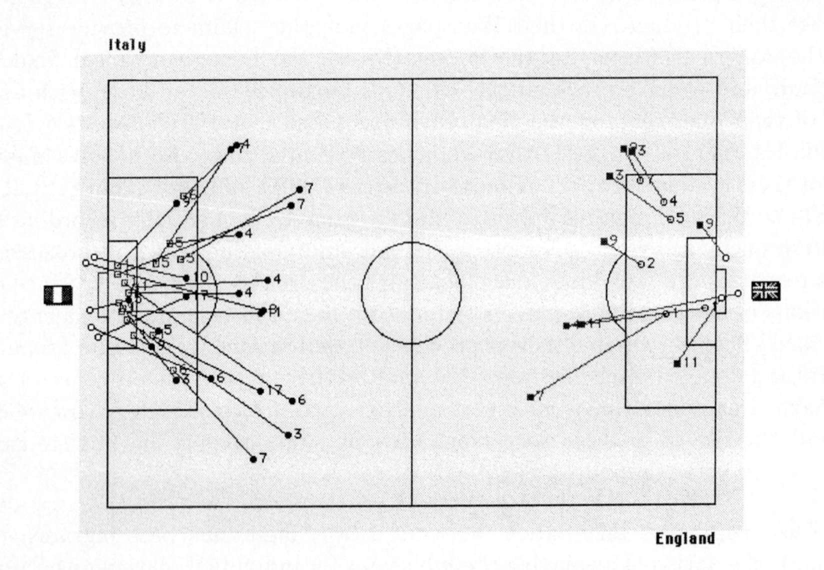

Fig. 6. Example screenshot from SECOND LOOK, showing all the shots on goal taken by Italy (left to right) and England (right to left) in their first World Cup qualifier in 1996. This image is reproduced by kind permission of SoftSport Inc.

team had a definite strategic shift to the left side of the Brazilian defense'. The number of completed passes within the opponent's third of the pitch (and in the penalty area) are also indicators of attacking strength. Further useful statistics to maintain are the percentage of each player's passes that are forward, and whether the passes received by attackers are long or short.
- **Turnover analysis**. SECOND LOOK monitors where the majority of the changes in possession occur.
- **Range**. SECOND LOOK can also produce a pitch diagram displaying every location that a player touched (kicked) the ball. This is very useful for examining how much of the pitch is covered by each player.
- **Corner kicks**. How many corner kicks did each player force? This is harder than simply counting the corner kicks taken by each team, since automatically identifying the player forcing the kick may not be trivial.

All the statistics produced by SECOND LOOK should be reproducible in the RoboCup environment. In fact, the level of information in the logs of the Soccer Server should provide even more opportunities for analysis (and also make the task of analysis still more challenging). As a first step, we plan to combine the MIKE and the SECOND LOOK expert analysis techniques described in this paper into a single program for automatically analyzing Soccer Server games. As we stated in §2, we believe that such a resource will be a powerful tool for researchers working on RoboCup.

5.2 SOCCER and ROCCO

In terms of academic research on automated soccer commentary, we are aware of only one other project: the SOCCER system described in [Andrè *et al* 88] (and more recently in [Andrè *et al* 94]). However, this system addresses a number of the research issues we identified in §2. For instance, the original version of SOCCER tackles machine vision and scene interpretation problems by using *geometrical scene descriptions* [Neumann 89] to interpret short sections of video recordings of real soccer games. It also employs multimedia presentation techniques to combine text, graphics and video in its presentations of these recordings. An updated version of SOCCER, called ROCCO, is being designed specifically for RoboCup, and is to be described in the proceedings of the 1997 RoboCup workshop (not yet in print at the time of writing this paper). Both ROCCO and SOCCER tackle the problem of incremental recognition of events, developing a model that allows the initiation of events to be recognized and commented upon before the event is actually completed.

For the task of simultaneous description, SOCCER has a control loop that selects events according to their *topicality*, where topicality is determined by the salience of an event and the time that has passed since its occurrence — a similar notion to the importance values used by MIKE. However, in SOCCER it seems that inferences are not used to identify higher-level, state-based explanations. MIKE not only does this, but also uses its importance values to guide its handling of natural language issues such as interruption, abbreviation and repetition.

Unlike MIKE, however, SOCCER can also carry out post-game analysis, for which the system has more complicated control structures, plus techniques for searching for events of certain types, modifying the viewpoint from which action is viewed, and enhancing an image to concentrate attention on important features, such as the superimposition of an arrow to indicate the forthcoming movements of ball or players. In terms of simple text, the system produces output such as 'Bommer, the midfield player, passes the ball... to Bosch the outside left... He is attacked by Müller, the outside right.'

5.3 BYRNE

Another commentary-related research project is the talking head BYRNE system described elsewhere in this volume [Binsted 98]. BYRNE uses the domain of soccer to investigate the effects of personality and emotion on the 'believability' of a designed character — in this case, a soccer commentator. BYRNE assumes that the identification of events in the domain is carried out externally and that they have 'relevance scores' that decay with time. It then generates *emotional structures* that describe the character's reactions to these events. These emotional structures have seven different possible types: fear, anger, sadness happiness, disgust and surprise, and interest. The process for generating these structures takes into account a set of *static characteristics* of the commentator, thus allowing different personalities to be explored. The final system output is then realized via a text-to-speech synthesizer and a facial animation system. BYRNE is due to be demonstrated at the 1998 RoboCup; we anticipate that it will provide an excellent example of how the domain of soccer commentary facilitates the investigation of a diverse range of research issues.

6 Conclusions

We have examined the role of automatic soccer commentary within the RoboCup initiative. Initially, we identified soccer commentary as allowing and requiring investigation of an unusually wide variety of research topics, many of which could not be addressed by the simple development of teams for the RoboCup leagues themselves. Next, we highlighted the crucial role of expert analysis, not just for the development of a commentary system, but also for the tackling of RoboCup challenges such as learning, opponent modeling and teamwork. We reviewed our own soccer commentary system, MIKE, paying special attention to its techniques for expert analysis. We also looked at related systems, highlighting the research issues that have been tackled to date, and also any further expert analysis techniques. Far from simply providing extra atmosphere in live games, then, we hope we have demonstrated that soccer commentary has the potential to play an indispensable role within the overall RoboCup initiative.

References

[Andrè *et al* 88] E. Andrè, G. Herzog, and T. Rist. On the simultaneous interpretation of real world image sequences and their natural language description: The system SOCCER. In *Proc. of the 8th ECAI*, pages 449–454, Munich, 1988.

[Andrè *et al* 94] E. Andrè, G. Herzog, and T. Rist. Multimedia presentation of interpreted visual data. In *Proceedings of AAAI-94, Workshop on Integration of Natural Language and Vision Processing*, pages 74–82, Seattle, WA, 1994.

[Binsted 98] K. Binsted. Character design for soccer commentary. In *Proc. of the Second International Workshop on RoboCup*, pages 22–36, 1998.

[Casti 97] John L. Casti. *Would-be Worlds: how simulation is changing the frontiers of science.* John Wiley and Sons, Inc, 1997.

[Grice 75] H. P. Grice. Logic and conversation. In P. Cole and J. L. Morgan, editors, *Syntax and Semantics: Speech Acts*, volume 3, pages 41–58. Academic Press, 1975.

[Jones 96] R. Jones. Quoted in Carlisle United fanzine, August 1996.

[Kitano *et al* 97a] H. Kitano, M. Asada, Y. Kuniyoshi, I. Noda, E. Osawa, and H. Matsubara. RoboCup: A challenge problem for AI. *AI Magazine*, pages 73–85, Spring 1997.

[Kitano *et al* 97b] H. Kitano, M. Tambe, P. Stone, M. Veloso, S. Coradeschi, E. Osawa, H. Matsubara, I. Noda, and M. Asada. The RoboCup synthetic agent challenge 97. In *Proceedings of IJCAI-97*, pages 24–29, Nagoya, Japan, 1997.

[Neumann 89] B. Neumann. Natural language description of time-varying scenes. In D.L. Waltz, editor, *Semantic Structures: Advances in Natural Language Processing*, pages 167–207. Lawrence Erlbaum, 1989. ISBN 0-89859-817-6.

[Noda *et al* 98] I. Noda, H. Matsubara, K. Hiraki, and I. Frank. Soccer Server: a tool for research on multi-agent systems. *Applied Atificial Intelligence*, 12(2–3):233–251, 1998.

[Oishi & Sugihara 95] Y. Oishi and K. Sugihara. Topology-oriented divide-and-conquer algorithm for Voronoi diagrams. *Graphical Models and Image Processing*, 57(4):303–314, July 1995.

[Sof] SECOND LOOK soccer analysis software. The makers, SoftSport Inc., can be found on the Web at `http://www.softsport.com/`.

[Taki *et al* 96] T. Taki, J. Hasegawa, and T. Furukawa. Development of motion analysis system for quantitative evaluation of teamwork in soccer games. In *Proc. of ICIP'96*, 1996.

[Tanaka-Ishii *et al* 98a] K. Tanaka-Ishii, K. Hasida, and I. Noda. Reactive content selection in the generation of real-time soccer commentary. In *Proceedings of COLING-ACL'98*, pages 1282–1288, Montreal, 1998.

[Tanaka-Ishii *et al* 98b] K. Tanaka-Ishii, I. Noda, I. Frank, H. Nakashima, K. Hasida, and H. Matsubara. MIKE: An automatic commentary system for soccer. In *Proceedings of ICMAS-98*, pages 285–292, 1998.

Rocco: A RoboCup Soccer Commentator System

Dirk Voelz, Elisabeth André, Gerd Herzog, and Thomas Rist

DFKI GmbH, German Research Center for Artificial Intelligence
D-66123 Saarbrücken, Germany
{andre,herzog,rist,voelz}@dfki.de

Abstract. With the attempt to enable robots to play soccer games, the RoboCup challenge poses a demanding standard problem for AI and intelligent robotics research. The rich domain of robot soccer, however, provides a further option for the investigation of a second class of intelligent systems which are capable of understanding and describing complex time-varying scenes. Such automatic commentator systems offer an interesting research perspective for additional integration of natural language and intelligent multimedia technologies.
In this paper, first results concerning the realization of a fully automated RoboCup commentator will be presented. The system called Rocco is currently able to generate TV-style live reports for arbitrary matches of the RoboCup simulator league. Based upon our generic approach towards multimedia reporting systems, step-by-step even more advanced capabilities are to be added with future versions of the initial Rocco prototype.

1 Introduction

The Robot World-Cup Soccer (RoboCup) challenge [8, 7] poses a common standard problem for a broad spectrum of specialized subfields in Artificial Intelligence and intelligent robotics research. Obviously, advanced techniques related to autonomous agents, collaborative multi-agent systems, real-time reasoning, sensor-fusion, etc. are required to enable robots to play soccer games. The rich domain of robot soccer, however, provides a further option for the development of a second class of intelligent systems that are capable of understanding and describing complex time-varying scenes. Such automatic commentator systems offer an interesting perspective for the additional integration of natural language and intelligent multimedia technology into the research framework.

In general, the combination of techniques for scene analysis and intelligent multimedia generation has a high potential for many application contexts since it will open the door to an interesting new type of computer-based information system that provides highly flexible access to the visual world.

A first approach towards the automated generation of multimedia reports for time-varying scenes on the basis of visual data has been introduced in our own previous work [2]. For the experimental investigations reported in [2] short sections of video recordings from real soccer games had been chosen as domain of discourse. In [3] our initial approach has been further elaborated and carried forward to the RoboCup domain. The system ROCCO (RoboCupCommentator) presented here constitutes a first practical result of our recent work related to the domain of robot soccer. The current ROCCO prototype is able to generate TV-style live reports for arbitrary matches of the RoboCup simulator league.

The exceptional research potential of automatic commentator systems is well reflected by the fact that at least two more related research activities have been started within the the context of RoboCup. Similar to the initial ROCCO version, the system MIKE (Multi-agent Interactions Knowledgeably Explained) described in [12] is designed to produce simultaneous spoken commentary for matches from the RoboCup simulator league. A specific focus of the MIKE approach is an elaborated analysis of agent behavior. The system aims to explain and classify interactions between multiple agents in order to provide an evaluation of team play and to generate predictions concerning the short-term evolution of a given situation.

The system BYRNE [4] employs an animated talking head as a commentator for matches of the RoboCup simulator league. The main focus of this work is on the generation of appropriate affective speech and facial expressions, based on the character's personality, emontional state, and the state of the play. Currently, BYRNE does not connect to the RoboCup soccer simulator program directly but uses pre-analysed game transcipts as input.

2 The Rocco System

A practical goal of our current experimental investigations in the area of multimedia reporting is the development and continuous advancement of ROCCO, a system for the automatic generation of reports for RoboCup soccer games.

As a first step, we restrict ourselves to matches of the simulator league which involves software agents only. This allows us to disregard the intrinsically difficult task of automatic image analysis instead of using real image sequences and starting from raw video material like in our previous work on the SOCCER commentator [1].

The basic architecture of the ROCCO system is sketched in Fig. 1 (cf. [3]). ROCCO relies on the RoboCup simulator called SOCCER SERVER [10], which is a network-based graphic simulation environment for multiple autonomous mobile robots in a two-dimensional space. The system provides a virtual soccer field and allows client programs (i.e. software robots) to connect to the server and control a specific player within the simulation. ROCCO utilizes the real-time game log that the SOCCER SERVER is able to supply to monitoring programs. The continously updated information includes the absolute positions of all mobile objects (i.e players and ball) as well as additional data concerning game score and play modes (like for example a *goal-kick* or a *throw-in*). On the basis of this material, ROCCO aims to provide a running report for the scene under consideration. The graphical user interface for monitoring object movements

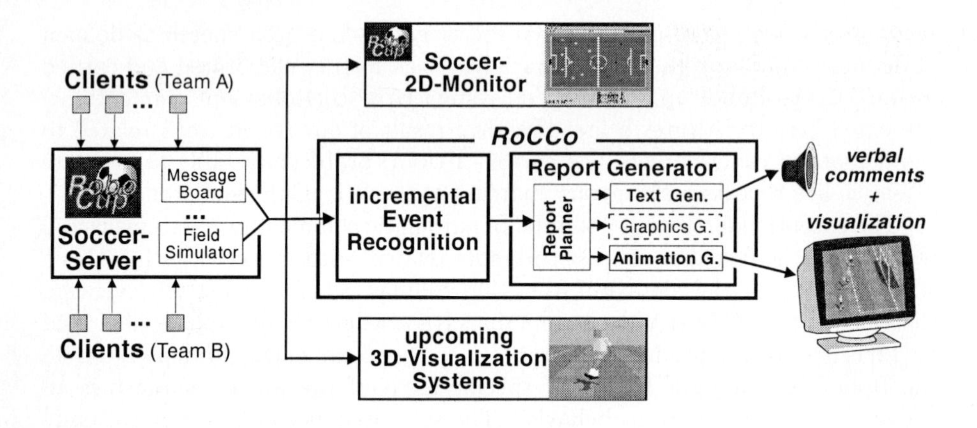

Fig. 1. Architecture of the ROCCO system

included within the SOCCER SERVER environment can be exploited as a simple visualization component. More advanced graphical displays become possible with already announced components for three-dimensional visualization (cf. [8]).

The automatic description of a time-varying scene can be thought of as a complex transformation process. Starting from a continuous flow of raw data, which basically describe simple geometric aspects of a scene, a more advanced analysis of the evolving scene is required in order to recognize relevant occurences that can be communicated to the user. ROCCO includes a dedicated component for the incremental recognition of interesting events in the time-varying scene. The resulting higher-level conceptual units form the input of the core report generator which exploits them for the construction of a multimedia report employing verbal comments and accompanying visualizations.

The initial Java-based ROCCO prototype, as it is shown in Fig. 2, is able to generate TV-style live reports for arbitrary matches of the RoboCup simulator league. The screenshot was taken during a typical session with the ROCCO system. In this case, the SOCCER SERVER logplayer is used to playback a previously recorded match. The window on the left contains a transcript of the spoken messages that have been generated for this example scene. Using the TRUETALK text-to-speech software, ROCCO is able to control intonation in order to generate a more lively commentary. The testbed character of the ROCCO system makes it easy to experiment with generation parameters to further explore the various possibilities for report generation.

3 High-level Scene Analysis

The continuously updated real-time game log from the RoboCup SOCCER SERVER forms the input for the ROCCO commentator system. This kind of *geometrical scene description* (cf. [9]) contains information concerning visible objects and

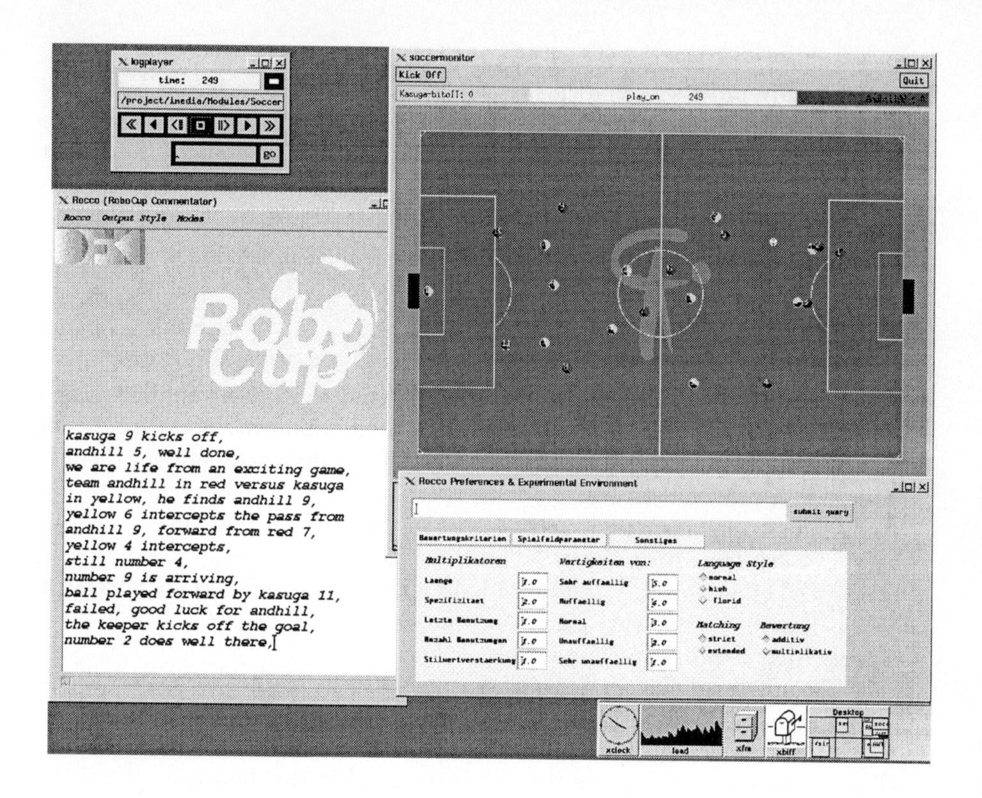

Fig. 2. The basic windows of the initial ROCCO system

their locations over time, together with additional world knowledge about the objects. High-level scene analysis provides an interpretation of this basicly visual information and aims at recognizing conceptual units at a higher level of abstraction. The information structures resulting from such an analysis encode a deeper understanding of the time-varying scene to be described. In general, they include spatial relations for the explicit characterization of spatial arrangements of objects, representations of recognized object movements, and further higher-level concepts such as representations of behaviour and interaction patterns of the agents observed (see [6]).

Within the ROCCO prototype we restrict ourselves to the incremental recognition of relevant events, i.e. interesting object movements involving players and the ball. The representation and automatic recognition of events is inspired by the generic approach described in [6], which has been adopted in ROCCO according to the practical needs of the application context. Declarative event concepts represent a priori knowledge about typical occurrences in a scene. These event concepts are organized into an abstraction hierarchy, grounded on specialization and temporal decomposition. Simple recognition automata each of which corresponds to a specific concept definition are used to instantiate events on the basis of the underlying scene data.

A peculiarity of the simulator league setting is the fact that the SOCCER SERVER already provides some elementary events related to selected actions of a single player (like *'kick'* or *'catch'*). Hence, ROCCO only needs to translate this data into a suitable representation format instead of doing a full event recognition. Elementary event and state predicates, which are oriented towards a single timepoint, can be utilized to define higher-level event concepts. Specific examples for these kind of elementary predicates are the following:

- *(Type:HasBall Time:time Agent:X)*
  Judging which agent actually is in possession of the ball, is an important part of event recognition in the soccer domain. As robots and softbots sometimes do not even perceive the ball when standing next to it, this kind of decision is not always an easy task in the RoboCup environment. For ROCCO, an agent X is considered to have the ball, if (1) the distance between them is less than a predefined constant value, (2) there is no one else closer to the ball, and (3) the agent is or was at least recently facing the ball.
- *(Type:Speed Time:time Object:X Speed:speed)*
  The speed of an object X (player or ball) will be interpreted as no movement if it is less than some very small constant value.
- *(Type:Angle Time:time Object:X Angle:angle)*
  The absolute angle associated with an object X depends on its current orientation or direction of movement.
- *(Type:Kick Time:time Agent:X)*
  The fact that agent X kicks the ball is recorded in the data from the simulator.
- *(Type:Region Time:time Object:X Region:region)*
  The playground is divided into several regions, like for example the penalty area. This predicate signifies that a certain object, i.e. the ball or a relevant player, enters a specific area of high interest.
- *(Type:Playmode Time:time Mode:mode)*
  A playmode change is to be interpreted as a game command by the referee (e.g. *'goal'*, *'upside'*) which also counts as an event that may be communicated.

From current and previous elementary predicates higher order events are calculated. Consider a *ball-transfer* event into the penalty area as an example:

(Type:BallTransfer
  StartTime:178 EndTime:185
  Spec:EnterPenaltyArea Agent:red9 Recipient:red7)

The recognition of a ball-transfer event depends on the recognition of all of its sub-events as listed below:

(Type:HasBall Time:start-time Agent:agent)
(Type:Kick Time:start-time Agent:agent)
(Type:Speed Time:time Object:Ball Speed:speed) with
  $speed > 0$ for all $start - time \leq time \leq end - time$.

(Type:Angle Time:time Object:Ball Angle:angle) with
$angle$ is constant for all $start - time \leq time \leq end - time$.
(Type:Region Time:start-time Object:Ball Region:NotPenaltyArea)
(Type:Region Time:end-time Object:Ball Region:PenaltyArea)
(Type:HasBall Time:end-time Agent:recipient)
As a further condition, there must be no event of the form
(Type:HasBall Time:time Agent:agent) with
$start - time < time < end - time$.

The other event concepts in the knowledge base of the ROCCO system are defined in a similar way. The initial ROCCO system includes about 15 event definitions covering essential soccer event concepts.

4 Report Generation

The initial ROCCO prototype is designed as a robust TV-style live commentator system for the RoboCup simulator league. The system generates simple multimedia reports which combine a 2-dimensional graphical animation of the soccer scene with a spoken commentary. ROCCO employs the monitor program provided with the RoboCup simulation environment for the graphical display. Currently, the central task of the reporting module is the generation of a verbal description of the scene under consideration.

One of the characteristics of a live report is the fact that temporal constraints makes it impossible to comment every detail within the time-varying scene. Given the large number of simultaneous occurances within a soccer scene, the verbalisation process continuously needs to carefully select those events that should be communicated. Selected events are passed on to the realisation component which uses a template-based approach to generate corresponding natural language utterances. For each event instance, an appropriate template is selected and instantiated. Then the templates are annotated with intonational information, and finally text-to-speech software is used to generate audio output.

4.1 Discourse Planning

Report generation in a system that is continuously provided with new data calls for a flexible content selection mechanism. At each point in time, the system has to decide which events should be verbalized. To solve this problem, ROCCO relies on an incremental discourse planning mechanism, paying special attention to timing constraints. Before generating a new comment, ROCCO performs the following steps:

Threshold calculation: A minimum importance value for the next comment is calculated first. It consists of a *basic threshold* which decreases during periods of inactivity of the commentator system, then allowing less important events to be commented. Furthermore, a longer time of inactivity will also increase the chance to state some background comment instead of describing the current activities.

Calculation of topicality: All new recognized events are evaluated for their topicality. Topicality is determined by the salience of an event and the time that has passed since its occurrence.

Content selection: The event with the highest topicality is selected for verbalization if the topicality value exceeds the current threshold for the minimum importance value. If no such event exists, either a comment containing background information is randomly selected, or the systems waits until more important events have been recognized or the threshold has lowered enough to select a less important event. Once an event is selected, the minimum importance value will be set back to the basic threshold.

Human soccer commentators often report an event while it is taking place, which leads to common phrases like: *"Miller passes in ⋯ to Meier"*. To meet the requirements of simultaneous scene description, the recognition component also provides information concerning only partly-recognized events. As a consequence, the verbalization component cannot always start from completely worked-out conceptual contents. Starting with the verbalization of *"Miller passes"* before complete information about the whole event is available, limits the possibilities of discourse planning and sometimes urges even human reporters to use gramatically incorrect sentences. The verbalization component in ROCCO contains various templates for most combinations of partial event instantiations. In the example given, ROCCO achieves to comment the complete event of Miller passing to Meier in the penalty area with a phrase well known from soccer life-reports: *"Miller passes ⋯ now in the penalty area ⋯ Meier gets it"*.

4.2 Generation of Verbal Comments

One objective of the ROCCO system is to provide a large variety of natural-language expressions that are typical of soccer live reports. A more flexible template-based approach is being employed since it is rather difficult to generate syntactic specifications for soccer slang expressions so that may be processed by a fully-fledged natural-language generator. The template knowledge base consists of about 300 text-fragments transcribed from about 12 hours of soccer matches during the WorldCup qualifications 1997. Each event specification is assigned to a list of templates, that might be applicable depending on further contextual constraints. A template consists of fixed strings and variables which are instantiated when a template is applied to a specific event instance. A separate *nominal phrase generator* is utilized to describe agents, teams, etc. with appropriate nominal phrases. To improve the variety of expressions some templates can be modified at random. For example the word *there* can be added to the phrase *well done* with a chance of 30 percent. Table 1 contains a list of some templates for the *ball-transfer* event. For the appropriate choice of a template for the verbalization of the selected event several aspects are taken into account:

Applicability Constraints: Constraints specify whether a certain template can be employed in the current context. For instance, a phrase like *"Now*

Template	Constraints	Vb	Floridity	Sp	Formality	Salience
(?x passes the ball to ?y)	None	8	dry	4	formal	normal
(?x playes the ball towards ?y)	None	8	dry	4	formal	normal
(?x towards ?y)	None	5	dry	3	slang	normal
(?x combines with ?y)	None	6	normal	3	slang	lower
(?x, now ?y)	(NotTop ?y)	5	dry	2	slang	low
(ball played towards ?y)	None	5	dry	3	colloquial	normal
(the ball came from ?x)	(NotTop ?x)	6	dry	3	colloquial	higher
(shot)	None	1	dry	1	colloquial	normal
(?x)	(NotTop ?x)	2	dry	1	slang	low
(?y was there {'again':WasTopic(?y)})	None	4	dry	1	slang	normal
(well done {'there':Random(30%)})	None	2	dry	0	colloquial	normal
(a lovely ball)	None	3	flowery	1	colloquial	higher

Table 1. Some verbalization templates for the event *ball-transfer*

Miller" should not be uttered if Miller is already topicalized. To avoid close similarities of templates in the knowledge base, constraints further control some details of instantiation. Consider the template *(?y was there {'again': WasTopic(?y)})*. If, for example, Miller has been mentioned not long ago, the template will be instantiated as *"Miller was there again"* instead of *"Miller was there"* otherwise.

Verbosity: The verbosity of a template depends on the number of words it contains. While instantiated slots correspond to exactly one word, non-instantiated slots have to be forwarded to the nominal phrase generator which decides what form of nominal phrase is most appropriate. Since the length is not known at the time of template selection a default word number is assumed in this case.

Floridity: We distinguish between dry, normal and flowery language. Flowery language is composed of unusual ad hoc coinages, such as *"a lovely ball"*. Templates marked as normal may contain metaphors, such as *(finds the gap)*, while templates marked as dry, such as *(playes the ball towards ?y)* just convey the plain facts.

Specifity: The specifity of a template depends on the number of verbalized deep cases and the specifity of the natural-language expression chosen for the action type. For example, the specifity of *(?x looses the ball to ?y)* is 4 since 3 deep cases are verbalized and the specifity of the natural-language expression referring to the action type is 1. The specifity of *(misdone)* is 0 since none of the deep cases occurs in the template and the action type is not further specified.

Formality: This attribute can take on the values *slang*, *colloquial* and *normal.* Templates marked as formal are grammatically correct sentences which are more common in newspaper reports. Colloquial templates, such as *"ball played towards Meier"*, are simple phrases characteristic of informal conversation. Slang templates are colloquial templates peculiar to the soccer domain, such as *"Miller squeezes it through"*.

Salience: To generate more interesting reports, unusual templates, such as *"Miller squeezes it through"* should not be used too frequently.

To select a template, ROCCO first determines all templates associated with the given event concept and checks whether the constraints are satisfied. For each template, a situation-specific preference value will be calculated and the template with the highest preference value is finally selected. If the system is under time pressure, a template will be punished for its length, but rewarded for its specifity. In phases with little activity, all templates will get a reward both for their length and specifity. In all other situations, only the specifity of a template will be rewarded. Templates that have recently or frequently been used are punished and unusual templates, i.e. templates that seldom occur in a soccer report, as well as very flashy comments get an additional punishment while inconspicuous phrases get a bonus.

To generate descriptions for a player or a team, the nominal phrase generator is applied which takes into account the discourse context to determine whether to generate a pronoun or a noun phrase (with modifiers). In some cases, the template structure allows to leave out the object description if the object is already topicalized. An example is: *"Here comes Miller ··· (he) combines with Meier"*.

4.3 Using Intonation to Convey Emotions

Intonation should not only correlate with the syntactical and semantical structure of an utterance, but also reflect the speaker's intentions. Furthermore, intonation is an effective means of conveying the speaker's emotions.

To generate affective speech, ROCCO first identifies the emotions of the assumed speaker. For instance, starting with a neutral observer, succesful actions of a team or player will lead to excitement, failed actions to disappointment. Excitement will increase with decreasing distance to the goal and find its peak when a goal is scored or, in the case of disappointment, when a shot misses the goal. As excitement and disappointment are main emotions occurring in a soccer report, these were selected as a starting point for the generation of affective speech in the ROCCO prototype.

Trying to map emotions onto instructions for the speech synthesizer, we realized that already the variation of only two parameters could lead to amazing results. In particular these parameters are *pitch accent* and *speed.* A pitch accent is applied to a word to convey sentential stress. In ROCCO we currently use two out of six pitch accents introduced in [11]:

1. *H** is a *'high'* accent which, roughly speaking, stresses the indicated syllable with a high tone.
2. *L** is a *'low'* accent which, is realized as a tone that occurs low in the speaker's pitch range.

Speed is another effective means of conveying emotions. ROCCO increases its talking speed to express excitement, and slightly slows down to express disappointment.

These parameters are modeled similar to the emotions *glad/indignant* and *sad/distraught* as described in [5] where a highly elaborated system of synthesizer instructions for the production of different emotions is introduced. The approach presented in [5] proposes a slightly faster speech rate and a high pitch range for gladness, and a very slow speech rate and a negative pitch for sadness.

Since we rely a template-based approach, it is possible to annotate natural-language utterances directly with intonational information. However, in some cases, ROCCO will override the default intonation to track the center of attention. Consider the phrase *"Miller got it"* as an example. If it has just been commented that he is trying to get the ball from Meier of the opposing team, and hence Miller is alread topicalized, ROCCO will accent the word "*got*" to indicate the new information. Alternatively if Miller is not topicalized, the player's name is to be emphasized. In this case, ROCCO will override the preset accent markers, and add a new H^* mark for the noun phrase "*Miller*". Depending on the current amount of excitement, ROCCO will then set speed and pitch range.

5 Conclusions and Further Work

The initial ROCCO prototype presented here constitutes a first promising step towards a multimedia reporting systems for RoboCup soccer matches. ROCCO is a robust TV-style live commentator system for the RoboCup simulator league. It combines emotional spoken descriptions of the running scene with the graphical display provided through the RoboCup simulation environment.

The approach described here is motivated by our vision of a novel type of computer-based information system that provides fully automated generation of multimedia reports for time-varying scenes on the basis of visual data. This work and related activities illustrate the high research potential for the integration of high-level scene analysis and intelligent multimedia presentation generation, especially in the context of the RoboCup challenge.

Our current experimental investigations concentrate on further improvements of the first ROCCO version. One interesting topic is the incorporation of a more elaborated model for the generation of affect in synthesized speech. A second line of research relates to the extension of multimedia presentation capabilities. We are currently working on our own 3D visualization component to enable intelligent control of the camera view as a novel feature of the multimedia presentation generator.

A third important aspect for us is the move from softbot games in the simulator league to RoboCup matches involving real robots. The main problem there is to obtain a suitable geometrical scene description that can be fed into our reporting system. Instead of relying on an external observer, the idea is to exploit the analyzed sensor information of the acting robots. Hence, no additional camera and no specific image analysis will be required. However, it can be expected that the geometric scene description to be obtained from one of the robot soccer teams will be less exact and less complete than in the case of the RoboCup simulator. Our plans are to further investigate these issues in close cooperation with one of the leading robot soccer teams at Universität Freiburg.

References

1. E. André, G. Herzog, and T. Rist. On the Simultaneous Interpretation of Real World Image Sequences and their Natural Language Description: The System SOCCER. In *Proc. of the 8th ECAI*, pages 449–454, Munich, Germany, 1988.
2. E. André, G. Herzog, and T. Rist. Multimedia Presentation of Interpreted Visual Data. In P. Mc Kevitt, editor, *Proc. of AAAI-94 Workshop on "Integration of Natural Language and Vision Processing"*, pages 74–82, Seattle, WA, 1994. Also available as Report no. 103, SFB 314 – Project VITRA, Universität des Saarlandes, Saarbrücken, Germany.
3. E. André, T. Rist, and G. Herzog. Generating Multimedia Presentations for Robocup Soccer Games. In H. Kitano, editor, *RoboCup-97: Robot Soccer World Cup I*, pages 200–215. Springer, Berlin, Heidelberg, 1998.
4. K. Binsted. Character Design for Soccer Commentary. In *Proc. of the Second International Workshop on RoboCup*, pages 25–35, Paris, 1998.
5. J. E. Cahn. The Generation of Affect in Synthesized Speech. *Journal of the American Voice I/O Society*, 8:1–19, 1990.
6. G. Herzog and P. Wazinski. VIsual TRAnslator: Linking Perceptions and Natural Language Descriptions. *AI Review*, 8(2/3):175–187, 1994.
7. H. Kitano, editor. *RoboCup-97: Robot Soccer World Cup I*. Springer, Berlin, Heidelberg, 1998.
8. H. Kitano, M. Asada, Y. Kuniyoshi, I. Noda, E. Osawa, and H. Matsubara. RoboCup: A Challenge Problem for AI. *AI Magazine*, 18(1):73–85, 1997.
9. B. Neumann. Natural Language Description of Time-Varying Scenes. In D. L. Waltz, editor, *Semantic Structures: Advances in Natural Language Processing*, pages 167–207. Lawrence Erlbaum, Hillsdale, NJ, 1989.
10. I. Noda. Soccer Server Manual Rev. 2.00 (for Soccer Server Ver. 3.00 and later). Internal document, ETL Communication Intelligence Section, Tsukuba, Japan, 1998.
11. J. B. Pierrehumbert. Synthesizing Intonation. *Journal of the Acoustical Society of America*, 70:985–995, 1981.
12. K. Tanaka, I. Noda, I. Frank, H. Nakashima, K. Hasida, and H. Matsubara. MIKE: An Automatic Commentary System for Soccer. In *Proc. of ICMAS'98, Int. Conf. on Multi-agent Systems*, pages 285–292, Paris, France, 1998.

The CMUnited-98 Champion Simulator Team *

Peter Stone, Manuela Veloso, and Patrick Riley

Computer Science Department,Carnegie Mellon University
Pittsburgh, PA 15213
{pstone,veloso}@cs.cmu.edu, priley@andrew.cmu.edu

Abstract. The CMUnited-98 simulator team became the 1998 RoboCup simulator league champion by winning all 8 of its games, outscoring opponents by a total of 66–0. CMUnited-98 builds upon the successful CMUnited-97 implementation, but also improves upon it in many ways. This chapter describes the complete CMUnited-98 software, emphasizing the recent improvements. Coupled with the publicly-available CMUnited-98 source code, it is designed to help other RoboCup and multi-agent systems researchers build upon our success.

1 Introduction

The CMUnited-98 simulator team became the 1998 RoboCup [4] simulator league champion by winning all 8 of its games, outscoring opponents by a total of 66–0. CMUnited-98 builds upon the successful CMUnited-97 implementation [8], but also improves upon it in many ways.

The most notable improvements are the individual agent skills and the strategic agent positioning in anticipation of passes from teammates. While the success of CMUnited-98 also depended on our previous research innovations including layered learning [9], a flexible teamwork structure [10], and a novel communication paradigm [10], these techniques are all described elsewhere. The purpose of this article is to clearly and fully describe the low-level CMUnited-98 agent architecture as well as the key improvements over the previous implementation.

Coupled with the publicly-available CMUnited-98 source code [11], this article is designed to help researchers involved in the RoboCup software challenge [5] build upon our success. Throughout the article, we assume that the reader is familiar with the soccer server [1].

The rest of the article is organized as follows. Section 2 gives an overview of the agent architecture. Section 3 describes the agents' method of keeping an accurate and precise world model. Section 4 details the agents' low-level skills. Section 5 presents the CMUnited-98 collaborative coordination mechanisms. Section 6 summarizes the RoboCup-98 results and Section 7 concludes.

2 Agent Architecture Overview

CMUnited-98 agents are capable of perception, cognition, and action. By perceiving the world, they build a model of its current state. Then, based on a set of behaviors, they choose an action appropriate for the current world state.

* This research is sponsored in part by the DARPA/RL Knowledge Based Planning and Scheduling Initiative under grant number F30602-95-1-0018. The views and conclusions contained in this document are those of the authors and should not be interpreted as representing the official policies or endorsements, either expressed or implied, of the U. S. Government.

A driving factor in the design of the agent architecture is the fact that the simulator operates in fixed cycles of length 100 msec. As presented in Section [1], the simulator accepts commands from clients throughout a cycle and then updates the world state all at once at the end of the cycle. Only one action command (dash, kick, turn, or catch) is executed for a given client during a given cycle.

Therefore, agents (simulator clients) should send exactly one action command to the simulator in every simulator cycle. If more than one command is sent in the same cycle, a random one is executed, possibly leading to undesired behavior. If no command is sent during a simulator cycle, an action opportunity has been lost: opponent agents who have acted during that cycle may gain an advantage.

In addition, since the simulator updates the world at the end of every cycle, it is advantageous to try to determine the state of the world at the end of the previous cycle when choosing an action for the current cycle. As such, the basic agent loop during a given cycle t is as follows:

- Assume the agent has consistent information about the state of the world at the end of cycle $t-2$ and has sent an action during cycle $t-1$.
- While the server is still in cycle $t-1$, upon receipt of a sensation (see, hear, or sense_body), store the new information in temporary structures. Do not update the current state.
- When the server enters cycle t (determined either by a running clock or by the receipt of a sensation with time stamp t), use all of the information available (temporary information from sensations and predicted effects of past actions) to **update the world model** to match the server's world state (the "real world state") at the end of cycle $t-1$. Then **choose and send an action** to the server for cycle t.
- Repeat for cycle $t+1$.

While the above algorithm defines the overall agent loop, much of the challenge is involved in updating the world model effectively and choosing an appropriate action. The remainder of this section goes into these processes in detail.

3 World Modeling

When acting based on a world model, it is important to have as accurate and precise a model of the world as possible at the time that an action is taken. In order to achieve this goal, CMUnited-98 agents gather sensory information over time, and process the information by incorporating it into the world model immediately prior to acting.

3.1 Object Representation

There are several objects in the world, such as the goals and the field markers which remain stationary and can be used for self-localization. Mobile objects are the agent itself, the ball, and 21 other players (10 teammates and 11 opponents). These objects are represented in a type hierarchy as illustrated in Figure 1.

Each agent's world model stores an instantiation of a stationary object for each goal, sideline, and field marker; a ball object for the ball; and 21 player

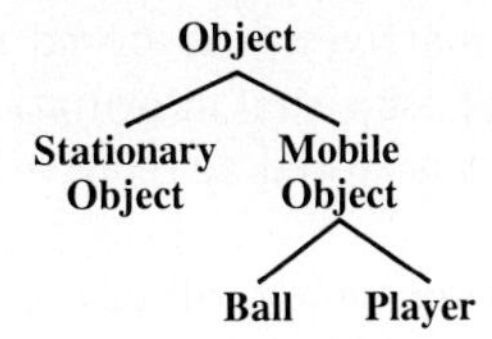

Fig. 1. The agent's object type hierarchy.

objects. Since players can be seen without their associated team and/or uniform number, the player objects are not identified with particular individual players. Instead, the variables for team and uniform number can be filled in as they become known.

Mobile objects are stored with confidence values within [0,1] indicating the confidence with which their locations are known. The confidence values are needed because of the large amount of hidden state in the world: no object is seen consistently [2].

The variables associated with each object type are as follows:

Object :
- Global (x, y) position coordinates
- Confidence within [0,1] of the coordinates' accuracy

Stationary Object : nothing additional

Mobile Object :
- Global (dx, dy) velocity coordinates
- Confidence within [0,1] of the coordinates' accuracy

Ball : nothing additional

Player :
- Team
- Uniform number
- Global θ facing angle
- Confidence within [0,1] of the angle's accuracy

3.2 Updating the World Model

Information about the world can come from

- Visual information;
- Audial information;
- Sense_body information; and
- Predicted effects of previous actions.

Visual information arrives as relative distances and angles to objects in the player's view cone. Audial information could include information about global object locations from teammates. Sense_body information pertains to the client's own status including stamina, view mode, and speed.

Whenever new information arrives, it is stored in temporary structures with time stamps and confidences (1 for visual information, possibly less for audial information). Visual information is stored as relative coordinates until the agent's exact location is determined.

When it is time to act during cycle t, all of the available information is used to best determine the server's world state at the end of cycle $t-1$. If no new information arrived pertaining to a given object, the velocity and actions taken are used by the predictor to predict the new position of the object and the confidence in that object's position and velocity are both decayed.

When the agent's world model is updated to match the end of simulator cycle $t-1$, first the agent's own position is updated to match the time of the last sight; then those of the ball and players are updated.

The Agent Itself: The following process is used to update the information about the agent:
- If new visual information has arrived:
  - The agent's position can be determined accurately by using the relative coordinates of one seen line and the closest stationary object.
- If no visual information has arrived:
  - Bring the velocity up to date, possibly incorporating the predicted effects of any actions (a dash) taken during the previous cycle.
  - Using the previous position and velocity, predict the new position and velocity.
- If available, reset the agent's speed as per the sense_body information. Assume velocity is in the direction the agent is facing.
- Bring the player's stamina up to date either via the sense_body information or from the predicted action effects.

The Ball: The ball information is updated as follows:
- If there was new visual information, use the agent's absolute position at the time (determined above), and the ball's temporarily stored relative position to determine the ball's absolute position at the time of the sight.
- If velocity information is given as well, update the velocity. Otherwise, check if the old velocity is correct by comparing the new ball position with the expected ball position.
- If no new visual information arrived or the visual information was from cycle $t-1$, estimate its position and velocity for cycle t using the values from cycle $t-1$. If the agent kicked the ball on the previous cycle, the predicted resulting ball motion is also taken into account.
- If the ball should be in sight (i.e. its predicted position is in the player's view cone), but isn't (i.e. visual information arrived, but no ball information was included), set the confidence to 0.
- Information about the ball may have also arrived via communication from teammates. If any heard information would increase the confidence in the ball's position or velocity at this time, then it should be used as the correct information. Confidence in teammate information can be determined by the time of the information (did the teammate see the ball

more recently?) and the teammate's distance to the ball (since players closer to the ball see it more precisely).

Ball velocity is particularly important for agents when determining whether or not (or how) to try to intercept the ball, and when kicking the ball. However, velocity information is often not given as part of the visual information string, especially when the ball is near the agent and kickable. Therefore, when necessary, the agents attempt to infer the ball's velocity indirectly from the current and previous ball positions.

Teammates and Opponents: In general, player positions and velocities are determined and maintained in the same way as in the case of the ball. A minor addition is that the direction a player is facing is also available from the visual information.

When a player is seen without full information about its identity, previous player positions can be used to help disambiguate the identity. Knowing the maximum distance a player can move in any given cycle, it is possible for the agent to determine whether a seen player could be the same as a previously identified player. If it is physically possible, the agent assumes that they are indeed the same player.

Since different players can see different regions of the field in detail, communication can play an important role in maintaining accurate information about player locations.

4 Agent Skills

Once the agent has determined the server's world state for cycle t as accurately as possible, it can choose and send an action to be executed at the end of the cycle. In so doing, it must choose its local goal within the team's overall strategy. It can then choose from among several low-level skills which provide it with basic capabilities. The output of the skills are primitive movement commands.

The skills available to CMUnited-98 players include kicking, dribbling, ball interception, goaltending, defending, and clearing. The implementation details of these skills are described in this section.

The common thread among these skills is that they are all *predictive, locally optimal skills* (PLOS). They take into account predicted world models as well as predicted effects of future actions in order to determine the optimal primitive action from a local perspective, both in time and in space.

One simple example of PLOS is each individual agent's stamina management. The server models stamina as having a replenishable and a non-replenishable component. Each is only decremented when the current stamina goes below a fixed threshold. Each player monitors its own stamina level to make sure that it never uses up any of the non-replenishable component of its stamina. No matter how fast it should move according to the behavior the player is executing, it slows down its movement to keep itself from getting too tired. While such behavior might not be optimal in the context of the team's goal, it is locally optimal considering the agent's current tired state.

Even though the skills are predictive, the agent *commits* to only one action during each cycle. When the time comes to act again, the situation is completely reevaluated. If the world is close to the anticipated configuration, then the agent will act similarly to the way it predicted on previous cycles. However, if the world is significantly different, the agent will arrive at a new sequence of actions rather than being committed to a previous plan. Again, it will only execute the first step in the new sequence.

4.1 Kicking

There are three points about the kick model of the server that should be understood before looking at our kicking style. First, a kick changes the ball's velocity by vector addition. That is, a kick accelerates the ball in a given direction, as opposed to setting the velocity. Second, an agent can kick the ball when it is in the "kickable area" which is a circle centered on the player (see Figure 2). Third, the ball and the player can collide. The server models a collision when the ball and player are overlapping at the end of a cycle. If there is a collision, the two bodies are separated and their velocities multiplied by -0.1.

As a first level of abstraction when dealing with the ball, all reasoning is done as a desired trajectory for the ball for the next cycle. Before a kick is actually sent to the server, the difference between the ball's current velocity and the ball's desired velocity is used to determine the kick to actually perform. If the exact trajectory can not be obtained, the ball is kicked such that the direction is correct, even if the speed is not.

In order to effectively control the ball, a player must be able to kick the ball in any direction. In order to do so, the player must be able to move the ball from one side of its body to the other without the ball colliding with the player. This behavior is called the *turnball* behavior. It was developed based on code released by the PaSo'97 team[7]. The desired trajectory of a turnball kick is calculated by getting the ray from the ball's current position that is tangent to a circle around the player (see Figure 3).

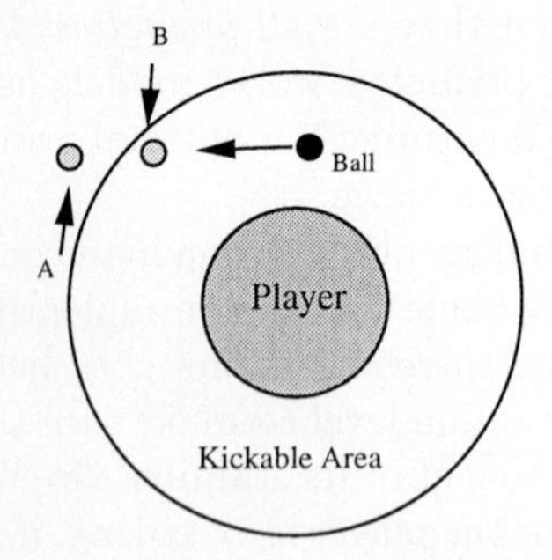

Fig. 2. Basic kicking with velocity prediction.

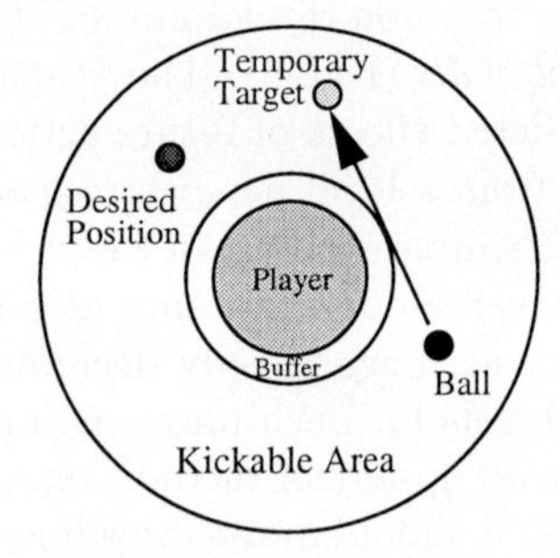

Fig. 3. The turnball skill.

The next important skill is the ability to kick the ball in a given direction, either for passing or shooting. The first step is to figure out the target speed of the ball. If the agent is shooting, the target speed is the maximum ball speed, but for a pass, it might be better to kick the ball slower so that the receiving agent can intercept the ball more easily.

In order to get the ball to the desired speed, several kicks in succession are usually required. By putting the ball to the side of the player (relative to the desired direction of the kick) the agent can kick the ball several times in succession. If a higher ball speed is desired, the agent can use the turnball kicks to back the ball up so that enough kicks can be performed to accelerate the ball.

This skill is very predictive in that it looks at future velocities of the ball given slightly different possible kicks. In some cases, doing a weaker kick one cycle may keep the ball in the kickable area so that another kick can be executed the following cycle. In Figure 2, the agent must choose between two possible kicks. Kicking the ball to position A will result in the ball not being kickable next cycle; if the ball is already moving quickly enough, this action may be correct. However, a kick to position B followed by a kick during the next cycle may result in a higher overall speed. Short term velocity prediction is the key to these decisions.

4.2 Dribbling

Dribbling is the skill which allows the player to move down the field while keeping the ball close to the player the entire time. The basic idea is fairly simple: alternate kicks and dashes so that after one of each, the ball is still close to the player.

Every cycle, the agent looks to see that if it dashes this cycle, the ball will be in its kickable area (and not be a collision) at the next cycle. If so, then the agent dashes, otherwise it kicks. A kick is always performed assuming that on the next cycle, the agent will dash. As an argument, the low-level dribbling code takes the angle relative to the direction of travel at which the player should aim the ball (see Figure 4). This is called the "dribble angle" and its valid values are $[-90, 90]$. Deciding what the dribble angle should be is discussed in Section 4.3.

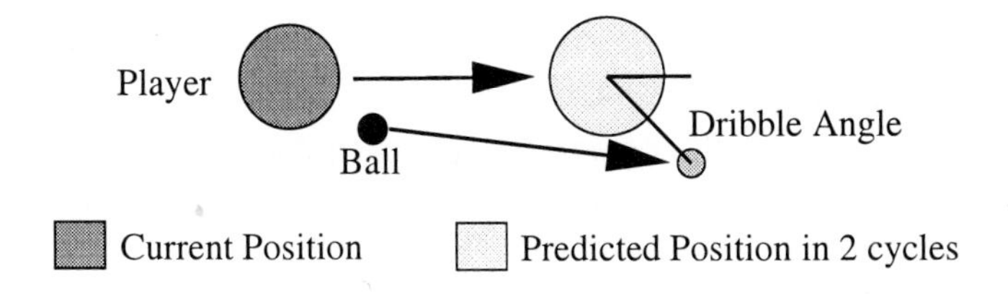

Fig. 4. The basic dribbling skill.

First the predicted position of the agent (in 2 cycles) is calculated:

$$p_{new} = p_{current} + v + (v * pdecay + a)$$

where p_{new} is the predicted player position, $p_{current}$ is the current position of the player, v is the current velocity of the player, $pdecay$ is the server parameter `player_decay`, and a is the acceleration that a dash gives. The a value is usually just the dash power times the `dash_power_rate` in the direction the player is facing, but stamina may need to be taken into account.

Added to p_{new} is a vector in the direction of the dribble angle and length such that the ball is in the kickable area. This is the target position p_{target} of the ball. Then the agent gets the desired ball trajectory by the following formula:

$$traj = \frac{p_{target} - p_{ball}}{1 + bdecay}$$

where $traj$ is the target trajectory of the ball, p_{ball} is the current ball position, and $bdecay$ is the server parameter `ball_decay`. This process is illustrated in Figure 4.

If for some reason this kick can not be done (it would be a collision for example), then a turnball kick is done to get the ball in the right position. Then the next cycle, a normal dribble kick should work.

As can be seen from these calculations, the basic dribbling is highly predictive of the positions and velocities of the ball and player. It is also quite local in that it only looks 2 cycles ahead and recomputes the best action every cycle.

4.3 Smart Dribbling

The basic dribbling takes one parameter that was mentioned above: the dribble angle. Smart dribbling is a skill layered on the basic dribbling skill that decides the best dribble angle based on opponent positions. Intuitively, the agent should keep the ball away from the opponents, so that if an opponent is on the left, the ball is kept on the right, and vice versa.

The agent considers all nearby opponents that it knows about. Each opponent is given a "vote" about what the dribble angle should be; each opponent votes for the valid angle $[-90, 90]$ that is farthest from itself. For example, an opponent at 45 degrees, would vote for -90, while an opponent at -120 degrees would vote for 60. Each opponent's vote is weighted by the distance and angle relative to the direction of motion. Closer opponents and opponents more in front of the agent are given more weight.

4.4 Ball Interception

There are two types of ball interception, referred to as active and passive interception. The passive interception is used only by the goaltender in some particular cases, while the rest of the team uses only the active interception. Each cycle, the interception target is recomputed so that the most up to date information about the world is used.

The *active interception* is similar to the one used by the Humboldt '97 team[3]. The active interception predicts the ball's position on successive cycles, and then tries to predict whether the player will be able to make it to that spot before the ball does, taking into account stamina and the direction that the player is facing. The agent aims for the earliest such spot.

The *passive interception* is much more geometric. The agent determines the closest point along the ball's current trajectory that is within the field. By prediction based on the ball's velocity, the agent decides whether it can make it to that point before the ball. If so, then the agent runs towards that point.

4.5 Goaltending

The assumption behind the movement of the goaltender is that the worst thing that could happen to the goaltender is to lose sight of the ball. The sooner the goaltender sees a shot coming, the greater chance it has of preventing a goal. Therefore, the goaltender generally uses the widest view mode and uses backwards dashing when appropriate to keep the ball in view to position itself in situations that are not time-critical.

Every cycle that the ball is in the defensive zone, the goaltender looks to see if the ball is in the midst of a shot. It does this by extending the ray of the ball's position and velocity and intersecting that with the baseline of the field. If the intersection point is in the goaltender box and the ball has sufficient velocity to get there, the ball is considered to be a shot (though special care is used if an opponent can kick the ball this cycle). Using the passive interception if possible (see Section 4.4), the goaltender tries to get in the path of the ball and then run at the ball to grab it. This way, if the goaltender misses a catch or kick, the ball may still collide with the goaltender and thus be stopped.

When there is no shot coming the goaltender positions itself in anticipation of a future shot. Based on the angle of the ball relative to the goal, the goaltender picks a spot in the goal to guard; call this the "guard point." The further the ball is to the side of the field, the further the goaltender guards to that side. Then, a rectangle is computed that shrinks as the ball gets closer (though it never shrinks smaller than the goaltender box). The line from the guard point to the ball's current position is intersected with the rectangle, and that is the desired position of the goaltender.

4.6 Defending

CMUnited-98 agents are equipped with two different defending modes: opponent tracking and opponent marking. In both cases, a particular opponent player is selected as the target against which to defend. This opponent can either be selected individually or as a defensive unit via communication (the latter is the case in CMUnited-98).

In either case, the agent defends against this player by observing its position over time and position itself strategically so as to minimize its usefulness to the other team. When *tracking*, the agent stays between the opponent and the goal at a generous distance, thus blocking potential shots. When *marking*, the agent stays close to the opponent on the ball-opponent-goal angle bisector, making it difficult for the opponent to receive passes and shoot towards the goal. Defensive marking and tracking positions are illustrated in Figure 5.

When marking and tracking, it is important for the agent to have accurate knowledge about the positions of both the ball and the opponent (although the ball position isn't strictly relevant for tracking, it is used for the decision of

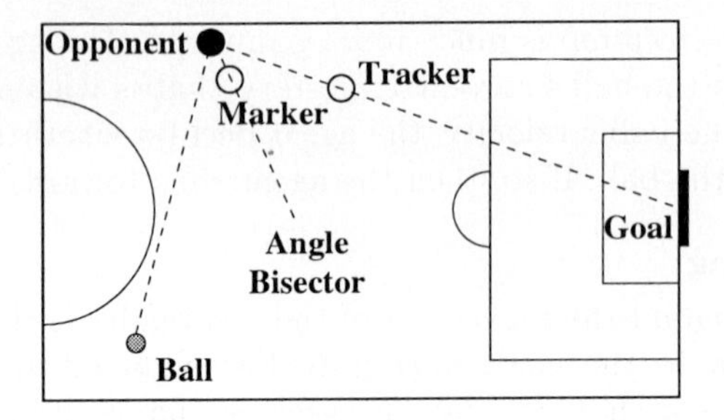

Fig. 5. Positioning for defensive tracking and marking.

whether or not to be tracking). Thus, when in the correct defensive position, the agent always turns to look at the object (opponent or ball) in which it is least confident of the correct position. The complete algorithm, which results in the behavior of doggedly following a particular opponent and glancing back and forth between the opponent and ball, is as follows:

- If the ball position is unknown, look for the ball.
- Else, if the opponent position is unknown, look for the opponent.
- Else, if not in the correct defensive position, move to that position.
- else, look towards the object, ball or opponent, which has been seen less recently (lower confidence value).

This defensive behavior is locally optimal in that it defends according to the opponent's current position, following it around rather than predicting its future location. However, in both cases, the defensive positioning is chosen in anticipation of the opponent's future possible actions, i.e. receiving a pass or shooting.

4.7 Clearing

Often in a defensive position, it is advantageous to just send the ball upfield, clearing it from the defensive zone. If the agent decides that it cannot pass or dribble while in a defensive situation, it will clear the ball. The important decision in clearing the ball is where to clear it to. The best clears are upfield, but not to the middle of the field (you don't want to center the ball for the opponents), and also away from the opponents.

The actual calculation is as follows. Every angle is evaluated with respect to its usefulness, and the expected degree of success. The usefulness is a sine curve with a maximum of 1 at 30 degrees, .5 at 90 degrees, and 0 at -90, where a negative angle is towards the middle of the field. The actual equation is (Θ is in degrees):

$$\text{usefulness}(\Theta) = \frac{sin(\frac{3}{2}\Theta + 45) + 1}{2} \quad (1)$$

The expected degree of success is evaluated by looking at an isosceles triangle with one vertex where the ball is, and congruent sides extending in the direction of the target being evaluated. For each opponent in the triangle, its distance from the center line of the triangle is divided by the distance from the player on that line. For opponent C in Figure 6, these values are w and d respectively. The expected success is the product of all these quotients. In Figure 6, opponent A would not affect the calculation, being outside the triangle, while opponent B would lower the expected success to 0, since it is on the potential clear line ($w = 0$).

By multiplying the usefulness and expected success together for each possible clear angle, and taking the maximum, the agent gets a crude approximation to maximizing the expected utility of a clear.

5 Coordination

Given all of the individual skills available to the CMUnited-98 clients, it becomes a significant challenge to coordinate the team so that the players are not all trying to do the same thing at the same time. Of course one and only one agent should execute the goaltending behavior. But it is not so clear how to determine when an agent should move towards the ball, when it should defend, when it should dribble, or clear, etc.

If all players act individually — constantly chase the ball and try to kick towards the opponent goal — they will all get tired, there will be nowhere to pass, and the opponents will have free reign over most of the field. Building upon the innovations of the CMUnited-97 simulator team [8], the CMUnited-98 team uses several complex coordination mechanisms, including reactive behavior modes, pre-compiled multi-agent plans and strategies, a flexible teamwork structure, a novel anticipatory offensive positioning scheme, and a sophisticated communication paradigm.

5.1 Behavior Modes

A player's top-level behavior decision is its behavior mode. Implemented as a rule-based system, the behavior mode determines the abstract behavior that the player should execute. For example, there is a behavior mode for the set of states in which the agent can kick the ball. Then, the decision of what to do with the ball is made by way of a more involved decision mechanism. On each action cycle, the first thing a player does is re-evaluate its behavior mode.

The behavior modes include:

Goaltend: Only used by the goaltender.
Localize: Find own field location if it's unknown.
Face Ball: Find the ball and look at it.
Handle Ball: Used when the ball is kickable.
Active Offense: Go to the ball as quickly as possible. Used when no teammate could get there more quickly.
Auxiliary Offense: Get open for a pass. Used when a nearby teammate has the ball.
Passive Offense: Move to a position likely to be useful offensively in the future.

Active Defense: Go to the ball even though another teammate is already going. Used in the defensive end of the field.
Auxiliary Defense: Mark an opponent.
Passive Defense: Track an opponent or go to a position likely to be useful defensively in the future.

The detailed conditions and effects of each behavior mode are beyond the scope of this article. However, they will become more clear in subsequent sections as the role-based flexible team structure is described in Section 5.3.

5.2 Locker-Room Agreement

At the core of the CMUnited-98 coordination mechanism is what we call the Locker-Room Agreement [10]. Based on the premise that agents can periodically meet in safe, full-communication environments, the locker-room agreement specifies how they should act when in low-communication, time-critical, adversarial environments.

The locker-room agreement includes specifications of the flexible teamwork structure (Section 5.3) and the inter-agent communication paradigm (Section 5.5). A good example of the use of the locker-room agreement is CMUnited-98's ability to execute pre-compiled multi-agent plans after dead-ball situations. While it is often difficult to clear the ball from the defensive zone after goal kicks, CMUnited-98 players move to pre-specified locations and execute a series of passes that successfully move the ball out of their half of the field. Such "set plays" exist in the locker-room agreement for all dead-ball situations.

5.3 Roles and Formations

Like CMUnited-97, CMUnited-98 is organized around the concept of flexible formations consisting of flexible roles [10]. Each role specifies the behavior of the agent filling the role, both in terms of positioning on the field and in terms of the behavior modes that should be considered.

A formation is a collection of roles, again defined independently from the agents. The entire team can dynamically switch formations. CMUnited-98 used a standard formation with 4 defenders, 3 midfielders, and 3 forwards (4-3-3) at the beginnings of the games. If losing by enough goals relative to the time left in the game (as determined by the locker-room agreement), the team would switch to an offensive 3-3-4 formation. When winning by enough, the team switched to a defensive 5-3-2 formation.

5.4 SPAR

The flexible roles defined in the CMUnited-97 software were an improvement over the concept of rigid roles. Rather than associating fixed (x, y) coordinates with each position, an agent filling a particular role was given a range of coordinates in which it could position itself. Based on the ball's position on the field, the agent would position itself so as to increase the likelihood of being useful to the team in the future.

However, by taking into account the positions of other agents as well as that of the ball, an even more informed positioning decision can be made. The idea of *strategic position by attraction and repulsion* (SPAR) is one of the novel contributions of the CMUnited-98 research which has been applied to both the simulator and the small robot teams [12].

When positioning itself using SPAR, the agent uses a multi-objective function with attraction and repulsion points subject to several constraints. To formalize this concept, we introduce the following variables:

- P - the desired position for the passive agent in anticipation of a passing need of its active teammate;
- n - the number of agents on each team;
- O_i - the current position of each opponent, $i = 1, \ldots, n$;
- T_i - the current position of each teammate, $i = 1, \ldots, (n-1)$;
- B - the current position of the active teammate and ball;
- G - the position of the opponent's goal.

SPAR extends similar approaches of using potential fields for highly dynamic, multi-agent domains [6]. The probability of collaboration in the robotic soccer domain is directly related to how "open" a position is to allow for a successful pass. Thus, SPAR maximizes the distance from other robots and minimizes the distance to the ball and to the goal, namely:

- *Repulsion* from opponents, i.e., maximize the distance to each opponent: $\forall i, \max \mathit{dist}(P, O_i)$
- *Repulsion* from teammates, i.e., maximize the distance to other passive teammates: $\forall i, \max \mathit{dist}(P, T_i)$
- *Attraction* to the active teammate and ball: $\min \mathit{dist}(P, B)$
- *Attraction* to the opponent's goal: $\min \mathit{dist}(P, G)$

This formulation is a multiple-objective function. To solve this optimization problem, we restate the problem as a single-objective function. As each term may have a different relevance (e.g. staying close to the goal may be more important than staying away from opponents), we want to apply a different weighting function to each term, namely f_{O_i}, f_{T_i}, f_B, and f_G, for opponents, teammates, the ball, and the goal, respectively. Our anticipation algorithm then maximizes a weighted single-objective function with respect to P:

$$\max\left(\sum_{i=1}^{n} f_{O_i}(\mathit{dist}(P, O_i)) + \sum_{i=1}^{n-1} f_{T_i}(\mathit{dist}(P, T_i)) - f_B(\mathit{dist}(P, B)) - f_G(\mathit{dist}(P, G))\right)$$

In our case, we use $f_{O_i} = f_{T_i} = x$, $f_B = 0$, and $f_G = x^2$. For example, the last term of the objective function above expands to $(\mathit{dist}(P, G))^2$.

One constraint in the simulator team relates to the position, or role, that the passive agent is playing relative to the position of the ball. The agent only considers locations that within one of the four rectangles, illustrated in Figure 7: the one closest to the position home of the position that it is currently playing. This constraint helps ensure that the player with the ball will have several different passing options in different parts of the field. In addition, players don't need to consider moving too far from their positions to support the ball.

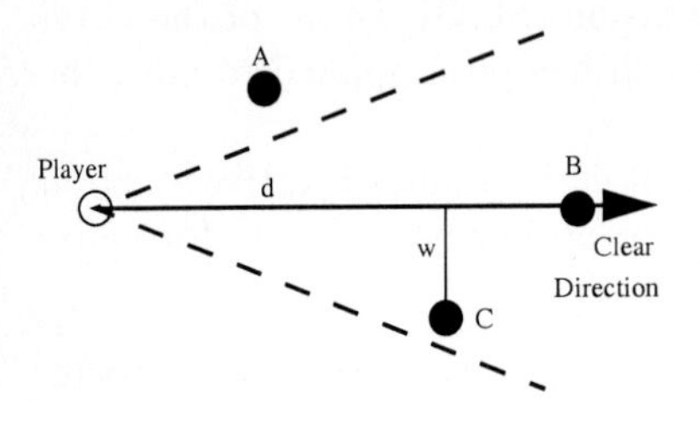

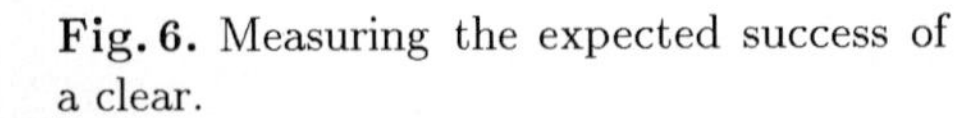
Fig. 6. Measuring the expected success of a clear.

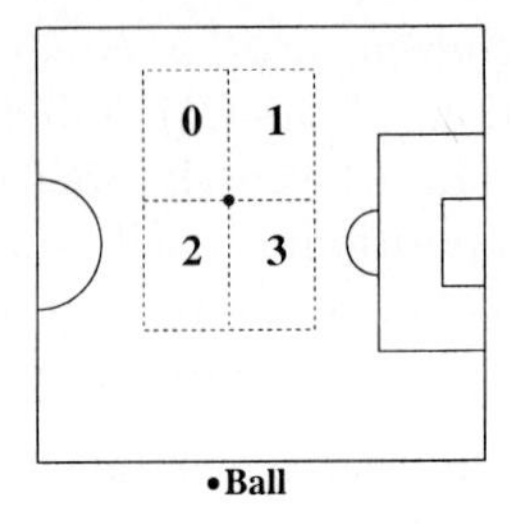

Fig. 7. The four possible rectangles, each with one corner at the ball's location, considered for positioning by simulator agents when using SPAR.

Since this position-based constraint already encourages players to stay near the ball, we set the ball-attraction weighting function f_B to the constant function $y = 0$. In addition to this first constraint, the agents observe three additional constraints. In total, the constraints in the simulator team are:

- Stay in an area near home position;
- Stay within the field boundaries;
- Avoid being in an offsides position;
- Stay in a position in which it would be possible to receive a pass.

This last constraint is evaluated by checking that there are no opponents in a cone with vertex at the ball and extending to the point in consideration.

In our implementation, the maximum of the objective function is estimated by sampling its values over a fine-grained mesh of points that satisfy the above constraints.

Using this SPAR algorithm, agents are able to *anticipate* the collaborative needs of their teammates by positioning themselves in such a way that the player with the ball would have several useful passing options.

5.5 Communication

The soccer server provides a challenging communication environment for teams of agents. With a single, low-bandwidth, unreliable communication channel for all 22 agents and limited communication range and capacity, agents must not rely on any particular message reaching any particular teammate. Nonetheless, when a message does get through, it can help distribute information about the state of the world as well as helping to facilitate team coordination. We identify a general communication environment for single-channel, low-bandwidth, unreliable communication environments [10].

All CMUnited-98 messages include a certain amount of state information from the speaker's perspective. Information regarding object position and teammate roles are all given along with the confidence values associated with this data. All teammates hearing the message can then use the information to augment their visual state information.

5.6 Ball Handling

One of the most important decisions in the robotic soccer domain arises when the agent has control of the ball. In this state, it has the options of dribbling the ball in any direction, passing to any teammate, shooting the ball, clearing the ball, or simply controlling the ball.

In CMUnited-98, the agent uses a complex heuristic decision mechanism, incorporating a machine learning module, to choose its action. The best teammate to receive a potential pass (called *potential receiver* below) is determined by a decision tree trained off-line [9].

6 Results

In order to test individual components of the CMUnited-98 team, it is best to compile performance results for the team with and without these components as we have done elsewhere [10]. However, competition against other, independently-created teams is useful for evaluating the system as a whole.

At the RoboCup-98 competition, CMUnited-98 won all 8 of its games by a combined score of 66–0, finishing first place in a field of 34 teams.

From observing the games, it was apparent that the CMUnited-98 low-level skills were superior in the first 6 games: CMUnited-98 agents were able to dribble around opponents, had many scoring opportunities, and suffered few shots against.

However, in the last 2 games, the CMUnited-98 strategic formations, communication, and ball-handling routines were put more to the test as the Windmill Wanderers (3rd place) and AT-Humboldt'98 (2nd place) also had similar low-level capabilities. In these games, CMUnited-98's abilities to use set plays to clear the ball from its defensive zone, to get past the opponents' defenders, and to maintain a cohesive defensive unit became very apparent. In addition, the fine points of the dribbling and goaltending skills came into play, with the players occasionally able to dribble around opponents for shots and with the CMUnited-98 goaltender making a particularly important save against the Windmill Wanderers.

Throughout the tournament, the CMUnited-98 software demonstrated its power as a complete multi-agent architecture in a real-time, noisy, adversarial environment.

7 Conclusion

The success of CMUnited-98 at RoboCup-98 was due to several technical innovations ranging from predictive locally optimal skills (PLOS) to strategic positioning using attraction and repulsion (SPAR). Building on the innovations of CMUnited-97, including flexible formation, a novel communication paradigm, and machine learning modules, CMUnited-98 successfully combines low-level individual and high-level strategic, collaborative reasoning in a single multi-agent architecture.

For a more thorough understanding of the implementation details involved, the reader is encouraged to study the algorithms described here in conjunction with the CMUnited-98 source code [11]. Other RoboCup researchers and multi-agent researchers in general should be able to benefit and build from the innovations represented therein.

References

1. David Andre, Emiel Corten, Klaus Dorer, Pascal Gugenberger, Marius Joldos, Johan Kummeneje, Paul Arthur Navratil, Itsuki Noda, Patrick Riley, Peter Stone, Romoichi Takahashi, and Travlex Yeap. Soccer server manual, version 4.0. Technical Report RoboCup-1998-001, RoboCup, 1998. At URL http://ci.etl.go.jp/~noda/soccer/server/Documents.html.
2. Mike Bowling, Peter Stone, and Manuela Veloso. Predictive memory for an inaccessible environment. In *Proceedings of the IROS-96 Workshop on RoboCup*, pages 28–34, Osaka, Japan, November 1996.
3. Hans-Diter Burkhard, Markus Hannebauer, and Jan Wendler. AT humboldt — development, practice and theory. In Hiroaki Kitano, editor, *RoboCup-97: Robot Soccer World Cup I*, pages 357–372. Springer Verlag, Berlin, 1998.
4. Hiroaki Kitano, Yasuo Kuniyoshi, Itsuki Noda, Minoru Asada, Hitoshi Matsubara, and Eiichi Osawa. RoboCup: A challenge problem for AI. *AI Magazine*, 18(1):73–85, Spring 1997.
5. Hiroaki Kitano, Milind Tambe, Peter Stone, Manuela Veloso, Silvia Coradeschi, Eiichi Osawa, Hitoshi Matsubara, Itsuki Noda, and Minoru Asada. The RoboCup synthetic agent challenge 97. In *Proceedings of the Fifteenth International Joint Conference on Artificial Intelligence*, pages 24–29, San Francisco, CA, 1997. Morgan Kaufmann.
6. Jean-Claude Latombe. *Robot Motion Planning*. Kluwer, 1991.
7. E. Pagello, F. Montesello, A. D'Angelo, and C. Ferrari. A reactive architecture for RoboCup competition. In Hiroaki Kitano, editor, *RoboCup-97: Robot Soccer World Cup I*, pages 434–442. Springer Verlag, Berlin, 1998.
8. Peter Stone and Manuela Veloso. The CMUnited-97 simulator team. In Hiroaki Kitano, editor, *RoboCup-97: Robot Soccer World Cup I*, pages 387–397. Springer Verlag, Berlin, 1998.
9. Peter Stone and Manuela Veloso. A layered approach to learning client behaviors in the RoboCup soccer server. *Applied Artificial Intelligence*, 12:165–188, 1998.
10. Peter Stone and Manuela Veloso. Task decomposition, dynamic role assignment, and low-bandwidth communication for real-time strategic teamwork. *Artificial Intelligence*, 1999. To appear.
11. Peter Stone, Manuela Veloso, and Patrick Riley. CMUnited-98 source code, 1998. Accessible from http://www.cs.cmu.edu/~ pstone/RoboCup/CMUnited98-sim.html.
12. Manuela Veloso, Michael Bowling, Sorin Achim, Kwun Han, and Peter Stone. The CMUnited-98 champion small robot team. In Minoru Asada and Hiroaki Kitano, editors, *RoboCup-98: Robot Soccer World Cup II*. Springer Verlag, Berlin, 1999.

The CMUnited-98 Champion Small-Robot Team

Manuela Veloso, Michael Bowling, Sorin Achim, Kwun Han, and Peter Stone

Computer Science Department
Carnegie Mellon University
Pittsburgh, PA 15213
{veloso,mhb,sorin,kwunh,pstone}@cs.cmu.edu
http://www.cs.cmu.edu/~robosoccer

Abstract. In this chapter, we present the main research contributions of our champion CMUnited-98 small robot team. The team is a multi-agent robotic system with global perception, and distributed cognition and action. We describe the main features of the hardware design of the physical robots, including differential drive, robust mechanical structure, and a kicking device. We briefly review the CMUnited-98 global vision processing algorithm, which is the same as the one used by the previous champion CMUnited-97 team. We introduce our new robot motion algorithm which reactively generates motion control to account for the target point, the desired robot orientation, and obstacle avoidance. Our robots exhibit successful collision-free motion in the highly dynamic robotic soccer environment. At the strategic and decision-making level, we present the role-based behaviors of the CMUnited-98 robotic agents. Team collaboration is remarkably achieved through a new algorithm that allows for team agents to anticipate possible collaboration opportunities. Robots position themselves strategically in open positions that increase passing opportunities. The chapter terminates with a summary of the results of the RoboCup-98 games in which the CMUnited-98 small robot team scored a total of 25 goals and suffered 6 goals in the 5 games that it played.

1 Introduction

The CMUnited-98 small-size robot team is a complete, autonomous architecture composed of the physical robotic agents, a global vision processing camera overlooking the playing field, and several clients as the minds of the small-size robot players. Fig. 1 sketches the building blocks of the architecture.

[0] This research is sponsored in part by the Defense Advanced Research Projects Agency (DARPA) and the Air Force Research Laboratory (AFRL) under agreement numbers F30602-97-2-0250 and F30602-98-2-0135, and in part by the Department of the Navy, Office of Naval Research under contract number N00014-95-1-0591. The views and conclusions contained in this document are those of the authors and should not be interpreted as necessarily representing official policies or endorsements, either expressed or implied, of the Air Force, of the Department of the Navy, Office of Naval Research or the United States Government.

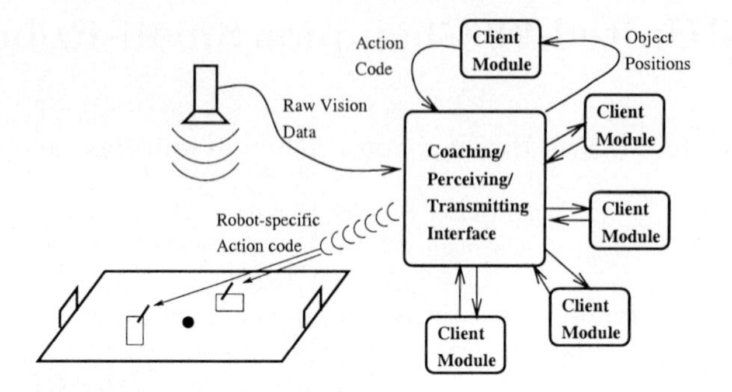

Fig. 1. The CMUnited architecture with global perception and distributed reaction.

The complete system is fully autonomous consisting of a well-defined and challenging processing cycle. The global vision algorithm perceives the dynamic environment and processes the images, giving the positions of each robot and the ball. This information is sent to an off-board controller and distributed to the different agent algorithms. Each agent evaluates the world state and uses its strategic knowledge to make decisions. Actions are motion commands that are sent by the off-board controller through radio frequency communication. Commands can be broadcast or sent directly to individual agents. Each robot has an identification binary code that is used on-board to detect commands intended for that robot. Motion is not perfectly executed due to inherent mechanical inaccuracies and unforeseen interventions from other agents. The effects of the actions are therefore uncertain.

The physical robots themselves are of size 15cm $\times$ 12cm $\times$ 10cm. Fig. 2 shows our robots. A differential drive mechanism is used in all of the robots. Two motors with integrated gear boxes are used for the two wheels. Differential drive was chosen due to its simplicity and due to the size constraints. The size of our robots conforms to RoboCup Competition rules[1]. Employing the differential drive mechanism means that the robot is non-holonomic, which makes the robot control problem considerably more challenging.

Although it may be possible to fit an on-board vision system onto robots of small size, in the interest of being able to quickly move on to strategic multiagent issues, the CMUnited-98 teams uses a global vision system. The fact that perception is achieved by a video camera overlooking the complete field offers an opportunity to get a global view of the world state. This setup may simplify the sharing of information among multiple agents, but it also presents a challenge for reliable and real-time processing of the movement of multiple mobile objects, namely the ball, five robots on our team, and five robots on the opponent's team.

This chapter presents the main technical contributions of our CMUnited-98 small robot team. It is organized as follows. Section 2 briefly overviews the hardware design of the robots. Section 3 describes the vision processing algorithm. Section 4 presents the motion planning approach for our robots including path planning to intercept moving targets and obstacle avoidance. Section 5 in-

[1] see http://www.robocup.org/RoboCup/

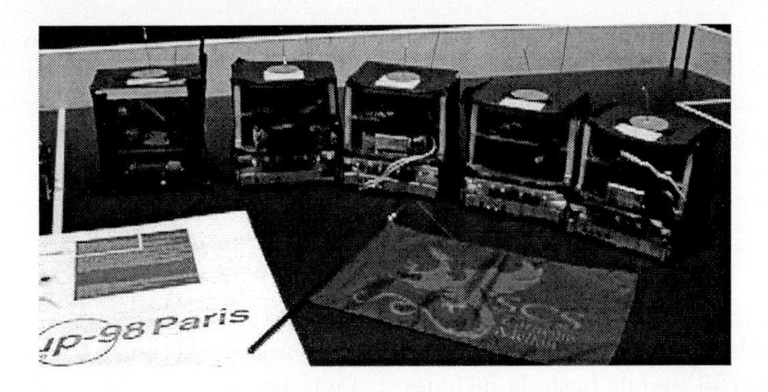

Fig. 2. The CMUnited-98 robots.

troduces the individual and team behaviors of the CMUnited-98 robots. We introduce the novel concept of *anticipation* which allows for the robots to effectively receive passes from teammates. Section 6 summarizes the results of the RoboCup-98 games and Section 7 draws conclusions.

2 Hardware

The CMUnited-98 robots are entirely new constructions built upon our experience in 1997. The new robots represent an upgrade of our own-built CMUnited-97 robots [8]. In this section, we present some of the ‘features of our robot design and refer to the differences with the CMUnited-97 robots when appropriate.

The robots of the CMUnited-98 team were designed according to the following principles: simple but robust, reliable electronics, and modularized mechanics. These goals were partially achieved in CMUnited-97 and some of the design was retained. We continue to use the Motorola HC11 8-bit micro-controller running at 8 MHz and the 40 KBd Radiometrix radio module. Improvements were made in two major areas: motors and control, and the mechanical chassis.

2.1 Motors and Control

CMUnited-98 uses two high-torque, 6V DC, geared motors, which are overpowered and use a simple PWM control. It is interesting to note that this represents a simpler design than our CMUnited-97 robots which used hardwired motion controllers where the control loop was implemented in hardware as a Proportional Integral Derivative (PID) filter and associated motor encoders. We realized that, although this design provided accurate navigation, it was not easily interruptible. We found that these interrupts were often needed in the highly dynamic robotic soccer environment. Hence, in the CMUnited-98 robots, the closed-loop motion control is achieved through software using visual feedback.

The new implementation only makes use of a "WRITE" operation to the corresponding 8-bit register for each requested change in the speed of the motors. Additionally, a "safety mechanism" guarantees that a "STOP" command is issued locally by on-board resources in case of a radio link malfunction (i.e., when the next valid command is not received within a preset time interval).

The main advantages obtained by implementing these changes are the simplification of the low level motion control, the simplification of the electronic circuitry design, and the reduction of the space needed by the motion control functions in the micro-controller's 2Kb EEPROM.

2.2 Mechanical Chassis

In designing the mechanical structure of the CMUnited-98 robots, we focused on modularity and robustness. The final design was a robot with a very compact and powerful mobile base. It includes a single battery module supplying three independent power paths (for the main-board, motors, and radio modules.) It also includes a single board containing all the required electronic circuitry, with multiple add-on capabilities. This was all combined in a layered design within an aluminum and plastic Frame. In addition, each of the modules within this design is completely interchangeable.

The mobile base module includes a kicking device driven by a DC motor. This motor is hardware activated by an array of four infrared sensors with adjustable sensing range which can be enabled or disabled by the software control. The circular motion of the kicker motor was transformed to linear motion using a rack-pinion system. The push-pull effect of the kicking device was implemented by alternating the voltage polarity at the DC motor. The push, stop and pull stages of the kicking action were all timed, which eliminated the need for control sensors. Other implementations for the kicking device were considered, like a dual front/back kicker driven by flexible strings, or a push-pull DC solenoid. Neither of these met the size, power and precision requirements for a reliable kicking system.

The hardware design proved to be challenging. By building upon our experience with CMUnited-97, CMUnited-98 successfully achieves a robust team of physical robots.

3 Vision Processing

The CMUnited-98 vision module is the same as the one used in the CMUnited-97 team. It successfully implements a very reliable and fast image processing algorithm. In this chapter, we focus mainly our detection and association algorithm which we present in more detail than in our description of the CMUnited-97 team [8]. In particular, in the setting of a robot soccer game, the ability to detect merely the locations of objects on the field is often not enough. We use a Kalman-Bucy filter to successfully predict the movement of the ball[8], which is very suitable since the detection of the ball's location is noisy.

3.1 Color-based Detection

The vision requirements for robotic soccer have been examined by different researchers [3, 4]. Both on-board and off-board systems have appeared in recent years. All have found that the response necessary for soccer robots requires a vision system with a fast processing cycle. However, due to the rich visual input, researchers have found that dedicated processors or even DSPs are often needed [1, 3]. Our current system uses a frame-grabber with frame-rate transfer from a 3CCD camera.

The RoboCup rules specify colors for different objects in the field and these are used as the major cue for object detection. The RoboCup rules specify a green color field with specific white markings. The ball is an orange golf ball. It also specifies a yellow or blue colored circle on the top of the robots, one color for each team (see Fig. 2.) A single color patch on the robot is not enough to provide orientation information. Thus, an additional pink color patch was added to each of our robots. These colors can be differentiated reliably in color-space.

The set of detected patches are initially unordered. The detected color patches on the tops of the robots are then matched by their distance. Using the constant distance between the team-color (blue or yellow) and the pink orientation patch, our detection algorithm matches patches that are this distance apart. Two distance-matched patches are detected as a robot.

Noise is inherent in all vision systems. False detections in the current system are often of a magnitude of 100 spurious detections per frame. The system eliminates false detections via two different methods. First, color patches of size not consistent with the ones on our robots are discarded. This technique filters off most "salt and pepper" noise. Second, by adding the distance matching mechanism described above, false detections are practically eliminated.

3.2 Data Association

The detection scheme described above returns an unordered list of robots for each frame. To be able to control the robots, the system must associate each detected robot in the field with a robot identification.

Each of the robots is fitted with the same color tops and no attempts are made to differentiate them via color hue. Experience has shown that, in order to differentiate 5 different robots by hue, 5 significantly different hues are needed. However, the rules of the RoboCup game eliminate green (field), white (markings), orange (ball), blue and yellow (team and opponent) from the list of possibilities. Furthermore, inevitable variations in lighting conditions over the area of the field and noise in the sensing system are enough to make a hue-based detection scheme impractical.

With each robot fitted with the same color, visually, all robots on our team look identical to the visual system. With visually homogeneous objects distinguishing between them in any given frame is not possible. Data association addresses this problem by retaining robot identification between subsequent frames. We devised an algorithm to retain association based on the spatial locations of the robots.

We assume that the starting positions of all the robots are known. This can be done trivially by specifying the location of the robots at start time. However, problems arise when subsequent frames are processed, the locations of the robots have changed due to robot movements. Association can be achieved by making two complementary assumptions: 1) Robot displacements over consecutive frames are local; 2) The vision system can detect objects at constant frame rate. By measuring the maximum robot velocity, we can know that in subsequent frames, the robot is not able to move out of a 5cm radius circular region. This provides the basis of our association technique.

3.3 Greedy Association

With these assumptions in mind, a minimum distance scheme can be used to retain association between consecutive frames. For each frame, association is maintained by searching for objects with a minimum displacement. Current robot positions are matched with the closest positions from the previous frame.

The following is the pseudo-code of a greedy association procedure:

let $prev[1..n]$ be the array of robot locations from the previous frame
let $cur[1..m]$ be the array of robot locations from the current frame
let ma be triangular array of size $n-1$ s.t.
$ma[i][j] = \text{dist}(prev[i], cur[j])$
for $i := 1$ to m **do**
  find smallest element $ma[i][j]$
  save (i, j) as a matched pair
  set all elements in row i and column j to be ∞
end
if $m < n$ **then**
  forall $prev[i]$ unmatched, save $(prev[i], prev[i])$
return the set of saved pairs as the set of matchings.

This algorithm searches through all possible matches, from the smallest distance pair upwards. Whenever a matched pair is found, it greedily accepts it as a matching pair. Due to noise, it is possible for the detection system to leave a robot or two undetected (i.e.. in the pseudo-code $m < n$). In this case, some locations will be left unmatched. The unmatched location will then be carried over to the current frame, and the robots corresponding to this location will be assumed to be stationary for this one frame.

This algorithm was implemented and was used in RoboCup-97 and RoboCup-98. Although the implementation was very robust, we present an improvement that allows for a globally optimal association.

3.4 Globally Optimal Association

The greedy association algorithm, as described above, fails in some cases. An example is the one illustrated in Fig. 3. In the figure, a greedy algorithm incorrectly matches the closest square and circular objects.

An improved algorithm was devised to handle the situation depicted above. The new algorithm generates all possible sets of matching and calculates the

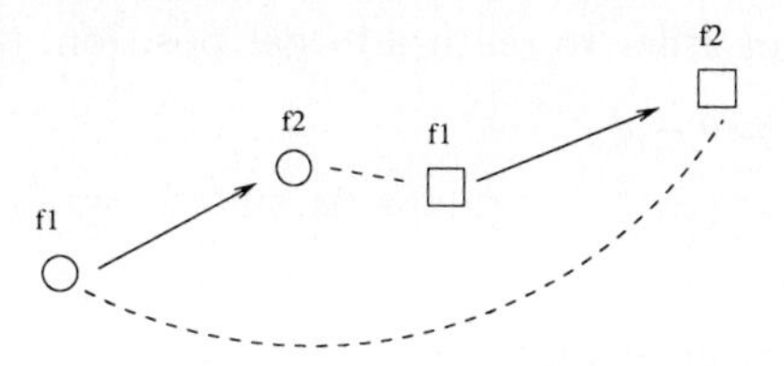

Fig. 3. A case in which greedy association fails but global optimal association performs well. The arrow indicates the actual object displacement over subsequent frames f1 and f2. The dotted lines indicate the wrong matches returned by greedy association.

fitness of each of the complete sets according to the following least square criteria:

$$\sum_{i=1}^{N}(dist(prev_i, cur_i))^2,$$

where $(prev_i, cur_i)$ are the ith matching pair. And the function $dist(x, y)$ is the Euclidean distance. The set of matches that minimizes the above criteria is selected.

While these algorithms do not theoretically guarantee perfect associations, in particular with noisy perception and cluttered environments, the implementation has proven to be very robust. Our vision processing approach worked perfectly during the RoboCup-98 games. We were able to detect and track 11 objects (5 teammates, 5 opponents and a ball) at 30 frames/s. Also the prediction provided by the Kalman-Bucy filter is an integral factor in our robots' strategic decisions.

4 Motion Control

The goal of our low level motion control is to be as fast as possible while remaining accurate and reliable. This is challenging due to the lack of feedback from the motors, forcing all control to be done using only visual feedback. Our motion control algorithm is robust. It addresses stationary and moving targets with integrated obstacle avoidance. The algorithm makes effective use of the prediction of the ball's trajectory provided by the Kalman-Bucy filter.

We achieve this motion control functionality by a reactive control mechanism that directs a differential drive robot to a target configuration. Though based on the CMUnited-97's motion control [8], CMUnited-98 includes a number of major improvements. The target configuration for the motion planner has been extended. The target configuration includes: (i) the *Cartesian position*; and (ii) the *direction* that the robot is required to be facing when arriving at the target position. Obstacle avoidance is integrated into this controller. Also, the target configuration can be given as a function of time to allow for the controller to reason about intercepting the trajectory of a moving target.

4.1 Differential Drive Control for Position and Direction

CMUnited- 98's basic control rules were improved from those used in CMUnited-97. The rules are a set of reactive equations for deriving the left and right wheel velocities, v_l and v_r, in order to reach a target position, (x^*, y^*):

$$\begin{aligned}\Delta &= \theta - \phi \\ (t,r) &= (\cos^2\Delta \cdot \mathrm{sgn}(\cos\Delta), \sin^2\Delta \cdot \mathrm{sgn}(\sin\Delta)) \\ v_l &= v(t-r) \\ v_r &= v(t+r),\end{aligned} \tag{1}$$

where θ is the direction of the target point (x^*, y^*), ϕ is the robot's orientation, and v is the desired speed (see Fig. 4(a))[2].

We extend these equations for target configurations of the form (x^*, y^*, ϕ^*), where the goal is for the robot to reach the specified target point (x^*, y^*) while facing the direction ϕ^*. This is achieved with the following adjustment:

$$\theta' = \theta + \min\left(\alpha, \tan^{-1}\left(\frac{c}{d}\right)\right),$$

where θ' is the new target direction, α is the difference between our angle to the target point and ϕ^*, d is the distance to the target point, and c is a clearance parameter (see Fig. 4(a).) This will keep the robot a distance c from the target point while it is circling to line up with the target direction, ϕ^*. This new target direction, θ', is now substituted into equation 1 to derive wheel velocities.

In addition to our motion controller computing the desired wheel velocities, it also returns an estimate of the time to reach the target configuration, $\hat{T}(x^*, y^*, \phi^*)$. This estimate is a crucial component in our robot's strategy. It is used both in high-level decision making, and for low-level ball interception, which is described later in this section. For CMUnited-98, $\hat{T}(x^*, y^*, \phi^*)$ is computed using a hand-tuned linear function of d, α, and Δ.

4.2 Obstacle Avoidance

Obstacle avoidance was also integrated into the motion control. This is done by adjusting the target direction of the robot based on any immediate obstacles in its path. This adjustment can be seen in Fig. 4(b).

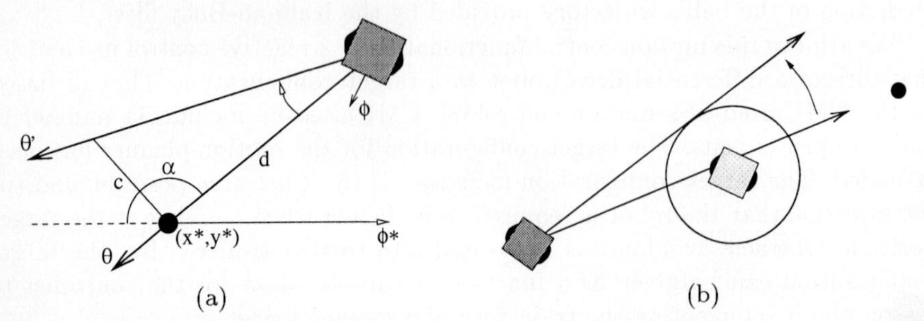

Fig. 4. (a) The adjustment of θ to θ' to reach a target configuration of the form (x^*, y^*, ϕ^*); (b) The adjustment to avoid immediate obstacles.

---

[2] All angles are measured with respect to a fixed coordinate system.

If a target direction passes too close to an obstacle, the direction is adjusted to run tangent to the a preset allowed clearance for obstacles. Since the motion control mechanism is running continuously, the obstacle analysis is constantly replanning obstacle-free paths. This continuous replanning allows for the robot to handle the highly dynamic environment and immediately take advantage of short lived opportunities.

4.3 Moving Targets

One of the real challenges in robotic soccer is to be able to control the robots to intercept a moving ball. This capability is essential for a high-level ball passing behavior. CMUnited-98's robots successfully intercept a moving ball and several of their goals in RoboCup-98 were scored using this capability.

This interception capability is achieved as an extension of the control algorithm to aim at a stationary target. Fig. 5(a) illustrates the control path to reach a stationary target with a specific direction, using the control mechanism described above. Our extension allows for the target configuration to be given as a function of time, where $t = 0$ corresponds to the present,

$$f(t) = (x^*, y^*, \phi^*).$$

At some point in the future, t_0, we can compute the target configuration, $f(t_0)$. We can also use our control rules for a stationary point to find the wheel velocities and estimated time to reach this hypothetical target as if it were stationary. The time estimate to reach the target then informs us whether it is possible to reach it within the allotted time. Our goal is to find the nearest point in the future where the target can be reached. Formally, we want to find,

$$t^* = \min\{t > 0 : \hat{T}(f(t)) \leq t\}.$$

After finding t^*, we can use our stationary control rules to reach $f(t^*)$. In addition we scale the robot speed so to cross the target point at exactly t^*.

Unfortunately, t^*, cannot be easily computed within a reasonable time-frame. We approximate this value, t^*, by discretizing time with a small time-step. We then find the smallest of these discretized time points that satisfies our estimate constraint. An example of this is shown in Fig. 5(b), where the goal is to hit the moving ball.

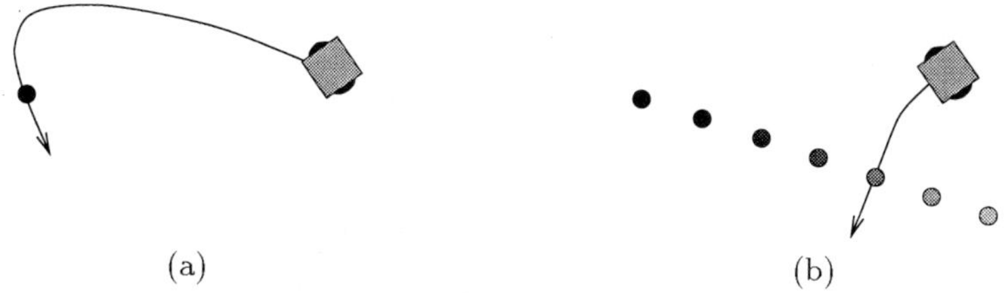

(a) (b)

Fig. 5. (a) Control for stationary target. (b) Control for moving target.

The target configuration as a function of time is computed using the ball's predicted trajectory. Our control algorithm for stationary points is then used to find a path and time estimate for each discretized point along this trajectory, and the appropriate target point is selected.

5 Strategy

The main focus of our research is on developing algorithms for collaboration between agents in a team. An agent, as a member of the team, needs to be capable of individual autonomous decisions while, at the same time, its decisions must contribute towards the team goals.

CMUnited-97 introduced a flexible team architecture in which agents are organized in *formations* and *units*. Each agent plays a *role* in a unit and in a formation [5, 8]. CMUnited-98 builds upon this team architecture by defining a set of roles for the agents. It also introduces improvements within this architecture to help address the highly dynamic environment.

CMUnited-98 uses the following roles: goalkeeper, defender, and attacker. The formation used throughout RoboCup-98 involved a single goalkeeper and defender, and three attackers.

5.1 Goalkeeper

The ideal goalie behavior is to reach the expected entry point of the ball in the goal *before* the ball reaches it. Assuming that the prediction of the ball trajectory is correct and the robot has a uniform movement, we can state the ideal goalie behavior: given the predicted v_g and v_b as the velocities of the goalie and of the ball respectively, and d_g and d_b as the distances from the goalie and the ball to the predicted entry point, then, we want $\frac{d_g}{v_g} = \frac{d_b}{v_b} - \epsilon$, where ϵ is a small positive value to account for the goalie reaching the entry point slightly before the ball.

Unfortunately, the ball easily changes velocity and the movement of the robot is not uniform and is uncertain. Therefore we have followed a switching behavior for the goalie based on a threshold of the ball's estimated trajectory.

If the ball's estimated speed is higher than a preset threshold, the goalie moves directly to the ball's predicted entry goal point. Otherwise, the goalie selects the position that minimizes the largest portion of unobstructed goal area. This is done by finding the location that bisects the angles of the ball and the goal posts as is illustrated in Fig. 6.

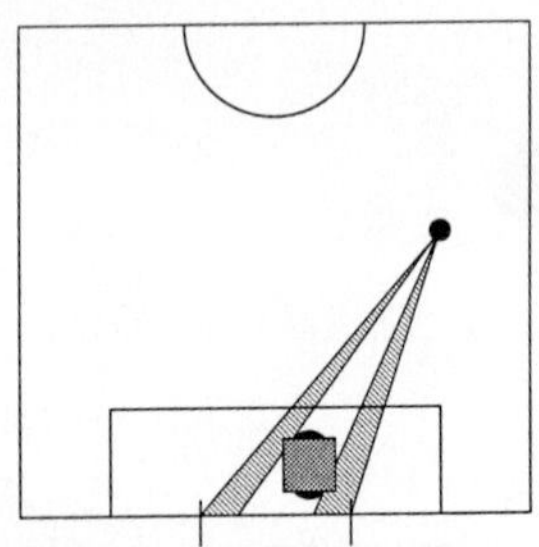

Fig. 6. The goalkeeper positions itself to minimize the unobstructed goal area.

The use of the predicted ball's velocity for the goalie's behavior was shown to be very effective in the RoboCup-98 games. It was particularly appropriate for defending a penalty shot, due to the accuracy of the predicted ball's trajectory when only one robot is pushing the ball.

5.2 Defender

The CMUnited-97's team did not have a well-specified defender's role, but our experience at RoboCup-97 made us understand that the purpose of a defending behavior is two-fold:

1. to stop the opponents from scoring in our goal; and
2. to not endanger our own goal.

The first goal is clearly a defender's role. The second goal comes as the result of the uncertain ball handling by the robots. The robots can easily push (or touch) the ball unexpectedly in the wrong direction when performing a difficult maneuver.

To achieve the two goals, we implemented three behaviors for the defender. *Blocking*, illustrated in Fig. 7(a), is similar to the goalkeeper's behavior except that the defender positions itself further away from the goal line. *Clearing*, illustrated in Fig. 7(b), pushes the ball out of the defending area. It does this by finding the largest angular direction free of obstacles (opponents and teammates) that the robot can push the ball towards. *Annoying*, illustrated in Fig. 7(c), is somewhat similar to the goalkeeping behavior except that the robot tries to position itself between the ball and the opponent nearest to it. This is an effort to keep the opponent from reaching the ball.

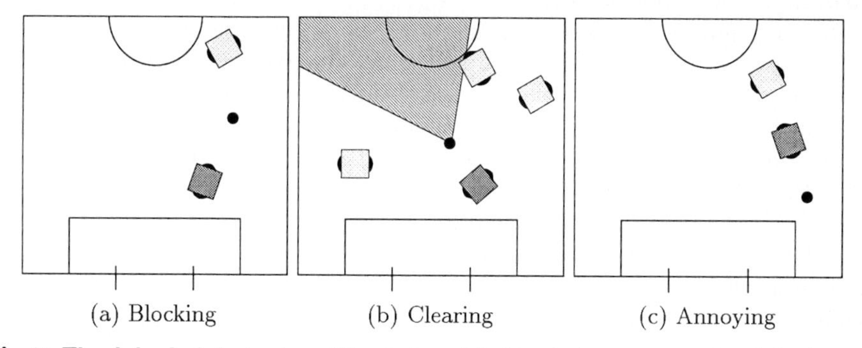

(a) Blocking (b) Clearing (c) Annoying

Fig. 7. The defender's behaviors. The dark and light robots represent the defender and the opponents respectively.

Selecting when each of these behaviors is used is very important to the effectiveness of the defender. For example, clearing the ball when it is close to our own goal or when it can bounce back off another robot, can lead to scoring in our own goal. We used the decision tree in Fig. 8 to select which action to perform based on the current state.

The two attributes in the tree, namely *Ball Upfield* and *Safe to Clear*, are binary. *Ball Upfield* tests whether the ball is upfield (towards the opponent's goal) of the defender. *Safe to Clear* tests whether the open area is larger than a preset angle threshold. If *Ball Upfield* is false then the ball is closer to the goal than the defender and the robot *annoys* the attacking robot. The CMUnited-98's annoying behavior needs to select one particular opponent robot to annoy. For example, when two opponent robots attack simultaneously, the current annoying behavior is able to annoy only one of them. We are planning on further improving this behavior for RoboCup-99.

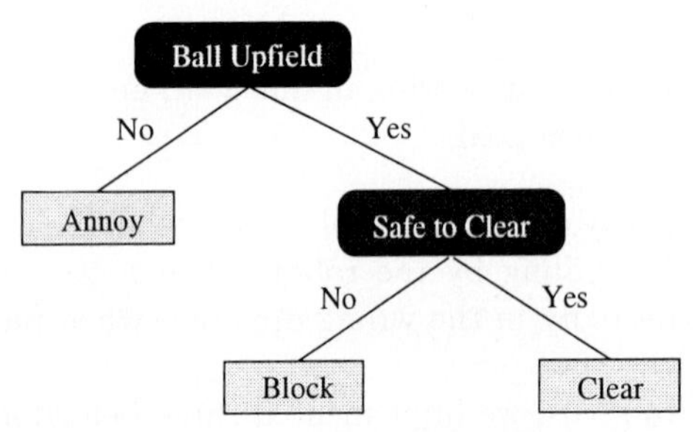

Fig. 8. The decision tree heuristic used by the defender to select its behavior.

If *Ball Upfield* is true, the defender clears or blocks, depending on the value of *Safe to Clear*. Clearing was shown to be very useful at RoboCup-98, with even a couple of our goals scored directly by a clearing action of the defender.

5.3 Attackers - Active Teammate and Anticipation

Attacking involves one of the best opportunities for collaboration, and much of the innovation of CMUnited-98 has been developing techniques for finding and exploiting these opportunities.

In many multi-agent systems, one or a few agents are assigned, or assign themselves, the specific task to be solved at a particular moment. We view these agents as the *active* agents. Other team members are *passive* waiting to be needed to achieve another task or assist the active agent(s). This simplistic distinction between active and passive agents to capture teamwork was realized in CMUnited-97. The agent that goes to the ball is viewed as the active agent, while the other teammates are passive.

CMUnited-98 significantly extends this simplistic view in two ways: (i) we use a decision theoretic algorithm to select the active agent; and (ii) we use a technique for passive agents *to anticipate* future collaboration. Passive agents are therefore not actually "passive;" instead, they actively *anticipate* opportunities for collaboration. In CMUnited-98 this collaboration is built on robust individual behaviors.

Individual Behaviors. We first developed individual behaviors for passing and shooting. Passing and shooting in CMUnited-98 is handled effectively by the motion controller. The target configuration is specified to be the ball (using its estimated trajectory) and the target direction is either towards the goal or another teammate. This gives us robust and accurate individual behaviors that can handle obstacles as well as intercepting a moving ball.

Decision Theoretic Action Selection. Given the individual behaviors, we must select an active agent and appropriate behavior. This is done by a decision theoretic analysis using a single step look-ahead. With n agents this amounts to n^2 choices of actions involving shooting or a pass to another agent followed by that agent shooting. An estimated probability of success for each pass and shot is computed along with the time estimate to complete the action, which is provided by the motion controller. A value for each action is computed,

$$\text{Value} = \frac{Pr_{\text{pass}} Pr_{\text{shoot}}}{\text{time}}.$$

The action with the largest value is selected, which determines both the active agent and its behavior. Table 1 illustrates an example of the values for the selection considering two attackers, 1 and 2.

		Probability of Success			
Attacker	Action	Pass	Shoot	Time(s)	Value
1	Shoot	–	60%	2.0	0.30
1*	Pass to 2	60%	90%	1.0	0.54
2	Shoot	–	80%	1.5	0.53
2	Pass to 1	50%	40%	0.8	0.25

Table 1. Action choices and computed values are based on the probability of success and estimate of time. The largest-valued action (marked with an *) is selected.

It is important to note that this action selection is occurring on each iteration of control, i.e., approximately 30 times per second. The probabilities of success, estimates of time, and values of actions, are being continuously recomputed. This allows for quick changes of actions if shooting opportunities become available or collaboration with another agent appears more useful.

Dynamic Positioning (SPAR). Although there is a clear action to be taken by the active agent, it is unclear what the passive agents should be doing. Although, in a team multiagent system such as robotic soccer, success and goal achievement often depends upon collaboration; so, we introduce in CMUnited-98, the concept that team agents should not actually be "passive."

CMUnited-97's team architecture allowed for the passive agents to flexibly vary their positions within their role only as a function of the position of the ball. In so doing, their goal was to *anticipate* where they would be most likely to find the ball in the near future. This is a first-level of single-agent anticipation towards a better individual goal achievement [7].

However, for CMUnited-98, we introduce a team-based notion of *anticipation*, which goes beyond individual single-agent anticipation. The passive team members position themselves strategically so as to optimize the chances that their teammates can successfully collaborate with them, in particular pass to them. By considering the positions of other agents and the attacking goal, in addition to that of the ball, they are able to position themselves more usefully: they *anticipate* their future contributions to the team.

This strategic position takes into account the position of the other robots (teammates and opponents), the ball, and the opponent's goal. The position is found as the solution to a multiple-objective function with repulsion and attraction points. Let's introduce the following variables:

- n - the number of agents on each team;
- O_i - the current position of each opponent, $i = 1, \ldots, n$;
- T_i - the current position of each teammate, $i = 1, \ldots, (n-1)$;
- B - the current position of the active teammate and ball;
- G - the position of the opponent's goal;
- P - the desired position for the passive agent in anticipation of a pass.

Given these defined variables, we can then formalize our algorithm for strategic position, which we call SPAR for *Strategic Positioning with Attraction and Repulsion.* This extends similar approaches using potential fields [2], to our highly dynamic, multi-agent domain. The probability of collaboration is directly related to how "open" a position is to allow for a successful pass. SPAR maximizes the repulsion from other robots and minimizes attraction to the ball and to the goal, namely:

- *Repulsion* from opponents. Maximize the distance to each opponent: $\forall i, \max dist(P, O_i)$.
- *Repulsion* from teammates. Maximize the distance to other passive teammates: $\forall i, \max dist(P, T_i)$.
- *Attraction* to the ball: $\min dist(P, B)$.
- *Attraction* to the opponent's goal: $\min dist(P, G)$.

This is a multiple-objective function. To solve this optimization problem, we restate this function into a single-objective function. This approach has also been applied to the CMUnited-98 simulator team [6].

As each term in the multiple-objective function may have a different relevance (e.g., staying close to the goal may be more important than staying away from opponents), we want to consider different functions of each term. In our CMUnited-98 team, we weight the terms differently, namely w_{O_i}, w_{T_i}, w_B, and w_G, for the weights for opponents, teammates, the ball, and the goal, respectively. For CMUnited-98, these weights were hand tuned to create a proper balance. This gives us a weighted single-objective function:

$$\max \left(\sum_{i=1}^{n} w_{O_i} dist(P, O_i) + \sum_{i=1}^{n} w_{T_i} dist(P, T_i) - w_B dist(P, B) - w_G dist(P, G) \right).$$

This optimization problem is then solved under a set of constraints:

- Do not block a possible direct shot from active teammate.
- Do not stand behind other robots, because these are difficult positions to receive passes from the active teammate.

The solution to this optimization problem under constraints gives us a target location for the "passive" agent. Fig. 9(a) and (b) illustrate these two sets of constraints and Fig. 9(c) shows the combination of these constraints and the resulting position of the anticipating passive teammate.

This positioning was very effective for CMUnited-98. The attacking robots very effectively and dynamically adapted to the positioning of the other robots.

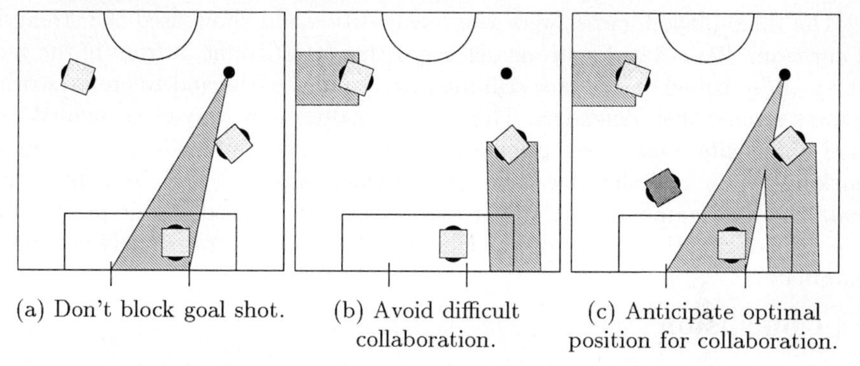

(a) Don't block goal shot. (b) Avoid difficult collaboration. (c) Anticipate optimal position for collaboration.

Fig. 9. Constraints for the dynamic anticipation algorithm are represented as shaded regions; (a) and (b) show three opponents and the current position of the ball; (c) illustrates the position of the passive agent - dark square - as returned by SPAR.

The SPAR anticipation algorithm created a number of opportunities for passes and rebounds that often led to goals and other scoring chances.

In general, we believe that our approach represents a major step in team multiagent systems in terms of incorporating *anticipation* as a key aspect of teamwork.

6 Results

CMUnited-98 successfully defended our title of the Small Robot Champion at RoboCup-98 in Paris. The competition involved 11 teams from 7 different countries. It consisted of a preliminary round of two games, followed by the 8 advancing teams playing a 3-round playoff. CMUnited-98 won four of five games, sweeping the playoff competition, scoring a total of 25 goals scored and only 6 suffered. The individual results of these games are in Table 2.

Phase	Opponent	Affiliation	Score (CMU - Opp.)
round-robin	iXS	iXs Inc., Japan	16 – 2
round-robin	5DPO*	University of Porto, Portugal	0 – 3
quarter-final	Paris-8*	University of Paris-8, France	3 – 0
semi-final	Cambridge	University of Cambridge, UK	3 – 0
final	Roboroos	University of Queensland, Australia	3 – 1

Table 2. The scores of CMUnited-98's games at RoboCup-98. The games marked with an * were forfeited at half time.

There were a number of technical problems during the preliminary rounds, including outside interference with our radio communication. This problem was the worst during our game against 5DPO, in which our robots were often responding to outside commands just spinning in circles. This led to our forfeit at half time and a clear loss against 5DPO, a very good team which ended in third place at RoboCup-98. Fortunately, the communication problems were isolated and dealt with prior to the playoff rounds.

The three playoff games were very competitive and showcased the strengths of our team. Paris-8 had a strong defense with a lot of traffic in front of the goal. Our team's obstacle avoidance still managed to find paths and to create scoring chances around their defenders. The final two games were very close against very good opponents. Our interception was tested against Cambridge, and included blocking a powerful shot by their goalie, which was deflected back into their goal. The final game against Roboroos demonstrated the dynamic positioning, especially during the final goal, which involved a pass to a strategically positioned teammate.

7 Conclusion

The success of CMUnited-98 at RoboCup-98 was due to several technical innovations, including robust hardware design, effective vision processing, reliable time-prediction based robot motion with obstacle avoidance, and a dynamic role-based team approach. The CMUnited-98 team demonstrated in many occasions its collaboration capabilities which resulted from the robots' behaviors. Most remarkably, CMUnited-98 introduces the concept of *anticipation*, in which passive robots (not going to the ball) strategically position themselves using attraction and repulsion (SPAR) to maximize the chances of a successful pass.

The CMUnited-98 team represents an integrated effort to combine solid research approaches to hardware design, vision processing, and individual and team robot behaviors.

References

1. Minoru Asada, Shoichi Noda, Sukoya Tawaratumida, and Koh Hosoda. Purposive behavior acquisition for a real robot by vision-based reinforcement learning. *Machine Learning*, 23:279–303, 1996.
2. Jean-Claude Latombe. *Robot Motion Planning*. Kluwer, 1991.
3. Michael K. Sahota, Alan K. Mackworth, Rod A. Barman, and Stewart J. Kingdon. Real-time control of soccer-playing robots using off-board vision: the dynamite testbed. In *IEEE International Conference on Systems, Man, and Cybernetics*, pages 3690–3663, 1995.
4. Randy Sargent, Bill Bailey, Carl Witty, and Anne Wright. Dynamic object capture using fast vision tracking. *AI Magazine*, 18(1):65–72, Spring 1997.
5. Peter Stone and Manuela Veloso. The CMUnited-97 simulator team. In Hiroaki Kitano, editor, *RoboCup-97: Robot Soccer World Cup I*, pages 387–397. Springer Verlag, Berlin, 1998.
6. Peter Stone, Manuela Veloso, and Patrick Riley. The CMUnited-98 champion simulator team. In Minoru Asada and Hiroaki Kitano, editors, *RoboCup-98: Robot Soccer World Cup II*. Springer Verlag, Berlin, 1999.
7. Manuela Veloso, Peter Stone, and Kwun Han. CMUnited-97: RoboCup-97 small-robot world champion team. *AI Magazine*, 19(3):61–69, 1998.
8. Manuela Veloso, Peter Stone, Kwun Han, and Sorin Achim. The CMUnited-97 small-robot team. In Hiroaki Kitano, editor, *RoboCup-97: Robot Soccer World Cup I*, pages 242–256. Springer Verlag, Berlin, 1998.

The CS Freiburg Robotic Soccer Team: Reliable Self-Localization, Multirobot Sensor Integration, and Basic Soccer Skills*

Jens-Steffen Gutmann, Wolfgang Hatzack, Immanuel Herrmann, Bernhard Nebel, Frank Rittinger, Augustinus Topor, Thilo Weigel, and Bruno Welsch

Albert-Ludwigs-Universität Freiburg, Institut für Informatik
Am Flughafen 17, D-79110 Freiburg, Germany
⟨*last name*⟩@informatik.uni-freiburg.de

Abstract. Robotic soccer is a challenging research domain because problems in robotics, artificial intelligence, multi-agent systems and real-time reasoning have to be solved in order to create a successful team of robotic soccer players. In this paper, we describe the key components of the CS Freiburg team. We focus on the self-localization and object recognition method based on using laser range finders and the integration of all this information into a global world model. Using the explicit model of the environment built by these components, we have implemented path planning, simple ball handling skills and basic multi-agent cooperation. The resulting system is a very successful robotic soccer team, which has not lost any game yet.

1 Introduction

Robotic soccer is a challenging research domain because problems in robotics, artificial intelligence, multi-agent systems and real-time reasoning have to be solved in order to create a successful team of robotic soccer players [11]. We took up the challenge of designing a robotic soccer team for two reasons. First of all, we intended to demonstrate the advantage of our perception methods based on laser range finders [7–9,19], which make *explicit world modelling* and *accurate and robust self-localization* possible. Secondly, we intended to address the problem of multirobot sensor integration in order to build a *global world model.* Of course, in order to demonstrate the usefulness of both concepts, we also had to implement basic ball handling skills, deliberation, and multi-agent cooperation that exploit the world model.

In a paper describing *challenge tasks* in robotic soccer, Asada *et al.* [1] conjectured that range finding devices are not sufficient for discriminating the ball, obstacles, and the goal [1, p.48]. Furthermore, it was conjectured in this paper

* This work has been partially supported by *Deutsche Forschungsgemeinschaft* (DFG) as part of the graduate school on *Human and Machine Intelligence*, by *Medien- und Filmgesellschaft Baden-Württemberg mbH* (MFG), and by *SICK AG*, who provided the laser range finders.

that a "conventional" approach to building an explicit world model, planning in this model, and executing the plan is not suitable for the dynamically changing environment in robotic soccer [1, p.49].

While we certainly agree that sonar sensors are not accurate and reliable enough, laser range-finders are definitely adequate for recognizing everything on the soccer field except the ball. Furthermore, this can be easily used to construct an explicit world model which can support sophisticated behaviors and some form of deliberation, provided deliberation is tightly coupled with observations.

As a matter of fact, we believe that building an explicit world model and using it for deliberation is a *necessary* prerequisite for playing an aesthetic and effective game of soccer. This conjecture is justified by the fact that the two winning teams in the simulation and the small size league in RoboCup'97 used this approach [4,20,21]. The performance of these two teams were in sharp contrast to the teams in the middle size league at RoboCup'97. Although much of the unsatisfying performance in the middle size league could be probably attributed to problems concerning radio communication and problems due to the lighting conditions [17], some of it was probably also caused by the lack of an explicit world model. Further evidence for our claim is the performance of our team at RoboCup'98, which won the competition in the middle size league.

The rest of the paper is structured as follows. In the next section, we give an overview of the robot hardware and general architecture of our soccer team. Section 3 focuses on our self-localization approach and Section 4 describes our player and ball recognition methods that are needed to create the local and the global world model. In Section 5 we describe the behavior-based control of the soccer agents and show how a basic form of multi-agent cooperation is achieved. Section 6 focuses on planning motion sequences that are needed to execute some of the behaviors. Finally, in Section 7 we describe our experience of participating in RoboCup'98 and in Section 8 we conclude.

2 Robot Hardware and General Architecture

The robot hardware components we used as well as our general architecture is very similar to that of the other teams. The main distinguishing points are the use of laser range-finders, the ball handling mechanism, and the global sensor integration leading to a global world model.

2.1 Hardware Components

Because our group is not specialized in developing robot platforms, we used an off-the-shelf robot—the *Pioneer 1* robot developed by Kurt Konolige and manufactured by *ActivMedia.*[1] In its basic version, however, the Pioneer 1 robot is hardly able to play soccer because of its limited sensory and effectory skills. For this reason, we had to add a number of hardware components (see Fig. 1).

[1] At this point we would like to thank *ActivMedia* for their timely support, resolving some of the problems which occured just a few weeks before RoboCup'98.

Fig. 1. Three of our five robots: Two field players and the goal keeper

On each robot we mounted a video camera connected to the *Cognachrome* vision system manufactured by *Newton Lab.*,[2] which is used to identify and track the ball. For local information processing, each robot is equipped with a *Toshiba* notebook *Libretto 70CT* running *Linux*. The robot is controlled using *Saphira* [13], which comes with the Pioneer robots. Finally, to enable communication between the robots and an off-field computer, we use the *WaveLan* radio ethernet.

2.2 Laser-Range Finders

In addition to the above components, we added *PLS200* laser range-finders manufactured by *SICK AG* to all of our robots. These range finders can give depth information for a 180° field of view with an angular resolution of 0.5° and an accuracy of 5 cm up to a distance of 30 m.[3] We used, however, only an angular resolution of 1° in order to reduce the data rate between the laser range-finder and the on-board computer. With an angular resolution of 1°, we can get five to eight scans per second using a 38,400 baud serial connection.

The information one gets from these laser-range finders is very reliable and there is hardly any noise. According to the specifications of *SICK AG*, one should

[2] *Newton Lab.* was also quite helpful in solving problems we had with their vision boards a few weeks before the tournament.

[3] The more recent *LMS 200* laser range-finders have an angular resolution of 0.25° and an accuracy of 1 cm.

not operate more than one laser range-finder in a given environment or one should make sure that the laser range-finders operate on different elevation levels. We operated them, however, on the same elevation level and there was no observable interference between the five laser scanners we used.

2.3 Ball-Handling Mechanism

Handling the ball with the body of the Pioneer 1 robot is not a very effective way of moving the ball around the field or pushing it into the opponent's goal. For this reason we developed a kicking device using parts from the *Märklin Metallbaukasten*. The kicker itself is driven by two solenoids and can kick the ball over a distance of approximately three meters. Furthermore, in order to steer the ball we used flexible flippers that have a length of approximately 35 % of the diameter of the ball as shown in Fig. 2.

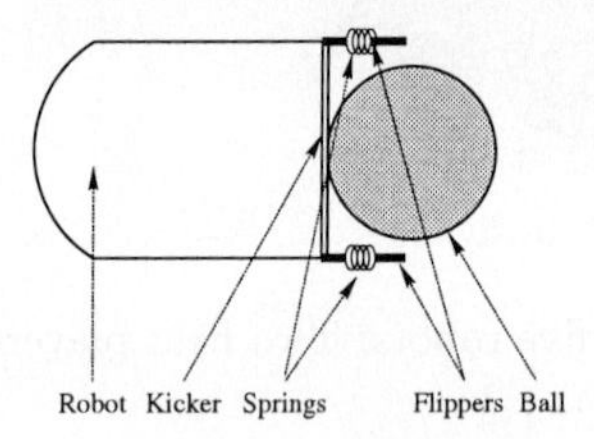

Fig. 2. Ball steering mechanism

These flippers led to some discussions before the tournament. However, it was finally decided that the use of such flippers does not violate the *RoboCup rules*. They do not take away all degrees of freedom from the ball and still allow for taking the ball away from the player. Nevertheless, whether such flippers should be allowed and, in particular, how long they may be are important issues for the further development of the RoboCup rules.

Taking the idea of *embodiment* seriously, we believe that such ball steering devices should be definitely allowed, since without it, it is virtually impossible to play soccer effectively. For example, without flippers, it is almost impossible to retrieve the ball from the wall, which means that the referee must relocate the ball, which is very annoying for everybody – in particular for spectators. Furthermore, without the ball steering mechanism the ball is very easily lost when running with the ball. In particular, our last goal in the final game, where the robot changed the direction in the last possible moment, would have been impossible without these flippers.

2.4 General Architecture

As mentioned above, our general architecture (see Fig. 3) is very similar to those of other teams in the middle size league. However, there are also some noticeable differences.

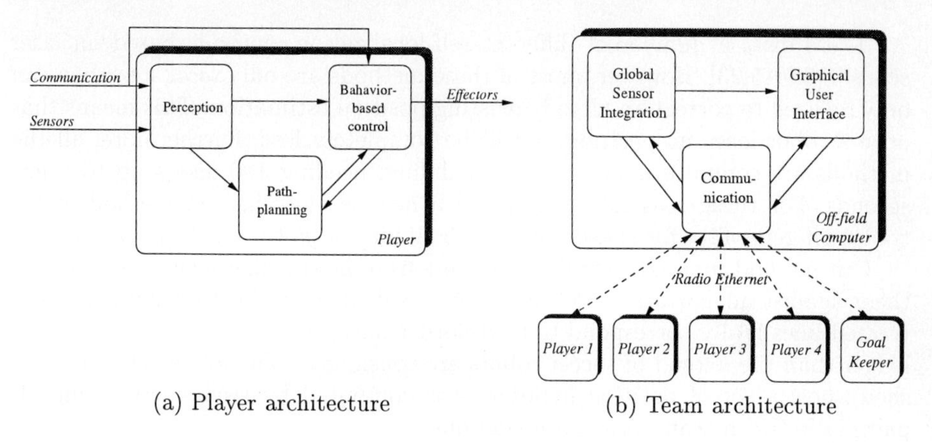

(a) Player architecture

(b) Team architecture

Fig. 3. Architecture

Our robots are basically autonomous robotic soccer players. They have all sensors, effectors and computers on-board. Each soccer agent has a *perception module* that builds a local world model (see Fig. 3 (a)). Based on the observed state of the world and intentions of other players communicated by the radio link, the *behavior-based control module* decides what behavior is activated. If the behavior involves moving to a particular target point on the field, the *path-planning* module is invoked which computes a collision-free path to the target point.

In order to initialize the soccer agents, to start and to stop the robots, and in order to monitor the state of all agents, we use a radio ethernet connection between the on-board computers and an off-field computer (see Fig. 3 (b)). If the radio connection is unusable, we still can operate the team by starting each agent manually. The *AGILO* team [12], the *ART* team [16], the *T-Team* [18], the *Uttori* team [25] and others use a very similar approach.

Unlike other teams, we use the off-field computer and the radio connection for realizing *global sensor integration*, leading to a *global world model.* This world model is sent back to all players and they can employ this information to extend their own local view of the world. This means that the world model our players have is very similar to the world model constructed by an overhead camera as used in the small size league by teams such as *CMUnited* [21]. It should be noted, however, that the information in our global world model is less accurate than the information obtained by direct observation (see Section 4).

3 Self-Localization

We started the development of our soccer team with the hypothesis that it is an obvious advantage if the robotic soccer agents know their position and orientation on the field. Based on our experience with different *self-localization* methods using laser range finders [7], we decided to employ such a method as one of the key components in our soccer agents.

There exist a number of different self-localization methods based on laser scans [3,6,9,15,23]. However, most of these methods are only *local*, i.e., they can only be used to correct an already existing position estimation. This means that once a robot loses its position, it will be completely lost. Furthermore, all the methods are computationally very demanding, needing 100 msecs up to a few seconds. For these reasons we designed a new self-localization method which trades off generality for speed and the possibility of *global self-localization*.

Our method first extracts line segments from laser range scans and matches them against an *apriori* model of the soccer field. In order to ensure that extracted lines really correspond to field-border lines, only scan lines significantly longer than the extend of soccer robots are considered. The following algorithm shows how a set of position hypotheses is computed by recursively trying all pairings between scan lines and model lines:

Algorithm 1. *PositionHypothesis(M, S, Match)*
Input: model lines M, scan lines S, correspondence-set $Match$
Output: set of positions hypotheses H

```
if |Match| = |S| then
    H := {FitMatch(M, S, Match)}
else
    H := {}
    s := SelectScanline(S, Match)
    for all m ∈ M do
        if VerifyMatch(M, S, Match, m, s) then
            H := H ∪ PositionHypothesis(M, S, Match ∪ {(m, s)})
return H
```

The *FitMatch* function computes a position hypothesis from the *Match* set, *SelectScanline* selects the next scan line that should be matched, and *VerifyMatch* verifies that the new (m, s) pairing is compatible with the *Match* set. This method is similar to the scan matching method described by Castellanos *et al.* [5]. In contrast to this approach, however, we only verify that the *global constraints* concerning translation and rotation as well as the length restrictions of scan lines are satisfied. This is sufficient for determining the position hypothesis and more efficient than Castellanos *et al.* approach.

Although it looks as if the worst-case runtime of the algorithm is $O(|M|^{|S|})$, it turns out that because of the geometric constraints the algorithm runs in $O(|M|^3|S|^2)$ time—provided the first two selected scan lines are not collinear or parallel [22]. For the RoboCup field the algorithm is capable of determining the global position of the robot modulo 180°—provided three field borders are visible.

For robust and accurate self-localization, the position information from odometry and scan matching is fused by using a Kalman filter. Therefore the probability that the robot is at a certain position l is modelled as a single Gaussian distribution:

$$l \sim N(\mu_l, \Sigma_l)$$

Here $\mu_l = (x, y, \alpha)^T$ is the mean value (the position with the highest probability) and Σ_l its 3×3 covariance matrix.

On robot motion $a \sim ((d, \theta)^T, \Sigma_a)$ where the robot moves forward a certain distance d and then rotates by θ, the position is updated according to:

$$\mu_l := E(F(l, a)) = \begin{pmatrix} x + d\ cos(\alpha) \\ y + d\ sin(\alpha) \\ \alpha + \theta \end{pmatrix}$$
$$\Sigma_l := \nabla F_l \Sigma_l \nabla F_l^T + \nabla F_a \Sigma_a \nabla F_a^T$$

Here E denotes the expected value of the function F and ∇F_l and ∇F_a are its Jacobians.

From scan matching a position estimate $s \sim N(\mu_s, \Sigma_s)$ is obtained and the robot position is updated using the formulas:

$$\mu_l := (\Sigma_l^{-1} + \Sigma_s^{-1})^{-1} \cdot (\Sigma_l^{-1} \mu_l + \Sigma_s^{-1} \mu_s)$$
$$\Sigma_l := (\Sigma_l^{-1} + \Sigma_s^{-1})^{-1}$$

To initialize the self-localization system, a pre-defined value for μ_l is chosen and the diagonal elements of Σ_l are set to infinity. For the specific RoboCup environment, this ensures global self-localization on the first scan match.

The self-localization algorithm can then be implemented in a straightforward way. From a set of position hypotheses generated by the *PositionHypothesis* algorithm, the most plausible one is selected and Kalman-fused with the odometry position estimate. The robot position is then updated taking into account that the robot has moved since the scan was taken and matched.

Our hardware configuration allows 5–8 laser scans per second using only a few milliseconds for computing position hypotheses and the position update. Although a laser scan may include readings from objects blocking the sight to the field borders, we didn't experience any failures in the position estimation process. In particular, we never observed the situation that one of our robots got its orientation wrong and "changed sides."

4 Building Local and Global World Models

Each soccer agent interprets its sensor inputs using the perception module shown in Fig. 4 in order to estimate its own position, the position of observed players and the ball position.

After the self-localization module matched a range scan, scan points that correspond to field lines are removed and the remaining points are clustered. For each cluster the center of gravity is computed and interpreted as the approximate position of a robot (see Fig. 5). Inherent to this approach is a systematic error depending on the shape of the robots.

Since our laser range finders are mounted well above the height of the ball, we cannot use it for ball recognition. In fact, even if it were mounted lower, it

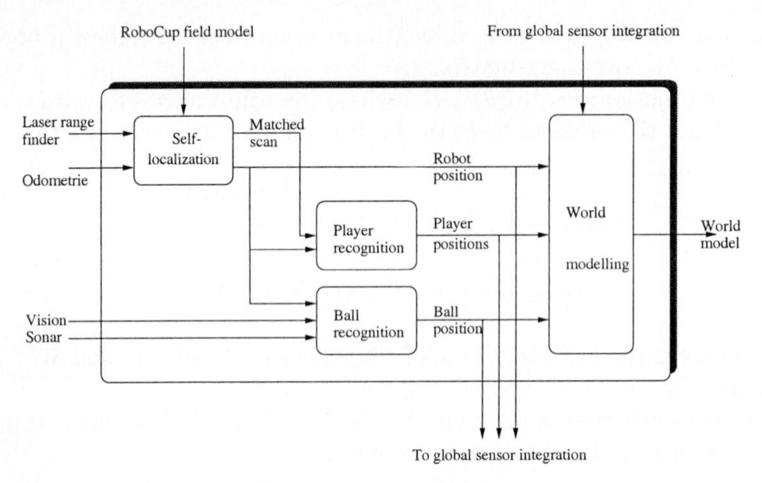

Fig. 4. Perception module

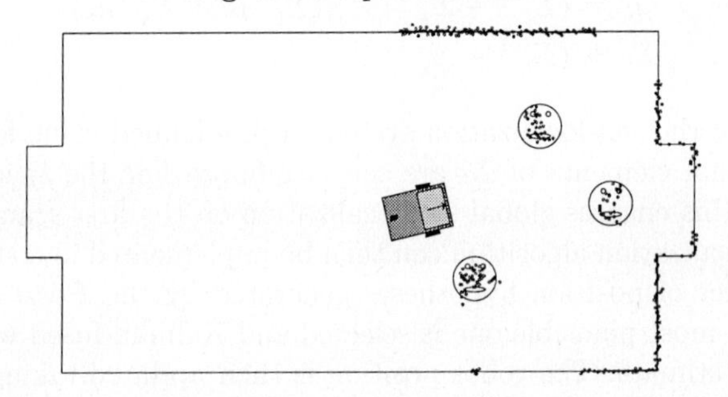

Fig. 5. Line segments are extracted from a range scan, matched against the field lines and three players are extracted from the scan.

is questionable whether it would be possible to distinguish the ball from the players by shape. For this reason, we use a commercially available vision system for ball recognition.

If the camera sees an object of a particular color (a so-called *blob*), the vision system outputs the pixel coordinates of the center of the blob, its width, height and area size. From these pixel coordinates we compute the relative position of the ball with respect to the robot position by mapping pixel coordinates to distance and angle. This mapping is learned by training the correspondence between pixel coordinates and angles and distances for a set of well-chosen real-world positions and using interpolation for other pixels. In order to improve the quality of the position estimation, we use the sonar sensors as a secondary source of information for determining the ball position.

From the estimated position of the player, the estimated position of other objects and the estimated position of the ball – if it is visible – the soccer agent constructs its own *local world model.* By keeping a history list of positions for all objects, their headings and velocities can be determined. To reduce noise, headings and velocities are low-pass filtered. Position, heading, and velocity estimates are sent to the *multirobot sensor integration module.*

In addition to objects that are directly observable, the local world model also contains information about objects that are not visible. First of all, if an object disappears temporarily from the robot's view, it is not immediately removed from the world model. Using its last known position and estimated heading and velocity, its most likely position is estimated for a few seconds. Secondly, information from the global world model is used to extend the local world model of a player. Objects of the global world model which don't correspond to any object of the local world model are added to the local world model, but marked as not really visible for the player. If an object of the global world model corresponds to an object of the local model the local information regarding exact position, heading end velocity is given priority because it is probably more recent and accurate. In this case the global information is only used to determine the objects identity.

The *global world model* is constructed from time-stamped position, heading, and velocity estimates that each soccer agent sends to the global sensor-integration module. Using these estimates, it is easy to tell whether an observed object is friend or foe (see Fig. 6). Knowing who and where the team members

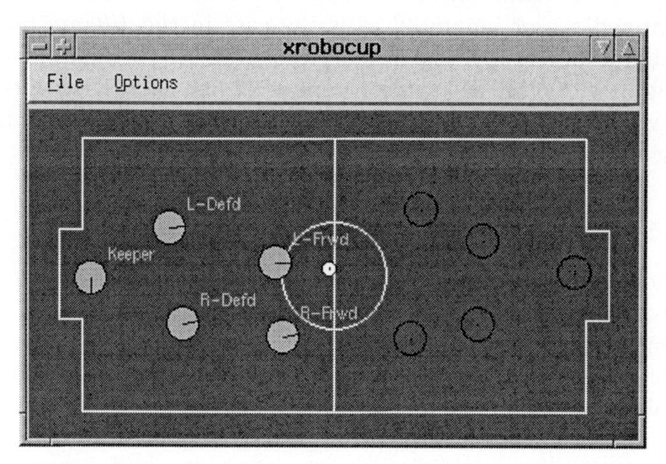

Fig. 6. Visualizing the results of global sensor integration

are is, of course, very helpful in playing a cooperative game.

Other information that is very useful is the global ball position. Our vision hardware recognizes the ball only up to a distance of 3–4 m. Knowing the global ball position even if it is not directly visible enables the soccer robot to turn its camera into the direction of where the ball is expected avoiding a search for the ball by turning around. This is important in particular for the goal keeper, which might miss a ball from the left while it searches for the ball on the right side.

It should be noted, however, that due to the inherent delay between sensing an object and receiving back a message from the global sensor integration, the information from the global world model is always 100–400 msecs old. This means that it cannot be used to control the robot behavior directly. However, apart from the two uses spelled out above, there are nevertheless a number of important problems that could be solved using this global world model – and we will work on these points in the future. Firstly, the global world model could be used to reorient disoriented team members. Although we never experienced such a disorientation, such a fall-back mechanism is certainly worthwhile. Secondly, it provides a way to detect unreliable sensor systems of some of the soccer agents. Thirdly, the global world model could be used for making strategic decisions, such as changing roles dynamically [20].

5 Behavior-based Control and Multi-Agent Cooperation

The soccer agent's decisions are mainly based on the situation represented in the explicit world model. However, in order to create cooperative team behavior, actual decisions are also based on the *role* assigned to the particular agent and on intentions communicated by other players.

Although the control of the execution can be described as behavior-based, our approach differs significantly from approaches where behaviors are activated by uninterpreted sensor inputs [2] as is the case in the *Ullanta* team [24]. In our case, high-level features that are derived from sensor inputs and from the communication with other agents determine what behavior is activated. Furthermore, behaviors may invoke significant *deliberation* such as planning the path to a particular target point (see Section 6).

5.1 Basic Skills and Behavior-Based Control

The behavior-based control module consists of a rule-based system that maps situations to actions. The rules are evaluated every 100 msecs so that the module can react immediately to changes in the world. Depending on whether the agent fills the *role* of the goal keeper or of a field player, there are different rule sets.

The goalie is very simple minded and just tries to keep the ball from rolling into our goal. It always watches the ball – getting its information from the global world model if the camera cannot recognize the ball – and moves to the point where the robot expects to intercept the ball based on its heading. If the ball is on the left or right of the goal, the goal keeper turns to face the ball. In order to allow for fast movements, we use a special hardware setup where the "head" of the goalie is mounted to the right as shown in Fig. 1. If the ball hits the goalie, the kicking device kicks it back into the field.

The field players have a much more elaborate set of skills. The first four skills below concern situations when the ball cannot be played directly, while the two last skills address ball handling:

Approach-position: Approach a target position carefully.

Go-to-position: Plan and constantly re-plan a collision-free path from the robot's current position to a target position and follow this path until the target position is reached.

Observe-ball: Set the robots heading such that the ball is in the center of focus. Track the ball without approaching it.
Search-ball: Turn the robot in order to find the ball. If the ball is not found after one revolution go to *home position* and search again from there.
Move-ball: Determine a straight line to the goal which has the largest distance to any object on the field. Follow this line at increasing velocity and redetermine the line whenever appropriate.
Shoot-ball: To accelerate the ball either turn the robot rapidly with the ball between the flippers or use the kicker-mechanism. The decision on which mechanism to use and in which direction to turn is made according to the current game situation.

The mapping from situations to actions is implemented in a decision-tree like manner as shown in Fig. 7. Taking into account the currently executed action

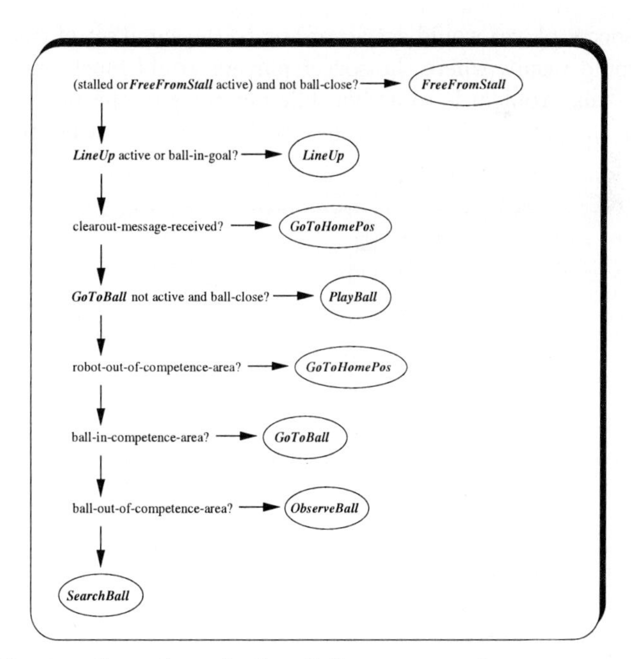

Fig. 7. Decision tree for action selection. Left arrow: yes, down arrow: no, circle: new state

and the current world model, the rules are permanently evaluated leading to a decision as to which action to take next. Possible actions include:

FreeFromStall: Select and follow a path to a clear position on the field.
GoToHomePos: Go to the home position using the *go-to-position* skill.
LineUp: "Be happy"(play music and turn on the spot), then go to the home position.
SearchBall: Invoke the *search-ball* behavior.

PlayBall: Attack using the *approach-position*, *move-ball* and *shoot-ball* skills.
ObserveBall: Track the ball using the *observe-ball* behavior.
GoToBall: Go to the ball using the *go-to-position* and *approach-position* behaviors.

It should be noted that details of tactical decisions and behaviors were subject to permanent modifications even when the competition in Paris had already started. As a reaction to teams which would just push the ball and opponents over the field we modified our stall behavior to not yield in such situations. Unfortunately the capability to recognize when a goal was shot and to line up to wait for game start was of no use since the field got too crowded with people repositioning their robots after our team scored a goal.

5.2 Multi-Agent Coordination

If all of the soccer player would act according to the same set of rules, a "swarm behavior" would result, where the soccer players would block each other. One way to solve this problem is to assign different *roles* to the players and to define *areas of competence* for these roles (see Fig. 8). If these areas would be

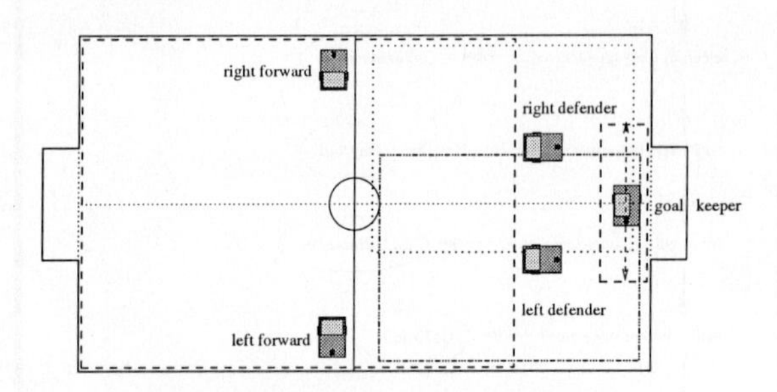

Fig. 8. Roles and areas of competence

non-overlapping, interference between team members should not happen, even without any communication between players. Each player would go to the ball and pass it on to the next area of competence or into the goal. In fact, this was our initial design and it is still the fall-back strategy when the radio communication is not working.

There are numerous problems with such a rigid assignment of competence areas, however. Firstly, players may interfere at the border lines between competence areas. Secondly, if a player is blocked by the other team, broken, or removed from the field, no player will handle balls in the corresponding area. Thirdly, if a defender has the chance of dribbling the ball to the opponent's goal, the corresponding forward will most probably block this run. For these reasons, we modified our initial design significantly. Even during the tournament in Paris we changed the areas of competence and added other means to coordinate attacks as a reaction to our experiences from the games.

If a player is in a good position to play the ball it sends a **clear-out** message. As a reaction to receiving such a message other players try to keep out of the playing robots way (see Fig. 7). This helps to avoid situations in which two team mates block each other. In other words, we also rely on *cooperation by communication* as the *Uttori* team [26]. However, our communication scheme is much less elaborate than Uttori's. Based on communicating intentions, areas of competence can be made overlapping as shown in Fig. 8. Now, the forwards handle three quarters of the field and attacks are coordinated by exchanging the intentions.

We do not have any special coordination for defensive moves. In fact, defensive behavior *emerges* from the behavior-based control described above. When the ball enters our half of the field, our defenders go to the ball and by that block the attack. Surprisingly, this simple defensive strategy worked quite successfully.

6 Path Planning

Some of the skills described in the last section concern the movement of the soccer robots to some target point on the field. While such movements could be realized in a behavior-based way, we decided to *plan* the motion sequence in order to avoid problems such as local minima.

Motion planning in the presence of moving obstacles is known to be a computationally very demanding problem [14]. Furthermore, because the movements of the opponents are hardly predictable, a motion plan would be probably obsolete long before it has been generated. For these reasons, we decided to approximate the solution to the motion planning problem with moving obstacles by solving the path planning problem with stationary obstacles. Although such an approach might seem to be inadequate in an environment that is as dynamic as robotic soccer, experience shows that often enough the opponent players can indeed be approximated as stationary obstacles. More importantly, however, our path planning method is so efficient – needing only a few milliseconds for 4–5 obstacles – that constant re-planning is possible.

To plan arbitrary paths around objects, we use the *extended visibility graph method* [14]. Objects in the world model are grown and the soccer field is shrunken allowing path planning for a robot shrunken to point. The actual planning is done by an A^* algorithm that finds shortest collision-free paths consisting of straight line and arc segments from the current robot position to the desired target position (see Fig. 9 (b))

To increase speed in planning, an iterative planning approach is used (see Fig. 9 (a)). Beginning with only the start and goal node, objects are only added to the graph if they interfere with a found path. To avoid oscillation between paths with similar costs, a distance penalty is added to paths which require large changes of the robot's heading in the beginning.

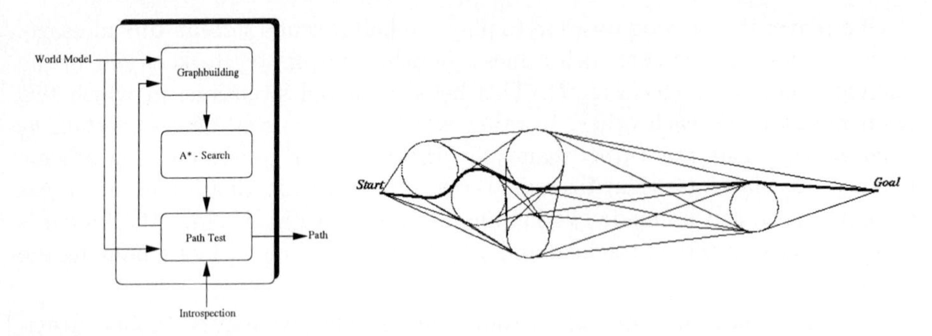

(a) Path-planning module (b) Path planning for a robot shrunken to a point

Fig. 9. Path Planning

7 Experience at RoboCup'98

Participating in the RoboCup'98 tournament was very beneficial for us in two ways. Firstly, we got the opportunity to exchange ideas with other teams and learned how they approached the problems. Secondly, we learned much from playing. As pointed out at various places in the paper, we redesigned tactics and strategy during the tournament incorporating the experience we made during the games.

Our experience with hard- and software reliability was mixed. The laser range finders and our self-localization worked without any problem, while the radio communication was sometimes jammed, perhaps by other teams playing at the same time on another field. The most fragile part was our vision system – not because of hardware failures, but because slightly changed lighting conditions led to serious problems in ball recognition. However, this seemed to be a problem for all teams.

The performance of our team at RoboCup'98 was quite satisfying. Apart from winning the tournament, we also had the best goal difference (12:1), never lost a game, and scored almost 25% of the goals during the tournament. This performance was not accidental as demonstrated at the national German competition *VISION-RoboCup'98* on the 1st of October 1998 in Stuttgart. Again, we won the tournament and did not lose any game.

The key components for this success are most probably the *self-localization* and *object-recognition* techniques based on laser range finders, which enabled us to create accurate and reliable *local* and *global world models*. Based on these world models, we were able to implement reactive path planning, fine-tuned behaviors, and basic multi-agent cooperation – which was instrumental in winning. Finally, the mechnical design of our kicker and the ball steering mechanism certainly also played a role in playing successful robotic soccer.

8 Conclusions and Future Directions

Robotic soccer is a challenging research domain. In this context, we addressed the problem of building an accurate and reliable world model for each soccer agent using laser range finders and integrated these into a *global world model.* Based on these explicit world models, simple soccer skills, path planning, and multi-agent cooperation were implemented. The resulting system is a very successful robotic soccer team, which has not been beaten yet in an official game. There are nevertheless a number of points where significant improvements are possible. For instance, we plan to improve

- the low-level sensor interpretation by exploiting more features of our hardware and by using real-time extensions of the Linux system for getting precise time-stamps of sensor measurements;
- the accuracy and robustness of multirobot sensor integration;
- the low-level control of the movements in order to get smoother behaviors;
- the soccer skills based on the above improvements, e.g., to accomplish ball passing;
- the strategic decision making by allowing for flexible role assignments and by using explicit deliberation based on the current global state.

Summarizing, we hope that we will be able to improve the level of our play and demonstrate that our robots are able to play even more effective and aesthetic robotic soccer at RoboCup'99.

References

1. M. Asada, P. Stone, H. Kitano, A. Drogoul, D. Duhaut, M. Veloso, H. Asama, and S. Suzuki. The RoboCup physical agent challenge: Goals and protocols for phase I. In Kitano [10], pages 42–61.
2. R. A. Brooks. A robust layered control system for a mobile robot. *IEEE Journal of Robotics & Automation*, RA-2(1), 1986.
3. W. Burgard, D. Fox, D. Hennig, and T. Schmidt. Estimating the absolute position of a mobile robot using position probability grids. In *Proceedings of the 13th National Conference of the American Association for Artificial Intelligence (AAAI-96)*, pages 896–901, Portland, OR, July 1996. MIT Press.
4. H.-D. Burkhard, M. Hannebauer, and J. Wendler. AT Humbold – development, practice and theory. In Kitano [10], pages 357–372.
5. J. A. Castellanos, J. D. Tardós, and J. Neira. Constraint-based mobile robot localization. In *International Workshop on Advanced Robotics and Intelligent Machines.* University of Salford, Apr. 1996.
6. I. J. Cox. Blanche: Position estimation for an autonomous robot vehicle. In I. J. Cox and G. T. Wilfong, editors, *Autonomous Robot Vehicles*, pages 221–228. Springer-Verlag, Berlin, Heidelberg, New York, 1990.
7. J.-S. Gutmann, W. Burgard, D. Fox, and K. Konolige. An experimental comparison of localization methods. In *Proceedings of the International Conference on Intelligent Robots and Systems (IROS'98).* IEEE/RSJ, 1998.

8. J.-S. Gutmann and B. Nebel. Navigation mobiler Roboter mit Laserscans. In P. Levi, T. Bräunl, and N. Oswald, editors, *Autonome Mobile System 1997*, Informatik aktuell, pages 36–47, Stuttgart, Germany, 1997. Springer-Verlag.
9. J.-S. Gutmann and C. Schlegel. Amos: Comparison of scan matching approaches for self-localization in indoor environments. In *Proceedings of the 1st Euromicro Workshop on Advanced Mobile Robots*, pages 61–67. IEEE, 1996.
10. H. Kitano, editor. *RoboCup-97: Robot Soccer World Cup I*, volume 1395 of *Lecture Notes in Artificial Intelligence*. Springer-Verlag, Berlin, Heidelberg, New York, 1998.
11. H. Kitano, M. Asada, Y. Kuniyoshi, I. Noda, E. Osawa, and H. Matsubara. RoboCup: A challenge problem for AI. *The AI Magazine*, 18(1):73–85, 1997.
12. M. Klupsch, M. Lückenhaus, C. Zierl, I. Laptev, T. Bandlow, M. Grimme, I. Kellerer, and F. Schwarzer. Agilo RoboCuppers: RoboCup team description. This volume.
13. K. Konolige, K. Myers, E. H. Ruspini, and A. Saffiotti. The Saphira Architecture: A Design for Autonomy. *Journal of Experimental and Theoretical Artificial Intelligence*, 9:215–235, 1997.
14. J.-C. Latombe. *Robot Motion Planning*. Kluwer, Dordrecht, Holland, 1991.
15. F. Lu and E. E. Milios. Robot pose estimation in unknown environments by matching 2D range scans. In *IEEE Computer Vision and Pattern recognition Conference (CVPR)*, pages 935–938, 1994.
16. D. Nardi, G. Clemente, and E. Pagello. Azzurra robot team. This volume.
17. I. Noda, S. Suzuki, H. Matsubara, M. Asada, and H. Kitano. Overview of RoboCup-97. In Kitano [10], pages 20–41.
18. M. Plagge, B. Diebold, R. Guenther, J. Ihlenburg, D. Jung, K. Zahedi, and A. Zell. Design and evaluation of the t-team of the university or tuebingen for robocup'98. This volume.
19. S. Thrun, J.-S. Gutmann, D. Fox, W. Burgard, and B. Kuipers. Integrating topological and metric maps for mobile robot navigation: A statistical approach. In *Proceedings of the 15th National Conference of the American Association for Artificial Intelligence (AAAI-98)*, Madison, WI, July 1996. MIT Press.
20. M. Veloso, P. Stone, and K. Han. The CMUnited-97 robotic soccer team: Perception and multiagent control. In *Autonomous Agents – Second Internation Conference (Agents'98)*. ACM Press, 1998.
21. M. Veloso, P. Stone, K. Han, and S. Achim. The CMUnited-97 small robot team. In Kitano [10], pages 242–256.
22. T. Weigel. Roboter-Fußball: Selbstlokalisation, Pfadplanung und Basisfähigkeiten. Diplomarbeit, Fakultät für Angewandte Wissenschaften, Freiburg, Germany, 1998.
23. G. Weiß and E. von Puttkamer. A map based on laserscans without geometric interpretation. In U. Rembold, R. Dillmann, L. Hertzberger, and T. Kanade, editors, *Intelligent Autonomous Systems (IAS-4)*, pages 403–407. IOS Press, 1995.
24. B. Werger, P. Funes, M. S. Fontan, R. Sargent, C. Witty, and T. Witty. The spirit of Bolivia: Complex behavior through minimal control. In Kitano [10], pages 348–356.
25. K. Yokota, K. Ozaki, A. Matsumoto, K. Kawabata, H. Kaetsu, and H. Asama. Omni-directional autonomous robots cooperating for team play. In Kitano [10], pages 333–347.
26. K. Yokota, K. Ozaki, N. Watanabe, A. Matsumoto, D. Koyama, T. Ishikawa, K. Kawabata, H. Kaetsu, and H. Asama. Cooperative team play based on communication. In M. Asada, editor, *RoboCup-98: Robot Soccer World Cup II. Proceedings of the second RoboCup Workshop*. Paris, France, July 1998.

The Survey of RoboCup '98: Who, How and Why

Igor M. Verner

Technion - Israel Institute of Technology, Dept. of Education in Technology & Science,
32000 Haifa, Israel
ttrigor@tx.technion.ac.il

Abstract. The need for educational research of robot competition programs is argued. A questionnaire for RoboCup '98 team members is presented. The survey data about the participants, their experience in robot competitions, activities, subjects and motivation for participating in the program are reported and discussed.

1 Introduction

Robotics competitions have become popular at universities, colleges and even schools throughout the world. They are arousing wide public interest. It becomes evident that people build robots and participate in competitions not only for research and fun but also in order „to learn design and engineering" [1].

Among other robot contests the Robot Soccer World Cup (RoboCup) program stands out for wide international representation of universities, a variety of competitions in leagues, the professional level of team projects and a strong aspiration to foster AI and robotics research. Attention is paid also to education.

Several arguments for the need to consider educational aspects of programs like the RoboCup have already been mentioned [2]. One other reason is that the RoboCup program provides a new learning environment in which new educational approaches may be developed and examined. This reason was the motivating factor for the author of this paper to participate as educator in RoboCup '97 and '98.

Though most of robot competitions are widely announced and reported, there is minimal information available about the educational aspects and values of these programs. Reports stating the success of the contests, creativity and enthusiasm of the participants (students) are not supported by concrete data and theoretical arguments as to the values of this form of education.

In fact, there is a lack of theoretical studies on didactics of design competitions in the engineering curriculum. This is currently a challenging problem for engineering education [3]. Therefore, educational research of programs like the RoboCup is currently required. This research can be carried out as case studies of specific team-projects such as [4] or as overall RoboCup statistical and analytic surveys [2].

In this paper statistical data from the participants of Robot Soccer World Cup II are summarized and discussed.

2 A Questionnaire for RoboCup '98 Team-Members

The idea of using a questionnaire was first raised at RoboCup '97 and was implemented operatively. Answers were obtained from 77 team members. This survey provided useful results [2] and it was decided to continue this study at RoboCup '98. A new questionnaire was prepared and approved by RoboCup president H. Kitano before the contest. It was presented during the competitions. Answers were obtained from 128 respondents - the majority of the team-members.

The questionnaire deals with three basic questions concerning the RoboCup activities: who, how and why. It includes five sections. The first section relates to personal data (name, country, team, institution, faculty/department and position). The second section describes personal experience in robot competition programs: period of participation, participation in competitions before RoboCup '98 and the form of participation (league). An additional question was formulated as follows: „To what extent is the participation in the RoboCup program important for your professional growth (from 1 to 5, with 5 indicating very important)?".

In the third section of the questionnaire the respondents were asked about the framework for their participation in RoboCup. In particular, they were asked to estimate (in percentage) the share of different activities in their RoboCup timetable. The mentioned activities are as follows: part of the course, extracurricular activities, hobby/voluntary activities, part of graduate/postgraduate research, research sponsored by grants, other activities.

In the fourth section, the RoboCup participants were asked to estimate (in percentage) the share of different subjects in his RoboCup activities. The mentioned subjects are: drive mechanism, control circuits, computer vision, sensor fusion, steering behaviors, learning behaviors, multiagent collaboration, computer graphics and animation and other subjects.

The last section of the questionnaire relates to respondents' motivation for participation in robot soccer. It includes a question that is similar to the one asked in the RoboCup '97 survey: „To what extent are the factors mentioned below important for your participation in the RoboCup". However the mentioned factors in the new questionnaire are different.

The factors listed in the '97 version do not provide comprehensive view of possible motivation, while the factors mentioned in the '98 version correspond one-to-one with Wlodkowski's six categories of learning motivation [5] summarized in [2]. The list of motivation factors is presented below in Table 4.

3 Questionnaire Results

3.1 Participants

Distribution of the team-members according to their positions in the universities is presented in Table 1.

Table 1. Team-members' positions (%)

Positions	RoboCup-97	RoboCup-98
Professors	23.4	14.3
Senior (PhD) researchers	3.9	8.7
Assistants & Ing.	14.3	8.7
Postgraduate students	15.6	23.8
Graduate students	23.4	17.5
Undergraduate students	19.5	27.0

It is obvious that the majority of team-members are students, they presented 58.5% at RoboCup '97 and 68.3% at RoboCup '98. There was a definite growth of postgraduate and undergraduate students. Most of the undergraduate students appearing in the table are juniors and seniors at the universities with direct master degree programs.

Possible reasons for relative decrease of representation of professors registered in the '98 survey are wider participation of students in the contest and the fact that some professors did not fill the survey form.

The distribution of the respondents according to the period of participation in the robot competitions is given in Table 2.

Table 2. Period of participation in RoboCup (%)

	Less than half a year	From half to 1 year	From 1 to 2 years	Over 2 years
RoboCup-97	48.1	32.5	14.3	5.2
RoboCup-98	29.5	32.1	17.9	20.5

Considerable growth of experience of the RoboCup '98 participants in relation to RoboCup '97 is observed. The part of respondents with less than half a year experience in robot competition programs dropped sharply. However, there is an increase of the part of respondents with experience of more than a year and especially more than two years. For the last group, robot soccer has become a subject of long-term research.

Data on participation of the respondents in robot competitions before RoboCup '98 are presented in Table 3.

Table 3. Experience in competitions before RoboCup '98 (%)

Lack of experience	One contest	Two or more contests
64.2	20.3	15.5

As follows from the table more than one third of the respondents already participated in competitions before RoboCup. A certain group of experienced „robocuppers" has been arisen. This evidence also indicates growing experience of the team members in robot soccer.

RoboCup program includes several forms of participation (leagues), with an absolute majority of participants competing in middle size, small size or simulation leagues. The number of respondents participating in each league is given in Table 4.

Table 4. Participation in leagues (%)

	Middle size	Small size	Simulation
RoboCup-97	37.6	18.2	46.8
RoboCup-98	61.3	24.2	14.5

One can mention a substantial change in division the RoboCup '98 participants into leagues in relation to RoboCup '97, due to several reasons whose discussion exceeds the limits of this paper.

3.2 Activities

The share of different types of activities in personal timetables for robot soccer, estimated by the respondents, is presented in Table 5.

The activities are listed in the first (left) column. The next columns contain data on specific groups of respondents: the second column – about all respondents; the t, forth and columns – about the participants of middle size, small size and simulation leagues; the sixth and seventh columns – about professors and students. The rows of

Table 5 contain data concerning specific types of activities appearing in the left column.

Table 5. Types of activities for robot soccer (%)

Type of activity	Share in timetable / Shaparticipants					
	All	Middle	Small	Simul	Prof.	Stud.
Part of the course	9.7 24.6	10.9 25.7	3.2 14.3	15.6 38.9	8.1 33.3	11.1 24.7
Extracurricular activities	5.9 18.0	7.6 24.3	0.7 3.6	6.7 11.1	8.9 22.2	5.9 18.8
Voluntary/Hobby activities	22.0 45.1	24.3 51.4	9.3 17.9	22.5 50.0	12.2 33.3	25.7 49.4
Part of the graduate or postgraduate research	37.7 57.4	35.0 58.1	53.6 67.9	35.6 50.0	28.9 50.0	45.3 65.9
Research work sponsored by grants	21.0 33.6	17.6 27.0	31.1 50.0	16.7 33.3	36.1 61.1	8.9 18.8
Other	2.7 8.2	2.9 8.1	2.1 7.1	3.1 11.1	5.8 16.7	1.7 7.1

Each cell contains two numbers. The top (first) number presents an average share of a certain activity in the timetable of a participant in a specific group of respondents. The bottom number (under the first one) shows the share of participants involved in a certain type of activity related to RoboCup. Both numbers are in percentage.

According to Table 5, an average RoboCup participant spares his time mainly in graduate/postgraduate research (37.7%), voluntary activities and hobbies (22.0%), research work sponsored by grants (21.0%) and as part of the course (9.7%). One can see that these values are lower than the values showing the share of participants involved in the appropriate activities. Thus, 57.4% of the respondents are involved in graduate or postgraduate research related to robot soccer. For 45.1% this is a subject of voluntary activities or hobbies, one third is doing research sponsored by grants and one fourth deals with robot soccer as part of the course.

There are several deviations from these average values for specific groups of respondents that can be considered characteristic of these groups. A considerable part of respondents from the simulation league (38.9%) participate in robot soccer as part of their course, while for the small size league this part is small (14.3%). The part of graduate/postgraduate research in the average timetable for RoboCup of the small size league members (53.6%) is larger than of the middle size league participants (35.0%). The percentage of professors involved in research sponsored by grants (61.1%) is substantially higher than the percentage of students (18.8%). Detailed discussion of the deviations exceeds the limits of this paper.

3.3 Subjects

Robot soccer requires dealing with various science and engineering subjects. On this occasion we asked the respondents to estimate (in percentage) the share of different subjects in their activities as members of the robot soccer teams. The answers are summarized in Table 6. The structure of Table 6 is similar to that of Table 5. The subjects are listed in the first (left) column. The next columns contain data concerning specific groups of respondents: all respondents, the participants of middle size, small size and simulation leagues, professors and students.

The top number in each cell of Table 6 presents an average share of a certain subject in the timetable of the participant from a specific group of respondents. The bottom number shows a share of participants dealing with a certain subject related to RoboCup. The numbers are in percentage.

As follows from Table 6, the average participants divide their time quite evenly between nine mentioned subjects with maximum of 18.5% spared for multiagent collaboration and minimum of 4.5% given for computer graphics and animation. More substantial differences are indicated in values that show the share of participants dealing with each subject.

More than half of all participants deal with multiagent collaboration (58.4%), computer vision (54.4%) and steering behaviors (51.2%). However, for every subject mentioned in Table 6 the number of participants dealing with it is significant.

Deviations from the average values by specific groups of respondents can characterize peculiar features of the groups.

For example, respondent from middle and small size leagues spare about 10% of their time with drive mechanisms and control circuits, while respondents from the simulation league spare for the first of these subjects only 2.6% of time and do not deal with the second subject. To counterbalance it, all respondents of the simulation league deal with multiagent collaboration, while for the other leagues the part of participants dealing with this subject is about 50%. One can see that professors, on average, are dealing with more subjects than students. Such comparison of the groups can be continued.

Table 6. Subjects related to robot soccer (%)

Subjects	Share in time-table / Share of participants					
	All	Middle	Small	Simul	Prof.	Stud.
Drive mechanism	8.6	9.8	10.4	2.6	4.2	7.8 38.8
	41.1	46.7	48.3	17.6	27.8	
Control circuits	8.9	9.6	13.5	0.0	5.3	8.2 31.4
	35.2	44.0	37.9	0.0	39.3	
Computer vision	17.0	18.3	20.7	3.2	15.1	17.7
	54.4	64.0	55.2	17.6	66.7	54.7
Sensor fusion	8.4	10.0	7.0	6.2	8.2	8.6 46.5
	48.0	57.3	37.9	35.3	50.0	
Steering behaviors	11.1	11.6	7.0	17.6	6.7	13.1
	51.2	53.3	48.3	58.8	33.3	57.0
Learning behaviors	8.3	6.7	6.4	20.9	13.9	7.9 36.0
	38.4	29.3	41.4	76.5	55.6	
Multiagent collaboration	18.5	15.6	18.7	34.7	21.4	18.5
	58.4	53.3	48.3	100.0	72.2	55.8
Computer graphics and animation	4.5	4.3	3.3	4.4	3.9	4.5 23.3
	23.2	22.7	20.7	23.5	27.8	
Other subjects	12.4	13.7	3.8	10.3	21.4	11.2
	28.0	29.3	10.3	35.3	44.4	25.6

3.4 Motivation

Motivation is a decisive factor in the processes of active learning, research and team work that are central for the robot soccer projects. The RoboCup '98 Survey included several questions related to personal motivation for participation in the program mentioned in Section 2 of the paper.

Answers to the question about the importance of participation in the RoboCup program for personal professional growth are summarized in Table 7.

Table 7. Importance for professional growth

AG all Imp (%)	AG Mid size Imp (%)	AG Small Imp (%)	AG Simul. Imp (%)	AG Stud Imp (%)	AG Prof Imp (%)
3.36	3.49	3.37	3.25	3.33	3.57
83.0%	87.9%	75.0%	76.5%	77.9%	100%

The table consists of six columns related to specific groups of respondents, namely, all respondents, the members of middle size, small size and simulation leagues, students and professors.

Each cell includes two numbers. The top one presents an average grade assigned by the respondents from certain groups to the importance of participation in the RoboCup for their professional growth (from 1 to 5, with 5 indicating very important). The bottom number indicates the part of respondents in the group (in percentage) that assigned grades from 3, 4 or 5 (quite important, important and very important) to their participation in RoboCup.

As follows from Table 7, the average grade given by all respondents is quite high - 3.36. The average grade in the middle size league (3.49) is higher than the grades given in other leagues, the average grade in the simulation league (3.25) is the lowest, though the difference is not sharp. The grade given by professors (3.57) is higher than that given by students (3.33). It should be mentioned that all professors assigned for this category a grade not less than 3.

Another question was related to the importance of various motivation factors for respondents' participation in RoboCup (see Section 2 of this paper). The answers are summarized in Tables 8 and 9.

Table 8 presents attitudes of all respondents, students and professors. It consists of four columns. Six motivation factors are listed in the first (left) column. The second, third and forth columns present data on the above mentioned specific groups of respondents. Each cell of the table includes two numbers. The top one shows the average grade assigned to importance ofa certain motifactor by respondents from a specific group. The bottom number indicates the part of respondents (in percentage) who assigned the grades 4 and 5 to each factor, i.e. consider it important or very important for their participation in the RoboCup.

The data presented in Table 8 indicate that high level motivation of the respondents to participate in the RoboCup is influenced by a combination of motivation factors.

Table 8. Motivation for participation in RoboCup

Motivation factors	AG all Imp (%)	AG Stud Imp (%)	AG Prof Imp (%)
1. A positive attitude towards the subject, the method and the framework for research and education suggested by the program	4.0 70.6%	4.0 69.9%	4.2 72.2%
2. Awareness of the practical need of knowledge and experience acquired through participation in the Program	3.5 49.6%	3.3 46.3%	4.0 61.1&
3. Extrinsic stimulation of your participation in the program (funds, scholarships, travel grants etc.)	2.5 24.8%	2.5 25.9%	2.7 27.8%
4. Taking pleasure in robot soccer gaming	3.6 56.3%	3.7 61.7%	3.3 50.0%
5. Ambition to cope with the RoboCup challenges and win a reward at this prestigious professional contest and forum	2.8 34.7%	2.7 31.7%	3.2 44.4%
6. Opportunity to apply your ideas, and reinforce practical and teaching/learning skills	3.9 71.1%	4.0 71.4%	4.4 88.9%

The highest average grades was assigned to the first factor (4.0) and the sixth factor (3.9) related to creative research, active learning and practical activities. Each factor is important to a certain group of the respondents. Indeed, even the factor of extrinsic simulation that got a minimal average grade between the motivation factors was mentioned as important or very important by the quarter of all respondents. One can see that motivation of professors is higher than that of students for all factors except one, namely, taking pleasure in robot soccer gaming.

Table 9 relates to attitudes of the participants in middle size, small size and simulation leagues towards the motivation factors listed above. The data included in Table 9 are similar to those in Table 8, but are presented in a more compact form. In the first (left) column of Table 9 the names of motivation factors are replaced by their ordinal numbers from M1 to M6.

The second, third and fourth columns of the table present attitudes of respondents from middle size, small size and simulation leagues. Each cell includes two numbers. The first (left) number presents an average grade, the second (right) number indicates the part of the respondents (in percentage) who assigned the grades 4 and 5 to each motivation factor.

Table 9. Motivation of Team Members in Leagues

Motivation factors	Middle size		Small size		Simulation	
	AG	Imp (%)	AG	Imp (%)	AG	Imp (%)
M1	4.1	73.6	3.6	64.0	4.1	66.7
M2	3.7	54.9	3.3	53.8	3.2	38.9
M3	2.6	25.4	3.0	36.0	1.9	16.7
M4	3.6	56.2	3.5	57.7	3.8	61.1
M5	2.9	34.2	2.8	44.4	2.8	27.8
M6	4.1	77.0	4.0	65.4	3.8	72.2

According to Table 9, there are some differences in evaluation of the motivation factors between the leagues. For example, respondents from the middle size league assign a higher average grade (3.7) to the factor M2 (related to awareness of the practical need of knowledge and experience) than their peers from other leagues. Respondents from the small size league grade the extrinsic motivation factor M3 (3.0), higher than other leagues. The part of respondents with strong ambition to win the RoboCup in the small league (44.4%) is larger than in other leagues.

4 Conclusions

1. Analysis of educational approaches and values of robot competition programs is currently a challenging problem for engineering pedagogy. Surveying the competitions and questioning the participants are suitable forms of educational research in this direction. Our paper summarizes answers to the questionnaire presented to the RoboCup '98 participants during the contest.
2. The three main questions under consideration are:
   - who are the participants;
   - how do they carry out their team-projects;
   - why do they participate in robot soccer competitions.
3. The significant population of the survey enables us to get down to studying characteristic features of specific groups of participants such as members of leagues, professors and students. The study of group differences requires more detailed analysis of the survey data, which will be dealt with in our future research.

References

1. Martin, F.: Building Robots to Learn Design and Engineering, 1992 Frontiers in Education Conference, 12C5 (1992) 213-217.
2. Verner, I. M.: The Value of Project-Based Education in Robotics. In: Kitano H. (ed.): RoboCup-97: Robot Soccer World Cup I, Lecture Notes in Artificial Intelligence 1395 (1998) 231-241.
3. Brandt, D., Ihsen, S.: Editorial: Creativity: How to Educate and Train Innovative Engineers, or Robots Riding Bicycles, European Journal of Engineering Education, 23(2) (1998) 131-132.
4. Coradeschi, S., Malec, J.: How to Make a Challenging AI Course Enjoyable Using the RoboCup Soccer Simulation System. *Proceedings of the Second RoboCup Workshop* (1998) 57-61.
5. Wlodkowski, R. J.: Enhancing Adult Motivation to Learn. San Francisco, CA: Jossey Bass Publ. (1984)

How to Make a Challenging AI Course Enjoyable Using the RoboCup Soccer Simulation System *

Silvia Coradeschi and Jacek Malec

Department of Computer and Information Science
Linköping University, Sweden
silco@ida.liu.se, jacma@ida.liu.se

Abstract. In this paper we present an AI programming organised around the RoboCup soccer simulation system. The course participants create a number of software agents that form a team, and participate in a tournament at the end of the course. The use of a challenging and interesting task, and the incentive of having a tournament has made the course quite successful, both in term of enthusiasm of the students and of knowledge acquired. In the paper we describe the structure of the course, discuss in what respect we think the course has met its aim, and the opinions of the students about the course.

1 Introduction

During the fall term 1997 we have been responsible for the course on Artificial Intelligence Programming (AIP, described on the home page `http://www.ida.liu.se/~jacma/official/aip.html`) at the Department of Computer Science, University of Linkping in Sweden. The main aim of this course is to let the students learn how to create nontrivial knowledge-based software systems. Students choosing the AIP course are expected to be taking the third year of Computer Science or the fourth year of Computer Engineering programmes. The prerequisites are an introductory artificial intelligence course and the Incremental Programming (read: Common Lisp) course. Credit for this course is given solely on the basis of approved lab assignments, differently from most other courses, where it depends mostly on examination results.

During the recent years the course has fought with bad rumour (was considered boring and time-consuming) and decreasing number of students. In 1996 only five students have chosen it, out of which just two have got all the lab assignments approved.

In order to remedy this situation, we have decided to introduce a challenging task so that the students feel interested in employing AI techniques they have learnt earlier and during the introductory lectures of AIP and are motivated for getting a working system at the end of the course, rather than quitting it in the middle of the term.

We have chosen to use RoboCup (for more information please confront `http://www.RoboCup.org/RoboCup/RoboCup.html`) as the primary lab assignment throughout the course. RoboCup is an initiative [1,2] to foster AI and intelligent robotics research by providing a standard problem (soccer) where wide range of

* A version of this paper in Japanese has appeared in the Journal of Robotic Society of Japan, special issue in Robotics and Education, May 98.

technologies can be integrated and examined. A number of international conferences and competitions are organized every year using both real robots and a simulated soccer environment. In our course we have used the simulation environment, as the main objective of AIP is to teach programming techniques. The RoboCup tournament organized at the end of the course gave the students an opportunity to present their solutions and confront them with those of others.

In the rest of the paper we describe in more detail the structure of the course, analyse whether the aims of this change have been met, summary the students' opinions and present our own conclusions about this experiment.

2 Structure of the course

The course spans two quarters of the academic year (approximately four months) and consists of 4 lectures 2 hours each, 10 hours of seminars and 52 hours of laboratory sessions.

The lectures are devoted to the introduction of the RoboCup framework and to the basic AI programming techniques we expect the students to use in their designs and implementations. The first lecture is an introduction to RoboCup and to the problem the students are expected to solve. In the second lecture several agent architectures are presented, ranging from Brooks' reactive architecture to the classical AI "sense-think-act" agent architecture. Learning is discussed during the third lecture. The final lecture introduces problems and important practical aspects related to the development of a team of agents. This lecture is given by a person that has actually developed a team for one of the previous competitions.

The seminars are mainly used by the students to discuss the architecture of their agents and to receive feedback from the teachers and from other students about their design. In the beginning of the course the students are given a number of articles describing earlier RoboCup teams to take inspiration for their design.

During the laboratory sessions a student has access to a number of Sun Sparcstation computers and can get help from the teachers about implementation and design issues.

In order to get credit the students are expected to implement agents fulfilling the specifications given in the beginning of the course and capable of taking part in the final competition. Also they have to write two reports: one after the first few weeks that gives a first idea about the agent architecture they intend to implement, and a second after the competition that explains, justifies and critically discusses the used architecture and the implementation.

3 Did the course meet its aims?

While evaluating the success or failure of the course we have to assume some criteria. They may be based on the formal definition of the course (as published in the course-book of the university), or may take into account satisfaction of the students and analyze whether they have learned something useful by taking the course.

Analysing the goals of the Artificial Intelligence Programming course, we can see that the students taking it are expected to learn a variety of implementation techniques commonly understood as AI-related, so that they can later employ them in whatever (software) system they create. From this point of view, using RoboCup only partially meets this goal, as the students make quite early decision about how their software agents will look like, what implementation techniques they would like to use and what are the system-design solutions (architecture, functionality, representational capabilities, finally the implementation language). On the other hand, in the beginning of the course they learn about various architectures that might be used, about several representation formalisms that might be employed and about different functional possibilities while implementing their agents. A number of papers is read and discussed by the students during this part of the course (all lectures and most seminars, plus a small number of lab sessions). Afterwards they are expected to file a design specification that states their design decisions and motivates them. This way we ensure that the expected amount of background knowledge has been assimilated, before proceeding to the more practical part of the course.

Another objection that might be raised against RoboCup as the practical task for AI Programming is that it requires a substantial amount of knowledge not related to AI, like real-time and process programming (there is a course on Process Programming given by our department, but we cannot expect that students choosing AIP have already taken it). Actually, this was the major complaint from the students - that they had to devote a lot of time to learning other things while creating their systems. Some of them have spent a substantial number of hours finding out how to make their agents communicate with the server via a socket. The fact that some of them used Lisp didn't make it easier.

However, the students were very enthusiastic until the very end of the course. We have received a number of evaluation questionaires; all of the participating students thought that RoboCup was a good idea to follow in this course. Moreover, we have managed to revert the decrease of the number of students taking the course: the lectures have been attended by eight participants and five teams have been presented at the final tournament. So from this point of view we consider this year's AI Programming course a success, too.

If we analyse intead the knowledge that the students have gathered during the course, then definitely the evaluation of the course must be positive. Besides learning some classical AI implementation techniques, they were forced to learn about reactivness vs. deliberativeness in complex autonomous systems, about multi-agent systems, agent cooperation and coordination (actually, the latter two topics are not covered during the basic course on Artificial Intelligence), last but not least about real-time programming and multi-language system implementations.

Thinking about the future of the course we might imagine two possible ways of proceeding: either keeping the AI Programming as the major focus, or shifting it towards agent-based programming. In the first case the course should be prolonged to three quarters of the year — the students feel that the time available for developping their systems is not sufficient. The first part of the course, lasting one quarter, would focus on general AI programming paradigms, while the next two quarters would give the opportunity to thoroughly design, implement and test a multi-agent team capable of participating in a RoboCup contest.

In the latter case, more information should be provided about the multi-agent systems, including theoretical foundations of cooperation and coordination, on the expense of techniques that are not relevant in this context. The course could then be given under the title "Agent programming" or similar.

4 Student opinions

We have asked the students a number of questions about the course. What follows is a summary of their answers.

The students were very happy about the course and felt that it was corresponding to their expectations. They found it stimulating to implement a "real system", and to see how it actually worked in practice, although they thought that implementing it was time-consuming. All students though that using RoboCup was a good idea. The main suggestions for improvement were in the direction of giving to the students more help in overcoming the practical problems related to the soccer simulator and in particular socket communication and multiprocessing. Also in order to run a team efficiently it could be necessary to use several Sparcstations and this can be impractical during the development of the team. A possible solution would be to slow down the simulator instead. The course was considered quite time consuming and this has been the main reason for the few drop-outs we had from the course.

5 Lesson learned and improvement of the course

The main lesson learned is that it is possible to have students design and develop quite a large system if the task is sufficiently interesting and appealing. During the course the learning process was mainly based on solving a specific problem. This has the disadvantage that the students learn less general knowledge about the subject, but on the other hand they practice their capability of actually solving a problem and assimilate much better the part of the knowledge directly related to the problem. As this course is taken by students that already have a basic knowledge in AI we think that the advantages of the problem-based learning (PBL) approach overcome the disadvantages.

For the future we intend to provide skeleton agents in several languages to help the students starting off their implementations. The soccer simulator could also be made run slower so that a number of practical problems could be removed. We have noticed that the team's performance was improving during the competition due to small changes made by the students. We think that it could be a good idea to have a pre-competition where the team could be tried out, and the real competition should happen some week after it. This way the student's satisfaction of their teams could be even larger.

6 Conclusions

We have described a course on Artificial Intelligence Programming given at the Department of Computer Science, University of Linköping. The course has much benefitted from incorporating the RoboCup challenge as the main laboratory task. Although the amount of knowledge students gathered during the course is hard to measure quantitatively, it is our firm belief that they have learned more than during previous years the course was given. The students' satisfaction is also a factor that shouldn't be neglected in this context.

We hope that our experience can help in preparing similar courses at other universities. All the information about AIP course is available at its home page (`http://www.ida.liu.se/~jacma/official/aip.html`), more data can be obtained form the authors.

References

1. H. Kitano, M. Asada, Y. Kuniyoshi, I. Noda, E. Osawa, and H. Matsubara. Robocup, a challenge problem for ai. *AI Magazine*, 18(1):73–85, 1997.
2. H. Kitano, M. Tambe, P. Stone, M. Veloso, S. Coradeschi, E. Osawa, H. Matsubara, I. Noda, and M. Asada. Robocup synthetic agent challenge 97. In *Proc. of IJCAI'97*. Nagoya, Japan, 1997.

A Quadruped Robot for RoboCup Legged Robot Challenge in Paris '98

Masahiro Fujita[1], Stephane Zrehen[1] and Hiroaki Kitano[2]

[1] D21 Laboratory, Sony Corporation,
6-7-35, Kitashinagawa,
Shinagawa-ku, Tokyo, 141 JAPAN
[2] Sony Computer Science Laboratory Inc.
Takanawa Muse Building, 3-14-13 Higashi-gotanda,
Shinagawa-ku, Tokyo, 141 JAPAN

Abstract. One of the ultimate dream in robotics is to create life-like robotics systems, such as humanoid robot and animal-like legged robot. We choose to build pet-type legged robot because we believe that dog-like and cat-like legged robot has major potential for future entertainment robotics markets for personal robots. However, numbers of challenges exists before any of such robot to be fielded in the real world. Robots have to be reasonably intelligent, maintains certain level of agility, and be able to engaged in some collaborative behaviors. RoboCup is an ideal challenge to foster robotics technologies for small personal and mobile robotics system. This paper, we present Sony's legged robots system which enter RoboCup-98 Paris as a special exhibition games.

1 Introduction

Robot systems with life-like appearance and behaviors are one of the ultimate dream in robotics and AI [Inaba, 1993, Maes, 1995]. Honda's Humanoid Robot announced in early 1997 clearly demonstrated that it is technically feasible and we have all recognized that social and psychological impacts of such technologies are far reaching. A robot system described in this paper presents our effort to build reconfigurable robot system that has high degree of design flexibility. While many robotics system uses wheel-based driving mechanisms, we choose to develop legged robot which can be converted into wheel-based robots easily by using reconfigurable physical components. Among various types of robot configurations attainable using our architecture, our immediate focus is a legged robot configuration with four legs and a head each of which has three degree of freedom. Given that our goal is to establish Robot Entertainment industry, this configuration is attractive because it resembles dogs or cats, so that people may view it as robot pets. It also entails numbers of technical issues in controlling its motion, while avoiding difficulties of biped robot systems.

We are interested in participating in RoboCup, Robot World Cup Initiative [Kitano, et al., 1997], using our legged robots, as a part of its exhibition program. RoboCup is an ideal forum for promoting intelligent robotics technologies because it offer challenging and exciting task of playing soccer/football games by multiple robots. In order for a team of robots to play soccer game, robots should be able to identify environment in real-time under less controlled world than the laboratory set up, move quickly, and has to have certain level of intelligence. For our legged robots, RoboCup would be even more challenging than wheel-based robots because participation to RoboCup requires sophisticated control system for 15 degree-of-freedom for multiple robots. In addition, body posture and head position moves constantly due to legged locomotion. This imposes serious problem for a vision system which acts as a central sensor system to identify ball position, opponent position, and position of goals.

All these challenges actually represents basic requirements for developing robot for personal use, especially for entertainment robots. For robot entertainment, each robot has to be fully autonomous, and it should be exhibit reasonable intelligence and maneuvering capability. For example, a robot dog should be able to identify the owner and chase a ball, and play with it. If unwanted person or objects show up in front of it, a robot dog should react quickly to escape from it. Often, few robot dogs may chase an emery as a team. Clearly, these basic elements can be found in RoboCup games. Of course, legged robot to play soccer game constitutes only a part of robot entertainment market. However, it can be a clear and present new market and it is a good starting point.

The legged robot described in this paper is based on OPEN-R standard [3], which is designed as a standard architecture and interfaces for Robot Entertainment [Fujita and Kageyama, 1997a]. OPEN-R aims at defining a standard whereby various physical and software components which meet the OPEN-R standard can be assembled with no further modifications, so that various robots can be created easily. We developed the legged robot platform based on the OPEN-R, and collaborating with three universities, we had an RoboCup Legged Robot Exhibition at Paris.

In this paper, first, we will describe the OPEN-R briefly, followed by description of the legged robot platform for RoboCup-98 Paris. Then, the field setup and the rule will be explained. Since the legged robot platform equipped with a color processing engine, the important items in the field were painted by different colors, which we selected carefully based on experiments. We will describe how we selected the colors, and how the robot can self-localize in the field. Finally, we will report the result of Legged Robot Exhibition at RoboCup-98 Paris with a brief introduction of features of each team.

2 OPEN-R Standard

2.1 Overview

OPEN-R is a standard for robot entertainment which enables reconfiguration of robot structure, as well as easy replacement of software and hardware components.

[3] **OPEN-R** is a trade mark of Sony Corporation.

One of the purposes of proposing OPEN-R is that we wish to further promote research in intelligent robotics by providing off-the-shelf components and basic robot systems. These robots should be highly reliable and flexible, so that researchers can concentrate on the aspect of intelligence, rather than spending a substantial proportion of their time on hardware troubleshooting.

(a) (b)

Fig. 1. MUTANT: (a) Fully autonomous pet-type robot and (b) remote-operated soccer game robots

For the feasibility study of OPEN-R, we developed an autonomous quadruped pet-type robot named MUTANT[Fujita and Kageyama, 1997a] (Fig.1 (a)). We also implemented and tested a remote-operated robot system for soccer game (Fig.1 (b)). Using these robots, software and application feasibility studies were carried out.

The legged robot for RoboCup described in is paper is a direct descendent of MUTANT and remote control soccer robots. The legged robot for RoboCup is a fully autonomous soccer robot.

2.2 Major Features

One of the salient feature of the OPEN-R standard is that is attains several critical dimensions of scalability. These are (1) size scalability, (2) category scalability, (3) time scalability, and (4) user expertise scalability.

Size Scalability (Extensibility): OPEN-R is extensible for various system configuration. For examples, in a minimum configuration, a robot may be composed of only few components, and perform a set of behaviors as a complete agent. This robot can be scaled up by adding additional physical components. It is possible to scale up such a system by having such robots as sub-systems of large robot systems.

Category Scalability (Flexibility): Category scalability ensures that various kinds of robots can be designed based on the OPEN-R standard. For example, two very different styles of robots, such as a wheel-based robot and a quadruped-legged robot, should be able to described by the OPEN-R standard. These robots may have various sensors, such as cameras, infra-red sensors, and touch sensors and motor controllers.

Time Scalability (Upgradability): OPEN-R can evolve together with the progress of hardware and software technologies. Thus, it must maintain a modular organization so that each component can be replaced with up-to-date modules.

User Expertise Scalability (Friendly Development Environment): OPEN-R provides a development environments, both for professional developers and for end-users who do not have technical knowledge. End users may develop or compose their own programs using the development environment. Thus, it is scalable in terms of the level of expertise that designers of the robot have.

2.3 OPEN-R Strategy

Our strategy to meet these requirements consists of the following:

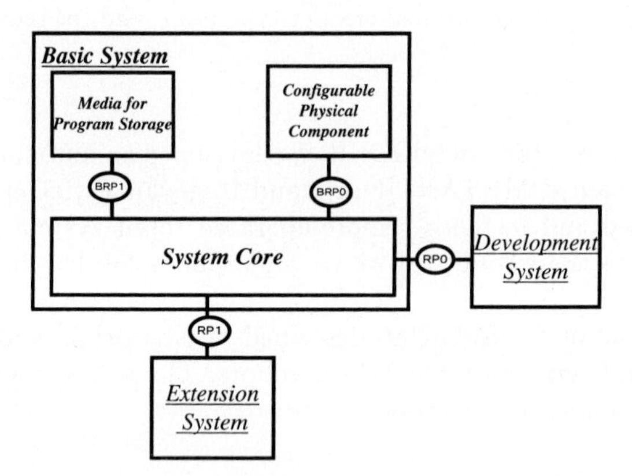

Fig. 2. Generic System Functional Reference Model and Examples of Derived System Architectures

Generic Reference Model: To meet the requirements of Extensibility and Friendly Development Environment, we define a generic system functional reference model (GSFRM) composed of Basic System, Extension System and Development System. By defining GSFRM, we are able to construct various kinds of robot systems with extensibility and development environments, as shown in Fig.2.

Configurable Physical Component: To meet the requirements of Flexibility and Extensibility, we devise a new idea of Configurable Physical Component (CPC). The physical connection between the robot components is done by a serial bus. In addition every CPC has non-volatile memory with (1) functional properties, such as an actuator and a two dimensional image sensor, and (2) physical properties, which help solve the dynamics of the robot consisting of these CPCs. Fig. 3 illustrate this concept.

Object-Oriented Programming: To meet the requirements of Up-gradability and Flexibility, we employ an object-oriented OS, **Aperios** [Yokote, 1992], which supports the Object-Oriented Programming (OOP) paradigm from the system level with several types of message passing among objects. In addition, Aperios is capable of customizing APIs by system designers.

Layering: To meet the requirements of Up-gradability and Friendly Development Environment, we utilize the layering technique which is often used for multi-vendor open architecture. OPEN-R divides each functional element into three layers, Hardware Abstraction Layer (HAL), System Service Layer (SSL), and Application Layer (APL).

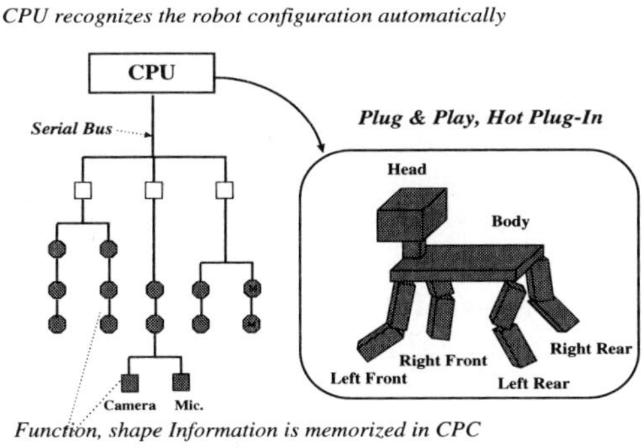

Fig. 3. Configurable Physical Component

In order to achieve the software component concept, an Agent Architecture for entertainment application was studied. The details is described in [Fujita and Kageyama, 1997a].

3 Legged Robot Challenge

A team of legged robot soccer players is a major challenge. It must address various issues involving control of robot with multiple degree of freedoms and stabilization and compensation of images obtained from continuously dislocating vision system due to legged locomotion, and other issues, in addition to all difficult problems offered in RoboCup Challenge for wheel-based robots.

3.1 Legged Robot Platform for RoboCup

Quadruped Legged Robot: For RoboCup-98 Paris, we will deploy legged robot with four legs and one head, each of which has three degree of freedom, and rich sensory channels, for example, a head has a color CCD camera, stereo microphone, touch sensors, and a loud speaker.

Most of the intelligent autonomous robots are implemented in wheel-based mechanical configuration. A wheel-based robot has advantage in their simplicity of motion control, so that researchers can concentrate on vision, planning, and other high-level issues. However, since our goal is robot entertainment, different emphasis shall be made. We believe that the capability of representation and communication using gesture and motion is very important in entertainment applications. Therefore, we choose a mechanical configuration of our robot as a quadruped-legged type, as shown in Fig.4.

The merits of the quadruped-legged configuration are, (1) walking control of a quadruped-legged is easier than that of a biped robot, and (2) when in a posture of sitting, two hands are free to move, therefore, they can be used to present emotions or to communicate with a human by the motions of the hands. Since each leg or hand has to be used for various purposes besides walking, we assign three degree of freedom (DoF) for each leg/hand. In addition, we add a tail and three DoF for neck/head so that the robot has enough representation and communication capabilities using motions.

During the RoboCup games, legs are not necessary used for expressing emotions. However, they can be used for sophisticated control of balls, such as passing ball to the side or back, or engaged in deceptive motions.

Disadvantages of using legged robot is their moving speed is not as fast as wheel-based robots. In future, speed issue may be resolved when galloping was made possible. For now, legged robot will be played within dedicated league. Although serious hardware limitation exists, teams with efficient leg motion coordination will have major advantages in the game.

Standalone System: In general, it is difficult for a standalone robot system to perform these tasks in real time in a real world environment because of its limited computational power. The remote operated robot system depicted in Fig.2 can solve the computational power problem; however, in general, much computational power is necessary for image processing tasks. This implies that it is necessary for each robot to be equipped with a video transmitter. This is sometimes difficult for regulation reason.

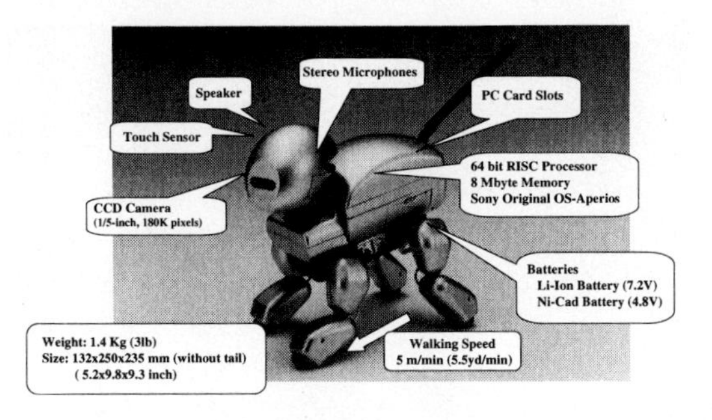

Fig. 4. Mechanical Configuration

Another solution to this problem is to set up a camera overlooking the entire field, and to distribute the image signal to all host computers. In this set up, each robot can use the huge computer power of the host computer, or a special engine such as image processing hardware. However, the image information taken by the overlooking camera is not the image taken from each robot's viewpoint.

We consider that technologies to process the image from each robot viewpoint without any global information will become very important in Robot Entertainment in future. Therefore, we decide to build RoboCup System with standalone robots under local communication constraint.

There are two hardware issues that need to be resolved to enable full onboard vision system for small size robots: (1) camera size, and (2) processor power. We solved these problem by actually manufacturing a dedicated camera and a processor chip.

Micro-Camera-Unit: In order to make a robot small in size and weight and to reduce cost, we developed a Micro-Camera-Unit (MCU) using multi-chip-module technology ([Ueda, 1996]), as shown in Fig.5. This MCU includes a lens in the same package to achieve a single thin camera module. The size of the MCU is $23 \times 16 \times 4mm$, with pixels 362×492.

OPEN-R Chip: To make the robot small in size and in power consumption, we employ MIPS architecture's R4000-series CPU with more than 100 MIPS performance. We have also developed the dedicated ASICs including the peripheral controllers of the CPU, as described before. Fig.6 shows the entire electrical block diagram, where we employ the Unified Memory Architecture with synchronous DRAMs (SDRAM) as a main memory. The features of the ASIC are as follows:

OPEN-R BUS controller: The OPEN-R bus controller controls all CPC

Fig. 5. Micro Camera Module

devices which are connected in tree structure fashion (Fig.3).

Image Signal Processing Engine: The Camera data is transferred to the SDRAM through the multi-resolution filter bank (FBK). This FBK consists of three layer resolution filters. The resolutions of the filters are 360×240, 180×120 and 90×60. The filtered image data through the FBK can be applied by the color detection engine (CDT) so that eight color can be detected in real-time.

DSP: For sound processing, the integer DSP with about 50MHz clock is integrated in the ASIC so that the FFT or filter bank processing for sound data can be done in real-time.

Although dedicated camera module and OPEN-R Chip enables on-board vision processing, it still requires major efforts to recognize natural images of balls and field. However, color marking regulation of RoboCup enable us to identify goals, a ball, and other robots using Color Detection Engine.

3.2 Soccer Field

The size of the soccer field for the quadruped robot competition is shown in Fig.7. This size is not the same as the official filed size of RoboCup Small Size League or Middle Size League. We considered three players by three players competition, and this size is considered based on the size of our robot.

Each wall is slanted by 45 degree with 10cm height. This slant keeps a distance between a ball and the wall so that a robot player can easily kick the ball. In addition, the corner has a triangle slant wall. This also avoids the ball stacking in the corner.

As we mentioned above, the items in the field are painted by carefully selected colors. The color assignment is shown in Fig.8.

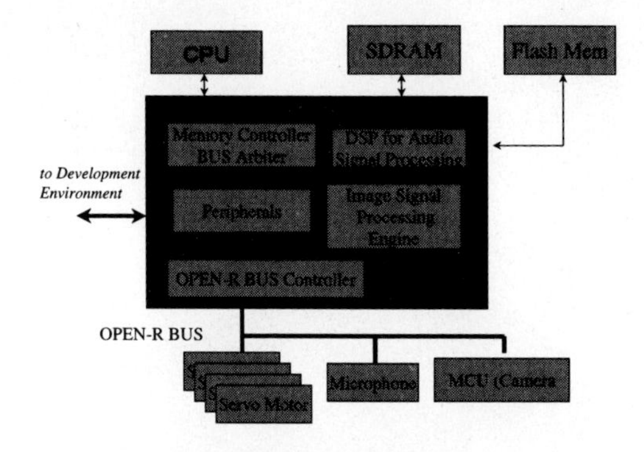

Fig. 6. Electrical Block Diagram including the ASIC

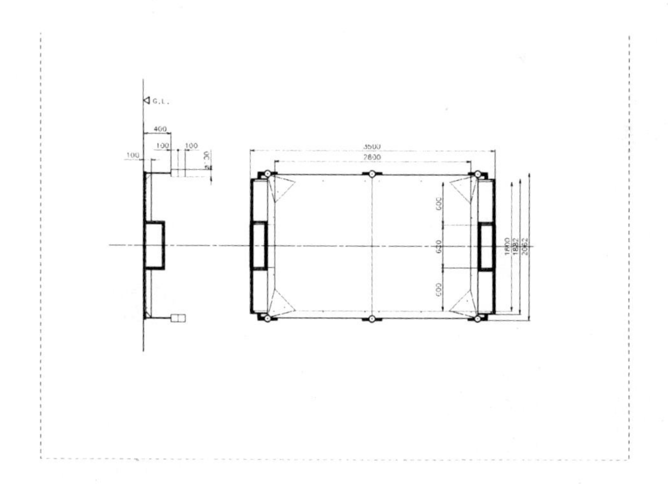

Fig. 7. A dimension of the Soccer Field

The six poles with two colors are considered to be used for landmarks, which can be used for self-localization in team collaboration strategy described in the following section.

The color samples used in the field is shown in Fig.9. The image was taken by a robot camera with a lighting condition of $580(lx)$ measured by an illuminometer, and a color temperature of $4500(K)$ measured by a CIE 1931 chromaticity meter.

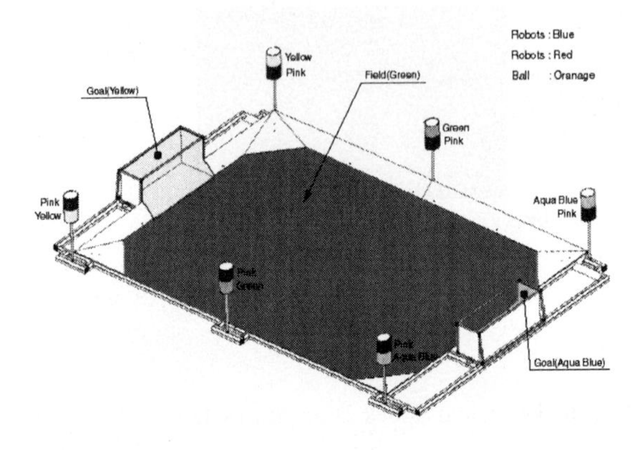

Fig. 8. Color assignment in the field

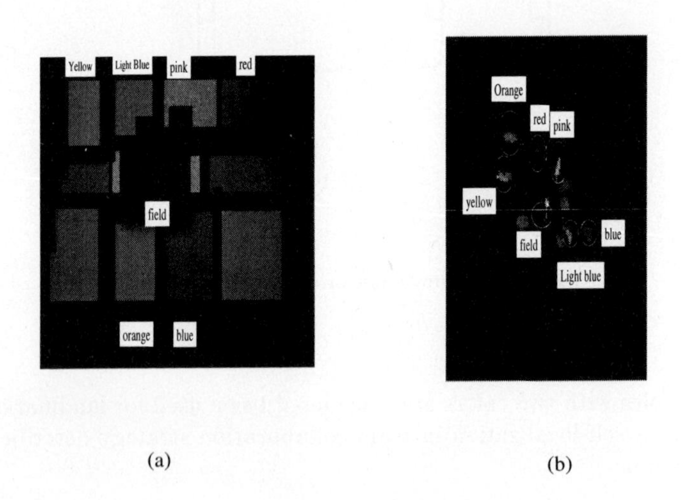

Fig. 9. (a) Color samples used in the field. and (b) its distribution in the UV plane

4 Vision Subsystem

While there are numbers of software issues exist, this paper briefly describe vision subsystem to detect and track objects, and to localize robot's own position.

4.1 Object Tracking and Distance Estimation

There are few objects which need to be identified in RoboCup games – goals, a ball, and other robots. Basically, each object can be identified with color. A ball is painted in a single color, and each goal is painted in a distinguishable color. Robots are basically painted in black or a dark color. We use color (red and blue) painted costume so that a robot can recognize opponent players. Obviously, the most important object is a ball. This can be done in rather simple manner, that to find a define color for the ball, e.g. red. To keep the ball in the center of the image, the robot head direction is feedback controlled, and the walking direction is also controlled using the horizontal head direction. The advantage of having a neck is that the robot can continue to keep the ball within visual field even when the robot do not walk toward the ball. These head position control is carried out by neck-subsystem using behavior-based approach [Brooks, 1986]. Our distributed layering of agent architecture, similar to [Firby, 1994] enables such an implementation.

Both the horizontal and vertical neck angle can be utilized to estimate the distance between the ball and the robot body (Fig.10). This information is also used as a queue to execute a kick motion.

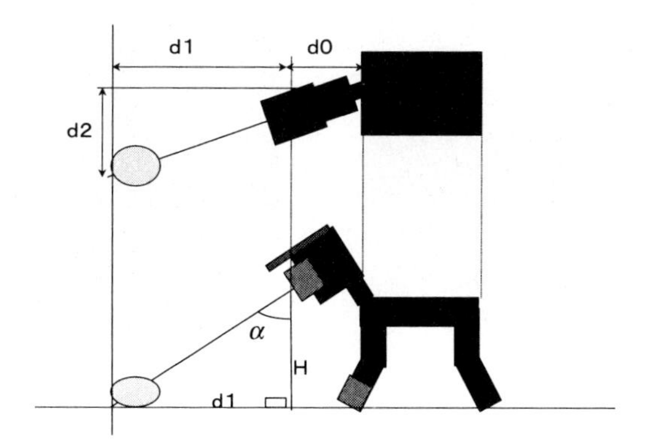

Fig. 10. Ball Positioning

Assume that the robot knows H, the height of its camera view point from the ground, following procedure identifies distance of the robot from the ball.

1. Find the ball.
2. Measure the neck's vertical angle facing the ball (α), and horizontal angle (θ).
3. $Htan^{-1}(\alpha)$ is the distance from the ball along the axis of body's moving direction (d1), and $Htan^{-1}(\theta)$ is the displacement of the ball from the axis of body direction (d2).

4.2 Self-localization through vision

Visual landmarks on the walls of the field can be used for self-localization purposes. Indeed, any x-y position can be computed by triangulation of three landmarks azimuths, assuming that the absolute position of landmarks is known. In order to limit the computational demands on the CPU for visual processing and to make exhaustive use of the color processing hardware provided with the camera module, we will paint visual landmarks regularly along the walls of the field. These landmarks are represented on figure 11. They are about 10cm high and 3cm wide. They are composed of three color strips, and are limited on their top and their bottom by bright yellow bands. The information carrying colors are chosen among four: blue, pink, yellow and indigo. The bottom color indicates the wall on which the landmark is located.

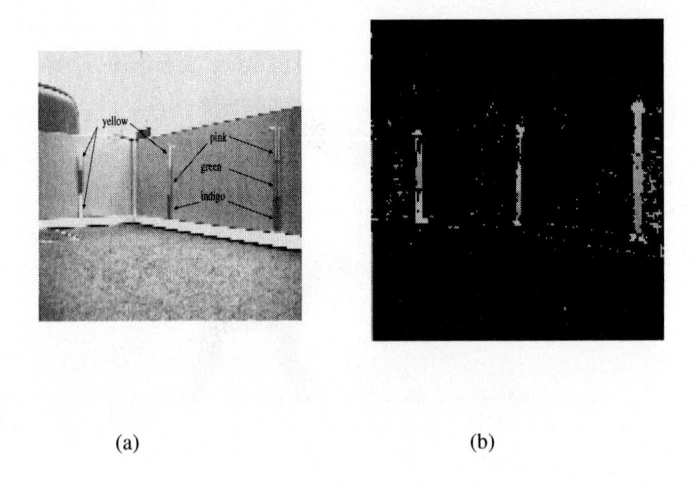

Fig. 11. (a) an image of two color landmarks pasted on the field walls, and (b) the result of color detection

The special color processing hardware allows the detection of 8 template colors and provides output in the form of 8 bit-planes, for every field, that is, every 50ms. Color detection proceeds as follows: The input image is represented in the YUV space, using 256 values for U and V and 32 values for Y. A lookup table is established for every template color in the form of a cube in the YUV space, the dimensions of which representing a matching threshold on color comparison. In our preliminary studies, using landmarks printed on a simple color printer and then pasted on the walls with yellow adhesive tapes, it appears that it is best to use two templates for each color, one for the lighter regions and one for the darker ones. This is notably due to the presence of shadows whose color is hard to model. An example of color detection is presented on Figure 11 (b). The image is noisy, due to the existence of reflections on the plastic walls and to the absence of any noise-removing routine. However, this is not a major problem since in the localization subsystem we are only trying to identify and localize the landmarks in the visual field, and not to exhaustively segment the image. Landmark detection and identification is performed by detecting three small color squares in a vertical arrangement. On a sequence of 10 seconds where a -90 to +90 degrees motion was performed, all the landmarks but one were correctly identified. In such cases, the information extracted from the sequence can help overcome the wrong detection.

Since the micro-camera module has a limited visual angle, a head rotation movement might be necessary to identify and localize distant landmarks, for good resolution in the triangulation computation. This is time consuming and may not always be necessary. Perspective information can be used to obtain qualitative information about the robot's position in the field. For instance, on Figure 11 (a), the relative size of the two landmarks on the right tells that the right wall of the field is on the right side of the robot. On the other hand the similar size of the two landmarks on the left tells that the robot is facing a corner since these two landmarks are known to be on different walls. In addition, it is possible to establish a lookup table of the visual size of a landmark as a function of its distance to the robot.

The use of this system will thus make self-localization possible and easy without requiring any global positioning system, which would need to be outside the robot.

5 Report of RoboCup 98 Paris

In this section we present some of the work developed with the legged robot platform at Carnegie Mellon University (CMU), in the Laboratoire de Robotique de Paris (LRP), and at Osaka University.

5.1 Participants

CMU: CMTrio: CMU had decomposed their work along the following aspects:

- Reliable detection of all of the relevant colors: orange (ball), light blue (goal and marker), yellow (goal and marker), pink (marker), light green (marker), dark blue (teammate/opponent), and drak red (opponent/teammate).
- Active ball chasing: the robot activity interleaves searching for the ball and localization on the field to evaluate both an appropriate path to the ball and final positioning next to the ball.
- Game-playing behaviors: robots play attacking and goal keeping positions.

They use supervised learning for color detection and Bayesian Estimation for Self-locarization. The main feature of CMTrio was stable visual perception. Since the basic behaviors of the player in the color painted field were based on the visual perception, the stability of the color detection is one of the most important points in the Legged Robot Exibition.

LRP: Les Titis Parisiens: The French team had focused on the three points:

- Vision Recognition System
- The design of walking patterns
- The strategy chosen to play soccer games

The most important feature of the French team was the development of walking pattern. Although Sony provided a basic walking generation software, only this team developed the walking pattern by themselves, which was very stable and fast.

Osaka University: Baby Tigers: The main feature of Osaka University team was the development of the metholodogy to acquire basic behaviors from the interactions between the legged robots through multi sensor-motor coordinations. The behaviors of Baby Tigers were not Top-down programmed, but teached by presenting many sets of visual patterns and behavior commands.

5.2 Results

The event schedule is shown in Fig.12.

From July 4th to 5th, we had non-official competitions, but only practical purpose games. On July 6th, it was not opened to public, but only to RoboCup participants. From July 7th to 8th, we had official competitions. The result of round-robin was shown in Fig.13.

6 Conclusion

Building a reliable and powerful robot platform meeting commercial-level product reliability is a challenging task. A small-size legged robot is even more challenging. We managed to produce such a robot platform by actually fabricating

Fig. 12. Schedule of RoboCup-98 Paris: Legged Robot Exhibition

RoboCup Legged Robot Exhibition Match

	CM-Trio CMU	Les Titis Parisiens LRP	Baby Tigers OSAKA	Total Game Point
CM-Trio CMU		2 - 1	1 – 1	5
Les Titis Parisiens LRP	1 – 2		1 - 0	3
Baby Tigers OSAKA	1 – 1	0 - 1		1

Fig. 13. Result of Competitions

dedicated components, such as a micro camera unit, a special chip, and mechanical/electrical joints. These specification of robot were now being compiled as OPEN-R standard. OPEN-R will be able to supply broad range of useful mechanical and electrical hardware and software components.

Nevertheless, software to control such a robot and to make them behave to carry out meaningful tasks is major research topic. In order to facilitate research in the control of legged robot with multiple degree of freedom and rich sensor systems, we choose to developed a physical robot platform for RoboCup based on OPEN-R, and to participate in RoboCup.

Acknowledgements

We thank to the RoboCup98 Legged Robot Challengers at CMU, LRP, and Osaka University. Especially we would like to thank to Dr. Veloso, Dr. Duhaut, and Dr. Asada. Without their efforts, we couldn't have exciting competition games during the short period of development. The authors also would like to thank to Ms. Aline Dahlke at Sony France and Ms. Asako Watanabe at D21 Laboratory, Sony Tokyo. Had it not been for their support and preparation, the event "RoboCup-98 Legged Robot Exhibition " would not have been a big success like that.

References

[Brooks, 1986] Brooks, R. A., 1986, A Robust Layered Control System for a Mobile Robot, *IEEE Journal of Robotics and Automation*, RA-2(1), March, pp.14–23.

[Firby, 1994] Firby, R. J., 1994, Task Networks for Controlling Continuous Processes, In *Proceedings on the Second International Conference on AI Planing Systems*, June.

[Fujita and Kageyama, 1997a] Fujita, M. and Kageyama, K., 1997, An Open Architecture for Robot Entertainment, In *Proceedings of the First International Conference on Autonomous Agents*, Marina del Ray, pp.234–239.

[Inaba, 1993] Inaba, M., 1993, "Remote-Brained Robotics: Interfacing AI with Real World Behaviors, In *Proceedings of the 6th International Symposium on Robotics Research (ISRR6)*, pp.335–344.

[Kitano, et al., 1997] Kitano, H., et al, 1997, RoboCup: A Challenge Problem for AI, *AI Magazine*, Spring, pp.73–85.

[Maes, 1995] Maes, P., 1995, Artificial Life meets Entertainment: Lifelike Autonomus Agents, *Communication of the ACM: Special Issue on New Horizons of Commercial and Industrial AI*, pp.108–114.

[Ueda, 1996] Ueda, K. and Takagi, Y., 1996, Development of Micro Camera Module, In *Proceedings of the 6th Sony Research Forum*, pp.114–119.

[Yokote, 1992] Yokote, Y., 1992, The Apertos Reflective Operating System: The Concept and Its Implementation, In *Proceeding of the 1992 International Conference of Object-Oriented Programing, System, Languages, and Applications.*

Robot Soccer with LEGO Mindstorms

Henrik Hautop Lund **Luigi Pagliarini**

LEGO Lab
University of Aarhus, Aabogade 34, 8200 Aarhus N., Denmark

hhl@daimi.aau.dk
http://www.daimi.aau.dk/~hhl/

Abstract We have made a robot soccer model using LEGO Mindstorms robots, which was shown at RoboCup98 during the World Cup in soccer in France 1998. We developed the *distributed behaviour-based approach* in order to make a robust and high performing robot soccer demonstration. Indeed, our robots scored in an average of 75-80% of the periods in the games. For the robot soccer model, we constructed a stadium out of LEGO pieces, including stadium light, rolling commercials, moving cameras projecting images to big screens, scoreboard and approximately 1500 small LEGO spectators who made the "Mexican wave" as known from soccer stadiums. These devices were controlled using the LEGO Dacta Control Lab system and the LEGO CodePilot system that allow programming motor reactions which can be based on sensor inputs. The wave of the LEGO spectators was made using the principle of *emergent behaviour*. There was no central control of the wave, but it emerges from the interaction between small units of spectators with a local feedback control.

1 Introduction

Before the LEGO Mindstorms Robotic Invention System was to be released on market, we wanted to make a large-scale test of the robot kit. We selected to make a LEGO Mindstorms robot soccer play with a distributed behaviour-based system to be demonstrated in Paris during the soccer World Cup France'98 at RoboCup'98. Robot soccer has been defined as a new landmark project for artificial intelligence [3], and its characteristics fitted our purpose. In contrast to previous artificial intelligence challenges such as computer chess, robot soccer is a dynamic and physical game, where real time control is essential. Further, where a game like chess might allow extensive use of symbolic representation, robot control put emphasis on embodiment and many aspects of this prohibits the use of symbolic representation. In general, participating in robot soccer is believed to provide both students and researchers with knowledge about the importance of embodiment and the problems that ungrounded abstractions might lead to [4].
However, we also found a number of problems that had to be solved before

robot soccer could be made appealing for a public audience. Robot soccer is a very young research field, so the performance of the robot soccer players might not look impressive enough from a public audience's point of view. Even though, we expected a huge public interest in our LEGO Mindstorms robot soccer demonstration[1], so it was important to alleviate this problem. The robot game had to be put into the right context. From an aesthetic point of view, some robot soccer players might be looked at as essentially cubic, metallic devices that move around in a pen and push a ball — it might not appear to be much like soccer to the public audience if the audience is not told in advance to look at this as soccer. Therefore, in our robot soccer game, we put much more emphasis on making a context that immediately would allow the public audience to recognise the game to be a soccer game. This was done by making a whole stadium (which we named Stade de Victor LEGO) out of LEGO with light towers, rolling commercials, and almost 1500 LEGO spectators who made the "wave", by providing sounds related to the game (tackling, kicking, spectator noise, etc.), and by giving the robot soccer players a face. Indeed, in the developing phase, we had a graphical designer to make huge colour drawings of possible scenarios, we had a technical designer to make appealing facial expression of the robots, and we made different scripts for games (how to enter the field, how to sing the national anthems, how to get into the kick off positions, what play strategies to use, etc.).

2 RoboCup and LEGO Mindstorms Robot Soccer

We wanted to construct LEGO Mindstorms robot soccer players to play a demonstration tournament during RoboCup'98. First, we held an internal fully autonomous LEGO robot soccer competition. In this tournament, our students were allowed to use nothing else than LEGO Mindstorms and LEGO sensors, which include light sensors, angle sensors, temperature sensors, and switch sensors. However, one has to think very carefully about the set-up to make an impressive robot soccer game with only these sensors. Since the initial results of the fully autonomous robot soccer experiment were not impressive enough to bring to RoboCup'98, we ran another experiment in parallel. In this experiment, we wanted to increase the sensing capabilities of the LEGO robot soccer players. This was done by using the approach taken in the RoboCup Small League, namely to use an overhead camera. Here, the idea is to increase the sensing capability by having a camera that can overview the whole scene (i.e. the field), interface this with a host computer, and then have the host computer to transmit information to the robots on the field.

Our set-up included an NTSC video-camera which was connected to a hardware vision system that extracted the position of the players and the ball in the robot soccer field. This information was processed to a host computer that would communicate information to the robot soccer players, and the information

[1] Indeed, our LEGO Mindstorms robot soccer demonstration was broadcasted to an estimated 200-250 million television viewers world-wide.

Figure1. The LEGO robot soccer set-up. There is one goalkeeper and two field players on each team (one red and one blue team). The stadium has light towers, scanning cameras that project images to large monitors, scoreboard, rolling commercials, and almost 1500 small LEGO spectators that make the "Mexican wave". ©H. H. Lund, 1998.

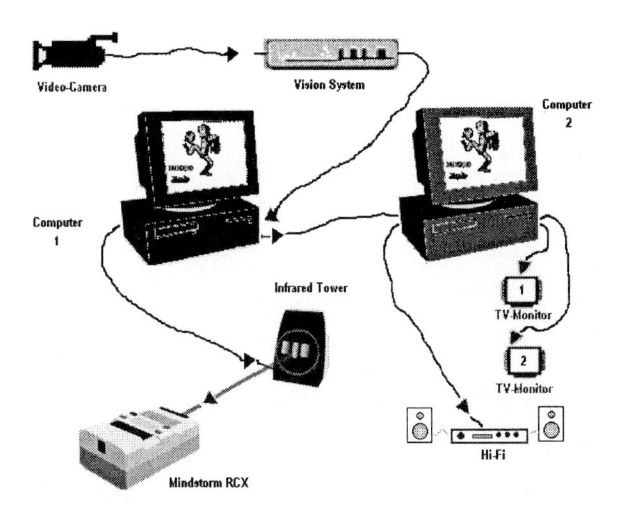

Figure2. The image from the video-camera was fed into a hardware vision system that extracted the position of players and ball in the field. This information was first sent to computer 1 that contained part of the distributed behaviour-based controller and communicated information via infra-red communication towers to the LEGO Mindstorms RCX (Robot Control System). Secondly, the information was forwarded to computer 2 that steered a couple of cameras from which images were sent to monitors, and it held the sound processing system. ©H. H. Lund and L. Pagliarini, 1999.

was passed on to a second computer that controlled sound and small cameras (for projecting the game from different angles to big screens) during the games. The set-up is shown on Figure 2. Again, it is important to note how we here use both monitoring and sound for making the robot soccer games more lively and pleasing for the public audience.
Each team consisted of one goalkeeper and two field players. The goalkeeper was controlled with a LEGO CodePilot, while the two field players were constructed around the LEGO Mindstorms RCX (Robot Control System). Each player had two independent motors to control two wheels to make the robot move around on the field, and one motor to control movement of the robot's mouth (so that it could "sing" the national anthem and "shout" when scoring a goal). A player had three angle sensors to detect the motion of wheels and mouth. All parts of the robots except for batteries and coloured hair to indicate the team were original LEGO elements (LEGO Dacta, LEGO Mindstorms, LEGO Technic).
Since we wanted to use LEGO Mindstorms only for the robot soccer demonstration, we had to communicate from host computer to robots with infra-red transmission. This is notoriously slow and unreliable, so most other RoboCup teams use a radio-link. Because of our infra-red link, the game would necessarily be a bit slow. But we solved a great deal of the problems by making a distributed behaviour- based system, in which only high levels of competence were implemented in the host computer, and low levels of competence ran directly on the robots.
When opening the mouth, a small bar on the back of the robot would move

Figure3. The LEGO Mindstorms robot soccer players have an appealing design with big eyes, a moving mouth, coloured hair, etc. ©H. H. Lund, 1998.

down on the ground, which made the robot tip slightly forward. This was made in order to make a shout look even more lively. However, the functionality turned out to provide a nice, unintentional use. The border of the field had a chamfered edge to allow the ball not getting stuck up against the wall. However, because of the slight slope, the robot soccer players could get stuck at the edge, since the distance between the wheels touch point on the ground and the base of the robot was quite small. The problem was solved by allowing the robot soccer player to scream after being stuck for a short time. Apart from opening the mouth, this would make the bar on the back go down on the ground and tip the robot, so that it would actually get free.
The LEGO Mindstorms robot soccer players were programmed using the Mindstorms OCX together with MicroSoft Visual Basic. The control system was a distributed system between the robots and the host computer. The host computer ran a program that collected co-ordinates from the hardware vision system, processed this information and sent angles to turn out via the infra-red transmitters. The robots had a control system that allowed them to collect the appropriate data, and react accordingly. In a sense, we implemented a control system that we can term a *distributed behaviour-based system*, after behaviour- based systems such as Brooks' subsumption architecture [1]. Some levels of competence were placed on the robot, while other higher levels of competence were placed on the host computer. The distributed behaviour-based system is described in full details in [7]. The division of levels of competence and their distribution on host computer and robot(s) is shown in Table 1.

Table1. The distributed behaviour based control system of the LEGO robot soccer players. The low levels of competence must necessarily be performed on the

	Competence	Level	Computational unit
Level 4	Position Planner	High level	Host computer
Level 3	Action selection	High level	Host computer
Level 2	Turn (and Scream)	Low level	Robot
Level 1	Forward	Low level	Robot
Level 0	Stop and wait	Low level	Robot

computational device that has direct access to sensors and actuators (here the robot), while higher levels of competence can be implemented in external computational devices (here the host computer).

3 The LEGO Stadium

In order to put the robot soccer play into a stimulating context, we built a whole LEGO stadium (see Figure 1). The stadium had light towers (with light) in each corner, and these towers also hold infra-red transmitters that could transmit information from a host computer to the RCXs. In one end, there was a scoreboard that could be updated when a goal was scored via an interface with the LEGO Dacta Control Lab (see Figure 4). Over each goal, there was a rolling commercial sign that held three commercials that were shown in approximately 30 seconds each before the sign would turn to the next commercial. The control of the two rolling commercial signs was made with the LEGO CodePilot. A camera-tower with a small b/w camera was placed in one corner. The camera (controlled from a CodePilot) could scan to the left and the right of the field, while displaying the image to the audience on a large monitor. Another camera was placed over the sideline on one side and should scan back and forth following the ball (see Figure 4). Also this camera image was displayed on a large monitors, and its control was made from LEGO Dacta Control Lab.

Figure4. The scoreboard and one of the small cameras. The camera runs up along the sideline, while projecting the images to a large monitor. ©H. H. Lund, 1998.

4 Emergent Behaviour for the Spectator Wave

When wanting to make a spectator wave (a "Mexican wave") at our stadium, we identified the spectator wave at stadiums to be an instance of emergent behaviour, rather than being with central control [5]. The "Mexican wave" that is made when spectators stand up and sit down does not have central control. The wave is initialised when a couple of spectators anywhere on the stadium decide to make the stand up + sit down movement, and some nearby spectators go with. There is no central control to tell the individual spectator to do a specific thing at a given time, rather it is an emergent behaviour.

Emergent behaviour is an interesting phenomenon that can be observed in natural systems. We define emergent behaviour as being the behaviour of a system that is the product of interaction between smaller sub-systems. The emergent

behaviour is of higher complexity than the sum of the behaviours of the smaller sub-systems. The reason that behaviour of higher complexity than the sum can emerge is the interaction between the sub-systems.
Emergent behaviour is known from flocks of birds, schools of fish and herds of land animals. When observing a flock of birds, we will notice that there is no apparent leader in the flock and there appears to be no central control of motion. The motion of the flock might seem complex and at times random, but on the other hand, it also appears synchronous. The motion of a flock of birds is an example of emergent behaviour and can be modelled as such. Reynolds [9] has made an impressive study of the general motion of flocks, herds, and schools in a distributed behavioural model with the goal of using this to model flocking in computer graphics. Recently, similar models have been used in the Disney movie *Lion King* for a wild-beast stampede and to produce photo-realistic imagery of bat swarms in the feature motion pictures *Batman Returns* and *Cliffhanger*. Reynolds calls his simulated bird-like organisms *boids* (bird-oids). The boids are controlled by three primary rules:

1. Collision Avoidance: avoid collision with nearby boids
2. Velocity Matching: attempt to match velocity with nearby boids
3. Flock Centering: attempt to stay close to nearby boids

The three rules are local in the sense that a boid only has knowledge about nearby boids and there is no global knowledge like size or centre of the flock. For instance, Flock Centering is achieved by having boids to perceive the centroid of nearby boids only. This actually gives the advantage of allowing for bifurcation: the flock can split around an obstacle in the moving direction, since the boids only tend to stay close to nearby flock-mates.
In general, the phenomenon of emergent behaviour is fundamental in a number of artificial life systems. Artificial life tries to synthesise life with a bottom-up approach by using small building blocks that emerge to a complex system by their interaction.
When using emergent behaviour in real world models, there are a number of pitfalls that we have to be aware of. When we look at Reynolds' boid model, we notice that only local knowledge of neighbours is used, so the model might appear appropriate for control tasks for autonomous agents in the real world. However, it is not clear how to obtain even the local knowledge that is available for the simulated boids. For instance, we have to solve the question of how to measure *nearby* (distance and direction). This demands an advanced sensor that can measure distance and direction, and at the same time identify an object as being a neighbour (and, for instance, not an obstacle). The task is worsened further by the demand for doing this in real time with moving objects.
There are other significant differences between a simulation model and a real world implementation that we have to take into account. For instance, the actuators will produce friction and there will be a whole range of noise issues that makes it very difficult to transfer an idealised model from simulation to the real

world. In some cases, it will be possible to transfer models from simulation to reality [8, 6, 2]. This is done by very careful building of a simulator that models the important characteristics of the real device and the way that real world noise interferes with this device.
In our emergent behaviour model, we work directly in the real world, so we avoid the problems of difficulties in transfer from an idealised model to the real world. Our model therefore has to work with the noise, friction, etc. that exists in the real world.

5 The LEGO spectator wave

Apart from robot soccer players, cameras, rolling commercials, and scoreboard, we had placed almost 1500 small LEGO spectators on the grandstands. Our idea was to have all these spectators make the "Mexican wave" as we see soccer fans make at real World Cup matches.
After first trying to make a prototype of the wave with a kind of open-loop con-

Figure5. The LEGO spectator wave. When the switch sensor is released, the next section of LEGO spectators will start to move upwards. ©H. H. Lund, 1998.

trol, the control of the wave was changed to a feedback control, and the idea was to allow the wave to *emerge* from the interaction between the different sectors of spectators that each had their own, local feedback control. In this case, the implementation of a system with emergent behaviour should be possible in the real world, since there would be no demand of advanced sensing. In fact, a switch sensor for each section of LEGO spectators turned out to be enough (see Figure

6). The idea was that the movement of one section should be dependent on sensing what the adjacent section was doing. If a section was moving upwards, then the section to the right should sense this and start to move upwards itself. The section would fall down when reaching the top position (in this way, we used the principle that *what goes up, must fall down*). This was built by placing a switch sensor under each section and connecting this switch sensor to the control unit (a LEGO CodePilot) of the next section. In resting mode, the switch sensor would be pressed by the section of spectators above it, but when this section started to move upwards, the switch sensor would no longer be pressed. This triggered the simple control program in the next section to start moving the section of LEGO spectators upwards. In pseudo-code, the control program of each section could be as follows:

```
Section N control:

            if (Switch(N-1)=false) then turn on motor
            else turn off motor
```

Figure6. The movement of a section is triggered by the switch sensor mounted underneath the adjacent section. The left section of spectators has risen, so the switch sensor is no longer pressed. The control of the right section will notice this, and start to move upwards (immediately after this photo was taken). ©H. H. Lund, 1998.

This very simple control allows the wave to emerge when one section is triggered from the external to move upwards. In the actual implementation in the LEGO CodePilot language, it was however necessary to use a timer, since the time slice of the CodePilot is so small, that the above pseudo-code program would result in the section barely moving upwards before the motor would be turned off. So

when it was sensed that the switch was no longer pressed, the control would turn the motor one direction for 0.5 seconds and then the other direction for 0.5 seconds. It would have been more sensible to have a switch sensor on the top position that the section should reach, and then base the time of upward and downward movement on the feedback from the top and the bottom sensors. But since the LEGO CodePilot has only one input channel (see Figure 7), we opted for the timing solution. However, it must be noted that this timing is very different from the timing in the first prototype, since here, even though the timing of a section might have been wrong, the section would still lift itself for some time and therefore the next section would be triggered. In a sense, we are setting the time-slice to 0.5 seconds and use the pseudo-code program.
The feedback control was used and the wave emerged from the interaction be-

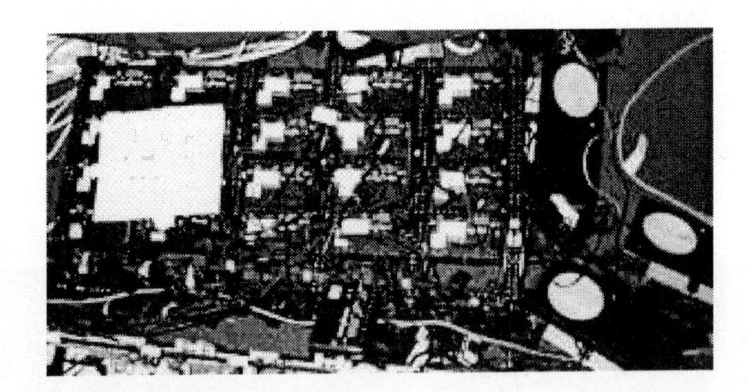

Figure7. The mixer. 16 LEGO CodePilots were used to construct the dynamics of the "Mexican wave". Each CodePilot had one switch sensor and one motor connected. ©H. H. Lund, 1998.

tween the simple units. It was run numerous times daily in Paris for a week without any need for refinements apart from a couple of changes of physical aspects (one or twice, a section got stuck).

6 Conclusion

The distributed behaviour-based control system gave a robot soccer play that allowed the LEGO Mindstorms robots to play a good robot soccer game, where goals were scored in most periods. In the demonstration tournament, we played games with five periods of up to two minutes (or until a goal was scored), and on average 3-4 goals were scored in each match. This is far higher average of goals/period than in most other robot soccer games, and it was essential for us to achieve this performance in order to provide a spectacular game for the public audience.
We used emergent behaviour to construct a "Mexican wave" of LEGO spectators

for the LEGO robot soccer demonstration. The wave emerged from the interaction between sections of LEGO spectators, each with its own simple control. The control was based on feedback from the local environment. Since sensing was straightforward with one switch sensor for each section, it was fairly easy to implement a real world emergent behaviour. Under other circumstances, it might be more difficult, since more advanced sensing might be necessary, and we cannot guarantee that the desired real world behaviour will emerge. Therefore, the study of real world emergent behaviour is important in order to identify the circumstances that will lead to successful results.

Further information, including photos and videos are available on the LEGO Lab web-site:
`http://legolab.daimi.au.dk/`

Acknowledgements

Jens Aage Arendt designed the LEGO stadium in collaboration with Jakob Fredslund. Students from the LEGO Lab made a highly valuable work in the LEGO robot soccer RoboCup'98 project. Lars Nielsen from LEGO Media worked as a very helpful partner who gave numerous valuable ideas. Ole Caprani participated in a number of inspiring discussions. Thanks to H. Kitano for collaboration on the Paris RoboCup'98 event.

References

1. R. A. Brooks. A robust layered control system for a mobile robot. *IEEE J. Robotics and Automation*, RA-2(1), 1986.
2. N. Jakobi. The Minimal Simulation Approach to Evolutionary Robotics. In T. Gomi, editor, *Proceedings of ER'98*, Kanata, 1998. AAI Books.
3. H. Kitano, M. Asada, Y. Kuniyoshi, I. Noda, E. Osawa, and H. Matsubara. RoboCup. a Challenge Problem for AI. *AI Magazine*, (spring):73–85, 1997.
4. H. H. Lund. Robot Soccer in Education. 1999. submitted to Advanced Robotics journal.
5. H. H. Lund, J. A. Arendt, J. Fredslund, and L. Pagliarini. Ola: What Goes Up, Must Fall Down. In *Proceedings of Artificial Life and Robotics 1999 (AROB'99)*, pages I9–I15, Oita, 1999. ISAROB.
6. H. H. Lund and O. Miglino. From Simulated to Real Robots. In *Proceedings of IEEE Third International Conference on Evolutionary Computation*, NJ, 1996. IEEE Press.
7. H. H. Lund and L. Pagliarini. LEGO Mindstorms Robot Soccer. a Distributed Behaviour-Based System. 1999. To be submitted.
8. O. Miglino, H. H. Lund, and S. Nolfi. Evolving Mobile Robots in Simulated and Real Environments. *Artificial Life*, 2(4):417–434, 1995.
9. C. W. Reynolds. Flocks, Herds, and Schools: A Distributed Behavioral Model. *Computer Graphics*, 21(4):25–34, 1987.

Ball-Receiving Skill Dependent on Centering in Soccer Simulation Games

Kazuaki Maeda kaz@solan.chubu.ac.jp
Akinori Kohketsu g94117@isc.chubu.ac.jp
Tomoichi Takahashi ttaka@isc.chubu.ac.jp

Department of Business Administration and Information Science,
Chubu University
1200 Matsumoto, Kasugai, Aichi 487-8501, JAPAN

Abstract. This paper describes an effective ball-receiving skill. When soccer games are played in real life, players generally must make consecutive actions in one play, for example, running, receiving, and shooting a ball. We believe that the same is true in the case of simulation soccer games. Therefore, we designed an experiment to check how changing ball-receiving methods which is dependent on the centering patterns influence scoring goals. The experiment shows that one ball-receiving method is more effective than the others. The result is embedded into our soccer team, Kasugai-bito II, and the effectiveness is discussed in games.

1 Introduction

In the last few years, many researchers have devoted their efforts toward RoboCup (The World Cup Robot Soccer)[2, 3, 4]. In RoboCup, robots or agents play a soccer game under given constraints. RoboCup was proposed as a standard problem to promote AI and intelligent robotics research. It consists of three competition tracks: Simulator League, Real Robot Small Size League, and Real Robot Middle Size Leage[1]. We have investigated only Simulator League.

RoboCup has features different from typical traditional AI problems such as computer chess. Features of computer chess are static and non-real-time. A computer with powerful and large parallel processors, enormous knowledge, and sophisticated software, can find the most effective move in a few minutes, and it defeated the world chess champion [6].

For RoboCup, however, situations change dynamically and in real-time. Therefore, we cannot apply the traditional techniques to RoboCup. To develop a high level robot soccer game, we must investigate intelligence (e.g. learning) for individual agents and teams.

The purpose of this paper is to confirm the effectiveness in changing the soccer agents' action dependent on situations. At this time, much of the work about intelligent agents is in the planning phase. However, it is important to confirm the effectiveness before we implement our intelligent agents. Our current soccer agents, Kasuga-bito II, do not have intelligent ability, but are hand-coded to change ball-receiving methods which are dependent on centering. After this confirmation of effectiveness, we will implement the new intelligent agents. If new intelligent agents are the same level as our hand-coded agents, they will be regarded as having good skill.

In section 2, we define centering patterns and ball-receiving methods. In section 3, we explain the outline of our experiment and evaluate the effectiveness for the changing of receiving methods. In section 4, we describe the results of the match games where we applied our skill. Finally, we summarize the paper.

2 Centering Patterns and Receiving Methods

2.1 Real Soccer Games

From the experience of playing soccer games in real life, players generally must make consecutive actions in one play, for example, running, receiving, and shooting a ball. In the consecutive actions, they must receive the ball by the most effective methods. That is, we should select one method to receive the ball. For example, when the centering ball moves toward the goal, we usually turn toward the goal and receive the ball, then shoot it. Another example is when the centering ball lands outside the goal. In this case, we usually turn back against the goal and receive the ball, then shoot it. Generally speaking, we decide the next action by considering current situations.

When human players shoot the ball toward the goal, it is said that as the number of times the ball is touched increases, the shooting success rate decreases[5]. Moreover, it is said that direct or one-trapping shots occupy 70–80% of the successful shoots. This points out the need for real-time processing for the soccer agents. To satisfy this need, the agents should shoot the ball toward the goal as soon as they receive it in the penalty area.

2.2 Classification of Centering Patterns

We can classify centering as the point of relationship between the player and the ball. We call this the centering patterns.

We introduce 4 centering patterns such as depicted in Figure 1. The patterns have different features relative to the position of the player (the "shooter") and the direction of the ball.

Our classification is as follows:

1. When the X-position of the ball is the same as the X-position of the shooter, the Y-position of the ball is
   - positive to the Y-position of the shooter (1 and 3 in Figure 1), or
   - negative to the Y-position of the shooter (2 and 4 in Figure 1).
2. The gradient of the ball-trajectory is
   - positive (3 and 4 in Figure 1), or
   - negative (1 and 2 in Figure 1).

These combinations make 4 centering patterns. To simplify our discussion, we will call the patterns Centering-1, Centering-2, Centering-3, and Centering-4.

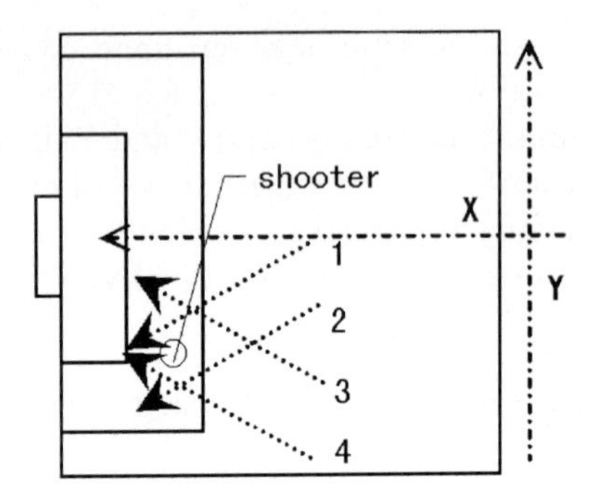

Fig. 1. Centering Patterns

2.3 Prediction and Receiving Methods

Prediction of the Ball Position: For receiving the ball, we need to design a prediction of ball movement by using see-information from the soccer server. We will describe our prediction in this section.

When we know the current ball position $\mathbf{B}(t)$ at time t, we can calculate the predicted ball position $\mathbf{B}(t + 100ms)$ at $100ms$ later below:

1. Calculate the distance, $\Delta(t)$, of the ball movement by using changes of relative distance and angle between the ball and the player.
$$\Delta(t) = \mathbf{B}(t) - \mathbf{B}(t - 100ms)$$
2. Multiply $\Delta(t)$ by the default Ball-Decay 0.96 and add the current position $\mathbf{B}(t)$ to it.

$$\mathbf{B}(t + 100ms) = \mathbf{B}(t) + 0.96\ \Delta(t)$$

When we predict the ball position at more than $100ms$ later, we can obtain it by repeating the above calculation.

Receiving Methods: From the experience of soccer games in real life, we have found some receiving methods from the point of relationship between the player and the ball. This naturally led us to introduce 3 methods for receiving the ball. The methods are different based on the relative position between the shooter and the ball such as depicted in Figure 2.

At the time t, we describe the shooter position as $\mathbf{S}(t)$ and the ball position as $\mathbf{B}(t)$. Moreover, we describe $t1$ as the moment when the shooter reaches on the ball-trajectory (maybe the predicted ball-trajectory), and $t2$ as the moment when the shooter receives the ball. This implies $\mathbf{S}(t2) = \mathbf{B}(t2)$. To simplify our discussion, we give the name to each method below:

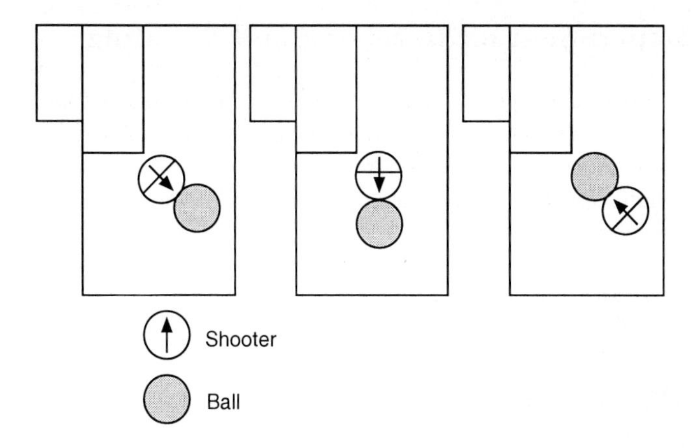

Fig. 2. Backward-Receiving, Simultaneous-Receiving, and Forward-Receiving

Backward-Receiving This method is used when the shooter reaches the predicted ball position in advance of the ball. The feature is

$$\begin{aligned} \mathbf{S}(t1) &= \mathbf{B}(t2) \\ t2 &= t1 + 300ms \end{aligned}$$

That is, the shooter waits for the ball before receiving it. Therefore, he can modify his position for the unpredictable ball movement. This method, however, has a drawback in that there is a delay in waiting for the ball.

Simultaneous-Receiving This method is used when the shooter receives the ball as soon as he reaches the predicted ball position. The feature is

$$t2 = t1$$

This method is the fastest among three methods. However, the shooter usually turns his back against the goal to receive the ball. Therefore, this method has a drawback in that he cannot see the situation around the goal.

Forward-Receiving This method is used when the shooter runs after the ball and receives it. The feature is

$$\mathbf{S}(t1) = \mathbf{B}(t1 - 100ms)$$

The shooter can easily see the situation around the goal (especially the goalkeeper). He utilizes his see-information around the goal and can shoot the ball effectively. This method, however, has a drawback of sometimes having to wait before receiving the ball.

These receiving methods make one action in consecutive actions to score. The next action is to shoot the ball. From the experience of soccer games in real life, our agents shoot the ball toward the goal as soon as they receive it[5].

3 An Experiment on Receiving and Shooting

We believe that soccer agent's skill of their shooting is related to their scoring ability. Therefore, we designed a skill experiment and examined the scores to evaluate our agents. If our well-designed agents improve their ability to score, we can conclude that the skill of our agents has improved.

We made an experiment to try 3 receiving methods from 4 centering patterns respectively. Figure 3 depicts an overview of the experiment. We describe here the experiment and the result.

Fig. 3. An Overview of the Experiment

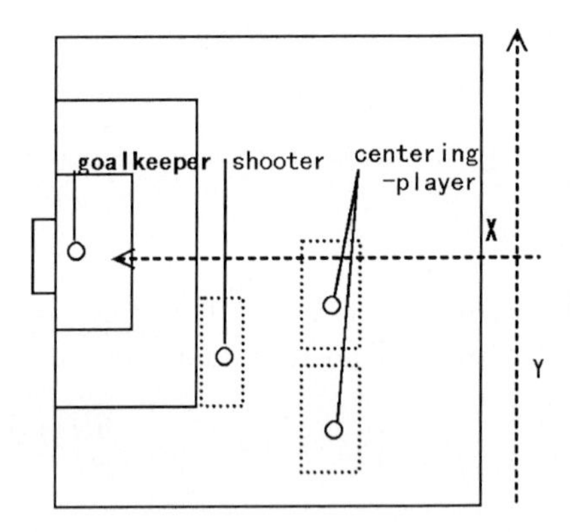

Fig. 4. Positions of the Players

3.1 The Condition of the Experiment

We designed the experiment with the following conditions:

- We prepared two players for offense. One kicked a ball according to a centering pattern. We called him the centering-player. The other was the shooter. We also prepared one goalkeeper for defense.
- We made the experiment in the coaching mode on the soccer server (version 3.17) to position the players and the ball.
- The shooter was positioned at random within a range of
  X-position: $33 \sim 37$ and
  Y-position: $-20 \sim -10$.
  The centering-player was positioned at the place where the direction from the shooter was
  150 degrees for Centering-1 and Centering-2, and
  -150 degrees for Centering-3 and Centering-4.
  Moreover, the distance between the centering-player and the shooter had a range of ± 3 around 17.5. The goalkeeper was positioned at 2.5 away from the goal. Figure 4 depicts the positions of players.

- The beginning position of the ball was at the centering-player's feet.
- The centering-player kicked the ball to the shooter by kick-power 100 toward the relative direction of
  $5 \sim 20$ degrees for Centering-1 and Centering-3, and
  $-5 \sim -20$ degrees for Centering-2 and Centering-4.
- The shooter shot the ball as soon as he received it.
- If it took less than 6 seconds to score the goal after the centering-player kicked the ball, the trial was regarded as success. Otherwise, the trial was regarded as a failure.

The number of trials was as follows:

- The centering-player kicked the ball to the shooter 300 times for each centering pattern. There were 4 centering patterns. Therefore, he kicked the ball 1200 times.
- The shooter received the ball 100 times by using each method for each centering pattern.

We adopted the goalkeeper of AT-Humboldt team (Germany) which won the championship at the 1st RoboCup Worldcup Simulation League held on 1997.

3.2 Experiment Result

The results of the experiment are shown in Table 1. The table shows the success rates of shooting the ball to 4 centering patterns by 3 receiving methods. The underlined rates are the most effective receiving methods for each centering pattern. We consider here the result for each centering pattern.

Table 1. Success Rate of Each Receiving Method (%)

	1	2	3	4
Backward-Receiving	42	40	39	<u>62</u>
Simultaneous-Receiving	<u>63</u>	46	<u>66</u>	50
Forward-Receiving	33	<u>53</u>	48	28

First, the table shows that the Simultaneous-Receiving is the most effective method for Centering-1. By comparing this with soccer games in real life, it is easily understood.

Secondly, the table shows that Forward-Receiving is the most effective method for Centering-2. In a real soccer game, however, we may find that Backward-Receiving is the most effective method. This is mainly because the shooter, when Forward-Receiving, can see the situation around the goal.

Thirdly, the table shows that Simultaneous-Receiving is the most effective method for Centering-3. In a real soccer game, however, we may find that

Forward-Receiving is the most effective method. This is mainly because it is likely that the goalkeeper will catch the ball if the shooter uses Forward-Receiving.

Finally, the table shows that Backward-Receiving is the most effective method for Centering-4. In a real soccer game, however, we may find that Forward-Receiving is the most effective method. This reason is the same as one given for Centering-3.

3.3 Statistical Testing

To examine the experiment results statistically, we tried hypothesis-testing. At first, we set a hypothesis:

> The receiving methods are not dependent on the success rate to score the goal.

Next, we made a 2×3 contingency table, where $\{$ *shoot success, failure* $\} \times \{$ *Backward-Receiving, Simultaneous-Receiving, Forward-Receiving* $\}$. Then we tested it, where the significance level is 1% and the rejection region is more than 9.28. Table 2 shows that our hypothesis was rejected to Centering-1, Centering-3, and Centering-4. That is, the receiving methods are dependent on the centering patterns except in the case of Centering-2.

Table 2. Result of Statistical Testing

	1	2	3	4
testing	reject	accept	reject	reject
	19.08	2.79	23.89	15.13

4 Application to a Soccer Game

As mentioned above, changing the receiving methods which is dependent on centering patterns is usually effective to score the goal. We will confirm here the effectiveness in a soccer simulation match game.

First, we prepared two teams for comparison:

Rand All players select one of three receiving methods at random.

Advc All players select the most effective method dependent on the centering patterns.

For both teams, the receiving methods are only used for the players which are positioned within 30 meters from the center of the goal (52.5, 0) such as depicted in Figure 5.

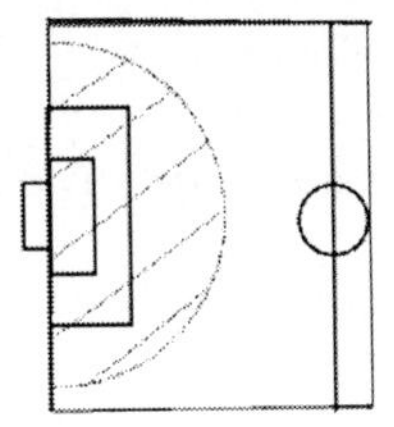

Fig. 5. Region to Change the Receiving Methods

We selected two opposing teams. One was AT-Humboldt team. The other was Andhill team (Tokyo Institute of Technology, Japan). Andhill won the second prize in the 1st RoboCup Worldcup Simulation League. The combinations were 4 match games. We tried each match game for 100 minutes[1].

The differences in the scores are presented in Table 3. When we changed teams from Rand to Advc, the difference in the scores was changed for both AT-Humboldt and Andhill. These results show the effectiveness of our changing methods in a soccer simulation match game.

Table 3. The Differences in the Scores

	Kasuga-bito II	AT-Humboldt	Difference	Kasuga-bito II	Andhill	Difference
Rand	31	33	-2	28	37	-9
Advc	36	31	+5	33	35	-2

5 Conclusion

The purpose of this paper was to examine the effectiveness of changing the ball-receiving methods in RoboCup simulation soccer game. For that purpose, we made an experiment to discover the most effective receiving methods which are dependent on the centering patterns. As a result, we found effective receiving methods except in the case of one centering pattern.

Acknowledgements

We wish to thank developers of AT-Humboldt and Andhill for freely distributing their soccer clients. We also thank the anonymous reviewers for some suggestions.

[1] We manipulated the length of games using the configuration file of soccer-server

References

1. The RoboCup Federation. *RoboCup: The Robot World Cup Initiative.* http://www.robocup.org, 1998.
2. Hiroaki Kitano, Minoru Asada, et al. The RoboCup: The Robot World Cup Initiative. In *Proceedings of IJCAI-95 Workshop on Entertainment and AI/Alife*, pages 19–24, 1995.
3. Hiroaki Kitano, Minoru Asada, et al. RoboCup: The Robot World Cup Initiative. In *Proceedings of First International Conference on Autonomous Agent*, 1997.
4. Hiroaki Kitano, Milind Tambe, et al. The RoboCup Synthetic Agent Challenge 97. In *Proceedings of IJCAI-97*, 1997.
5. Jiro Ohashi, Kozo Tajima, and Takashi Kakemizu. *Science of Soccer (in Japanese).* Tokyo Denki University Press, 1997.
6. Herbert A. Simon and Toshinori Munakata. AI Lessons. *Communications of the ACM*, 40(8):23–25, 1997.

Appendix: Team Description

Our team, Kasugabito-II, was runner-up in the Japan Open 98 Simulation League. The most important feature of Kasugabito-II is, we mentioned it in this paper, changing ball-receiving methods which is dependent on the centering patterns. We describe other features here.

Formation and Roles

Kasugabito-II is composed of 3 Defenders (DFs), 5 Midfielders (MFs), 2 Forwards (FWs) and 1 Goalkeeper (GK) such as depicted in Fig.6.

DFs defend away a little bit from their goal to trap in off-side. MF*s in Fig.6 sometimes participate in an attack by dribbling. FWs stand nearby the penalty area and wait for centering from a MF.

In Kasugabito-II, each player has the same algorithm for predicting the ball position (see section 2.3). About the movement, however, each player's action is dependent on the position. For instance, a GK and DFs wait for the ball until the ball enters within a distance, and players move to receive it. This is because they should receive the ball safely by using more reliable prediction of the ball position. MFs and FWs move earlier toward the ball to rival opposite players to receive the ball.

Implementation

Kasugabito-II is written in C and it consists of about 3,000 lines. We are developing it on multiple SPARCstations which are running under Solaris 2.5.

We were using two workstations (SPARCstation 5/110, memory 64M bytes) in the early development. In the recent development, we are using five workstations;

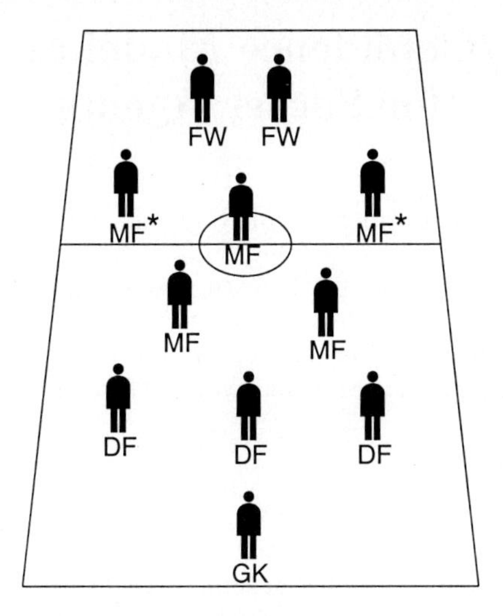

Fig. 6. Formation

- One Sun Ultra 1/140 (Memory 96M bytes)[2]
- Four Sun Ultra 1/170 (Memory 128M bytes)[3]

We would like to thank people for lending these workstations to us.

[2] This machine is supported by Nippon Steel, Inc. and Nissho Electronics, Inc.
[3] These machines are supported by Nihon Sun Microsystems, Inc.

The Priority/Confidence Model as a Framework for Soccer Agents

Jan Lubbers* and Rogier R. Spaans

Department of Computer Science, University of Amsterdam
Kruislaan 403, NL-1098 SJ Amsterdam, The Netherlands
E-mail: {jlubbers, spaans}@wins.uva.nl
URL: http://carol.wins.uva.nl/~spaans/RoboCup/

Abstract. We propose the *Priority/Confidence model* as a reasoning model for agents. Decisions are made according to a *confidence measure* which is based on the importance of actions (priority) and the satisfaction of a priori preconditions. We implemented the Priority/Confidence model for a robotic soccer domain, namely the RoboCup. Our team, AIACS, has demonstrated the feasibility of this model by beating the world champion of 1997, *AT-Humboldt* in training matches and reaching a 9th place in the RoboCup'98 tournament.

1 Introduction

Multiple Agent Systems (MAS) are environments in which several autonomous individuals (agents) co-exist. These agents might be robots in the real world or computer programs in a virtual world. Since actions of one agent may effect the world of other agents it is necessary that agents interact, ranging from collision avoidance to communication and cooperation. In this paper we propose a novel reasoning model for agents based on object oriented concepts which we call the *Priority/Confidence model.* The Priority/Confidence model provides a framework for reasoning. Decisions are made according to a *confidence measure* which is based on the importance of actions (priority) and the satisfaction of a priori conditions.

We implemented the Priority/Confidence model for a robotic soccer domain, namely the RoboCup [Kitano et al., 1997]. The RoboCup domain is a popular testbed to evaluate the performance of agents. It offers a dynamic environment which must be dealt with in real-time under strict rules of behaviour. It offers the possibility to test individual behaviour (dribble, pass, intercept or goal attempt) as well as collective behaviour, like coordinating an offensive action. The RoboCup competition is divided into leagues (leagues for real robots, in several size classes, and a separate simulator league). Our team, AIACS[1], is one of the two teams developed at the University of Amsterdam; the other team

* This research is carried out as part of the authors' Master's thesis.

[1] The name 'AIACS' is derived from the well-known Amsterdam soccer team AJAX, and is an acronym for *A*msterdam *I*ntelligent *A*gents for *C*hallenging *S*occer

is named *Windmill Wanderers* [Corten & Rondema, 1998]. Both teams plan to participate in the simulator league. AIACS has demonstrated the feasibility of the Priority/Confidence model by beating the world champion og 1997, *AT-Humboldt* and reaching a 9th place in the RoboCup'98 tournament.

1.1 Outline of this Paper

The remainder of this paper is organised as follows: Sect. 2 discusses related work. After that, Sect. 3 and 4 present the basic principles of the Priority/Confidence model. Section 5 describes how to evaluate the performance. Section 6 reports on aspects of the implementation and presents preliminary results. Section 7 concludes and gives suggestions for future work.

2 Related Work

Sahota [Sahota et al., 1995] developed a decision making technique called *reactive deliberation*, which makes a choice between several hard-wired behaviours. This is muxh like the *layered learning* paradigm is proposed in [Stone & Veloso, 1997, Stone & Veloso, 1998] propose the, in which low-level skills are learned first, and then, in subsequent layers, gradually more and more higher-level skills and behaviours are added. Their approach differs from Sahota's, in the way that they do not use hard-wired behaviours, and that their system is suited for team play, while reactive deliberation was mainly developed for a 1 vs. 1 scenario.

[Tambe et al., 1998] describes the principles behind the *ISIS* team. *ISIS* uses an explicit model of teamwork called *STEAM*. Although we do not use such an explicit model, we expect that the proper choice of actions and the pursuit of a collective goal leads to implicit teamwork. This approach makes our agent more robust, since it is neither dependent on information it receives from other agents, nor on the number of agents in the team.

We developed a model that is much like the approaches of Stone and Veloso and of Sahota et al. We make a clear distinction between two layers in our architecture: acting and reasoning. The reasoning layer is domain independent and decides between several tasks. These tasks can be implemented in any way: whether they are learned, as in the approach of Stone and Veloso or fully hard-wired like in Sahota's system, is not important to our model. The actions are fully domain dependent. The information on which the reasoning layer bases its decisions is also domain dependent, but the information is translated into conditions which are consulted by the reasoning layer.

3 The Priority/Confidence Model Architecture

The Priority/Confidence model is a two level architecture, in which the reasoning is completely separated from the execution. On the top level resides the *Action Manager*, which tasks are to select and coordinate actions based on a rational analysis of the current situation. On the bottom level the *Action Executor* is located, where it performs the actual actions. Figure 1 sketches the basic architecture.

The Action Manager is the component which directs the Action Executor by suggesting which action to perform. Its functionality is further discussed in Sect. 4. The Action Executor consists of two components, the *SkilledPlayer* and the *BasicPlayer*.[2] The BasicPlayer is a small, low-level component which facilitates the communication with the simulator in a clean and transparent way. It is an interface similar to Noda's libsclient [Noda, 1997b], and is used by the SkilledPlayer to execute the basic actuator commands (e.g. `kick`, `turn`, and `dash`) and to receive and process messages coming from the simulator.

We want to stress the importance of low-level skills. In order to win a game of soccer, a player must have precise control over its own movement and at the same time be able to handle the ball (e.g. to intercept a moving ball, to dribble with the ball without losing it, to avoid other players when running across the field). The possession of these skills is an essential prerequisite for playing soccer: no matter how sophisticated the reasoning is, without sufficient skills a team can never win.

The above can be summarised by the following principle: in the game of soccer, each player possesses a number of skills (such as dribbling and passing) which correspond to a set of simple actions that are necessary and sufficient to play soccer. This principle is used as a guideline in the design of the SkilledPlayer.

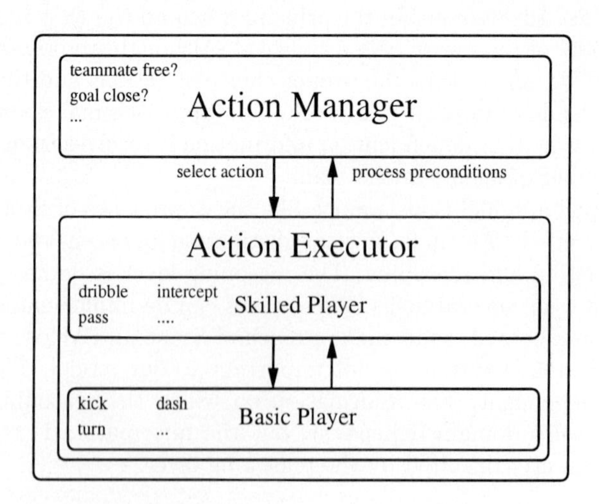

Fig. 1. The Priority/Confidence model architecture.

[2] See [Corten & Rondema, 1998].

3.1 Actions and Preconditions

Actions are short sequences of the basic actuator commands (like `kick`, `dash` and `turn`). We have identified a few standard actions that each player should master to play the game of soccer. A sample list of actions is shown in Table 1. All actions are executed sequentially: although it is possible to abort one action to start another one, it is not possible to execute two actions at the same time. Furthermore, to start an action, it is required that its preconditions are satisfied (or at least, that the confidence in the action is high enough). Actions can be seen as the means to satisfy the preconditions of a higher priority action (e.g. dribbling towards the goal satisfies the precondition that the goal should be within a certain distance before an attempt to score can be made).

The preconditions fall into two classes: (a) *essential* preconditions, and (b) *beneficial* preconditions. The former class includes all preconditions that are a *sine qua non.* The latter class encompasses the remaining preconditions, which merely increase the probability of success. If one or more essential preconditions are unsatisfied, an action cannot be executed successfully, and will therefore not get selected by the Action Manager. Beneficial preconditions return a satisfaction value, depending on the degree of satisfaction (e.g. for passing the ball there must be a free teammate. The more opponent players surround a teammate, the less free it becomes, resulting in a lower satisfaction value).

Table 1. A sample list of actions.

Action	**Example preconditions**
Pass	Teammate free
Dribble	Free course, stamina high
Goto Ball	No teammate closer to ball
Intercept Ball	Ball moving towards me
Go Home	Not already at home position
Score Goal	Close enough to opponent's goal
Follow Ball	Ball almost out of sight

4 The Action Manager

The Action Manager bases its selection of actions on two distinct criteria. On the one hand, each action has a *priority*, which indicates its usefulness or desirability. On the other hand, each action is associated with a *confidence* measure, which indicates its feasibility in a given situation. In this section we motivate the underlying ideas and present the Action-Selection algorithm.

It is our hypothesis that the ordering of priorities is related to the current strategy (e.g., when the strategy is offensive, scoring a goal is a high-priority action). To express this ordering, we prioritize all actions according to their importance for each strategy. In this way, we obtain a number of priority lists, each of which is related to a specific strategy (e.g. offensive or defensive). The initial values for the priorities are chosen by analysis of the (relative) desirability of each action.

In a way, each ordered list of actions defines an *implicit plan.* With 'implicit' we mean that there is no fixed "first, do this... then, do that" structure, but the ordering still specifies the strategy.

4.1 Definition of Terms

We define a precondition c to be a tuple of a unique descriptive *name*[3] and a *type* $\in$ {*essential, beneficial*}:

$$c =< name, type > \tag{1}$$

Let N be the total number of preconditions. Then, we define C_{total} to be the set of all preconditions:

$$C_{\text{total}} = \{c_1, \ldots, c_N\} \tag{2}$$

We define a state of the world W, to be the complete description of the current state.

For each precondition $c_i \in C_{\text{total}}$, we define a function $f^i_{\text{sat}}(c_i, W)$, which calculates its *satisfaction*, given world state W. In case of an *essential* precondition, $f^i_{\text{sat}}(c_i, W)$ is a binary function which evaluates to 0 (false) if the precondition is not satisfied, and 1 (true) if it is. In case of a *beneficial* precondition, the satisfaction is a value in the range $[0, 1]$ (continuous).

We define an action a to be a tuple of a unique descriptive *name* and a set of preconditions $C_{\text{action}} \subseteq C_{\text{total}}$:

$$a =< name, C_{\text{action}} > \tag{3}$$

The strategy $f_{\text{str}}(W)$ is defined as a function over the state W, and has a range of {*offensive, defensive*}. The priority of an action a_i is defined as a function over the strategy:

$$f_{\text{pri}}(a_i, f_{\text{str}}(W)) \tag{4}$$

Finally, to express the confidence an agent has in an action, we introduce the *confidence measure*, which acts as an evaluation function for the action to be chosen. We define the confidence measure $f_{\text{conf}}(a_i)$ of action a_i to be a function over its preconditions and over its priority:

$$\begin{aligned} f_{\text{conf}}(a_i) = f_{\text{pri}}(a_i, f_{\text{str}}(W)) \cdot \prod_{j=1}^{N_{ess}} f^j_{\text{sat}}(c_j, W) \cdot \frac{\sum_{k=1}^{N_{ben}} f^k_{\text{sat}}(c_k, W)}{N_{ben}}, \\ \text{where } c_j, c_k \in C_{\text{action}}, \\ N_{ess} \text{ is the cardinality of essential preconditions, and} \\ N_{ben} \text{ is the cardinality of beneficial preconditions} \end{aligned} \tag{5}$$

[3] Which allows us to refer to actions by their label.

Rationale:
The confidence measure can be seen as a way to incorporate a variable robustness into the team: with low confidence measures, the probabilities of success are also low and the team plays a risky game (which might be advantageous against weak opponents). To win from strong opponents, it might be better to play more conservative and commit only to actions that are bound to be successful.

4.2 The Action-Selection Algorithm

In Fig. 2 we state the Action-Selection algorithm, which is used by the Action Manager to select the actions. Following this, we give a short example that makes use of the Action-Selection algorithm to choose an action.

1. *Initialisation*:
   - Set action a_{best} to $< none, \emptyset >$
   - Set confidence $f_{\text{conf}}(a_{\text{best}}) = 0$
2. *For each* action a_i:
   - Calculate its confidence $f_{\text{conf}}(a_i)$
   - *If:* $f_{\text{conf}}(a_i) > f_{\text{conf}}(a_{\text{best}})$
   - *Then:* set action a_{best} to a_i
3. Selected action is a_{best}

Fig. 2. The Action-Selection algorithm.

Example 1. Imagine a situation in which you are an attacker close to the opponent's goal. The goal is defended by a goal-keeper. There is a teammate nearby to assist you, but he is marked by an opponent. What should you decide to do? Let us assume for now that there are only two relevant options, to *pass* the ball to your teammate, or to attempt to *score* by yourself. See Fig. 3.

We have the following conditions:
$C_{\text{total}} = \{\ c_1 =< \textit{ball kickable, essential} >,$
$c_2 =< \textit{teammate free, beneficial} >,$
$c_3 =< \textit{shoot course free, beneficial} >\ \}.$

And the actions: $a_1 =< \textit{pass}, \{c1, c2\} >$ and $a_2 =< \textit{score}, \{c1, c3\} >$.

With the strategy $f_{\text{str}}(W) = \textit{offensive}$, the priorities are given as follows: $f_{\text{pri}}(a_1, \textit{offensive}) = 0.5$ and $f_{\text{pri}}(a_2, \textit{offensive}) = 0.6$.

For the purpose of this example, the satisfaction values for the three preconditions are:
$f_{\text{sat}}(c1, W) = 1.0$, $f_{\text{sat}}(c2, W) = 0.6$ and $f_{\text{sat}}(c3, W) = 0.8$.

By applying the Action-Selection algorithm and calculating the confidence measures, we arrive at: $f_{\text{conf}}(a1) = 0.4$ and $f_{\text{conf}}(a2) = 0.54$. Thus, $a_{\text{best}} = a_2$, and the player would choose to *score*.

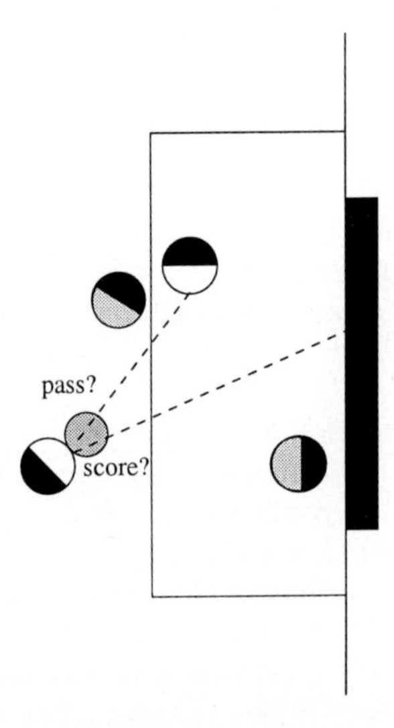

Fig. 3. An example situation: pass or score?

5 Evaluating the Performance

Since the selection of actions is based on their priority values, it is essential that these values are optimised. In this section we present three evaluation functions and discuss their advantages and disadvantages. We tested these and other functions as fitness functions in an optimisation with a genetic algorithm (GA). There are however a few problems we encounter; First, due to the dynamic nature of the domain, the results are variable. The randomness factor in the simulator is one cause of this problem, another cause is the unpredictability of opponents. We can disable the randomness to obtain a more steady evaluation. Second problem is the time it takes to play a match. A possible solution is parallelisation, which is especially useful if we use a genetic algorithm. It is possible to use embarrassing parallelisation because we can play every single match on a different computer. As long as we have enough processors available we can obtain an ideal speedup. The following list discusses three possible evaluation functions.

1. Goal difference
   - **Estimated duration of evaluation**
     To obtain a fair estimation of the performance, a trial should last at least

3000 simulations cycles, since this is the official duration of one half of a match. A trial should last long enough to show the effects of player fatigue (implemented in the simulator as decrease of stamina).

   - **Possible function**
   $\alpha \cdot (goals\ for) - (1 - \alpha) \cdot (goals\ against)$
   With the term α we are able to stress the importance of defense or attack. We can then evaluate two sets of priorities, one for an offensive and one for a defensive strategy, which we can use depending on player task or depending on team strategy. If $\alpha = 0.5$, both attack and defense are equally important. If $\alpha < 0.5$, we evaluate the quality of the defense, otherwise we evaluate the quality of the attack.
   It is important that the evaluation is reliable. To ensure this, we play all evaluations against one opponent only. Furthermore we set the randomness variable of the simulator to zero, so there will be no luck involved in an evaluation match.

2. Duration of ball possession
   - **Estimated duration of evaluation**
   Like the first method, this should last one half of a match.
   - **Possible function**
   An important issue of this method is defining 'ball possession'. We define the duration of ball possession as:
   The time between the first contact of a player with the ball and the moment that a player of the other team touches the ball, where 'touch' means that the ball is within kicking distance.
   A disadvantage of this function is that a good estimation of ball possession requires multiple measurements to minimize the error. This function can be useful to give extra information in combination with other evaluation functions.

3. Average distance between the ball and opponent's goal
   - **Estimated duration of evaluation**
   Like the first method, this should last one half of a match.
   - **Possible function**
   If a team has a high percentage of ball possession, it does not mean that the team is a winner. Although the champion of RoboCup'97 (*AT-Humboldt*) has a great ability to score, it does not stay in ball possession very long. Including the average distance between ball and opponent's goal gives a more accurate prediction of the actual outcome of a match.

We use a coach to monitor the match and record statistics. The statistics are written to a result file, which is used by a GA to calculate the fitness. Experiments with the fitness function have shown that a combination of criteria one (goal difference) and three (average distance between the ball and opponent's goal) currently gives the most reliable evaluation of the performance of a team.

6 Implementation and Results

The Priority/Confidence model has been implemented in C++. We started by constructing an empty framework for the control architecture and its components. Next, single actions from the SkilledPlayer were added and tested in an incremental manner. Every addition enabled the players to perform a little better.

6.1 Environment

For the evaluation of our team we ensured a fair comparison with the teams that competed in RoboCup'97. All matches were played on Soccer Server version 3.28 with the official parameter settings of RoboCup'97 [Noda, 1997a].

The Soccer Server simulator runs on a Sun Microsystems UltraSPARC-2 workstation clocked at 200-MHz with 128 Megabytes of memory. All 22 client programs run on a single UltraSPARC-1 clocked at 143-MHz with 160 Megabytes of memory. Both machines are running the Solaris 2.5 operating system.

6.2 Results

6.3 Results

The results of the matches played by AIACS in the RoboCup'98 competition is given in Table 6.1.AIACS won 3 out of 4 matches in the qualification round, scoring a total of 40 goals (6 goals were scored against AIACS, all by AT-Humboldt'98). AIACS finished second in its group after AT-Humboldt'98[4], which meant it qualified for the next round. In the championship rounds AIACS won 1 out of 3 matches, scoring a total of 15 goals (11 goals against). AIACS finished on an equal 9th place.

At RoboCup'98, apart from the competition, there was also an evaluation session in which teams could play against the world champion of RoboCup'97, AT-Humboldt'97. The purpose of this session was to evaluate the robustness of teams by disabling a number of players of one team, while AT-Humboldt'97would play with a full team. Four half matches were played, with 0, 1, 2, and 3 players disabled[5]. The results (see Table 3)demonstrate the robustness of the Priority/Confidence model. No games were lost, even with the goalkeeper and two defenders disabled.

7 Conclusion and Future Work

In this paper we have proposed the Priority/Confidence model as a framework for agent control in a dynamic real-time environment. Its feasibility has been demonstrated by several victories against leading teams of the RoboCup'97 competition. The results offer enough perspectives to proceed with our work. The obvious next step is to continue the work on the optimisation.

[4] AT-Humboldt'98 reached the finals and lost against CMUnited-98.

[5] The inactive players stayed on-field, so they could still be perceived by all others.

Round	Opponent	Result
Qualification 1	AT-Humboldt'98	0-6
Qualification 2	TU-Cluj	7-0
Qualification 3	ERIKA	17-0
Qualification 4	Dartbotics	16-0
Championship 1	Mainz Rolling Brains	2-7
Championship 2	Miya-2	11-0
Championship 3	ISIS	2-4

Table 2. Results of all competition matches played by AIACS at RoboCup'98.

match	disabled	score
1	0	7-0
2	1	6-0
3	2	6-1
4	3	4-4

Table 3. Results of the evaluation matches against AT-Humboldt'97 at RoboCup'98. Match nr. 4 was played with the goalkeeper disabled.

Acknowledgments

We would like to thank our supervisor Prof. Frans Groen for his valuable help in the preparation of this paper. We also thank Emiel Corten and Erik Rondema for stimulating discussions. Our special gratitude goes to Emiel Corten for providing the BasicPlayer and SkilledPlayer components.

References

[Corten & Rondema, 1998]
Emiel Corten and Erik Rondema, Team Description of the Windmill Wanderers, In this proceedings, 1998.

[Kitano et al., 1997]
Hiroaki Kitano, Milind Tambe, Peter Stone, Manuela Veloso, Silvia Coradeschi, Ei-Ichi Osawa, Hitoshi Matsubara, Itsuki Noda, and Minoru Asada, The RoboCup Synthetic Agent Challenge 97, In Martha E. Pollack, editor, *Proceedings of the 15th International Joint Conference on Artificial Intelligence (IJCAI-97), Volume 1*, pages 24–29, 1997.

[Kitano et al., 1998]
Hiroaki Kitano, editor, *RoboCup-97: Robot Soccer World Cup I, Proceedings of the First International Workshop on RoboCup*, Lecture Notes in Artificial Intelligence (LNAI), Volume 1395, Springer-Verlag, Berlin, Germany, 1998.

[Luke et al., 1998]
Sean Luke, Charles Hohn, Jonathan Farris, Gary Jackson, and James Hendler, Co-Evolving Soccer Softbot Team Coordination with Genetic Programming, In [Kitano, 1998], pages 398–411, 1998.

[Noda, 1997a]
Itsuki Noda, Regulations of the Simulation Track in RoboCup-97, available at: http://ci.etl.go.jp/~noda/soccer/regulations.1997/reg1997.html, March 1997.

[Noda, 1997b]
Itsuki Noda, libsclient software, available at: ftp://ci.etl.go.jp/pub/soccer/client/libsclient-3.03.tar.gz, May 1997.

[Sahota et al., 1995]
Michael K. Sahota, Alan K. Mackworth, Rod A. Barman, and Stewart J. Kingdon, Real-time Control of Soccer-playing Robots using Off-board Vision: the Dynamite Testbed, In *IEEE Transactions on Systems, Man, and Cybernetics*, 1995.

[Stone & Veloso, 1997]
Peter Stone and Manuela Veloso, A Layered Approach to Learning Client Behaviors in the RoboCup Soccer Server, In *Applied Artificial Intelligence (AAI) Journal, Volume 12*, 1997.

[Stone & Veloso, 1998]
Peter Stone and Manuela Veloso, Using Decision Tree Confidence Factors for MultiAgent Control, In [Kitano, 1998], pages 99–111, 1998.

[Tambe et al., 1998]
Milind Tambe, Jafar Adibi, Yaser Al-Onaizan, Ali Erdern, Gal A. Kaminka, Stacy C. Marsella, Ion Moslea, and Marcello Tallis, Using an Explicit Model of Teamwork in RoboCup, In [Kitano, 1998], pages 123–131, 1998.

A User Oriented System for Developing Behavior Based Agents

Paul Scerri, Silvia Coradeschi and Anders Törne

Department of Computer and Information Science
Linköping University, Sweden
Email: pausc@ida.liu.se, silco@ida.liu.se, ato@ida.liu.se

Abstract. Developing agents for simulation environments is usually the responsibility of computer experts. However, as domain experts have superior knowledge of the intended agent behavior, it is desirable to have domain experts directly specifying behavior. In this paper we describe a system which allows non-computer experts to specify the behavior of agents for the RoboCup domain. An agent designer is presented with a Graphical User Interface with which he can specify behaviors and activation conditions for behaviors in a layered behavior-based system. To support the testing and debugging process we are also developing interfaces that show, in real-time, the world from the agents perspective and the state of its reasoning process.

1 Introduction

Intelligent agents are used in a wide variety of simulation environments where they are expected to exhibit behavior similar to that of a human in the same situation. Examples of such environments include RoboCup[10], air combat simulations[14] and virtual theater[16].

Defining agents for simulation environments is a very active research area. The research has resulted in a large number of agent architectures being proposed. Many of the proposed architectures have accompanying languages for defining the behaviors, for example [1, 3, 5, 6, 11, 15, 16]. However many of these methods for specifying behavior are oriented towards a computer experts way of thinking, rather than to a domain experts, i.e. they use logic or other kinds of formalisms.

The quality of the behavior exhibited by an agent is closely related to the quality and quantity of the knowledge held by the agent. As domain, rather than computer, experts are likely to have superior knowledge of intended agent behavior, it seems to be advantageous to develop methods whereby domain experts can directly specify the behavior of the agents. It may be the case, especially in simulation environments, that parts of the behavior of an agent change often over the life of a system, in which case it is even more desirable to enpower domain experts to define and update behavior.

When developing agents with complex behavior for real-time complex environments it is often hard to debug and tune the behavior in order to achieve the desired result[8, 13]. When incomplete and uncertain information are added to the cocktail, as occurs in many domains including RoboCup, determining the reason for unwanted behavior can become extremely difficult.

The goal of allowing non-computer experts to specify complex behavior of simulated agents quickly and easily is a lofty one. In this paper we present the design of a system that addresses three aspects of the problem as it relates to RoboCup, namely: vertical rather than horizontal decomposition of behaviors; specification of conditions and behaviors in a high level abstract natural language manner; and a short and simple design-specify-debug cycle. The underlying ideas are not new, we have mainly pieced together existing ideas simplifying or adapting where necessary in order to create an environment that is simple for non-computer experts. Where possible we have tried to make the way a user specifies behavior as close as possible to the way human coaches would explain behavior to their players.

The system we are developing allows a user to specify the behaviors for a layered behavior based system via a Graphical User Interface(GUI). Activation conditions for behaviors are in the form of abstract natural-language like statements, which we refer to as predicates. The runtime system maps the natural language statements to fuzzy predicates.

The behavior specification interface presents the user with a window where they can define behaviors for each layer of a behavior based decision making system. The user can specify an appropriate activation predicate and a list of lower level behaviors, with associated activation information, that implements the functionality of the behavior. At runtime the behavior specification is used as input to a layered behavior based controller which uses the specification, along with world information abstracted from the incoming percepts, to turn low level control routine skills on and off as required. A GUI is provided to show in real time the way an agent perceives the field.

The choice of a behavior based architecture as the underlying architecture seems to be a natural choice as the structure of the architecture seems to correspond well to the way a human coach would naturally explain behavior. [1] A behavior-based architecture uses vertical decomposition of overall behavior, i.e. into defend and attack, rather than horizontal decomposition, i.e. into navigate and plan. For example a coach is likely to divide his discussion of team tactics into discussions on attacking and defending - this would directly correspond to attacking and defending behaviors in a behavior based system.

We use natural language statements which map to fuzzy predicates as a way of allowing an agent designer to specify conditions for behavior in his/her own language. The use of abstract natural language statements about the world as behavior activation mechanisms attempts to mimic the way a human might describe the reason for doing something. For example, a coach may tell a player

---

[1] This may or may not correspond to the way human decisions are actually made. However, the relevant issue is trying capture the experts explanation of the behavior rather than copying the decision making process.

to call for the ball when he is in `free space`. The idea of *free space* is a vague one, hence the natural choice of fuzzy predicates as an underlying implementation.

An artifact of behavior based systems is that they are difficult to predict before testing and, usually, also difficult to explain when observed. The process is further complicated in environments where incoming information is incomplete and uncertain. Consequently behavior based systems must go through a design-test cycle many times. To assist in the development process we have developed a real-time GUI interface that shows the world as the agent sees it and we have developed an interface that shows the state of the agents reasoning process i.e. which behaviors are executing and the activation level of non-executing behaviors.

Developing systems which allow non-computer experts to define agents is an active research area. Different approaches are often successful at simplifying the specification of behavior for a particular domain. Strippgen has developed a system for defining and testing behavior-based agents called INSIGHT [13]. It is claimed that a graphical representation of the internal state of an agent coupled with a visualization environment aids in testing and debugging agents. The INSIGHT system also includes an interface for incremental development of behaviors. Firby's RAP's system uses Reactive Action Packages to turn low level control routines on and off[6]. Our system is similar in that it provides an abstract method for deciding which skills to turn on and off, however we believe that RAP's is more suited to relatively static domains where the activities to be performed involve sequential tasks, with possibly a number of different available methods for achieving the goal, whereas our system is more suited to very dynamic domains. Moreover the usability aspect is not especially considered in RAPS. Harel has developed an extension of state machines called Statecharts[7] which is a powerful formalism for the representation of system behavior. However Statecharts are usually only used for the specification of complex physical system behavior. HCSM [4] is a framework for behavior and scenario control which uses similar underlying ideas to Statecharts. Like Statecharts HCSM is a very powerful way of representing behavior however it is not designed for easy specification by non-expert users. At the other end of the ease-of-use spectrum is KidSim [12] which allows specification of only simple behavior. As the name suggests, KidSim allows children to specify the behavior of agents in a dynamic environment via a purely graphical interface. An alternative way of specifying the behavior of agents is to have agents follow scripts like actors in a theater [16]. In [3] a system is presented that also has the aim of making behavior specification easier, however in that system no GUI is present and a different decision making mechanism is used.

Currently we are applying our system to the specification of RoboCup agents. However, we intend in the future to adapt it for specifying agents for air-combat and rescue simulations.

2 How a User Perceives the System

We are developing a system which allows users who are not necessarily programmers to specify the complex behavior of an agent for a complex domain. An agent designer can define complex agents by defining layers of behaviors, specifying activation conditions and testing agents all without having to write or compile any code.

An agent definition is in the form of an arbitrary number of layers of behaviors where behaviors on higher levels are more complex and abstract. At any time a single behavior on each level is executing. In higher levels the executing behavior implements its functionality by specifying the lower level, less abstract behaviors that should be considered for execution. The bottom level behaviors execute by sending commands to an interface which in turn turns on or off low level *skills*. The selection of the behavior to execute is determined by finding the behavior with the highest *activation* at that point in time. The activation level of a behavior is a time dependent function that depends on the truth of the fuzzy predicate[2] underlying the natural language statement and user specified activation parameters associated with the behavior.

The development process consists of developing behaviors for lower levels, testing the partially specified agent, then using previously developed behaviors to specify behaviors for higher layers. Lower layers, even incomplete lower layers, can be fully tested and debugged before higher level behaviors are specified.

A behavior is defined by specifying a name, the level of the system the behavior is at, a predicate for the behavior and a list of, possibly parameterized, lower level behaviors with associated activation parameters that together implement the functionality of the behavior. This is all done via a graphical user interface. At runtime the user can observe, via another GUI, the interactions between behaviors.

Another window provides real time feedback on exactly how the player perceives the world. This enables an agent developer to better understand the information upon which the agents reasoning is being done and therefore design a better set of behaviors. The GUI is relatively decoupled from the rest of the system and is in our intention to make it publicly available for other developers to use.

2.1 Creating Behaviors

When the users first starts up the specification system a main window opens up. This window gives the user the opportunity to open previously saved agent behavior specifications or to start a new behavior specification. The main window shows the behaviors that have been created previously for this agent. These behaviors can be used to implement the functionality of new higher level behaviors.

[2] We use this term very loosely to indicate a function that returns a value between `true` and `false` to indicate the perceived truth of some statement.

The user can choose to edit an existing behavior or create a new behavior. Clicking on either the *New Behavior* or the *Edit Behavior* button pops up a second window (see Figure 1). Here the user specifies information about the nature of the behavior.

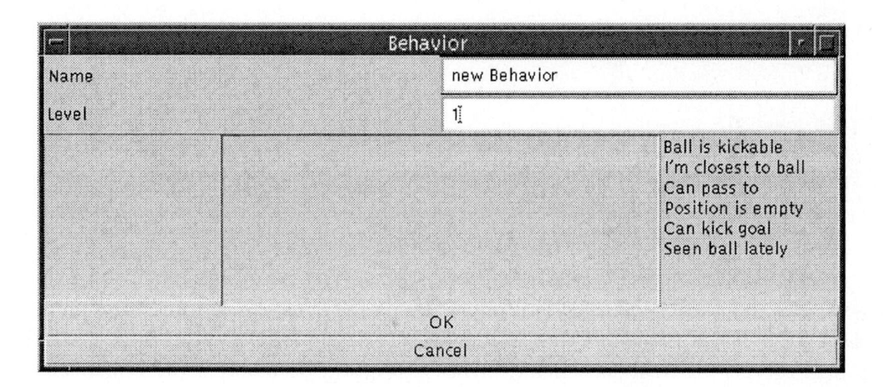

Fig. 1. The Behavior Specification Window. This is how the window appears when the *New Behavior* button is pressed. The list on the right shows the predicates that can be used by the behavior. The, initially empty, list on the left shows names of behaviors that can be used to implement the functionality of the behavior. The middle of the window will show selected behaviors and their associated activation parameters.

Once the user has finished specifying the behavior he can click OK and the behaviors name appears in the main window. The specification can then be saved and tested in the RoboCup simulator. Alternatively the user may directly begin work on other behaviors or a new higher level behavior which uses the previously defined behavior. There is no compilation required when the behaviors for an agent are changed.[3] However, real-time performance is still achieved as the control system that executes the behaviors is compiled and acts sufficiently quickly.

The way a user perceives different elements of a behavior specification does not necessary correspond to the actual underlying implementation. The intention is that the users are presented with an interface that allows them to express their ideas as naturally as possible and the underlying system takes the specification and uses it to make decisions.

The information required to fully specify a behavior is the following:

- Name
- Level
- Predicate
- List of Behaviors

[3] It is anticipated that eventually the behaviors will be able to be changed on line. However, at present the agent must be restarted when behaviors are modified.

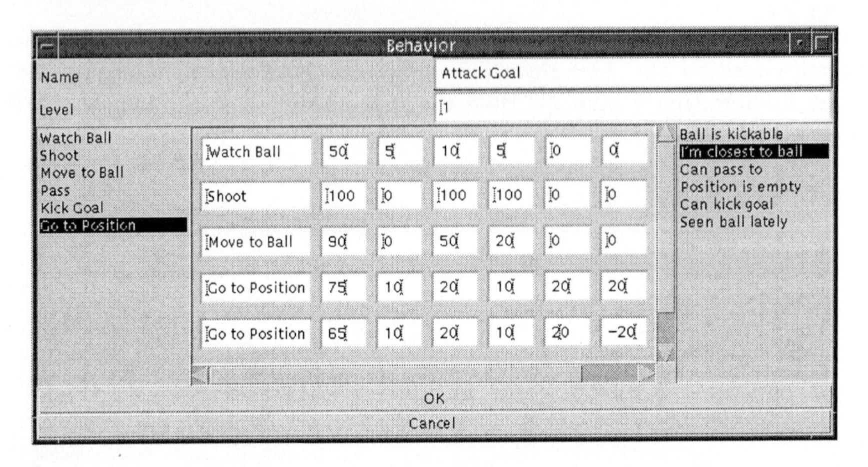

Fig. 2. The Behavior Specification Window with a behavior almost fully defined. In the middle is the list of behaviors that implements the functionality of `Attack Goal`. The numbers associated with each behavior are activation parameters (described below). In the list there are two instantiations of `Go to position`. The first `Go to position` is preferred as its maximum applicability, 75, is higher.

Each of these elements is discussed seperately below.

Name

Each behavior has a unique *name*. A designer specifying an agent can use names that abstractly represent the idea the behavior is trying to capture. Examples of behavior names for the RoboCup domain are `kick goal` - at a relatively low level of abstraction, `attack down left` - at a higher level of abstraction or `our free kick` - at an even higher level of abstraction. The names of behaviors on low levels are then used to specify behaviors on the next level up. The idea is that an agent designer can implement a behavior such as `our free kick` in terms of behaviors like `attack left` and `kick to open space`. The underlying system uses the behavior name as an identifier for the behavior.

Level

The *level* specifies which layer of the behavior based structure the behavior is to be used at. Different agents may have different numbers of levels depending on the design. The agent designer uses the level number to capture the intuitive idea that behaviors occur at different levels of abstraction - `moving to the ball` is at a low level of abstraction whereas `attacking` is at a high level of abstraction. Perhaps slightly less intuitive is the idea that very abstract behaviors are a result of interactions between a number of slightly less abstract behaviors. [4]

Predicate

To the agent designer a *predicate* is a statement about the world for which the

---

[4] This is an underlying concept in behavior based systems that has yet to be conclusively shown to be correct. However, as the underlying agent architecture for this system is a behavior based one, it is necessary that the agent designer uses this idea.

level of truth changes as the world changes. These statements occur at different levels of abstraction. Some example predicates for the RoboCup domain are `ball close enough to kick` - at a low level of abstraction, `good position to shoot at goal`, and `attacking position` - at a higher level of abstraction. The activation level of the behavior increases as the truth of the predicate statement increases. To the underlying system a predicate is merely the name of an abstraction of information received from the server. It is implemented as a function that maps data from the environment to a fuzzy truth value according to a programmer definition. An example of a mapping is a predicate `close to ball` which is implemented as a polynomial function of the last seen distance to the ball.

List of Behaviors

The *list of behaviors* is a list of behaviors less abstract than the one being defined that together implement the functionality of the behavior. Effectively the list of behaviors forms a hierarchical decomposition of the behaviors functionality. The behaviors in the list should interact in such a way that the intended complex behavior emerges. The process of choosing the behaviors and activation conditions is a difficult one. The short design-test cycle and the interfaces to aid analysis of the interactions can make the process of choosing appropriate behaviors simpler.

Each of the behaviors in the list may have some associated parameters which determine, along with the predicate that was specified when the less abstract behavior was created, the activation characteristics of the behavior. In order to influence the activation of each of the behaviors the user specifies four values: *Maximum Activation*, *Minimum Activation*, *Activation Increment*, and *Activation Decrement.*

To a user Maximum Activation is the highest activation a behavior can have. Intuitively, when more than one behavior are applicable the applicable behavior with the highest Maximum Activation will be executed. This allows representation of priorities between behaviors. To the system Maximum Activation is a hard limit above which the controller does not allow the runtime activation level of the behavior above.

To the user Minimum Activation is the lowest activation a behavior can have. Intuitively when no behaviors are applicable the behavior with highest Minimum Activation is executed. To the system Minimum Activation is a hard limit below which the controller does not let the runtime activation level of the behavior below.

To a user Activation Increment is the rate at which the activation level of the behavior increases when its predicate is true.[5] The Activation Decrement is the rate at which the activation of the behavior decays over time. These two values are closely related. High values for both the Decrement and Increment create a very reactive behavior, i.e. it quickly becomes the executing behavior when its predicate statement is true and quickly goes off again when the statement becomes false. Relatively low values for the Increment and Decrement result in

[5] As the predicate is actually a fuzzy predicate the activation increase is actually a function of the "truth" of the predicate.

a behavior that is not activated easily but temporarily stays active even after its predicate has become false. In the RoboCup domain a behavior such as `Kick Goal` may have high Increment and Decrement values, i.e quick reaction, so that when the predicate statement `ball close enough to kick` becomes true the behavior immediately starts executing and quickly stops executing when the predicate becomes false, i.e the ball is not close enough to kick. Behaviors such as `Attack` may have low values for Increment and Decrement so that the player does not start executing the `Attack` behavior until the appropriate predicate, possibly something like `We are in an attacking position`, has been consistently true for some time. However, it maintains the `Attack` behavior even if the predicate becomes false for a short time - perhaps due to temporarily incorrect information. It was the authors experience with previous behavior based systems for RoboCup that much instability is caused by temporarily incorrect information mainly due to incomplete and uncertain incoming information.

The use of a list of behaviors and corresponding activation parameters allows multiple uses of the same lower level behavior in a single higher level behavior specification. For example, an attack behavior may use two instances of the lower level behavior `move to position` (having predicate `Position is empty`) with different activation parameters and position to move to. The result may be that the agent "prefers" to move to one position over another.

2.2 Debugging

Behavior based agents interact very closely with their environment. Interactions between the world and relatively simple behaviors combine in complex ways to produce complex observed overall behavior[2]. Although the resulting behavior may exhibit desirable properties the complex interactions that occur make behavior based systems extremely difficult to analyze and predict especially when they exist in dynamic, uncertain domains[9]. It can often even be difficult to determine the reasons for unwanted behavior simply by observing the overall, complex behavior of the agent. Therefore an important part of any system for creating behavior based agents is a mechanism for allowing the user to quickly test and debug agents. To this end we have developed a graphical interface which shows in real-time the world as the agent see it. We have also developed an interface which graphically shows the reasoning of the agent, i.e. the currently selected behaviors at each level and the activation levels of all behaviors.

The world information interface draws the soccer ground as the agent sees it, displaying information such as the agents calculated position, the calculated position of the ball, the position and team of other players and the status of the game. This interface is intended to make it easier for developers to determine the causes for unwanted behavior in an agent.

The designers can make more informed behavior designs when they have a better understanding of the information the agent has available.[6] For example,

[6] During the overall system development the interface has also proved useful in determining errors in the way the agent processes information.

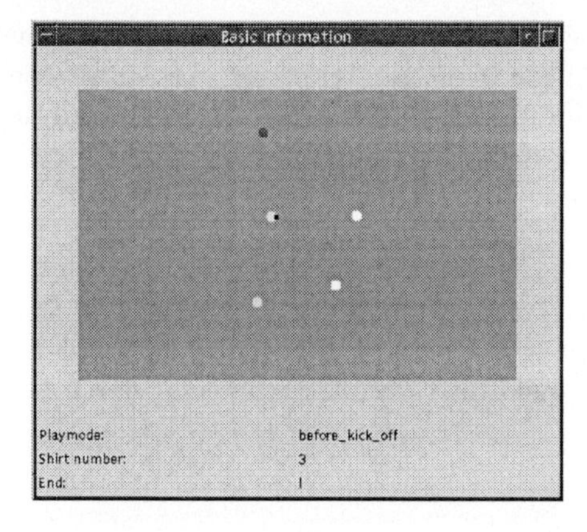

Fig. 3. The GUI showing how the agent perceives the world. The dark colored circle near the top of the screen is the agent whose world view is shown. Other circles represent the players in view. In the middle are team mates and on the right are opponents. Player of unknown team are shown in a different color (all players are known on the above diagram). Notice that the player that appears directly next to the agent of interest in the SoccerMonitor (See Figure 4) window does not appear in the agents perception of the world.

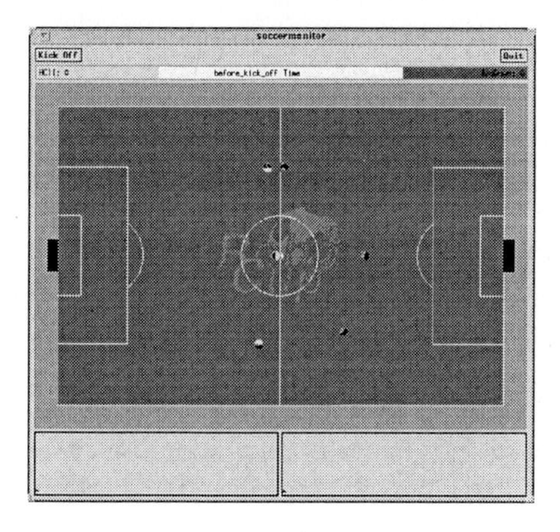

Fig. 4. The RoboCup Simulator showing the actual state of the world. Notice the two players close together near the top of the field. The player on the right does not appear in the player on the lefts view of the world - see Figure 3.

in Figure 3 the darker player near the top of the window perceives that he is in an empty space on the field although it is not, as can be seen from the RoboCup Soccermonitor (see Figure 4). The teammate with the ball can see that the player is not alone. This may possibly indicate to a designer that it is better for a player with the ball to look around for someone to pass to rather than relying on team mates to communicate that they would like to get the ball.

Also developed is an interface that displays in real-time the currently executing behavior on each level and the activation levels of all available behaviors (see Figure 5). This interface will allow designers to quickly determine the particular interactions that are resulting in undesirable behavior.

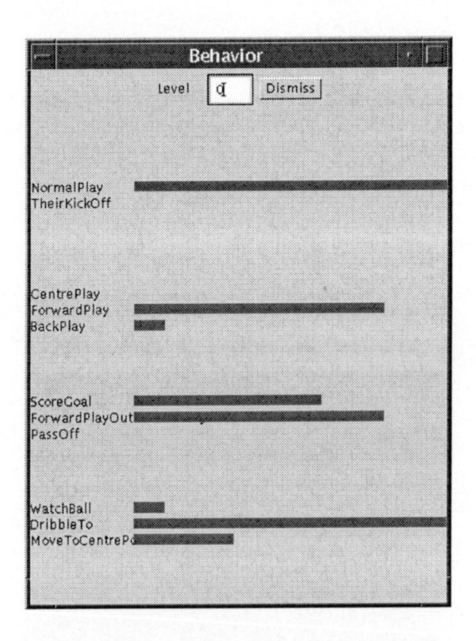

Fig. 5. A snapshot of the behavior activation window for an attacking player. Horizonatal lines represent the relative activation of the behaviors. Behaviors near the top of the window are higher level behaviors.

3 Underlying Agent Architecture

Many agent architectures have been developed, each with properties that make them suitable for some type of domain or some type of activity. For this system we use a layered behavior oriented architecture. The behavior activation mechanisms and behavior specifications are designed to allow the agents behavior to be specified without programming. Furthermore the entire agent system architecture is designed in such a way that it can accommodate a behavior based decision making system that uses abstract predicates and acts by turning skills on and off.

The system is divided into seven sub-systems (see figure 6):

- *Information Processing:* The information processing sub-system is responsible for receiving incoming perceptual information and creating and maintaining an accurate view of the world for the agent.
- *Predicates:* The Predicate sub-system forms the interface between Information Processing sub-system and Behavior Based Decision Making sub-systems and consists of a number of different fuzzy predicate objects. Predicates abstract away the details of the incoming information so that behaviors, and therefore agent designers, can use high level information for decision making.Rather than being precisely `true` or `false` predicates have a value that ranges between `true` and `false`.
- *Skills:* The Skills sub-system consists of a number of low level control routines for achieving particular simple tasks.The skills can have a very narrow scope such as moving towards a ball that has been kicked out of play.
- *Behavior Based Decision Making:* The Behavior based Decision Making system is responsible for the decision making of the agent. When the agent is started up an agent behavior description is loaded. At each layer of the system there is a controller which continually executes the following loop:
  – *Check if the layer above has specified a new set of behaviors. If so remove the old set of behaviors and get the new set.*
  – *Increase the* `activation level` *of all currently available behaviors by the value of the behaviors predicate times the* Activation Increment *value for the behavior.*
  – *Decrease the* `activation level` *of all currently available behaviors by the* `Activation Decrement` *value for the behavior.*
  – *If any behaviors* `activation level` *has gone above its* `Maximum Activation` *or below its* `Minimum Activation` *adjust the* `activation level` *so it is back within the legal range.*
  – *Find the behavior with the highest* `activation level` *and send the behavior list for this behavior to the next layer down (or in the case of the bottom level send a command to the interface).*
  – *Sleep until next cycle.*
- *Interface:*Interfaces between symbolic decision making systems and continuous control routines are an active area of research, e.g. [6]. We have implemented a very simple interface that may need to be extended in the future. The Interface receives strings representing simple commands from the decision making sub-system and reacts by turning on an appropriate skill in the Skills sub-system.
- *Debugging:* The Debugging sub-system acts as an observer of the rest of the agent architecture. The debugging sub-system works by periodically checking predefined information in the agent and representing the information graphically.

- *Server Interface:* The Server Interface is the sub-system responsible for communicating with the Soccer Server. The Server Interface sends incoming percepts to the Information Processing sub-system. Commands from the skills come to Server Interface to be sent to the Soccer Server.

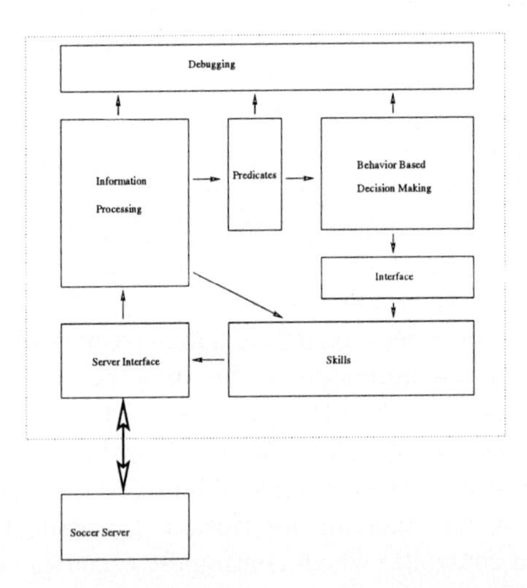

Fig. 6. An abstract view of the system architecture. The architecture of the agent is shown inside the dotted bow. Each box inside the agent represents a separate sub-system of the architecture. Arrows between sub-systems represent information flows.

The system was implemented in Java. Object Oriented techniques have been used in a way that allows new Skills and new Predicates to be quickly added by a programmer without changes being required to other parts of the system. For example the creation of a new predicate simply requires creating a subtype of an abstract Predicate class.

4 Evaluation

At the time of RoboCup98 in Paris the system was not sufficiently complete to allow evaluation with end users. However the authors used the GUI to specify a team which reached the quarter finals of the World Cup. During the development of the team and its subsequent use an evaluation based on observations was made. During the evaluation there were two main areas which were focused on, namely: the overall behavior of a finished agent; and the development process.

The overall observed behavior of the agent was reasonable. The agents consisted of around 40-50 different behaviors arranged into five or six levels. The agents were smoothly able to handle a range of different siutations at different levels of abstraction, for example agents exhibited different behavior in reponse to almost all referee calls. A plesantly surprising aspect of the behavior of the agents was the way they handled uncertainty in information. During testing bugs in the timing of updates on the agents memory were noticed. However with some tuning of the activation increment and activation decrement parameters the bugs made little difference to the players behavior. A considerable weakness in the behavior was the inability of agent to combine output of active behaviors. The winner-take-all arbitration strategy of the layer controller means that only one behavior can act at any time. Although careful tuning of the specification avoided oscillations between behaviors, behaviors that should have been concurrently acting, most notably *move to position* and *avoid obstacle*, required extra programming at the skill level.

The development system, although promising, had a number considerable weaknesses. The GUI made it relatively simple and fast to make farily considerable changes to the behavior of an agent. The major problem with the interface was the need for an expert programmer to define predicates. Early in development almost every new behavior required low level coding of a new predicate. As the list of predicates became longer the need for the creation of new predicates became rarer but managing the list became more difficult. The requirement that each behavior be given a fixed level turned out to be rather inconvienient often requiring that *dummy* behaviors were created so that lower level behaviors could be used by higher layers.

The testing and debugging interfaces proved valuable when debugging a single agent however they were inadequate for debugging teams of agents. The interface showing what the player sees was mainly used for debugging information processing aspects of the agents rather than for debugging behavior of agents. The interface showing the state of the agents reasoning proved extremely valuable for determining the reason for incorrect behavior of an agent. The debugging interfaces turned out to be inadequate when trying to determine reasons for undesirable behavior in teams or team situations. The problem seemed to stem from the fact that it was impossible for a user to monitor the reasoning of 11 agents in real-time. Some system for focusing a users attention on important aspects of the reasoning process, or more simply record and playback facilities would be required to make the interfaces useful for multiagent debugging.

5 Conclusion

In this paper we have described a system that allows non-computer experts to specify the behavior of agents for the RoboCup domain. We also describe an interface that shows in real-time the world as the agent sees it and an interface that shows the state of the agents reasoning process. Future research will look at user acceptance of the system and work towards making the interface more intuitive to designers. Agents developed with this system competed at RoboCup'98 reaching the quarter finals.

Acknowledgments

Paul Scerri has been supported by the NUTEK project "Specification of agents for interactive simulation of complex environments". Silvia Coradeschi has been supported by the Wallenberg Foundation project "Information Technology for Autonomous Aircraft".

References

1. Bruce Blumberg and Tinsley Galyean. Multi-level control of autonomous animated creatures for real-time virtual environments. In *Siggraph '95 Proceedings*, 1995.
2. Rodney Brooks. Intelligence without reason. In *Proceedings 12th International Joint Conference on AI*, pages 569–595, Sydney, Australia, 1991.
3. Silvia Coradeschi and Lars Karlsson. *RoboCup-97: The First Robot World Cup Soccer Games and Conferences*, chapter A Role-Based Decision-Mechanism for Teams of Reactive and Coordinating Agents. Springer Verlag Lecture Notes in Artificial Intelligence, Nagoya, Japan, 1998.
4. James Cremer, Joseph Kearney, and Yiannis Papelis. HCSM: A framework for behavior and scenario control in virtual environments. *ACM Transactions on Modeling and Computer Simulation*, 1995.
5. Kieth Decker, Anandeep Pannu, Katia Sycara, and Mike Williamson. Designing behaviors for information agents. In *Autonomous Agents '97 Online Proceedings*, 1997.
6. James Firby. Task networks for controlling continuous processes. In *Proceedings of the Second International Conference on AI Planning Systems*, June 1994.
7. D. Harel. Statecharts: A visual formalism for complex systems. *Sci. Comput. Program*, 8:231–274, 1987.
8. Maja Mataric. Behavior-based systems: Main properties and implications. In *IEEE International Conference on Robotics and Automation, Workshop on Architectures for*, pages 46–54, Nice, France, May 1992.
9. Maja Mataric. *Interaction and Intelligent Behavior*. PhD thesis, Massachusetts Institute of Technology, 1994.
10. Itsuki Noda. Soccer server: A simulator of RoboCup. In *Proceedings of AI Symposium'95*, Japanese Society for Artificial Intelligence, December 1995.
11. Itsuki Noda. Agent programming in Gaea. In *RoboCup '97 Proceedings*, 1997.
12. David Smith, Allen Cypher, Jim Spohrer, Apple Labs, and Apple Computer. *Software Agents*, chapter KidSim: Programming Agents without a Programming Language. AAAI Press/The MIT Press, 1997.
13. Simone Strippgen. Insight: A virtual laboratory for looking into behavior-based autonomous agents. In *Autonomous Agents '97 Online Proceedings*, 1997.
14. Milind Tambe, W. Lewis Johnson, Randolph Jones, Frank Koss, John Laird, Paul Rosenbloom, and Karl Schwamb. Intelligent agents for interactive simulation environments. *AI Magazine*, 16(1), Spring 1995.
15. Sarah Thomas. *PLACA, An Agent Oriented Programming Language*. PhD thesis, Dept. Computer Science, Standford University, 1993.
16. Peter Wavish and David Connah. Virtual actors that can perform scripts and improvise roles. In *Autonomous Agents '97 Online Proceedings*, 1997.

From Play Recognition to Good Plays Detection - Reviewing RoboCup 97 Teams from Logfile -

Tomoichi Takahashi and Tadashi Naruse

Chubu University, 1200 Matsumoto Kasugai-shi Aichi 487, JAPAN
Aichi Prefecture University, 3-28 Takada-cho Mizuho-ku Nagoya-shi Aichi 467, JAPAN

Abstract. This paper describes an attempt to review the teams participating in RoboCup simulator leagues. RoboCup simulator games are played through communication between soccer player clients and the soccer server. The server simulates the player's requirements and sends the result to the soccer monitor. We enjoy the games by seeing the players' actions and the ball's movement displayed on monitors. A method is proposed to recognize actions of a player such as shooting, kicking, etc., or the ball's movement from the log files, which are equivalent to images displayed on the monitor. Action recognition is a necessary technique for scoring, commenting and judging a game. The games in RoboCup '97 are reviewed from log files and the analysis results are discussed.

1 Introduction

RoboCup simulator games are played through communication between soccer player clients and the soccer server. The server receives the player's requirements and simulates the game. The server sends the game information to the soccer monitor. We enjoy the games by seeing the actions of the players and the ball's movement displayed on CRT monitors.

It was pointed out in RoboCup '97 workshop that other agents can join the soccer games through network without changing the simulator game frameworks. The agents are outside the field and receive the same information as displayed on CRT monitors. They score the game or comment on the plays. Their outputs indicate who passed the ball, who received the ball, and shot, or which opposing player interrupted a pass. It means that they recognize the soccer clients play as a human would see the game.

In this paper, methods of recognizing soccer actions from time sequence position data of the players and the ball are proposed. All games in RoboCup '97 are objectively reviewed using the numbers of the actions. The analysis result of all teams shows the necessary conditions for a strong team.

2 Information outside field in simulation track

Fig. 1 shows the data flows in simulation track. The data between clients agents and the soccer server is bi-directional and the flow is drawn in solid lines. Each client controls the corresponding player basically by repeating the cycles - sensing the environment around it, planning the next actions, and sending the corresponding commands to the server. The soccer server receives the requests from clients and simulates the games.

The server sends the game information to the soccer monitor periodically during the game. The data from the soccer server to the soccer monitor is unidirectional and contains the positions of all player/clients and the ball. The flow is drawn in dotted lines in the figure. Upon receiving the data, the soccer monitor displays the clients and the ball on CRTs. The clients and the soccer server are agents inside on the field and make up the game. The soccer monitor is an outside agent which does not participate in the game.

The other agents which scores the game or comment on the plays are also outside agents. They see the game by receiving the data from the soccer server.

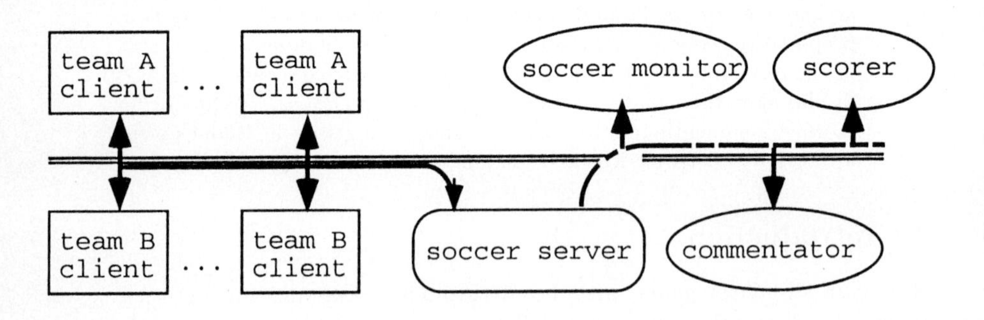

Fig. 1. data flows among agents

3 Action recognition from time sequence data

A viewer enjoy the client's plays by seeing the images displayed by the soccer monitor, and can comment on the game by indicating which agent passed the ball, whether the pass was very nice, etc. After games, the viewer becomes a programmer and improves his own client program manually by reflecting on the comments.

Our aims are to make outside agents able to detect good plays by seeing the game, and to make use of the detected plays to improve client agent ability. To do this, it is necessary to recognize the player's action and the ball movement from the time sequence data sent from the soccer server.

The issues in recognizing the player's actions and the ball movements are (1) detection of changes in games, and (2) which agent causes the changes. To make use of them as learning data, (3) division of the games into parts which are sufficient to replay the changes is necessary, (4) as well as transforming them into symbols. The ideal set of actions recognized by agents is the same one as the humans recognize by seeing the game.

We assume that the outside agents receive the same data as the soccer monitor receives them from the soccer server. The following is the data format sent to the soccer monitor. `showinfo_t` contains the positions of 22 players and the ball at time t which is represented by $\boldsymbol{p}_t$ and $\boldsymbol{b}_t$ respectively. `msginfo_t` contains the referee's command.

```
typedef struct {
   short enable;
   short side;
   short unum;
   short angle;
   short x;
   short y;
 } pos_t;

typedef struct {
   short  board;
   char   message[2048];
 } msginfo_t;

typedef struct {
   char  name[16];
   short score;
 } team_t;

typedef struct {
   char    pmode;
   team_t  team[2];
   pos_t   pos[23];
   short   time;
 } showinfo_t;

typedef struct {
   short mode;
   union {
      showinfo_t  show;
      msginfo_t   msg;
   } body;
 } dispinfo_t;
```

(from sserver-3.28[1]/server/type.h)

The actions recognized at present are *kick* and *shoot*. The recognized ball movements are *goal*, *own-goal* and the player who *assist*ed the goal is recognized. The followings are the methods used to recognize actions using a time sequence of $\boldsymbol{p}_t$ and $\boldsymbol{b}_t$.

kick:
1. the ball direction is changed. (the angle between $\boldsymbol{b}_{t_i} - \boldsymbol{b}_{t_{i-1}}$ and $\boldsymbol{b}_{t_{i-1}} - \boldsymbol{b}_{t_{i-2}}$ is larger than a specified value.)
2. the ball speed is increased. ($|\boldsymbol{b}_{t_i} - \boldsymbol{b}_{t_{i-1}}|$ is larger than $|\boldsymbol{b}_{t_{i-1}} - \boldsymbol{b}_{t_{i-2}}|$.)
3. at least one player is within kickable area when the 1st condition or 2nd one is satisfied.

shoot:
1. the ball moves to the goal.
2. the ball speed is sufficient to reach the goal. (the distance of a ball is estimated using $|\boldsymbol{b}_{t_i} - \boldsymbol{b}_{t_{i-1}}|$.)
3. the goal's team is different from the team of a player who kicked the ball.

goal:
1. the ball reached the goal.
2. the same as the 3rd condition in shoot.

the last player who kicked the ball is the goal player.

[1] http://ci.etl.go.jp/~noda/soccer/server.html

own-goal: 1. the same as the 1st condition in goal.
2. the opposite one to the 3rd condition in shoot.

assist: 1. goal occurred.
2. the player who kicked the ball before the shooter and was the same team of the shooter is assigned as an assist player.

4 Analysis of RoboCup 97's teams

We see soccer games implicitly on assumptions that:

1. players of a strong team move more actively than players of a weak team,
2. a team which controls a game kicks the ball more times than the other team.

We checked to see if these assumptions held true in soccer simulation games. 29 teams participated in RoboCup '97. The data sent from the soccer server during a game was stored as a Logfile. The Logfiles have been made public through the internet.

Table 1 shows the result of comparing teams by seven items - score, distances, kick, shoot, goal, own-goal and assist -. The first column contains the team names who participated in RoboCup '97. The second column contains game numbers that a team played. "Score" is the points the team gained in RoboCup. The others are recognized items. "Distance" is the sum of distance all players moved in a game. The numbers are the average per game. The numbers in parentheses show the order in each item.

From Table 1, the followings are shown,

1. The top four teams, andhill, AT_Humbolt, CMUnited, ISIS, happen to be in the middle in distance ranking from 16 to 19.
2. The top two teams, andhill, AT_Humbolt, are highly ranked in kick and shoot.
3. Team C4 is the highest ranked in score, goal and assist. However, it is also the lowest ranked in own-goal.

Table 2 shows the result of the game between andhills vs. Kasugabito in preliminary league Group D. The score was 23 to 0 and andhills won. The rows data are each player's data. In looking at the activity of a player, distance ranking and shoot ranking seems to be related.

From these, conditions of strong teams are (1) numbers of kicks or shots are numerous, (2) number of own-goal is small. These conditions match our human assumptions.

The score is the sum of its goal and the opponent's own goal. However, in Table 1, the score is not equal to the goal. The reason is that the number is an average over different opponents.

Table 1. Comparison among RoboCup'97 teams

	team name	game	score	distance	kick	shoot	goal	own-goal	assist
1	andhill(2)	7	12.7 (3)	11330 (17)	191.7 (3)	30.5 (1)	8.1 (2)	0.5 (4)	3.2 (2)
2	AT_Humboldt(1)	7	13.2 (2)	11315 (18)	204.5 (2)	25.7 (2)	8.0 (3)	0.8 (6)	3.0 (3)
3	C4	5	14.4 (1)	12291 (15)	166.2 (6)	0.0 (4)	10.6 (1)	3.8 (22)	4.8 (1)
4	CAT_Finland	5	2.8 (16)	9351 (23)	142.8 (13)	3.8 (20)	1.8 (15)	1.8 (13)	1.0 (11)
5	CMUnited(4)	7	8.8 (4)	11852 (16)	160.5 (7)	13.7 (9)	5.8 (5)	1.4 (10)	1.8 (7)
6	CTH	3	4.0 (13)	7581 (26)	133.3 (16)	16.3 (6)	3.0 (12)	1.3 (9)	1.3 (10)
7	FCMellon	6	8.8 (5)	8691 (25)	142.5 (14)	14.1 (8)	5.6 (6)	0.6 (5)	1.5 (8)
8	gamma	2	0.0 (26)	13475 (10)	136.0 (15)	0.5 (28)	0.0 (24)	6.5 (29)	0.0 (22)
9	Georgia_Tech	3	1.0 (23)	17037 (4)	104.0 (23)	4.3 (18)	0.3 (23)	3.3 (21)	0.0 (22)
10	HAARLEM	3	2.0 (17)	18019 (2)	124.0 (19)	7.6 (14)	1.3 (18)	1.6 (12)	0.3 (19)
11	HChick	3	1.6 (18)	12916 (12)	93.6 (27)	2.6 (22)	0.6 (20)	2.3 (14)	0.0 (22)
12	Inoue &Wilkin	2	0.0 (26)	9614 (22)	52.5 (29)	1.0 (25)	0.0 (24)	2.5 (18)	0.0 (22)
13	ISIS(3)	6	6.0 (11)	10664 (21)	147.6 (9)	13.3 (12)	3.5 (10)	2.6 (19)	0.8 (12)
14	kasuga-bito	3	1.3 (21)	15113 (8)	149.3 (8)	3.6 (21)	1.0 (19)	6.0 (27)	0.6 (13)
15	Maryland	4	6.7 (10)	15691 (6)	175.7 (4)	18.2 (5)	3.5 (10)	2.2 (14)	0.5 (16)
16	MICROB	3	7.0 (8)	12452 (14)	120.6 (21)	23.6 (3)	2.3 (13)	2.3 (14)	0.6 (13)
17	miya	4	8.7 (5)	11200 (19)	144.7 (11)	13.5 (11)	4.5 (7)	0.2 (2)	2.7 (4)
18	NIKEN	4	7.0 (8)	16326 (5)	120.0 (22)	7.0 (15)	3.7 (9)	1.0 (7)	1.5 (9)
19	NIT-stones	3	0.3 (24)	13464 (11)	78.6 (28)	1.0 (25)	0.0 (24)	4.6 (24)	0.0 (22)
20	Ogalets	4	8.5 (7)	17065 (3)	128.5 (17)	16.3 (6)	6.0 (4)	0.5 (3)	2.5 (5)
21	Orient	6	6.1 (19)	13478 (29)	117.3 (24)	12.1 (19)	4.5 (17)	0.33 (1)	1.8 (16)
22	PaSo	3	3.6 (14)	13776 (9)	173.6 (5)	6.0 (17)	2.3 (13)	1.0 (7)	0.3 (19)
23	ProjectMAGI	5	6.0 (11)	10993 (20)	144.6 (12)	8.6 (13)	4.2 (8)	3.0 (20)	2.0 (6)
24	R.M.Knights	3	0.0 (26)	6924 (28)	124.6 (18)	1.0 (25)	0.0 (24)	4.3 (23)	0.0 (22)
25	sicily	3	0.3 (24)	9153 (24)	94.3 (25)	1.3 (24)	0.0 (24)	5.6 (26)	0.0 (22)
26	TeamGC	3	1.3 (21)	12666 (13)	216.3 (1)	1.6 (23)	0.6 (20)	5.6 (25)	0.6 (13)
27	TUT11	2	0.0 (26)	7427 (27)	123.0 (20)	0.5 (28)	0.0 (24)	6.0 (27)	0.0 (22)
28	UBC_Dynamo	2	1.5 (20)	15446 (7)	94.0 (26)	6.5 (16)	0.5 (22)	1.5 (11)	0.5 (16)
29	uc3m	3	3.6 (14)	18374 (1)	146.6 (10)	13.6 (9)	1.6 (16)	2.3 (14)	0.3 (19)

note: The numbers are rounded at first decimal points.
So, the ranks are different for the same score.

5 Discussion and Summary

The methods to recognize the player's action and events related to goal are described. The games in RoboCup '97 were reviewed and the teams were analyzed quantitatively.

The analysis shows that the necessary conditions to be a strong team are to increase the number of kicks, and not to do the own-goal.

This paper is a step to make agents who can score or comment on the game. The next steps are (1) to segment data record into a sequence of a player's

Table 2. Comparison between Andhills and Kasuga-bito

	andhil:23						Kasuga-bito:0					
player	distance	kick	shoot	goal	own-goal	assist	distance	kick	shoot	goal	own-goal	assist
1	223	8	0	0	0	0	500	22	0	0	7	0
2	1276	27	0	0	0	1	1030	11	0	0	1	0
3	785	9	0	0	0	0	1201	10	0	0	0	0
4	641	12	0	0	0	0	920	13	0	0	0	0
5	502	7	0	0	0	0	823	9	0	0	0	0
6	1022	24	1	0	0	0	806	8	0	0	0	0
7	991	17	1	0	0	2	1174	55	3	0	0	0
8	1392	13	5	1	0	2	713	1	0	0	0	0
9	1516	21	4	1	0	1	732	5	0	0	0	0
10	1918	23	13	7	0	0	881	6	0	0	0	0
11	1871	16	7	6	0	0	953	9	0	0	0	0
Total	12144	177	31	15	0	6	9738	149	3	0	8	0

actions, (2) relate the actions between and among players, (3) estimate them from the viewpoint of using learning samples. In RoboCup '98, a special session was held to evaluate simulation teams year by year. Quantitative game scores are one of fundumentals for evaulations.

The program which analyzes the logfiles is in http://www.bais.chubu.ac.jp/˜ttaka/.

References

1. http://ci.etl.go.jp/˜noda/soccer/client.html
2. Noda Itsuki, Kuniyoshi Yasuo. *Simulator track and Soccer Server.* bit (in Japanese), Vol.28, No.5, pp.28-34.

Inductive Verification and Validation of the KULRoT RoboCup Team

Kurt Driessens, Nico Jacobs,
Nathalie Cossement, Patrick Monsieurs, Luc De Raedt

Dept. of Computer Science, K.U.Leuven,
Celestijnenlaan 200A, B-3001 Heverlee, Belgium
http://www.cs.kuleuven.ac.be/~nico/robocup

Abstract. As in many multi-agent applications, most RoboCup agents are complex systems, hard to construct and hard to check if they behave as intended. We present a technique to verify multi-agent systems based on inductive reasoning. Induction allows to derive general rules from specific examples (e.g. the inputs and outputs of software systems). Using inductive logic programming, partial *declarative* specifications of the software can be induced. These rules can be readily interpreted by the designers or users of the software, and can in turn result in changes to the software. The approach outlined was used to test the KULRoT RoboCup simulator team, which is briefly described.

1 Introduction

The RoboCup simulator offers a rich environment for the development and comparison of multi-agent systems (MAS). These systems are often very complex. An agent must cope with incomplete and partially incorrect data about the world it acts in, the goal of the agent isn't formally defined ('play good football') and the system operates in real-time as a consequence of which the timing-issue is very important. Because of all this, the design and programming of a single agent is a difficult task; verifying that the agent behaves as described in the design is even harder as is the verification and validation (V&V) of MAS (see section 3).

In the area of verification and validation of knowledge based systems there have been attempts to adjust the V&V systems towards the problem of verifying and validating MAS (see for instance [13]). However most of these systems assume that the agents act upon a knowledge base and that one can specify the intended behavior of the knowledge base, which is not the case for many MAS.

The V&V method we propose is based on induction. Rather than starting from the specification and testing whether it is consistent with an implementation, inductive reasoning methods start from an implementation, or more precisely, from examples of the behaviour of the implementation, and produce a (partial) specification. Provided that the specification is declarative, it can be interpreted by the human expert. This machine generated specification is likely to give the expert new insights into the behaviour of the system he wants to verify. If the induced behaviour is conform with the expert's wishes, this will (partly) validate the system. Otherwise, if the expert is not satisfied with the induced specification, he or she may want to modify the knowledge based system and repeat the verification or validation process.

This paper addresses the use of inductive reasoning for verification and validation of MAS. The employed techniques originate from the domain of inductive logic programming, as these methods produce declarative specifications in the form of logic programs. The sketched techniques are tested in the RoboCup domain. This paper is organized as follows: in section 2, we introduce inductive learning through inductive logic programming; in section 3 we show how this technique can be used in verification. In section 4 we show how we used our technique to verify our agents, after which we discuss related work and conclude. The agents are described in appendix A. For the remainder of this article we assume the reader is familiar with Prolog [18].

2 Inductive Logic Programming

Inductive logic programming [15] lies at the intersection of machine learning and computational logic. It combines inductive machine learning with the representations of computational logic. Computational logic (a subset of first order logic) is a more powerful representation language than the classical attribute-value representation typically used in machine learning. This representational power is necessary for verification and validation of knowledge based systems, because such knowledge based systems are in turn written in expressive programming languages or expert system shells. Another advantage of inductive logic programming is that it enables the use of background knowledge (in the form of Prolog programs) in the induction process.

An ILP system takes as input examples and background knowledge and produces hypotheses as output. There are two common used ILP settings which differ in the representation of these data: learning from entailment ([7] compares different settings) and learning from interpretation [10]. In this paper we will use the second setting. In learning from interpretations, an example or observation can be viewed as a small relational database, consisting of a number of facts that describe the specific properties of the example. In the rest of the paper, we will refer to such an example as a model. Such a model may contain multiple facts about multiple relations. This contrasts with the attribute value representations where an example always corresponds to a single tuple for a single relation.

The background knowledge takes the form of a Prolog program. Using this Prolog program, it is possible to derive additional properties (through the use of Prolog queries) about the examples. If for instance we are working in a domain where family-data is processed, possible background knowledge would be:

parent(X,Y) ← mother(X,Y).   parent(X,Y) ← father(X,Y).
grandmother(X,Y) ← mother(X,Z), parent(Z,Y).

There are also two forms of induction considered here: predictive and descriptive induction. Predictive induction starts from a set of classified examples and a background theory, and the aim is to induce a theory that will classify all the examples in the appropriate class. On the other hand, descriptive induction starts from a set of unclassified examples, and aims at finding a set of regularities that hold for the examples. In this paper, we will use the TILDE system [1] for predictive induction, and the CLAUDIEN system [9] for descriptive induction.

TILDE induces logical decision trees from classified examples and background theory. Consider for example this background knowledge:

```
replaceable(gear).  replaceable(wheel).  replaceable(chain).
not_replaceable(engine).  not_replaceable(control_unit).
```

and a number of models describing worn parts and the resulting action (in total 15 models were used):

```
begin(model(1)).    begin(model(2)).    begin(model(3)).    ...
sendback.           fix.                keep.
worn(gear).         worn(gear).         end(model(3)).
worn(engine).       end(model(2)).
end(model(1)).
```

TILDE will return this classification tree:

```
worn(A) ?
+--yes: not_replaceable(A) ?
|       +--yes: sendback
|       +--no:  fix
+--no:  keep
```

CLAUDIEN induces clausal regularities from examples and background theory. E.g. consider the single example consisting of the following facts and empty background theory:

```
human(an).   human(paul).   female(an).   male(paul).
```

The induced theory CLAUDIEN returns is:

$$\text{human}(X) \leftarrow \text{female}(X). \qquad \text{human}(X) \leftarrow \text{male}(X).$$
$$\text{false} \leftarrow \text{male}(X) \wedge \text{female}(X). \qquad \text{male}(X) \vee \text{female}(X) \leftarrow \text{human}(X).$$

Notice that this very simple example shows the power of inductive reasoning. From a set of specific facts, a general theory containing variables is induced. It is not the case that the induced theory deductively follows from the given examples. Details of the TILDE and CLAUDIEN system can be found in [1, 9].

3 ILP for Verification and Validation

Given an inductive logic programming system, one can now verify or validate a knowledge based or multi-agent system as follows. One starts constructing examples (and possibly background knowledge) of the behaviour of the system to be verified. E.g. in a knowledge based system for diagnosis, one could start by generating examples of the inputs (symptoms) and outputs (diagnosis) of the system. Alternatively, in a multi-agent system one could take a snapshot of the environment at various points in time. These snapshots could then be checked individually and also the relation between the state an agent is in and the action it takes could be investigated.

Once examples and background knowledge are available one must then formulate verification or validation as a predictive or descriptive inductive task. E.g. in the multi-agent system, if the aim is to verify the properties of the states of the overall system, this can be formulated as a descriptive learning task. One then starts from examples and induces their properties. On the other hand, if the aim is to learn the relation among the states and the actions of the agent, a predictive approach can be taken.

After the formulation of the problem, it is time to run the inductive logic programming engines. The results of the induction process can then be interpreted by the human verifiers or validators. If the results are in agreement with the wishes of the human experts, the knowledge based or multi-agent system can be considered (partly) verified or validated. Otherwise, the human expert will get insight into the situations where his expectations differ from the actual behaviour of the system. In such cases, revision is necessary. Revision may be carried out manually or it could also be carried out automatically using knowledge revision systems (see e.g. Craw's KRUST system [5], or De Raedt's Clint [6]). After revision, the validation and verification process can be repeated until the human expert is satisfied with the results of the induction engines.

4 Experiments in RoboCup

In this section we describe some verification experiments. For this, a preliminary version of the agents described in appendix A was used. The most important differences are that the agents in the experiments did not yet used predictions about the future position of ball and players and that the decision-network used (figure 2) was much simpler.

4.1 Modeling the Information

The first tests were run to study the behavior of a single agent. We supplied the agents with the possibility to log their actions and the momentary state of the world as perceived by them. This way we were able to study the behavior of the agents starting from their beliefs about the world. Because all the agents of one team were identical except for their location on the playing field, the log files were joined to form the knowledge base used in the experiments. A sample description of one state from the log-files looks as follows :

```
begin(model(e647)).
   player(my,1,-43.91466,5.173167,3352).
   player(my,2,-30.020395,7.7821097,3352).
```

```
    ...
    player(other,10,14.235199,15.192206,2748).
    player(other,11,0.0,0.0,0).
    ball(-33.730022,10.014952,3352).
    mynumber(5).
    bucket(1).
    rctime(3352).
    moveto(-33.730022,10.014952).
    actiontime(3352).
end(model(e647)).
```

The different predicates have the following meaning :

player(T, N, X, Y, C) the agent has last seen the player with number N from team T at location (X, Y) at time C.
ball(X, Y, C) the agent has last seen the ball at location (X, Y) at time C.
mynumber(N) this state was written by the agent with number N. It thus corresponds to the observation of agent N.
bucket(N) the bucket used for bringing the agent back to its home position. The bucket-value was lowered every input/output cycle and forced the agent to its home-location and reset when it reached zero.
rctime(C) the time the state was written.
actiontime(C) the time the action listed was executed.
moveto(X, Y), *shoottogoal*, *passto*(X, Y), *turn*(X), *none* the agent's action.

The *rctime*(C) predicate was used to judge the age of the information in the model as well as to be able to decide how recent the action mentioned in the model is. This was done by comparing the *rctime*(C) with *actiontime*(C). Logging was done at regular time intervals instead of each time an action was performed, so we could not only look at why an agent does something, but also why an agent sometimes does nothing. The time-units used were the simulation-steps from the soccer-server.

Some of the actions that were used while logging were already higher level actions than the ones that can be sent to the soccer-server. However these actions, such as *shoottogoal* for example, were trivial to implement.

To make the results of the tests easier to interpret an even higher level of abstraction was introduced in the background knowledge used during the experiments. Actions that are known to have a special meaning were renamed. For instance a soccer-player that was looking for the ball always used the *turn*(85) command, so this command was renamed to *search_ball*. An other example of information defined in the background knowledge is the following rule :

```
action(movetoball):- validtime, moveto(X1,Y1), ball(X2,Y2),
                     distance(X1,Y1,X2,Y2,Dist), Dist =< 5 .
```

in which the *moveto*(X, Y) command was merged with other information in the model to give it more meaning. For instance, *moveto*$(-33.730022, 10.014952)$ and *ball*$(-33.730022, 10.014952, 3352)$ in the model shown above, would be merged into *movetoball* by this rule. Often a little deviation was permitted to take the dynamics and noise of the environment into account. The actions used to classify the behavior of the agent were : *search_ball*, *watch_ball*, *moveto*, *movetoball*, *moveback*, *shoottogoal*, *passto*, *passtobuddy* and *none*. Some of these actions were not used in the implementation of the agent but were included anyway for verification purposes. For instance, although — according to specifications — an agent should always "move to the ball" or "move back", the possible classification *moveto* was included in the experiments anyway, to be able to detect inconsistencies in the agent's behavior..

Other background knowledge included the playing areas of the soccer-agents and other high level predicates such as *ball_near_othergoal*, *ball_in_penaltyarea*, *haveball* etc. Again, not all of these concepts were used when implementing the agent. This illustrates the power of using background knowledge. Using background knowledge, it is possible for the verifier to focus on high-level features instead of low-level ones.

4.2 Verifying Single Agents

The first tests were performed with TILDE, which allowed the behavior of the agent to be classified by the different actions of the agent. The knowledge base used was the union of the eleven log-files of the agents of an entire team. The agents used in the team all had the same behavior, except for the area on the field. The area the agent acted in depended on the number of the agent and also was specified in the used background knowledge. The resulting knowledge base consisted of about 17000 models (14 Megabyte), collected during one test-game of ten minutes. The first run of TILDE resulted in the following decision tree

```
seeball ?
+--yes: ball_in_my_area ?
|       +--yes: haveball ?
|       |       +--yes: ball_near_othergoal ?
|       |       |       +--yes: action(shoottogoal) [15 / 15]
|       |       |       +--no:  action(passtobuddy) [122 / 124]
|       |       +--no:  action(movetoball) [1007 / 1015]
|       +--no:  bucket_was_empty ?
|               +--yes: action(moveback) [342 / 347]
|               +--no:  action(watch_ball) [2541 / 3460]
+--no:  action(search_ball) [7770 / 7771]
```

Only about 12000 models were classified. We did not include the action *none* as a classification possibility because although a lot of models corresponded to this action, it's selection was a result of the delay in processing the input information instead of depending on the state of the agent's world. Most of the classifications made by TILDE were very accurate for the domain. However, the prediction of the *action*(*watch_ball*) only reached an accuracy of 73,4%.

To get a better view on the behavior of the agent in the given circumstances CLAUDIEN was used to describe the behavior of the agent in case "*seeball*, *not*(*ball_in_my_area*), *not*(*bucket_was_empty*)." CLAUDIEN found two rules that describe these circumstances. The first rule was the one TILDE used to predict the *watch_ball* action.

```
action(watch_ball) if  not(action(none)), seeball,
                       not(ball_in_my_area), not(bucket_was_empty).
```

CLAUDIEN discovered the rule had an accuracy of 73%. The other rule that was found by CLAUDIEN was the following :

```
action(moveback) if  not(action(none)), seeball ,
                     not(ball_in_my_area), not(bucket_was_empty).
```

which reached an accuracy of 26%. It states that the agent would move back to its home location at times it was not supposed to. Being forced to go back to its home-location every time the bucket was emptied, this behavior was a result of the bucket getting empty while the player was involved in the game and therefore not paying immediate attention to the contents of the bucket.

To gain more consistency in the agents behavior, the bucket mechanism was removed and replaced by a new behavior where the agent would move back when it noticed that it was to far from its home location. The new behavior, after being logged and used in a TILDE-run resulted in the following tree :

```
seeball ?
+--yes: ball_in_my_area ?
|       +--yes: haveball ?
|       |       +--yes: ball_near_othergoal ?
|       |       |       +--yes: action(shoottogoal) [48 / 48]
|       |       |       +--no:  action(passtobuddy) [85 / 85]
|       |       +--no:  action(movetoball) [796 / 810]
|       +--no:  at_place ?
|               +--yes: action(watch_ball) [3826 / 3840]
|               +--no:  action(moveback) [384 / 394]
+--no:  action(search_ball) [7180 / 7318]
```

in which the *action*(*watch_ball*) was predicted with an accuracy of 99,6 %. The increase in consistency in the behavior in the agent, improved its performance in the RoboCup environment. Because the agent only moved back to its home-location when necessary it could spend more time tracking the movement of the ball and fellow agents and therefore react to changing circumstances faster.

4.3 Verifying Multiple Agents

In agent applications it is often important that not only all agents individually work properly, the agents also have to cooperate correctly. One important point in this is to check if the beliefs of the different agents more or less match. In the case of our RoboCup agents we want to know for instance if there is much difference between the position where player A sees player B and the position where player B thinks it is[1]. So we used CLAUDIEN to find out how often agents have different believes about there positions, and how much their beliefs differ.

[1] it is impossible to know what the real position of a player is, so we can only compare the different believes the agents have.

To do these tests, we transformed the datafile so that one model contains the believes of multiple agents at the same moment in time. CLAUDIEN found multiple rules like the one below:

```
Dist < 2 if mynumber(A,Nr), vplayer(A,my,Nr,X1,Y1), vplayer(B,my,Nr,X2,Y2),
            mynumber(B,Nr2), vplayer(B,my,Nr2,X3,Y3), not(A=B),
            distance(X1,Y1,X2,Y2,Dist),distance(X2,Y2,X3,Y3,Dist2),Dist2<10.
```

This rule, which has an accuracy of 78% states that if two players are less then 10 units apart, the difference in the believes of the position of one of those two players is less then 2 units. From all the rules we could conclude useful information, for instance, we found out that our agents can best estimate a team mate's position from distance 10. All rules found were 'acceptable' rules (e.g. for distances larger than 10, the error is positively correlated with the distance between the players), so we can conclude from the observed behavior that the beliefs of the different agents do not differ much.

5 Related work

This work builds upon earlier ideas on combining verification and validation with inductive logic programming [8]. It is also related to other approaches applying machine learning with validation and verification. This includes the work of Susan Craw on her KRUST system for knowledge refinement [5], the work by Bergadano et al. and the work by De Raedt et.al. [11]. The approach taken in KRUST is complementary to ours. Rather than starting from examples of the actual behaviour of the system, KRUST starts from examples of the desired behaviour of the system. Whenever the two behaviours do not match, KRUST will automatically revise the knowledge based system. It is clear that the KRUST approach could also be applied within our methodology, at the point where the human discovers inconsistencies between the two behaviours. If the human then specifies examples of the intended behaviour, KRUST might help revising the original knowledge base. The approach of Bergadano et. al. and De Raedt et. al. using inductive machine learning to automatically and systematically generate a test set of examples that can be used for verification or validation. Finally, our work is also related to the work by William Cohen [3] on recovering software specifications from examples of the input-output behaviour of the program.

6 Conclusions

We sketched a novel approach to verification and validation, based on inductive reasoning rather than deduction. We reported a number of experiments in the domain of MAS (RoboCup) which prove the concept of the approach.

Further work on this topic could involve applying the verification and validation technique also to other multi-agent systems (such as e.g. DESIRE [2]), and also to extend the inductive method to other representations. For instance, it seems very well possible to apply inductive techniques in order to automatically construct decision tables starting from the knowledge base. Such decision tables are already popular in V&V, but they are typically made by the human expert (in collaboration with the machine), see e.g. [19].

Acknowledgements: The authors wish to thank Hendrik Blockeel and Luc Dehaspe for their help with the TILDE and CLAUDIEN system. Nico Jacobs is financed by a specialisation grant of the Flemish Institute for the promotion of scientific and technological research in the industry (IWT). Luc De Raedt is supported by the Fund for scientific research, Flanders. This work is supported by the European Community Esprit project no. 20237 (ILP 2).

A Team Description

A.1 Introduction

In this appendix we describe the KULRoT team for the RoboCup '98 simulator league [12]. It is the result of some preliminary experiments in the domain, and the main emphasis is on detecting problems related to building multi agent systems for the RoboCup task and using machine learning techniques to overcome these problems. To simplify the building of a RoboCup team the complete team consists of identical players[2] which only differ in their field position.

The agents are implemented in Java for different reasons. The main reason is that Java is platform independent, an important aspect for code being simultaneously developed on different operating systems. Moreover using Java it's easy to use multi-threading. The use of the UDP protocol [17] is also embedded in Java.

This description is structured as follows: in section A.2 a general overview of the structure of the soccer agent is presented, in section A.3 we discuss the beliefs held by an agent. The acting of the agent is split up in low level skills (section A.4) and high level skills (section A.5). Finally in section A.6 we describe some timing problems and how we tackled these.

A.2 General overview of the soccer agent

The general structure of each player is presented in figure 1. It consists of five main parts:

- **Communicator**: this module acts as an intermediate layer between the real agent and the soccer server; it mainly manages the sockets for communication.
- **Sensors**: this thread parses the incoming information about what the agent sees and hears. At the moment the full input string gets parsed and the information stored in the world model.
- **World model**: the information received from the sensors is used to update the world model. This model is however an active model in the sense that it is able to predict the (approximate) position of objects in the future using the formulae the server uses to calculate the trajectory of objects. This is explained in more detail in section A.3.

[2] except for the goalie.

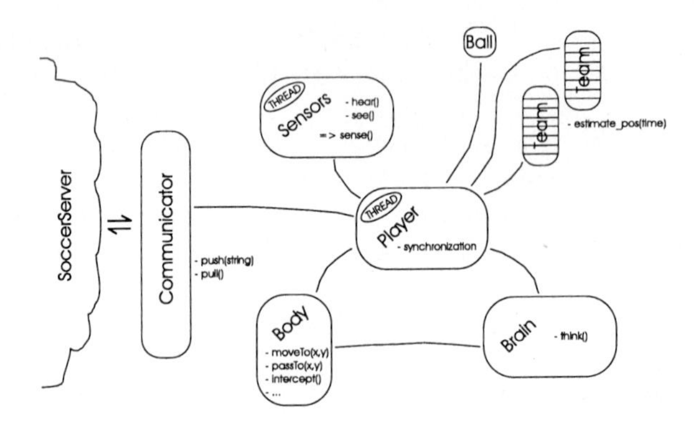

Fig. 1. General overview of the soccer agent

- **Body**: this module contains the low level skills of the agent. These are skills like turning with a ball, moving to a certain position or intercepting a ball. It uses the world model for this. See section A.4 for a description of these low level skills.
- **Brain**: this is the module in which the real decision making occurs. This module decides based on the world model which actions need to be executed and translates these to low level commands, which are then sent to the body module. This is explained in more detail in section A.5.

The player class only starts up the other modules and is used to let the other modules communicate with each other.

A.3 Beliefs and World Model

The information about the world stored within the agent is modeled using absolute coordinates. The see-messages arriving from the soccer server are parsed by the Sensors class and transformed to absolute coordinates.

Although the modeling in absolute coordinates requires extra work during the processing of the sensor information, it limits the updates required when changing the position of the agent and enables easier reasoning about future locations of moving objects.

The algorithm available in the libsclient library [16] was translated and used to calculate the absolute coordinates and facing direction of the agent. This information is then used to triangulate the absolute coordinates of the other objects present in the see-string.

The memory of the player consists of field objects that represent the twenty-two players and the ball. They hold the information needed to calculate their position at a given time in the near future, i.e. the coordinates they were last seen by the agent, the speed and direction they were traveling in at that moment and the simulator time they were last seen at. This, together with the mathematical formulas that represent the course of the object that are used within the soccer server allows the field objects to estimate their future positions. This information will become more and more inaccurate when the object hasn't been seen a long time or when looking far into the future. Also the estimations will be wrong when the object deviates from its course at which it was last seen.

Because every object on the field has its own representation as an object of the field object class, it is not easy to use information about players of which the number or even the team is not visible. As a consequence, this information is not used. Seen objects are only considered by the player when all identifying information is available. This of course limits the long range view of the agent and may be changed in the future. For the same reason, the low quality setting for the sensor information which is available in the soccer server is not used by the agent at the moment, because it supplies the sensory information without the identifying information necessary to use it.

A.4 Low Level Skills

The low level skills of the agent are represented by, and implemented in the Body class. Most of the low level skills implemented in the agent are concerned with transforming actions in the used coordinate system to actions which can be performed by the agent through the soccer-server.

Such actions are $moveTo(x,y)$, $passTo(x,y)$ or $shootToGoal()$. Starting from the agents own position and direction, either as observed or predicted, the necessary actions are calculated and performed. During these calculations, consideration is also paid to the fact whether the player stands between the ball and the target. For timing issues discussed later, the time the action will end is also calculated and returned as a result of the action.

Based on these actions, it was possible to supply the agent with a bit more complicated actions on a low level. Such actions as $markPlayer(number, team)$, $dribble()$, $moveAndCheck(x,y)$ or $turnWithBall()$ which are independent of the current field- or team-situation were implemented at this level were checks about own and target position could be evaluated half way during the action.

Successfully intercepting the ball requires a combination of three things the agent must performs. First it must estimate the time it will take to reach the ball. Then the player must estimate the position of the ball at that time. To complete the intercept, it must move to that position. The last two actions are already discussed above. The tricky part is estimating the best future time-instance to intercept the ball at.

Different strategies were tried to accomplish this. The first implementation looked a fixed number of simulator steps ahead and moved the agent to the estimated position of the ball at that time. Because neither the distance to the ball, nor its speed, nor the direction it is traveling in are considered this way, the method was not very accurate nor successful.

The second strategy calculated the intercept-time by starting from an underestimation and increasing this value by one simulator step until a time-instance was found at which the soccer-agent could reach the estimated position of the ball at that same time. This method was more successful but had computation requirements too large to be useful in the real time environment of RoboCup.

The solution used by the KULRoT team was obtained by generating a large set of examples of intercept-times together with seven relevant values : distance to the ball, the relative view-angle of the ball, the relative travel direction of the ball, the travel direction of the player, the speed of the ball, the speed of the player and the player's current effort. This set of examples was then used to generate a decision tree with TILDE [1] which predicted the intercept time based on the values given. The success rate of the intercept using this decision tree was comparable with the one that calculated the correct time — 59% vs. 63%, the reason for the failing of the intercept with the correct calculations being the error in the estimations of the ball speed and location — but the calculation time was much lower.

Preliminary tests using neural networks resulted in a lower success rate. Other strategies (e.g. using perpendicular intercept trajectories) were tested as well, but with bad results. More information can be found in [4].

A.5 High Level Skills

In the previous section we discussed some actions that the agent can perform. In this section we discuss how we decide which action to perform at a certain moment in time. The basic decision structure is based on a network structure depicted in figure 2, which can also be seen as a tree with the rightmost node as the root.

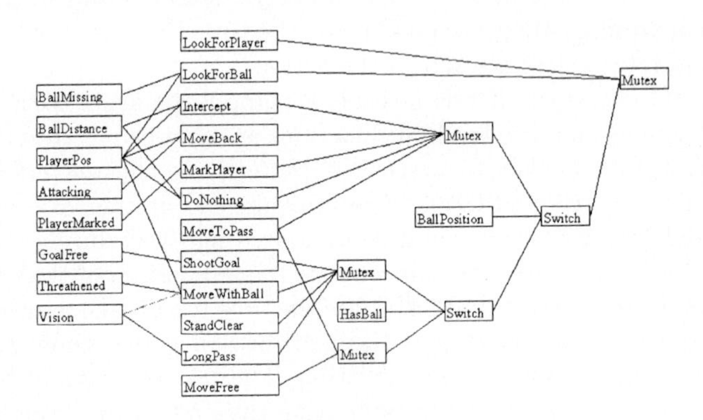

Fig. 2. Selecting high level actions

The network consists of two types of nodes: value-nodes and conditional nodes. A basic value node returns a value based on the world model. All leaves of the tree are basic nodes. For instance the basic node GoalFree returns a higher value if there are less 'obstacles' between itself and the opponents' goal. Besides basic value nodes there are also combined value nodes. These nodes combine values from other nodes into a new value. For instance the combined value node LookForBall combines the values of BallMissing and PlayerPosition: if the value for PlayerPosition is low (this means the player is close to its standard field position) and the value for BallMissing is high (which means it has been a long time since the player last saw the ball) the value for LookForBall will be high.

Results of value-nodes can also be the inputs for conditional nodes, of which there are two types. A mutex node takes multiple values as input and returns the highest value as output. A switch node takes three value-nodes as input and returns either the first or the third value, based on whether the value of the second node is below some threshold.

Everytime the agent can act, the tree is evaluated. The value nodes in the second column are each related to one high level action. The root node will return the value of one of those value nodes, and the high level action corresponding to this value node will be executed. This high level action (for instance looking for the ball) has to be translated to low level actions (e.g. multiple turn commands) which are then sent to the body of the player to be executed.

An important question is how to weigh all the values in combined value nodes. For instance: is the negative influence of PlayerPosition larger than the positive influence of BallMissing on the combined value node LookForBall? To solve this problem all input links to combined value nodes are given a weight (either positive or negative), which makes this a parameter optimization problem. This is solved by implementing a genetic algorithm, with the weight vector as elements of the population, and the result of a RoboCup simulator game with a team of players using these weights in their network as the fitness function. More information on the subject of high level skills can be found in [14].

A.6 Timing the Actions

One of the most difficult tasks for the agent was making sure it was working with up-to-date information. Because all the actions of the agent depend on the correct estimation of the agents own position and other field objects position, it is important to have the agents the world information correct and up-to-date at the time a new action is chosen.

Because of the non continuous way the see-information is provided to the agent, a possibility of delay between the actions performed and the related visual information exists. This delay can originate from visual information which originated during the agent's action and was not interpreted as such.

To take care of this problem, the agent now estimates the end-time of every action it performs. Then, if it wants to be sure the information about the world is accurate and up to date, it can wait for visual information that originated in the soccer server after the end-time of its last action.

To make this possible the current simulator step was kept in a clock within the agent which updated itself every 100 milliseconds and synchronized itself with the time given in see- or sensebody-messages. This enabled the agent to know the precise time an action was chosen and the time it was started by sending it to the soccer server and as a consequence to calculate the time the action should be performed.

This solution was preferred above the more obvious one of comparing the position of the agent or other field objects — such as the ball — to the target position of the action. The choice was based on the fact that actions cannot be guaranteed to succeed so an agent moving to a field location (x, y) could wait indefinitely for its action to end if, for instance, another player was standing in its way.

References

1. H. Blockeel and L. De Raedt. Lookahead and discretization in ILP. In *Proceedings of the 7th International Workshop on Inductive Logic Programming*, volume 1297 of *Lecture Notes in Artificial Intelligence*, pages 77–85. Springer-Verlag, 1997.
2. F. Brazier, B. Dunin-Keplicz, N. R. Jennings, and J. Treur. Desire: Modelling multi-agent systems in a compositional formal framework. *International Journal of Cooperative Information Systems*, 6:67–94, 1997. Special Issue on Formal Methods in Cooperative, Information Systems.
3. W. Cohen. Recovering Software Specifications with ILP. In *Proceedings of the 12th National Conference on Artificial Intelligence (AAAI-94)*, pages 142–148, 1994.
4. N. Cossement. Robocup: developing low level skills. Master's thesis, Department of Computer Science, Katholieke Universiteit Leuven, 1998.
5. S. Craw and D. Sleeman. Knowledge-based refinement of knowledge based systems. Technical Report 95/2, The Robert Gordon University, Aberdeen, UK, 1995.
6. L. De Raedt. *Interactive Theory Revision: an Inductive Logic Programming Approach.* Academic Press, 1992.
7. L. De Raedt. Logical settings for concept learning. *Artificial Intelligence*, 95:187–201, 1997.
8. L. De Raedt. Using ILP for verification, validation and testing of knowledge based systems, 1997. invited talk at EUROVAV 1997.
9. L. De Raedt and L. Dehaspe. Clausal discovery. *Machine Learning*, 26:99–146, 1997.
10. L. De Raedt and S. Džeroski. First order jk-clausal theories are PAC-learnable. *Artificial Intelligence*, 70:375–392, 1994.
11. L. De Raedt, G. Sablon, and M. Bruynooghe. Using interactive concept learning for knowledge-base validation and verification. In *Validation, Verification and Test of Knowledge-based Systems*, pages 177–190, 1991.
12. H. Kitano, M. Veloso, H. Matsubara, M. Tambe, S. Coradeschi, I. Noda, P. Stone, E. Osawa, and M. Asada. The robocup synthetic agent challenge 97. In *Proceedings of the 15th International Joint Conference on Artificial Intelligence*, pages 24–29. Morgan Kaufmann, 1997.
13. N. Lamb and A. Peerce. Verification of multi-agent knowledge-based systems. In *Proceedings of the ECAI-96 Workshop on Validation, Verification and Refinement of Knowledge Based Systems*, 1996.
14. P. Monsieurs. Developing high level skills for robocup. Master's thesis, Department of Computer Science, Katholieke Universiteit Leuven, 1998.
15. S. Muggleton and C. D. Page. A learnability model for universal representations. In S. Wrobel, editor, *Proceedings of the 4th International Workshop on Inductive Logic Programming*, pages 139–160, Sankt Augustin, Germany, 1994. GMD.
16. I. Noda. Libsclient (for c language). URL: http://ci.etl.go.jp/~noda/soccer/client/index.html.
17. J. Postel. RFC 768: User datagram protocol, 1980.
18. Leon Sterling and Ehud Shapiro. *The art of Prolog.* The MIT Press, 1986.
19. J. Vanthienen, C. Mues, and C. Wets. Inter-tabular verification in an interactive environment. In *Proceedings of the '97 European Symposium on the Validation and Verification of Knowledge Based Systems (EUROVAV-97)*, pages 155–165, 1997.

Layered and Resource-Adapting Agents in the RoboCup Simulation

Christoph G. Jung*

GK Kogwiss. & MAS Group, FB Inform., Univ. des Saarlandes & DFKI GmbH
Im Stadtwald, D-66123 Saarbrücken, Germany
jung@dfki.de

Abstract. Layered agent architectures are particularly successful in implementing a broad spectrum of (sub-)cognitive abilities, such as reactive feedback, deliberative problem solving, and social coordination. They can be seen as special instances of *boundedly rational* systems, i.e., systems that trade off the quality of a decision versus the cost of invested resources. For sophisticated domains, such as the soccer simulation of RoboCup, we argue that a generalised framework that combines a layered design with explicit, resource-adapting mechanisms is reasonable. Based on the InteRRaP model, we describe a prototypical setting that is to guide and to evaluate the development of reasoning about *abstract resources*. These are representations of general interdependencies between computational processes. The realised soccer team, CosmOz Saarbrücken, participated successfully in the RoboCup-98 competition and confirmed that abstract resources are an appropriate modelling device in layered and resource-adapting agents.

1 Introduction

In the nineties, the complementary AI paradigms of deliberative, perfect rationality and of reactive, myopic emergence have found their reconciliation in the more and more prominent principle of *bounded rationality* [20]. Boundedly rational systems trade off the quality of a solution versus the cost of invested computation and interaction. On the one hand, this implies turning away from purely complex[1] decision making into a more tractable, thus situated form of intelligence. On the other hand, bounded rationality still demands optimality with respect to given domain constraints.

We adopt the term *resource* to describe these mostly quantitative constraints that are imposed onto the agent either by its environment (external resources, such as tools, fuel, workspace, etc.) or by its own computation device (internal resources, such as time and memory). Along Zilberstein [22], there are three options to realise bounded rationality based on this notion. *Resource-adapted* systems, e.g., [18], are built with

* supported by a grant from the "Deutsche Forschungsgemeinschaft" (DFG).

[1] We speak of *complex* decisions as long-term intentions that are composed of several primitive system operations in order to produce an optimal answer. Generally, the corresponding decision procedures turn out to be complex, too: Their computational needs increase exponentially with problem size. Complementary, *simple* decisions denote primitive measures that maximise the system's performance just for a single step in time. Often, they can be computed using fairly undemanding, therefore simple procedures.

a pre-designed off-line reflection of the resource characteristics of a specific domain. Secondly, *resource-adaptive* systems are "somehow" able to react on-line to changes in domain-specific resources. However, generic agent models should be applicable to a range of demanding and dynamic environments. They should cope with a great variety of resources. This renders the construction of adapted or adaptive mechanisms highly difficult. Thus, domain-independent agent architectures favour the third option of *resource-adapting* designs, e.g., [19], that incorporate explicit resource representations and reasoning[2].

Hybrid agents, especially the three-*layered* InteRRaP [16] design, are resource-adaptive systems since integrating the different computational expenses and subsequently the different decision qualities of reactive feedback, deliberative problem solving, and even social reasoning. Compared to monolithic agent models incorporating only a single form of inference, hybrid agents provide advantages in domains in which a whole spectrum of (sub-)cognitive abilities is required.

Along with recent developments to define a formal methodology for hybrid systems [4, 12, 10], we have proposed a more elaborate notion of *layering*. Our investigations identify a meta-object relationship in InteRRaP, i.e., the deliberative *Local Planning Layer* (LPL) monitors and configures the computations inside the reactive *Behaviour-Based Layer* (BBL). Similarly, the *Social Planning Layer* (SPL) negotiates about LPL-goals and LPL-intentions, commits to change them, and adjusts the LPL accordingly.

This "layering as meta-reasoning" perspective is very close to the resource-adapting framework of Russell & Wefald [19] and leads to a generalised InteRRaP architecture [7, 8] in which resources of a lower layer are explicitly represented and reasoned about by its upper companion. The proposed representation is called *abstract resource* and denotes both internal, computational as well as external, environmental interdependencies between computational processes.

To guide the refinement of such boundedly rational models, we regard the definition of challenging domains, such as the soccer simulation (Figure 1) of the RoboCup initiative [13], to deliver an appropriate experimental and empirical basis. Simulated robot soccer offers a controllably continuous, dynamic, inaccessible, and noisy environment. It provides for a concise success criterion and introduces a competitive, realistic background for cross-evaluation.

The explicit management of internal as well as external resources is an important topic in RoboCup. As exemplified in Figure 1, a wide range of rapidly — during the fraction of a second — and unpredictably changing situations occur. Therefore, an immediate and flexible motivational shift of the computational engine is essential, e.g., for implementing a sudden retreat if your team instantly looses possession of the ball. Furthermore, the simulated physics of the soccer agents is also imposed some far reaching constraints. Especially the recently revised model of *stamina* severely penalises constant, exhaustive dashing.

---

[2] Russell & Subramanian argue that any such reasoning is itself subject of resource consumption and thus prevents optimality [18]. We rather define the task of a generic, resource-adapting system to *approximate* optimality — as mentioned in [19], experiences of explicit resource reasoning can be compiled into simpler, implicit control.

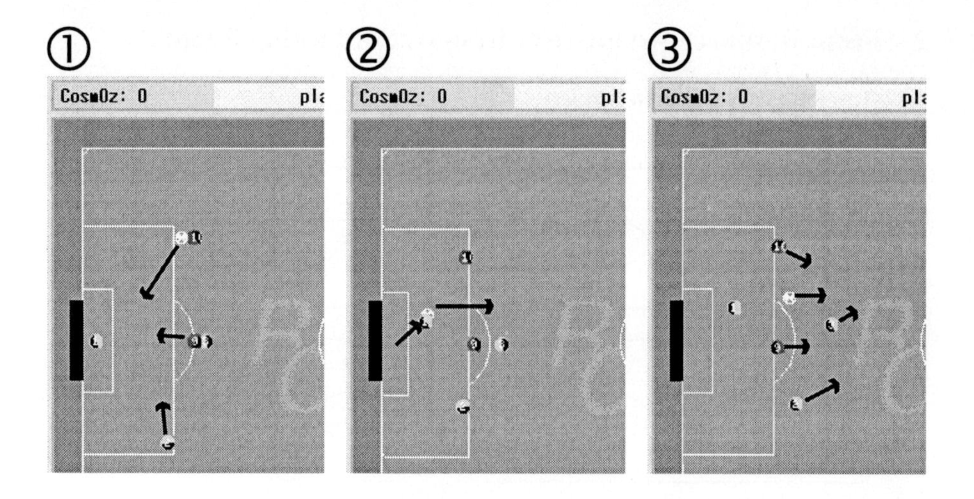

Fig. 1. The Rapid Shift of Motivations in the RoboCup Simulation

We also recognise the aspect of layering inside soccer agents. Positioning, orientation, ball handling, aiming, and tracking are reasonably defined as reactive patterns of behaviour. The strategic composition of "moves" (defending in your own half, dribbling in the middle-field, attacking from the right flank) can be assigned to deliberative planning. Hereby, each move can be realised by configuring the reactive subsystem accordingly, in particular by influencing the resource allocation to the behaviour patterns. Finally, coherent team play is the ultimate requirement to score goals and to prevent your opponent from doing so. Social reasoning including negotiation has to coordinate the deliberative planning in order to obtain combined moves (double-pass, offside-trap), assign roles (attacker, defender, goalie), and implement tactics (offensive, defensive).

Contribution. To gain experience with the generalised InteRRaP model (Section 2), in particular with abstract resources and respective decision algorithms, the present paper specifies prototypical layered and resource-adapting agents (Section 3) for the demanding RoboCup simulation. Our team, CosmOz Saarbrücken, has successfully participated the RoboCup-98 competition (quarter final; 9th rank) and will guide future research on the complete architecture (Section 4).

Opposed to, e.g., Burkhard et al. [3], our aim is to study mainly domain-independent mechanisms for intelligent real-time systems which addresses the RoboCup *Synthetic Teamwork Challenge* [14]. Abstract resources provide advantages with respect to both typical control-of-search and typical control-of-behaviour techniques. Russell & Wefald [19], for example, focus on a single and *sequential* time resource. Controlling real-time systems, but, requires a more flexible treatment of general interdependencies between *concurrent* processes. Using *a priori* estimations in conflict resolution, our approach is able to prevent the redundant computations found in *a posteriori* arbitration used by, e.g., Riekkie & Röning [17]. Furthermore, our meta-control is smoothly integrated with three important styles of inference (reactive, deliberative, social) only found separately in current RoboCup systems.

2 From Resource-Adaptive to Resource-Adapting Agents

2.1 InteRRaP: A Layered Model

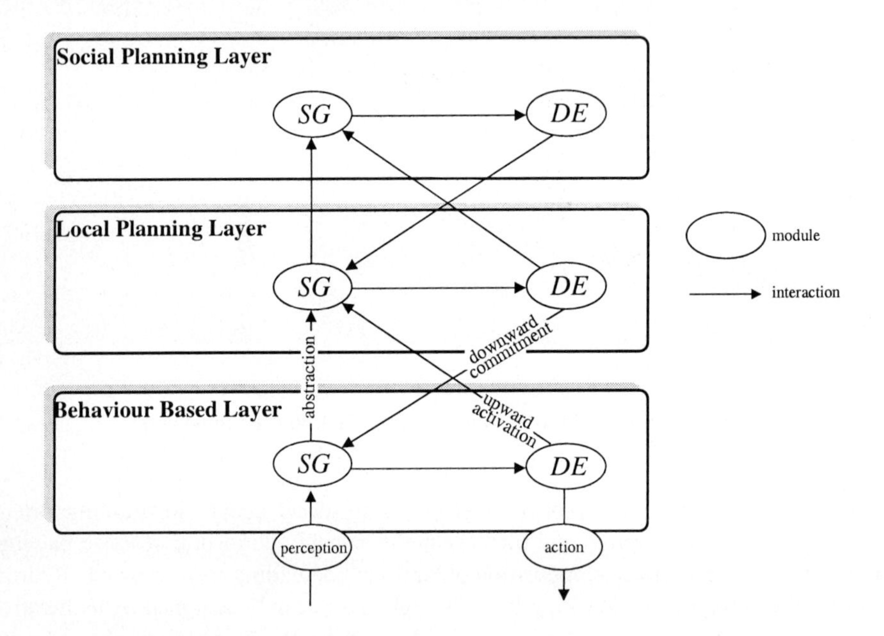

Fig. 2. The Layered InteRRaP Architecture

Hybrid architectures, such as InteRRaP [16] (Figure 2), integrate the functionality of separate *modules* by determining their interactions in a rather pragmatic manner. InteRRaP models the smooth transition from sub-symbolic reactivity to symbolic deliberation and even social capabilities by realising three different *layers*. Each layer internally follows a common flow of control: the *Situation Recognition and Goal Activation* module (SG) spawns new goals out of the maintenance of belief. The *Decision Making and Execution* module (DE) decides about how to meet these goals and thus obtains plans or intentions to execute.

The most concrete *Behaviour-Based Layer* (BBL) provides a short feedback loop with the environment by applying procedural routines, so-called *patterns of behaviour*. BBL decision making is quite fast and simple, because behaviour patterns are reactively triggered (*reflex*) by recognised situations. Stacked on top, the *Local Planning Layer* (LPL) reasons (plans) about how to meet long-term, abstract goals. Similarly, social decisions (at the *Social Planning Layer* — SPL) that involve negotiating and coordinating with other agents are also expressed as a planning problem.

These informal architectural considerations have been formalised by means of a detailed formal specification [12, 10] that bridges the gap to verifiable implementations and also to theorical issues. Our *computational model* of InteRRaP applies the principle of fine-grained concurrency between (sub-)cognitive *processes* located **inside** the

SG and DE modules, such as perception, reflexes, patterns of behaviour, or planning. Processes encapsulate logical inferences and compute continuously in an independent manner. Communication between processes happens (explicitly) via *signals* and (implicitly) via logical data structures in a *shared memory*. The Oz language [21,9] turns out to be a highly suitable implementation platform for both the inference engines and the process model, because Oz combines a logical background with modern programming features, such as concurrency, object-orientation, and transparent distribution[3].

The formalisation of a computational model also requires the conscientious reinvestigation of layering. Originally ([16]), layers interact via upward activation, e.g., a behaviour pattern "calls" the planner as a subroutine, and downward commitment, e.g., the planner activates additional patterns of behaviour (see Figure 2). Thus each layer represents an optional, possibly more useful, but expensive path of computation. Decisions of any layer have basically the same status and are arbitrated in between. This resembles many other layered designs, such as the Subsumption architecture [2].

In the extended specification [12, 10], the layer itself is realised as a designated *control process* that supervises the communication of encapsulated (sub-)cognitive processes, such as the BBL desires, reflexes, and patterns of behaviour. The inter-layer interaction is now installed as a special form of meta-object relationship. Hereby, a lower layer, such as the BBL, does indeed implement **all** the functionality of the agent. To be guided towards exhibiting a specific rational function, it is monitored, reasoned about, and reconfigured by its super-layer by means of the control process, e.g., for approaching objects, the LPL demands to suppress an avoid-collision reflex.

The BBL is thus no more subsumed, but supported by its super-layer LPL. Layers do no more stand in competition, but in a structured, cooperative relation with their superlayers. This perspective decouples the higher-level reasoning from the critical timing constraints of dynamic environments, especially the social reasoning in the SPL has the state of the LPL (current goals, future intentions) as its topic and its commitments non-monotonically affect the goals (adding or removing goals) and intentions (adopt or drop intentions) in the LPL.

2.2 A Layered and Resource-Adapting Model

By the different computational needs of its layers, InteRRaP so far represents an instance of resource-adaptive rationality. Changing resources in the environment and in the computation device influence the quality of actually executed decisions: If the environment becomes more calm, it is more likely that the deliberative module can timely influence the fast decisions of the reactive module. If the environment becomes more dynamic, the reactive module will constantly act without the planner being able to intervene. This adaption is implicitly encoded into the model.

Demanding environments are full of changing constraints and require to make such relevant design-time conventions rather a part of the run-time decisions of the agent. Russell & Wefald [19], for example, develop a resource-adapting architecture where

[3] In CosmOz, the distribution facilities are used to neatly manage a computer pool for running agents in parallel. In concordance with the official simulation rules, inter-player communication uses the soccer server as the only medium.

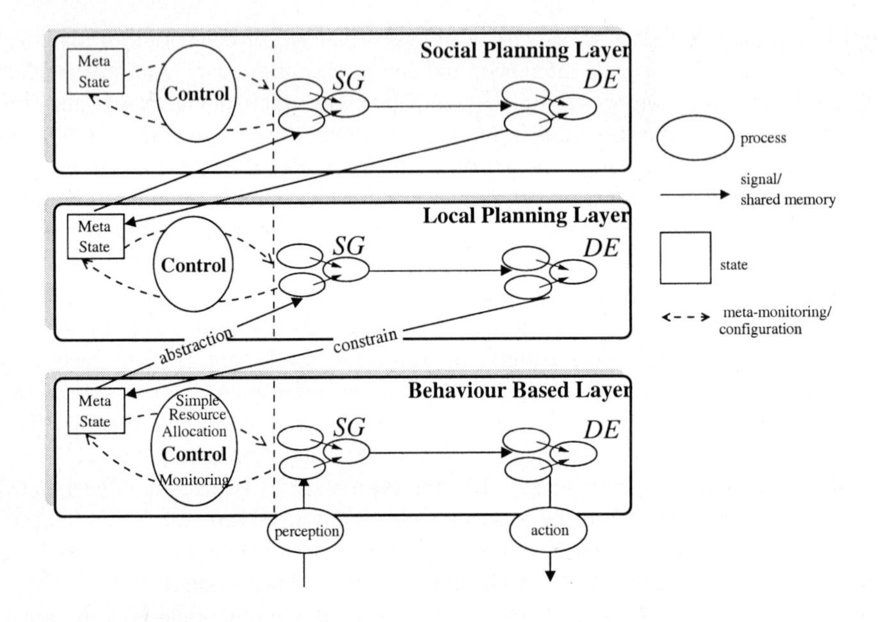

Fig. 3. Extending InteRRaP to a Resource-Adapting Scheme

complex object-level reasoning about external resources is guided by simple meta-level allocation of a single, internal resource (time). They report its successful application to non-trivial game-playing. In [7, 8], we have argued that this architecture is not immediately applicable to real-world, multi-agent domains such as RoboCup. One reason is the complexity of the single object-level. Another problem relates to the short-sight of the meta-level caused by the inherent object-level dependencies which go beyond the pure consumption of time. Generally, a clear separation of internal and external resources is not possible since they are substitutable, i.e., by withdrawing computation time from a particular decision process, taking actions can be prevented.

Consequently, a generalisation of both the layered InteRRaP model and the design of Russell & Wefald [19] into a multi-staged resource management is envisaged (Figure 3). Each layer hereby realises a complete meta-functionality by guiding (constraining) the resource allocation to the computations on its subordinate layer. LPL and SPL planning is thus established as a form of explicit and complex reasoning about resources. To uphold tractability, the control processes of any layer (see Section 2.1) are additionally equipped with a simple and situation-oriented mechanism to refine higher-level guidelines into a concrete resource assignment to the supervised processes. The control process and its simple resource allocation are thus already a part of the meta-interface; the complex higher-level reasoning can focus on rather abstract, long-term conflict resolution. In both forms of (resource) decision making, the representation device of *abstract resources* is employed. It denotes general interdependencies between (or constraints on) the supervised processes, be they of internal or external nature.

3 A Prototypical Soccer Agent

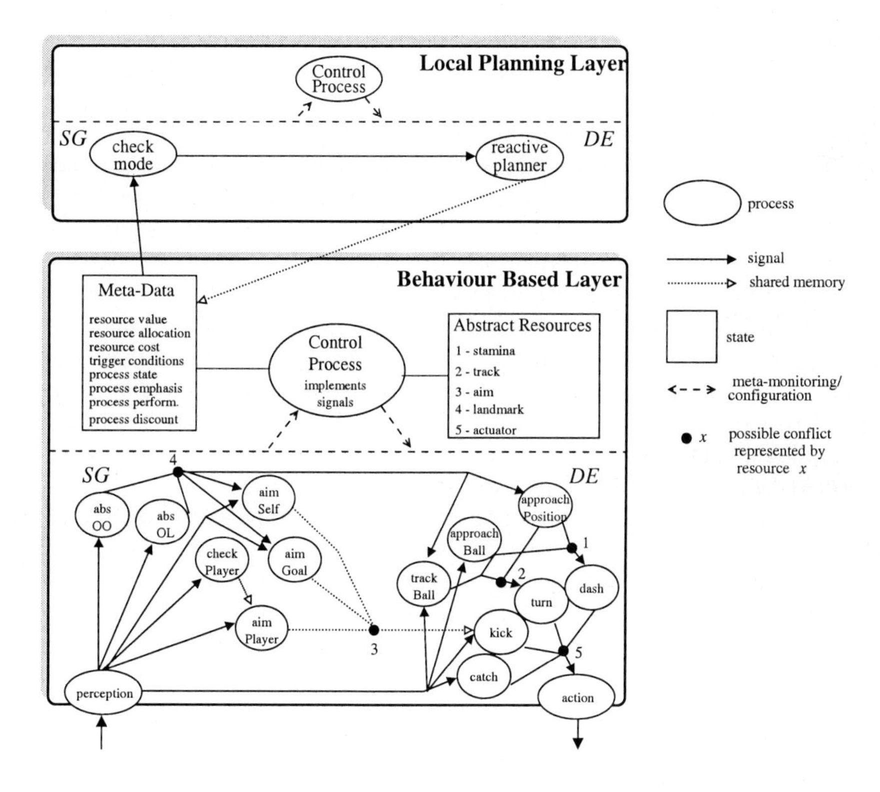

Fig. 4. The Lean Model of Resource-Adapting InteRRaP Agents in the RoboCup

Making the ideas of the preceding Section 2.2 more concrete is not trivial since building on an already matured agent framework which incorporates many design decisions. Therefore, we first step back to a leaner agent model consisting of BBL and simplified LPL instantiated to the already motivated RoboCup simulation (Figure 4). This model is used to gain experience with the important representational and algorithmic issues of abstract resources. Our detailed presentation in the following sections focuses on the interface between deliberative and reactive facilities, thus the control process of the BBL and its parameterisation by LPL plan operators. We discuss our presumptions, such as synchronous concurrency, process-built-in self-evaluation, and discrete resource values which are statically, i.e., in fixed amounts, assigned to active processes. As Section 4 concludes, the lean model has proven successful in RoboCup-98 and appears to be incrementally extendible to the whole layered and resource-adapting design.

3.1 Reactive, Sub-Cognitive Processes

The computational model of InteRRaP [12, 10] extends the ideas of reactive control systems [2, 15, 5] to a dynamic network of concurrent computation spread over the whole layered agent. Concurrency in the model hereby ensures the responsiveness[4] of any inference within the agent. This is of course particularly important for the reactive BBL processes of RoboCup agents depicted in Figure 4.

The `perception` process senses perceptual data from a UDP datagram connection to the simulation server. Its activity is further distributed by outgoing signals triggering other processes, such as `catch` or `kick`. Signals are emitted upon specific trigger conditions in a process state (see Section 3.2), e.g., if `perception` indicates that the ball appears to be near the player, `kick` will be invoked. Because `kick` additionally uses defaults for determining where and how hard to kick, this signal path thus implements a highly reactive reflex which does not involve much computation.

The direction and power defaults inside `kick` are accessible via shared memory to the `aimGoal`, `aimSelf`, and `aimPlayer` processes in order to allow more sophisticated goal-kicks, dribbling, and passes. These computations determine their decisions both from the relative data in `perception` and from other preprocessing steps, such as the derivation of absolute coordinates via triangulation (`absOO`: use two landmarks, `absOL`: use a landmark and an adjacent line) and such as checking team mates with respect to the usefulness of their position (`checkPlayer`).

Navigation is encoded into the `approachPosition`, the `approachBall`, and the `trackBall` patterns of behaviour. They position and orientate the player with respect to landmarks and to the ball by triggering *virtual* `turn` or `dash` actions. Approaching and tracking mainly depend on relative perception, but additionally require absolute data in the case that the envisaged objects are not visible. *Virtual actions* (including the aforementioned `kick` and `catch`) signal activity to the `action` process that finally sends primitive, external actions (*commands*) over the datagram connection to the simulation server.

Indeed, this reactive network already incorporates the whole functionality of soccer agents, e.g., the complete motivational basis for all different situations and for all different player roles in Figure 1, **at the same time**. This eventually raises conflicts because processes are unaware of their side-effects. For example, simultaneously haunting the ball and keeping a certain position results in a "paranoid" floundering of the soccer agent. This is a conflict with external grounding in the simulated "body" of the player. Similarly, commands could fail due to such restrictions (`catch` is useless for non-goalies; limited number of commands per cycle; exhausted stamina). Conflicts also have internal grounding, such as redundancy (`absOO` and `absOL`) or mutual overwriting of output (the `aim` processes). In Figure 4, we have marked five such sources of interdependencies between the reactive processes in our soccer agents.

---

[4] This is not to say that other models could not exhibit interactivity. To obtain a similar degree of responsiveness from a sequential model, but, the programmer has to put additional interaction and scheduling knowledge into the domain-dependent part of the agent. A good model, however, already integrates those facilities required in most domains, thus eases the programmer's task. Therefore, we regard both *Turing Machines* as well as models with hidden concurrency as bad agent models.

3.2 The Control Process, Synchronous Concurrency, and Abstract Resources

The BBL control process (Figure 4) is responsible for a reasonable mid-term interaction of the supervised behaviour patterns. The control process influences their computation by implementing their communication. This amounts to a synchronous form of concurrency, because each reactive process is sliced into subsequent computation *chunks* of flexible, but limited size. For example, such a chunk once takes aim (in `aimGoal`) or once produces a limited trajectory by firing a number of turn and dash actions (in `approachPosition`).

To trigger and determine the next, ready-to-run chunk of a process, the control process generates *signals* in each of its cycles. This is guided by *trigger conditions* that have to be satisfied by the state of the signal-emitting process, e.g., the relative position to the opponent's goal has to change in `perception` to trigger `aimGoal` anew.

In a second step, the set of generated signals is filtered to obtain an approximately optimal set of chunks, i.e., to minimise the conflicts between active computations. Therefore, we introduce the discrete, quantitative representation of *abstract resources* for each of these interdependencies: Natural numbers indicate the available amount of each resource. The dependent processes "consume" this amount by each computation chunk according to its length and its type. Once a resource is exhausted, signals to the "applying" processes are suppressed. In each cycle, resources recreate according to a recreation function.

Interestingly, the interdependencies in Figure 4 fit very well into this generic scheme. For example, the limited `landmark` resource (range $\{0, 1\}$) restricts the concurrent and redundant operation of the two optional methods for obtaining absolute coordinates. Both `absOO` and `absOL` apply and only one is able to get a hold on it. The `aim` resource (range $\{0, 1\}$) introduces a similar restriction with respect to `aimPlayer`, `aimSelf`, and `aimGoal`. So to speak, the agent has only a limited aiming capacity[5].

`landmark` and `aim` denote purely internal interactions inherent in the computational design. We do also find abstract resources with combined internal and external grounding. For example, the `stamina` resource (range $\{0, \ldots, 2000\}$[6]) matches the external model of the soccer agents' power and also represents the internal dependency between the possibly conflicting positioning behaviours. Thus `approachPosition` and `approachBall` apply for it. By assigning larger portions of `stamina`, a respective chunk is able to implement a longer trajectory.

The `actuator` resource (range $\{0, 1\}$) describes the limited amount of commands allowed in each simulation cycle: the virtual actions `kick`, `catch`, `turn`, and `dash` access it. Finally, the `track` resource (range $\{0, 1\}$) mediates between the different needs in orientation of `approachPosition` and `trackBall`.

---

[5] As discussed in [7], abstract resources provide an attractive device for cognitive science as well. They can be used to emulate cognitive restrictions of the human mind, such as limited short-term memory, focus of attention, etc. Indeed, cognitive architectures, such as ACT-R [1], also take a bounded rationality perspective towards such constraints.

[6] The range depends on the relation of inner-agent and simulator scale. The stamina resource is not purely traced internally, because RoboCup allows for "mental" actions that deliver the exact, external value. We have omitted this fact for reasons of simplicity.

Performance Monitoring and Simple Resource Allocation. We now develop the control process as a simple resource allocation procedure which optimises the choice of admissible signals, thus maximises the expected utility of a particular allocation choice. However, for such a short-term decision, sporadic external performance measures, such as the score in soccer, are not suitable. We therefore propose a form of internal profiling which is surprisingly applicable to many difficult cases: supervised processes incorporate self-evaluation functions. During each chunk, they produce a performance report which is then used as a prediction for the allocation decision.

An example is the `aimGoal` process: In order to adjust the `kick` defaults, this process has to analyse distance and direction to the opponent's goal, anyway. The greater the distance, the less likely the next kick will score a goal and the less useful the current chunk of aiming has been. This way, straightforward performance mappings (also of `aimPlayer` and `aimSelf` chunks; similarly for chunks of all other processes) onto a normalised scale ($[0, 1]$) can be found.

If we additionally assume the possible allocations of abstract resources to processes to be static, the simple cycle of the control process can now be formalised[7] as in Figure 5: After having updated the values for resources with a recreation function (1.), we obtain a set of signals using the trigger conditions (2.). This set is now sorted according to the utility of the destination processes (3.). Hereby, the utility of running a process depends on its expected performance reported by its last chunk minus the cost of the resource allocation. Here, we use independent cost functions for each resource. As long as the utility is greater than zero (4.), the static allocations are tried to be granted to the processes in order. If possible, resource values are updated (4.(a)) and the respective signals are transmitted (4.(b)). Otherwise, if any related resource would get exhausted, the selected set of signals is deleted (6.). A particular problem is posed by inactive processes not being able to adjust their evaluation to the current setting. Therefore, a discount of frequently selected "winners" (3.;4.(c);5.) improves the chance of steadily probing also the supposedly bad candidates. Furthermore, we introduce for each process an emphasis parameter.

3.3 Deliberative Influence: Plan Operators

In the RoboCup simulation, the BBL already delivers a reasonable mid-term behaviour: motivations and decisions are smoothly interpolated and appropriate parameterisations of the control process implement interesting "moves" of the agent: An example for such a move is the right-wing attack of player 10 in Figure 1 which, by the `aim` resource, steadily mediates between shooting to the goal, dribbling, and passing. Mediation hereby depends on an appropriate set of trigger conditions monitoring the distance to the goal, the free space of the agent, and the position of team mates.

Long-term motivations and strategic intentions composed out of particular moves are however not exhibited at the BBL level. This does also exclude projecting the evolution of resources into the future, thus a complex resource optimisation. The `stamina`

[7] In the present paper, we do not want to go into the detailed decision-theoretic assumptions, such as abstract resources describing conflicts in an independent and exhaustive manner. This will be the topic of a separate paper.

processes $\mathcal{P} = \{P_i\}$; resources $\mathcal{R} = \{R_j\}$; signals S; trigger condition $T : 2^P \times P \to S$;
resource value $V : R \to \mathbf{N}_0$; expected process performance $\pi : P \to [0, 1]$;
allocation $A : P \times R \to \mathcal{N}_0$; resource cost $C : R \times N_0 \times [0, 1]$;
recreation function $\rho : R \times \mathbf{N}_0 \to \mathbf{N}_0$; process emphasis $E : P \to [0, 1]$;
discount $D : P \to [0, 1]$; discount factors $\{d_i\}$.

1. recreate resource values $V(R_j) \leftarrow \rho(R_j, V(R_j))$.
2. out of processes $\mathcal{P} = \{P_i\}$ and trigger conditions T, generate signals $\mathcal{S} = \{S_i = T(\mathcal{P}, P_i)\}$.
3. choose a P_i with $\mathcal{S} \ni S_i \neq \{\}$ and highest utility

$$U(P_i) = D(P_i) * E(P_i) * \pi(P_i) - \sum_j C(R_j, A(P_i, R_j))$$

4. if $U(P_i) > 0$ and for each R_j, $V(R_j) - A(P_i, R_j) \geq 0$
   (a) consume resources $V(R_j) \leftarrow V(R_j) - A(P_i, R_j)$.
   (b) start P_i with signals S_i and allocation $A(P_i)$.
   (c) discount $D(P_i) \leftarrow \frac{D(P_i)}{d_i}$.
5. else reset discount $D(P_i) \leftarrow 1$.
6. delete signals $\mathcal{S} \leftarrow \mathcal{S} \setminus \{S_i\}$.
7. if $\mathcal{S} \neq \{\}$ then goto step 3
8. else goto step 1.

Fig. 5. Cycle of the Control Process

resource, for example, desperately needs such a projection because it does not allow freely dashing across the field. The required reasoning is appropriately described as a planning task of the LPL. Primitive moves are the *means* to be concatenated into plans that probably install the *ends*, such as to score a goal or to prevent your opponent from doing so. Our setting implies a form of decision-theoretic planning, because the planner additionally has to minimise the (resource) costs of the plan. Yet, these considerations are not in the focus of the lean model in Figure 4 concentrating on the interface between plan operators and the control process of the BBL.

At the moment, a simple situation recognition, `checkMode`, is tied to the state of the BBL control process and in turn triggers the `reactivePlanner`. `checkMode` infers the agent's role depending on the shirt number, derives the possessor of the ball, and determines a coarse play mode. The planner thereafter activates a particular corresponding move, such as what has to be done for a "kickoff", for "defending with the ball in own half", for "passing the ball into the middle field", or for "attacking from the right wing". Section 4 comes back to strategic, goal-oriented planning upon these moves.

As already anticipated, moves are implemented by appropriate configurations of the BBL control process. The configurable meta-data (Figure 4) comprises the trigger conditions of signals, the possible allocation of resources, emphasis of processes, and discount factors.

Let us discuss the impact of these parameters at hand of the defensive behaviour of the goal keeper (shirt number 1) in Figure 1 which is invoked once the opponent possesses the ball (one of the opponent players is recognised to have the least distance to it) and once the ball is residing within the goalie's half of the field. Herein, the emphasis parameters are set to prioritise catching before kicking before turning before dashing. This allows the goalie to quickly intercept the movement of the ball (the installed trigger condition of `catch` monitors the ball to fall short off a velocity-dependent distance), afterwards shooting it away, possibly to a team mate nearby. Catching and kicking are vital; their discount is chosen to be small ("no experiments in the emergency case"). Since `aimGoal` and `aimSelf` do not make sense for the goalie in this mode, they are completely deemphasised ($E(P_i) = 0$).

Besides, it is important for the goalie to frequently adjust its central position while tracking the ball (`approachPosition` triggers on leaving the goal). Hereby, the `stamina` resource is only available in minimal amounts while the sudden interception of `approachBall` triggered upon the ball entering the penalty area should be a continuous trajectory, thus a greater amount of `stamina` is granted if activated.

4 Conclusion and Outlook

Although we have yet implemented a lean version of the generalised framework for hybrid, resource-adapting systems, our CosmOz team already went into the quarter finals of the RoboCup-98 competition. Its performance demonstrated reactive, rational, and even implicitly cooperative behaviour. This supports our claim that the connection between layered and resource-adapting mechanisms can be made and that the representation of abstract resources applies well. The current prototype will be used to empirically study the influence of the resource-adapting mechanisms and its parameterisation onto the efficiency of a soccer team. Besides the close investigation of the decision-theoretic assumptions within our model (partially discussed in [7]), open issues for the future are:

Reactive Processes. The modularisation of reactive soccer facilities into processes has to be improved. For example, `aimPlayer` and `aimSelf` were only poorly implemented at RoboCup-98 not taking the trajectory of opponent players into account. Additional processes are needed to leave bad positions for actually receiving a pass or avoiding off-side. We also think of how to pursue particular opponent players in one-to-one defensive strategies. These changes will eventually introduce new conflicts whose management is straightforwardly supported by our generic resource representation.

Resource Allocations and On-line Learning. In the current model, possible allocations of resources and the priority of processes are just guided by LPL operators and otherwise remain static. Our experiences in RoboCup-98 have shown that the overall team behaviour is extremely sensitive to minimal changes in those parameters. Since this furthermore depends on the actual simulator settings, it poses a major design problem. [6] proposes to use local search which frequently probes allocation variants at runtime in order to improve the overall sum of process utilities to a satisficing level.

We also plan to adjust process priorities on the fly by a combination of reinforcement learning and memory-based reasoning.

Long-Term Deliberation and Social Abstract Resources. The goal-oriented, partial-order planner of InteRRaP is described in [11]. Its latest version includes hierarchical planning and on-line facilities in a logical setting. We are keen to integrate this planner into the lean model in order to obtain reasonable strategies out of complex intentions of the agent. We are exploring the role of BBL resources in representing the effects of LPL moves. Furthermore, the LPL is also subject of adaption due to its control process and the SPL. It is not clear up to now, how abstract resources can be used to describe the computational interdependencies within the LPL. We expect that this will change the discrete representation of resources from numbers to sets of items, such `roles` to incarnate, or different motivational `foci` of planning in order to coordinate moves (double pass, off-side trap) or even constrain the behaviour of the whole team (tactics, role assignment, etc.).

Acknowledgements

The author is grateful for the fruitful discussions with his colleagues of the MAS group. In particular, the concept of abstract resources has been developed in cooperation with Christian Gerber. The author would like to thank Jörg Müller for his path-setting thesis on InteRRaP. CosmOz owes much to the wonderful Oz programming language developed at the PS Lab at the Universität des Saarlandes and the Swedish Institute for Computer Science. Finally, thanks to the RoboCup initiative for designing a highly motivating and entertaining domain.

References

1. J. R. Anderson. *Rules of the mind.* Lawrence Erlbaum Associates, Hilldsdale, NJ, 1993.
2. R. A. Brooks. A robust layered control system for a mobile robot. In *IEEE Journal of Robotics and Automation*, volume RA-2 (1), pages 14–23, April 1986.
3. H. D. Burkhard, M. Hannebauer, and J. Wendler. AT humboldt — development, practice and theory. In *RoboCup-97: Robot Soccer World Cup I*, volume 1395 of *Lecture Notes in Artificial Intelligence*, pages 357–372. Springer, 1998.
4. B. Dunin-Keplicz and J. Treur. Compositional formal specification of multi-agent systems. In *Intelligent Agents*, volume 890 of *Lecture Notes in Artificial Intelligence*, pages 102–117. Springer, 1994.
5. R. James Firby. Task networks for controlling continuous processes. In *Proc. of the 2nd International Conference on Artifical Intelligence Planning Systems*, 1994.
6. C. Gerber. An Artificial Agent Society is more than a Collection of "Social" Agents. In *Socially Intelligent Agents - Papers from the 1997 AAAI Fall Symposium.* Technical Report FS-97-02, AAAI, 1997.
7. C. Gerber and C. G. Jung. Towards the bounded optimal agent society. In C. G. Jung, K. Fischer, and S. Schacht, editors, *Distributed Cognitive Systems*, number D-97-8 in DFKI Document, Saarbrücken, 1997. DFKI GmbH.

8. C. Gerber and C. G. Jung. Resource management for boundedly optimal agent societies. In *Proceedings of the ECAI'98 Workshop on Monitoring and Control of Real-Time Intelligent Systems*, pages 23–28, 1998.
9. S. Haridi, P. Van Roy, P. Brand, and C. Schulte. Programming languages for distributed applications. *New Generation Computing*, 1998. To appear.
10. C. G. Jung. On the Role of Computational Models for Specificying Hybrid Agents. In *Cybernetics And Systems'98 — Proceedings of the 14th European Meeting on Cybernetics and System Research*, pages 749–754, Vienna, 1998. Austrian Society for Cybernetic Studies.
11. C. G. Jung. Situated abstraction planning by abductive temporal reasoning. In *Proc. of the 13th European Conference on Artificial Intelligence ECAI'98*, pages 383–387. Wiley, 1998.
12. C. G. Jung and K. Fischer. A Layered Agent Calculus with Concurrent, Continuous Processes. In *Intelligent Agents IV*, volume 1365 of *Lecture Notes in Artificial Intelligence*, pages 245–258. Springer, 1998.
13. H. Kitano, M. Asada, Y. Kuniyoshi, I. Noda, and E. Osawa. Robocup: The robot world cup initiative. In *Proc. of The First International Conference on Autonomous Agent (Agents-97)*, Marina del Ray, 1997. The ACM Press.
14. H. Kitano, M. Tambe, P. Stone, M. Veloso, S. Coradeschi, E. Osawa, H. Matsubara, I. Noda, and M. Asada. The robocup synthetic agent challenge. In *RoboCup-97: Robot Soccer World Cup I*, volume 1395 of *Lecture Notes in Artificial Intelligence*, pages 62–73. Springer, 1998.
15. D.M. Lyons and A.J. Hendricks. A Practical Approach to Integrating Reaction and Deliberation. In *Proceedings of the 1st International Conference on Artifical Intelligence Planning Systems*, 1992.
16. J. P. Müller. *The Design of Intelligent Agents: A Layered Approach*, volume 1177 of *Lecture Notes in Artificial Intelligence*. Springer-Verlag, December 1996.
17. J. Riekki and J. Röning. Playing soccer by modifying and combining primitive reactions. In *RoboCup-97: Robot Soccer World Cup I*, volume 1395 of *Lecture Notes in Artificial Intelligence*, pages 74–87. Springer, 1998.
18. S J. Russell and D. Subramanian. Provably Bounded Optimal Agents. *Journal of Artificial Intelligence Research*, 2, 1995.
19. S J. Russell and E. Wefald. *Do the Right Thing*. MIT Press, Cambridge Mass, 1991.
20. H. A. Simon. *Models of Bounded Rationality*. MIT Press, Cambridge, 1982.
21. G. Smolka. The Oz Programming Model. In Jan van Leeuwen, editor, *Computer Science Today*, Lecture Notes in Computer Science, vol. 1000, pages 324–343. Springer-Verlag, Berlin, 1995.
22. S. Zilberstein. Models of Bounded Rationality. In *AAAI Fall Symposium on Rational Agency*, Cambridge, Massachusetts, November 1995.

A Description-Processing System for Soccer Agents and NIT Stones 98

Nobuhiro Ito, Kouichi Nakagawa, Xiaoyong Du, and Naohiro Ishii

Department of Intelligence and Computer Science, Nagoya Institute of Technology, Gokiso-cho, Showa-ku, Nagoya 466, JAPAN

Abstract. Many conventional object-oriented models suffer some problems in the representation of the multiple objects. A multiple object is an object with multiple aspects, autonomy, and pro-activeness. This paper proposes a new agent model called EAMMO, which applies the agent-oriented paradigm to represent multiple objects. EAMMO consists of three types of agents as follows: (1) an upper-agent describes autonomous objects, (2) a lower-agent describes reflective objects, and (3) an environmental-agent describes the environment including agents. We design a description-processing system for EAMMO to confirm the efficiency of our model. We describe soccer games in EAMMO. A soccer player is a good example for EAMMO, because a soccer player is a kind of multiple objects. We define only basic skills and strategies in the soccer agent. However, we found that soccer agents can corporate in a more complex strategy generated by the system. As the result, we confirm that EAMMO is an efficient model for describing multiple objects.

1 Introduction

It is an important problem in AI how an object collaborates with others to accomplish a goal in a dynamic environment. A good example is soccer games, where a soccer player need to collaborates with other teammates to win a game. Recently, much attention is paid to simulate soccer games on computer. A contest called the Robot Soccer World Cup (RoboCup) has been held two times. It has seen as a new landmark of AI research instead of chess games. In this paper, we pay special attention to such kinds of objects like soccer players. They have the following features:

1. **multiple aspects:** The object has different aspects (roles) in different time, places, and cases. For example, a soccer player has several possible roles like passer, shooter, or dribbler.
2. **autonomy:** The object can decide one of its multiple aspects by itself according to time, places, and cases. For example, a soccer player passes the ball if he keeps the ball and is near to a teammate. However, he shoots the ball if he is near enough to the opponent goal.
3. **pro-activeness:** The object can initial some behaviors by itself in a dynamic environment. For example, a soccer player can change his strategies according to time left.

We call the objects with the above three features "multiple objects" in this paper. Informally,

Multiple object = multiple aspects + autonomy + pro-activeness

To represent multiple aspects of objects, many concepts like views[2], aspects[3], roles[4], deputy[5], are proposed in existing object-oriented models. However, these models can not represent the autonomous and pro-active features of multiple objects. On the other hand, the existing agent models[6, 7, 8, 9] can represent autonomy and pro-activeness, appropriately. However, they omit the needs of multiple aspects of multiple objects.

In this paper, we propose a new agent model, called EAMMO(an Environmental Agent Model for Multiple Objects), to represent multiple objects. Our model consists of three type agents.

1. An upper-agent describes an autonomous object with multiple aspects. It has a status processor to decide its next state, and an action processor to decide actually its operation.
2. A lower-agent describes a reactive object, and has only one action processor to realize reactive behaviors.
3. An environmental-agent describes an autonomous and pro-active object. It manages the environment information which affects all agents in the environment.

We also implement an interpreter for the description language of EAMMO to confirm the effectiveness of our model. Furthermore, we describe soccer players in the language and simulate soccer games on the computer. From experiments,we found that the soccer players can corporate in a more complex strategy generated by the system itself than those strategies we defined initially for the soccer players. Thus, we confirm that EAMMO is an effective model for describing multiple objects (soccer players).

The paper is organized as follows: In Section 2, and Section 3, we introduce some basic concepts and describe the EAMMO and its description language. In Section 4, we explain an implementation of an interpreter for EAMMO. For evaluating the effectiveness of our model, we define three types of soccer players (multiple objects) in the language, and simulate the soccer games in Section 5. We also introduce our team(NIT Stones) of RoboCup-98 in the section. Finally, we conclude the paper in Section 6.

2 Basic Concepts

We discuss basic concepts of our model in this section. The definition of EAMMO is given in the next section.

As we know, an object in conventional object-oriented models (we call it a regular object in this paper) consists of an identifier, a list of attributes, and a set of methods. The attributes show the internal status of a regular object. They are defined as follows.

Definition 1 . (Attribute) An attribute is a pair of attribute_name and attribute_value. That is.

$$\langle\texttt{attribute}\rangle ::= \langle\texttt{attribute\_name}\rangle(\langle\texttt{attribute\_value}\rangle)$$

The list of attributes of an object can be specified as

$$\texttt{Attributes}\{\langle\texttt{attribute}\rangle, \ldots, \langle\texttt{attribute}\rangle\}$$

In the definitions, the words being from an uppercase are reserved words.

Definition 2 . (Identifier) An object always has a unique identifier.

$$\langle\texttt{identifier}\rangle ::= \texttt{Object}(\langle\texttt{object\_name}\rangle)$$

Definition 3 . (Method) A method is a pair of method_name and method_body.

$$\langle\texttt{method}\rangle ::= \langle\texttt{method\_name}\rangle(\langle\texttt{method\_body}\rangle)$$

where $\langle\texttt{method\_body}\rangle$ is a procedure/function written in some languages. The set of methods of an object can be specified as

$$\texttt{Methods}\{\langle\texttt{method}\rangle, \ldots, \langle\texttt{method}\rangle\}$$

By using these definitions, a regular object can be defined as follows.

Definition 4 . (Regular Object) A regular object consists of an identifier, a list of attributes, and a set of methods. That is,

$$\begin{aligned}\langle\texttt{regular\_object}\rangle ::= \; & \texttt{Object}(\langle\texttt{object\_name}\rangle), \\ & [\texttt{Attributes}\{\langle\texttt{attribute}\rangle, \ldots, \langle\texttt{attributes}\rangle\}, \\ & [\texttt{Methods}\{\langle\texttt{method}\rangle, \ldots, \langle\texttt{method}\rangle\}]]\end{aligned}$$

A message is an import concept in communication among regular objects in the object-oriented paradigm. Our agents also communicate each other through messages.

Definition 5 . (Message) A message consists of a content, sender, and receiver of the message.

$$\langle\texttt{message}\rangle ::= \texttt{Message}(\langle\texttt{a\_content\_of\_the\_message}\rangle, \langle\texttt{sender}\rangle, \langle\texttt{receiver}\rangle)$$

where $\langle\texttt{a\_content\_of\_message}\rangle$ is a method name or a keyword for invoking a method.

The multiple objects are defined by adding autonomy and pro-activeness to regular objects.

Our way is to define a status processor and an action processor, for a multiple object. By using these processors, a multiple object is able to decide its behavior autonomously through following two steps.

1. By using the status processor, the object decides one of its multiple aspects by itself.
2. It then decides a method by the action processor.

Definition 6 . (Status Processors) The status processor of an object is a process that modifies the attributes and methods of the object according to the relation with others, the environment, and its current status.

$$\langle\texttt{status\_processor}\rangle ::= \texttt{Status}\{\langle\texttt{relations}\rangle, \langle\texttt{roles}\rangle\}$$

$$\langle\texttt{relations}\rangle ::= (\langle\texttt{relation}\rangle, \ldots, \langle\texttt{relation}\rangle)$$

$$\langle\texttt{relation}\rangle ::= \texttt{Relation}(\langle\texttt{relation\_name}\rangle) \leftarrow \texttt{Object}(\langle\texttt{object\_name}\rangle)$$

$$\langle\texttt{roles}\rangle ::= (\langle\texttt{role}\rangle, \ldots, \langle\texttt{role}\rangle)$$

$$\langle\texttt{role}\rangle ::= \langle\texttt{attribute\_name}\rangle(\langle\texttt{attribute\_value}\rangle) \leftarrow \texttt{Relation}(\langle\texttt{relation\_name}\rangle), \texttt{Own}(\langle\texttt{attributes}\rangle), \texttt{Env}(\langle\texttt{attributes}\rangle)$$

$$\texttt{Own}(\langle\texttt{attributes}\rangle) ::= \texttt{Own}\{\langle\texttt{attribute}\rangle, \ldots, \langle\texttt{attribute}\rangle\}$$

$$\texttt{Env}(\langle\texttt{attributes}\rangle) ::= \texttt{Env}\{\langle\texttt{attribute}\rangle, \ldots, \langle\texttt{attribute}\rangle\}$$

where `Status` shows that the followings are rules to decide its own status. `Relation` shows that the followings are rules to restrict its relations between the object and others. `Own` shows that the attributes are the current own status(attributes). Env shows that the attributes are the current status of the environment. `Own` and Env are represented by a list of attributes.

The arrow in ⟨`relation`⟩ means that if the object in the right-hand is appeared, the relation denoted by ⟨`relation_name`⟩ occurs. Similarly, the arrow in ⟨`role`⟩ means that if the conditions in the right-hand are satisfied the attribute ⟨*attribute_name*⟩ will be assigned a new value ⟨`attribute_name`⟩.

Example 1 . We define a status process for the object "`Ito`".

```
Status { (Relation(offense) ← Object(Nakagawa),
          Relation(offense) ← Object(Du),
          Relation(defense) ← Object(Kawai),
          Relation(defense) ← Object(Matsui),
                  :              :           :       ),
         (role(passer)←Relation(offense),
                      Own(ball_keeping(true)),
                      Env(forward_sight(no_player_of_opponent_team)),
              :      :             :                          ),
       }
```

It means that "`Ito`" is in the relation "`offense`" if the object "`Nakagawa`" or "`Du`"(who is the member of own team) is detected by himself. He is in relation "defense" if the object "Kawai" or "Matsui" (who is the member of opponent team)is detected. If "Ito" is in the relation "offense", and he is keeping the ball and no player of the opponent team is in his forward sight, then he will change his attribute "role" into "passer".

Definition 7. (Action Processor) The action processor of an object is a process that select a proper method for the object according to decided status by the status processor, the environment, and the message that the object received.

$$\langle \texttt{action\_processor} \rangle ::= \texttt{Actions}\{\langle \texttt{action} \rangle, \ldots, \langle \texttt{action} \rangle\}$$
$$\langle \texttt{action} \rangle ::= (\langle \texttt{method\_name} \rangle \leftarrow \texttt{Own}(\langle \texttt{attributes} \rangle), \texttt{Env}(\langle \texttt{attributes} \rangle), \texttt{Msg}(\langle \texttt{a\_content\_of\_the\_message} \rangle))$$

where Msg shows that the following ⟨a_content_of_the_message⟩ is a message defined in Definition 5 . When the conditions in the right-hand of the arrow are satisfied, the method ⟨*method_name*⟩ is invoked.

Example 2 . One possible action processor for the object "Ito" is

$$\begin{array}{l} \texttt{Actions}\{\texttt{pass} \leftarrow \texttt{Own}(\texttt{role}(\texttt{passer})) \\ \qquad \texttt{Env}(\texttt{distance\_to\_the\_goal}(50)) \\ \qquad \texttt{Msg}(\texttt{pass}), \\ \qquad \vdots \quad \vdots \qquad \vdots \\ \} \end{array}$$

It means that "Ito" will kick the ball to the goal if he has a role of passer and he is near to the goal, and he gets a message including pass in the content of the message.

3 EAMMO

EAMMO(an Environmental Agent Model for Multiple Objects) is a multi-agent model that we proposed for representing multiple objects.

By using a status processor and an action processor, we extend a regular object with autonomy and pro-activeness. These processors require "the environment" in their descriptions. In other words, representing a multiple object requires an environment. The environment has pro-activeness (like time) and autonomy (like up/down temperature). So, we can regard the environment as one kind of agents. In real-world, entities have different powers. Some behave complicatedly such as human beings, and others behave simply such as tools. Hence, when we use agents to model these entities, different types of agents with different powers are necessary. For example, the agents for tools do not always have autonomous and pro-activeness features, and the agents for men, to be opposite, need more power. Therefore, we classify agents in our model

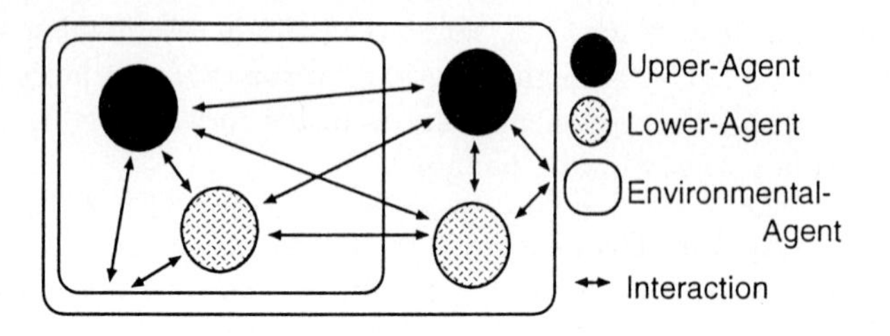

Fig. 1. Relationship among Three Types of Agents

to upper-agents who behave autonomously and pro-actively, and lower-agents who behave reactively, as well as environmental-agents who send messages to the other agents pro-actively. Figure.1 shows the relationship among these three types of agents. In Fig. 1, the environmental-agents include upper-agents and lower-agents. These three types of agents interact with each other and all of them are used to represent multiple objects.

3.1 Lower-Agents

Definition 8 Lower-Agent.
A lower-agent consists of an identifier, a list of attributes, a set of methods, and an action processor.

$$\langle \texttt{lower-agent} \rangle ::= \texttt{Agent}(\langle \texttt{agent\_name} \rangle), \\ [\texttt{Attributes}\{\langle \texttt{attribute} \rangle, \ldots, \langle \texttt{attributes} \rangle\}, \\ [\texttt{Methods}\{\langle \texttt{method} \rangle, \ldots, \langle \texttt{method} \rangle\}, \\ [\texttt{Actions}\{\langle \texttt{action} \rangle, \ldots, \langle \texttt{action} \rangle\}]]]$$

This definition shows that a lower-agent consists of a regular object and an action processor. The concept of the lower-agent is shown in Fig. 2. The lower-agent only behaves with messages from other agents, reactively. So, a single action processor deals with the attributes and the methods in Fig. 2.

3.2 Upper-Agents

Definition 9 . (Upper-Agent) An upper-agent is defined as an object that has multiple aspects, and behaves autonomously and pro-actively. An upper-agent consists of an identifier, a list of attributes, a set of methods, a status processor, and an action processor.

$$\langle \texttt{upper-agent} \rangle ::= \texttt{Agent}(\langle \texttt{agent\_name} \rangle), \\ [\texttt{Attributes}\{\langle \texttt{attribute} \rangle, \ldots, \langle \texttt{attribute} \rangle\}, \\ [\texttt{Methods}\{\langle \texttt{method} \rangle, \ldots, \langle \texttt{method} \rangle\}$$

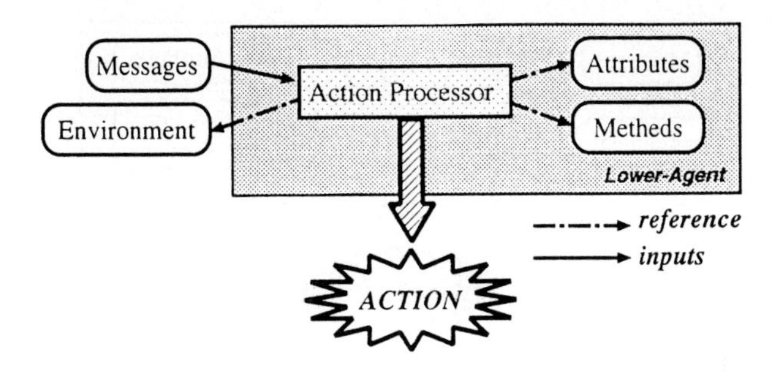

Fig. 2. A Lower-Agent

$$
\begin{aligned}
&[\texttt{Sequential\_methods}\{\langle\texttt{sequential\_method}\rangle,\\
&\qquad\qquad\qquad\qquad \ldots,\langle\texttt{sequential\_method}\rangle\}]]],\\
&[\texttt{Status}\{\langle\texttt{relatiosn}\rangle,\langle\texttt{roles}\rangle\}],\\
&[\texttt{Actions}\{\langle\texttt{aciton}\rangle,\ldots,\langle\texttt{aciton}\rangle\}]
\end{aligned}
$$

where $\langle\texttt{sequential\_method}\rangle$ is a sequence of methods. It is defined as

$$
\begin{aligned}
&\langle\texttt{sequential\_method}\rangle\\
&::= (\langle\texttt{sequential\_method\_name}\rangle(\langle\texttt{method\_name}\rangle,\ldots,\langle\texttt{method\_name}\rangle)\\
&\qquad\qquad \leftarrow \texttt{Own}(\langle\texttt{attributes}\rangle),\\
&\qquad\qquad\quad \texttt{Env}(\langle\texttt{attributes}\rangle),\\
&\qquad\qquad\quad \texttt{Msg}(\langle\texttt{a\_content\_of\_the\_message}\rangle))
\end{aligned}
$$

When a sequential method $\langle\texttt{sequential\_method\_name}\rangle$ is detected, the following $\langle\texttt{method_name}\rangle$s are invoked orderly.

A lower-agent plus a status processor constructs an upper-agent. The concept of the upper-agent is shown in Fig. 3. The status and action processors modify attributes and sequential methods, when the upper-agent receives messages. Since upper-agents have autonomous and pro-active features, these agents must modify attributes and sequential methods at runtime. "Messages" in Fig. 3 are the messages from other agents, including the environmental-agents. The status processor modifies the attributes and the sequential methods according to the status of the environment, the messages, and the attributes before modification. By modifying the attributes and the sequential methods with the status processor, the upper-agent decides next own status. Then, the action processor actually decides its behavior(a sequential method), using an output of the status processor.

3.3 Environmental-Agents

Definition 10 . (Environmental-Agent) An environmental-agent consists of an identifier, a list of attributes, a set of methods, a status processor, an action processor and an environmental database.

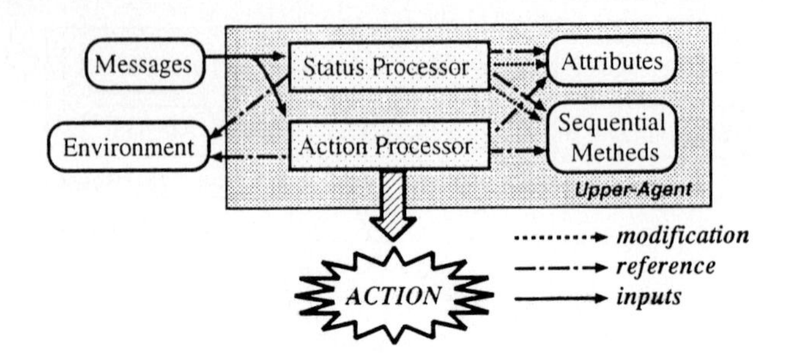

Fig. 3. An Upper-Agent

⟨environmental-agent⟩
::= Agent(⟨agent_name⟩),
    [Attributes{⟨attribute⟩, ..., ⟨attribute⟩},
    [Methods{⟨method⟩, ..., ⟨method⟩},
    [Sequential_methods{⟨sequential_method⟩, ..., ⟨sequential_method⟩}]]],
    [Status{⟨relations⟩, ⟨roles⟩}],
    [Actions{⟨action⟩, ..., ⟨action⟩}],
    [Env_db{⟨data⟩, ..., ⟨data⟩]

where Env_db is a reserved words, and means an environmental database. It is defined as

⟨environmental_database⟩ ::= Env_db{⟨data⟩, ..., ⟨data⟩}

⟨data⟩ ::= (Agent(⟨agent_name⟩), Attributes{⟨attribute⟩, ..., ⟨attribute⟩})

An environmental-agent consists of an upper-agent and an environmental database. The concept of an environmental-agent is shown in Fig. 4. Upper-agents and lower-agents do not have visible sensors. The environmental-agent plays a visual role for the other agents, and it provides other agents the visual data. The visual data are stored in the environmental database. The environmental-agent can multi-cast and broadcast to the agents with the same attribute values, by searching the environmental database for these attributes. When agents move from an environmental-agent to another environmental-agent, the result of the movement is registered in and deleted from the two environmental-databases respectively.

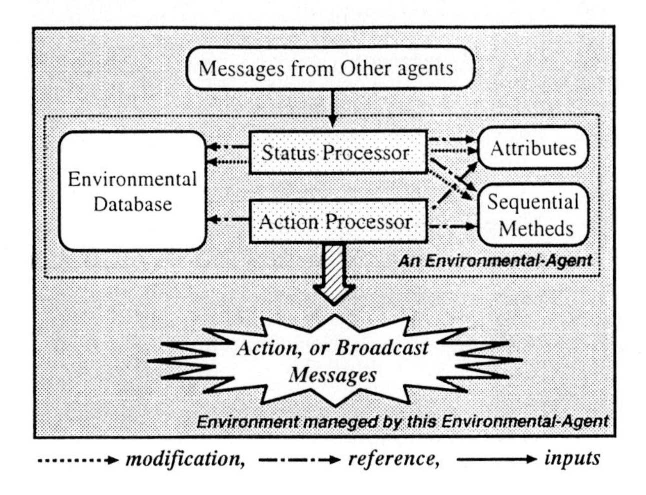

Fig. 4. An Environmental-Agent

Table 1. Strategies for A Soccer Player

Relation	offense				defense
Location	in goal area		not in goal area		—
Ball	○	×	○	×	—
Strategy	shoot	go to goal	pass	go to goal	track ball
exceptional strategy	—	—	Short distance to the goal → shoot	—	—

3.4 An Example

Here, we consider a soccer player "`Ito`" who behaves under the strategies in Table 1. This is a good example to represent multiple objects, because a soccer player require some aspects with surroundings in the game. We use "—" to show that the strategy is not restricted by the condition (Location, Ball and etc.) in Table 1.

We define this player as an upper-agent. It is described in Fig. 5.

Assume that this player keeps a ball and his team is in offense. Besides, this player is out of goal area (`where(not_goal_area)`). And when the distance between him and the opponent goal becomes short, he behaves actually. In this case, this player decides his status (`role(pass)`). Then, he decides his behavior (`shoot`) based on decided status (`role(pass)`) and status of environment (`Dist:short`). You should notice that this player can shoot the ball. Our upper-agent behaves through two steps, deciding a status and deciding an action. For

```
Agent(Ito),
Attributes{team(A), number(10),role(track_ball), ball(true)},
Methods{shoot(...),pass(...),goto_goal(...),track_ball(...)} ,
Status{(Relation(offense) -> offense,
        Relation(defense) -> defense),
       (role(shoot) -> Relation(offense),
                       Env(where(goal_area)),
                       Own(ball(true)),
        role(pass) -> Relation(offense),
                      Env(where(not_goal_area)),
                      Own(ball(true)),
        role(goto_goal) -> Relation(offense),
                           Env(),Own(ball(false)),
        role(track_ball) -> Relation(defense),
                            Env(),Own()}
Actions{shoot -> Own(role(shooter)),Env(),Msg(doing),
        shoot -> Own(role(passer)),Env(dist(short)),Msg(doing),
        pass -> Own(role(passer)),Env(),Msg(doing),
        goto_goal -> Own(role(goto_goal)),Env(),Msg(doing),
        track_ball -> Own(role(track_ball)),Env(),Msg(doing)}
```

Fig. 5. A Description of A Soccer Player

the second step, he can change his behavior from passing the ball to shooting the ball. It implies that our agent model can represent exceptional behaviors.

4 A Description-Processing System

In this section, we describe an implementation of multiple objects. A multiple object consists of some types of agents. Hence, the key is to implement agents. In the current version of the system, it is implemented as an interpreter which may result in a modification of the original specification of the agent.

4.1 A Common Architecture of Agents

As we stated in Definition 8 , 9 , 10 , each agent has some descriptions and one or two processors. Each agent has an interpreter which interprets the descriptions of the agent. Figure. 6 illustrates the common architecture of agents in our system. Each agent basically consists of three units as follows:

1. Communication Unit
   This unit is an interface that communicates with other agents, and translates messages coming from other agents into a form that the agent can understand.
2. Specification Unit
   This unit is a file that includes attributes, methods, and rules which the agent required in behaving autonomously and pro-actively.

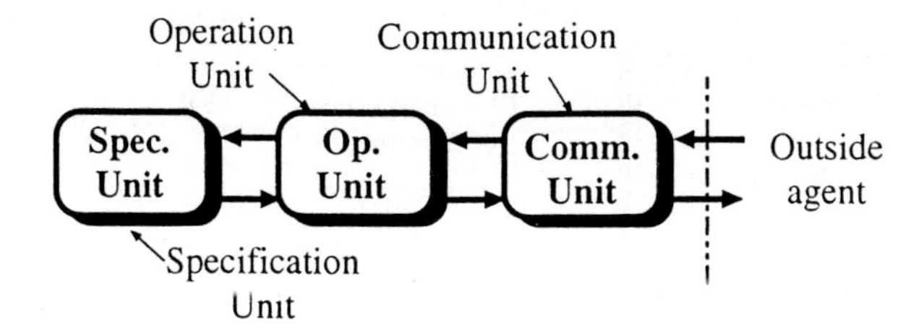

Fig. 6. Common Architecture of Agents

3. Operation Unit
   This unit is an interpreter that interprets(executes) contents of the specification unit. It also modifies contents of the specification unit and invokes a proper method, possibly.

The operation unit is a kind of interpreter and interprets contents of the specification unit. An agent communicate with other agents via the communication unit in behaving.

4.2 Class Hierarchy

Each agent is an instance of a class. Our system contains the following classes:

1. Agent Class
   This class owns common properties of agents, which are an agent identity, attributes, and methods.
2. Lower-Agent Class
   This class inherits the properties from Agent Class and has an additional property which is an action processor(interpreter).
3. Upper-Agent Class
   This class inherits the properties from Lower-Agent Class, and has an additional property which is a status processor(interpreter).
4. Environmental-Agent Class
   This class inherits the properties from Upper-Agent Class, and has properties which are method to broadcast and multi-cast in the environmental-agent.

These classes forms an inheritance hierarchy rooted at Agent Class.
For implementing soccer agents easily, a set of subclass are designed.

1. SoccerEAgent Class
   This class inherits properties from Environmental-Agent Class. As we use this class as interface of the soccer server, this class require method of Input/Output to the soccer server.

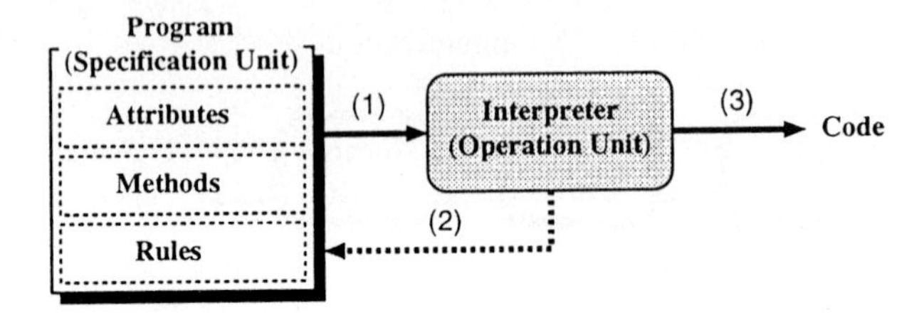

Fig. 7. A Behavior of the operation unit(interpreter)

2. SoccerUAgent Class
   This class inherits properties from Upper-Agent Class. This class has communication method to SoccerEAgent Class.
3. SoccerLAgent Class
   This class inherits properties from Lower-Agent Class. This class also has communication method to SoccerEAgent Class.

In the current implementation, we do not allow communications between upper-agents and lower-agents directly. They communicate through the environmental-agent.

4.3 The Interpreter of agents

The interpreter(operation unit) of agents is implemented in Java. It plays the role of a status processor and an action processor. Each agent has an only interpreter, even if the agent like an upper-agent, and an environmental-agent has both of these processors. In this case, the unit behaves according to the description of the agent in two steps. Therefore, this interpreter behaves as illustrated in Fig. 7. The unit of an upper-agent and an environmental-agent behaves in order of (1) $\rightarrow$ (2) $\rightarrow$ (1) $\rightarrow$ (3) in Fig. 7. In (2), the (operation) unit modifies the specification unit(program) according to the rules. Then the operation unit interprets the program again. On the other hand, The unit of an lower-agent behaves in order of (1) $\rightarrow$ (3).

5 Experiments

To confirm the efficiency of our model and system, two sets of experiments are designed. One is to compare the power of different teams which are consisted of different type players. The other is to let our team attend the RoboCup Japan Open held in Tokyo, Japan, recently.

5.1 Designs and Experiments in Our System

First, three primary types of soccer players are designed.

Table 2. Hybrid-Player Team(HPT) vs. Upper-Player Team(UPT)

Hybrid-Player Team kicks off.						Upper-Player Team kicks off.					
HPT	UPT	result	HPT	UPT	result	HPT	UPT	result	HPT	UPT	result
3	1	HPT	4	3	HPT	4	1	HPT	4	0	HPT
3	1	HPT	3	1	HPT	4	1	HPT	7	1	HPT
2	0	HPT	6	3	HPT	4	1	HPT	3	4	UPT
4	1	HPT	5	4	HPT	2	2	DRAW	5	0	HPT
3	4	UPT	4	5	UPT	8	1	HPT	4	2	HPT
3	1	HPT	4	4	DRAW	4	1	HPT	3	2	HPT
3	4	UPT	3	1	HPT	5	1	HPT	4	2	HPT
4	3	HPT	5	6	UPT	3	3	DRAW	2	2	DRAW
2	2	DRAW	8	1	HPT	6	4	HPT	2	1	HPT
5	1	HPT	4	6	UPT	4	2	HPT	6	2	HPT
5	1	HPT	3	2	HPT	5	3	HPT	4	3	HPT
2	3	UPT	4	1	HPT	2	3	UPT	3	3	DRAW
6	1	HPT.				5	2	HPT			

Winning ratio of HPT : 0.72
Winning ratio of UPT : 0.16

1. Lower-Player
   This player consists of lower-agent and environmental-agent, and behaves reflectively using lower-agent.
2. Upper-Player
   This player consists of upper-agent and environmental-agent, and behaves based on status decided by status processor.
3. Hybrid-Player
   This player consists of upper-agent, lower-agent, and environmental-agent. This agent autonomously behaves based on status decided by status processor or behaves reflectively.

We then generate three teams using these players: Lower-Player Team, Upper-Player Team, and Hybrid-Player Team. We simulates games using these three teams.

Table 2 shows the results of matches Hybrid-Player Team (HPT) and Upper-Player Team (UPT). HPT wins the games at winning ratio 0.72. In HPT, when the upper-agent does not behaves in 100ms intervals, the lower-agent behaves reflectively. The number of behaviors of the upper-agent is 5 times of that of the lower-agent in the games. On the other hand, only the upper-agent behaves autonomously in UPT. The cause of win of HPT is that HPT behaves faster than UPT exactly.

Table 3 shows the results that Hybrid-Player Team (HPT) competes Lower-Player Team (LPT). HPT does not lose the games. All Lower-Players traced the

Table 3. Hybrid-Player Team(HPT) vs. Lower-Player Team(LTP)

Hybrid-Player Team kicks off.						Lower-Player Team kicks off.					
HPT	LPT	result	HPT	LPT	result	HPT	LPT	result	HPT	LPT	result
6	3	HPT	5	1	HPT	5	2	HPT	6	1	HPT
6	3	HPT	5	2	HPT	5	3	HPT	4	2	HPT
7	3	HPT	5	2	HPT	5	1	HPT	5	2	HPT
6	3	HPT	5	3	HPT	4	4	DRAW	4	2	HPT
7	1	HPT	5	2	HPT	5	1	HPT	5	3	HPT
5	3	HPT	7	2	HPT	6	2	HPT	4	3	HPT
5	2	HPT	4	4	DRAW	5	3	HPT	4	1	HPT
6	3	HPT	5	1	HPT	5	3	HPT	5	2	HPT
6	3	HPT	7	2	HPT	4	1	HPT	5	3	HPT
5	2	HPT	6	2	HPT	6	3	HPT	4	2	HPT
7	2	HPT	5	2	HPT	4	4	DRAW	7	3	HPT
4	2	HPT	5	2	HPT	6	2	HPT	5	1	HPT
6	3	HPT				4	2	HPT			

Winning ratio of HPT : 0.94
Winning ratio of LPT : 0.00

ball, and gathered around the ball. On the other hand, Players of HPT spread in the soccer field and behave autonomously.

We do not show that Upper-Player Team competes Lower-Player Team. The result of the games is similar to the result of the matches HPT and LPT. UPT wins the games in a high rate.

From the above results, we conclude that HPT is the strongest team in our experiments. It implies that the ability of agents who can behave appropriately in a fixed interval is very important in RoboCup.

Our system has another more important advantage. From a real point of view, it is necessary to modify the strategies of the upper/lower-agents dynamically in the game. In our system, the environmental-agent are behaved as the interface between the soccer server and upper/lower-agents. In the other words, the environmental-agents can be viewed as soccer players, and the upper/lower-agents are the inference engine of the environmental-agents. Therefore, it is possible to modify upper/lower-agents dynamically in the game. It implies that our system is very useful when modifying soccer players was repeated to improve team plays and strategies.

5.2 NIT Stones 98

For comparing with the other teams that are implemented in different model and architecture, we took part in in the RoboCup Japan Open which held on April 9–11, 1998 in Tokyo, Japan. We implemented our team, NIT Stones, as a hybrid-player team. We qualified in the trial heats and finally entered into semifinals. As the result, we are the third winner among all 10 teams. The detail of the result are shown in the following URL(in Japanese):

Trial http://www.er.ams.eng.osaka-u.ac.jp/robocup/jnc/events/jopen98/jopen-result-sim.html

Final http://www.er.ams.eng.osaka-u.ac.jp/robocup/jnc/events/jopen98/jopen-result-sim2.html

Our team played faster and more exact than other teams in the competitions. It implies that our model and system are effective in dynamic environment like soccer games.

5.3 Results of RoboCup-98

JDK-1.2β3 was installed in the Workstations of RoboCup-98. However, we implemented our team with JDK-1.1.x. To our great regret, our players froze in the games frequently, although they had never frozen before RoboCup-98. As stated above, our team had a good performance and was awarded third prize in Japan Open 98. The β version system may have been the cause that our players froze.

In these games, we have observed that our players behaved faster and more accurate than other teams. It shows that our model is efficient for the given challenge. In the next RoboCup, we must confirm that our players can behave well in various systems. Besides, we look forward to build up better team plays.

6 Conclusion

In this paper, we applied the agent-oriented paradigm to represent multiple objects and a new agent model, EAMMO, is proposed. Besides, we also proposed a description-processing system for EAMMO. We confirmed that our agent model and system are applicable to soccer agents as multiple objects in RoboCup through our simulations. We also confirmed that EAMMO can represent real-time multi-agents systems in our simulations (Hybrid-Player Team). Our system has another more important advantage. From a real point of view, it is necessary to modify the strategies of the upper/lower-agents dynamically in the game. In our system, the environmental-agent are behaved as the interface between the soccer server and upper/lower-agents. In other words, the environmental-agents can be viewed as soccer players, and the upper/lower-agents are the inference engine of the environmental-agents. Therefore, it is possible to modify upper/lower-agents dynamically in the game. It implies that our system is very useful when soccer players are modified frequently to improve team plays and strategies. This effect is also confirmed by the result of our team in the competitions of the RoboCup Japan Open.

Our current system is an interpreter-based system implemented in Java. It makes our system behavior a little bit slowly. To improve the performance of our system, we plan to introduce a translator which generates Java code from the specification units. Moreover, we will use threads more in implementing our future system.

References

1. J. Runbaugh, M. Blaha, W. Premerlani, F. eddy, W. Lorensen: Object-Oriented Modeling and Design, Prentice Hall(1991).
2. J. J. Shilling: "Three Steps to Views: Extending the Object-Oriented Paradigm", OOPSLA '89 Proceedings, (1989) 353–361.
3. J. Richardson, P. Schwarz: "Aspects: Extending Objects to Support Multiple, Independent Roles", ACM SIGMOD '91, (1991) 298–307.
4. A. Albano, R. Bergamini, G. Ghelli, R. Orsini: "An Object Data Model with Roles" Proceedings of the 19th VLDB Conference, (1993) 39–51.
5. Y. Kambayashi, Z. Peng: "Object Deputy Model and Its Applications" Proceedings of Fourth International Conference on Database System for Advanced Applications DASFAA '95, (1995) 1–15. 1995.
6. Y. Shoham: "Agent Oriented Programming" Artificial Intelligence, 60(1), (1993) 51–92.
7. F. G. McCabe, K. L. Clark: "April — Agent PRocess Interaction Language" Intelligent Agents, Lecture Notes in Artificial Intelligence 890, (1995) 324–340.
8. M. J. Wooldridge, N. R. Jennings (Eds.): Intelligent Agents, Lecture Notes in Artificial Intelligence **890**, (1995).
9. S. Janson, S. Haridi: Programming Paradigms of the Andrra Kernel Language, SICS Research Report R**91**:**08**, (1991).

Using an Explicit Teamwork Model and Learning in RoboCup: An Extended Abstract

Stacy Marsella, Jafar Adibi, Yaser Al-Onaizan, Ali Erdem, Randall Hill
Gal A. Kaminka, Zhun Qiu, Milind Tambe

Information Sciences Institute and Computer Science Department
University of Southern California
4676 Admiralty Way, Marina del Rey, CA 90292, USA
robocup-sim@isi.edu

1 Introduction

The RoboCup research initiative has established synthetic and robotic soccer as testbeds for pursuing research challenges in Artificial Intelligence and robotics. This extended abstract focuses on teamwork and learning, two of the multi-agent research challenges highlighted in RoboCup. To address the challenge of teamwork, we discuss the use of a domain-independent explicit model of teamwork, and an explicit representation of team plans and goals. We also discuss the application of agent learning in RoboCup.

The vehicle for our research investigations in RoboCup is **ISIS (ISI Synthetic)**, a team of synthetic soccer-players that successfully participated in the simulation league of RoboCup'97, by winning the third place prize in that tournament. In this position paper, we briefly overview the ISIS agent architecture and our investigations of the issues of teamwork and learning. The key novel issues for our team in RoboCup'98 will be a further investigation of agent learning, and further analysis of teamwork related issues.

2 The ISIS Architecture

An ISIS agent uses a two-tier architecture, consisting of a higher-level that makes decisions and a lower-level that handles various time critical functions tied to perception and action.

ISIS's lower-level, developed in C, communicates inputs received from the RoboCup simulator (after sufficient pre-processing), to the higher level. The lower-level also rapidly computes some recommended directions for turning and kicking, to be sent to the higher-level. For instance, a group of C4.5 rules compute a direction to intelligently shoot the ball into the opponents' goal (discussed further in Section 4). The lower-level also computes a plan to intercept the ball consisting of turn or dash actions.

The lower-level does not make any decisions with respect to its recommendations however. Instead, all such decision-making rests with the higher level, implemented in the Soar integrated AI architecture[11, 14]. Once the Soar-based higher-level reaches a decision, it communicates with the lower-level, which then sends the relevant information to the simulator.

The Soar architecture involves dynamic execution of an operator (reactive plan) hierarchy. An operator begins execution (i.e., the operator is activated), when it is selected for execution, and it remains active until explicitly terminated. Execution of higher-level abstract operators leads to subgoals, where new operators are selected for execution, and thus a hierarchical expansion of operators ensues. Operators in Soar are thus similar to reactive plans in architectures such as RAP[4].

3 Teamwork

There are two key aspects of ISIS's approach to teamwork. The first is the explicit representation of team activities via the use of explicit representation of *team operators* (reactive team plans). Team operators explicitly express a team's joint activities, unlike the regular "individual operators" which express an agent's own activities. Furthermore, while an individual operator applies to an agent's private state (an agent's private beliefs), a team operator applies to an agent's *team state*. A team state is the agent's (abstract) model of the team's mutual beliefs about the world, e.g., the team's currently mutually believed strategy. An ISIS agent can also maintain subteam states for subteam participation. Each team member maintains its own copy of the team state, and any subteam states for subteams it participates in. That is, there is no shared memory among the team members.

The second key aspect of teamwork in ISIS is its novel approach to coordination and communication via the use of a general-purpose teamwork model. In particular, to surmount uncertainties that arise in dynamic environments and maintain coherence in teamwork, team members must be provided the capability of highly flexible coordination and communication. To this end, general-purpose explicit models of teamwork have recently been proposed as a promising approach to enhance teamwork flexibility[6, 19]. Essentially, teamwork models provide agents with the capability of first principles reasoning about teamwork to provide teamwork flexibility. Such teamwork models also enable code reuse.

We investigate the use of STEAM[17, 18, 19], a state-of-the-art general-purpose model of teamwork. STEAM models team members' responsibilities and commitments in teamwork in a domain-independent fashion. As a result, it enables team members to autonomously reason about coordination and communication, improving teamwork flexibility. Furthermore, due to its domain-independence, STEAM has been demonstrated to be reusable across domains. STEAM uses the formal *joint intentions* framework[3, 7] as its basic building block, but it is also influenced by the SharedPlans theory[5], and includes key enhancements to reflect the constraints of real-world domains. For instance, the Joint intentions theory requires that agents attain mutual belief in establishing and terminating joint intentions, but does not specify how mutual belief should be attained. STEAM uses decision-theoretic reasoning to select the appropriate method for attaining mutual belief. Thus, it does not rely exclusively on explicit communication (e.g, "Say" in RoboCup simulation) for attaining mutual belief in the team; instead, it may rely on plan-recognition.

A typical example of STEAM in operation is the DEFEND-GOAL team operator executed by the defender subteam. In service of DEFEND-GOAL, players in this subteam normally execute the SIMPLE-DEFENSE team operator to position themselves properly on the field and to try to be aware of the ball position. Of course, each player can only see in its limited cone of vision, and particularly while repositioning itself, can be unaware of the approaching ball. Here is where teamwork can be beneficial. In particular, if any one of these players sees the ball as being close, it declares the SIMPLE-DEFENSE team operator to be irrelevant. Its teammates now focus on defending the goal in a coordinated manner via the CAREFUL-DEFENSE team operator. Should any one player in the goalie subteam see the ball move sufficiently far away, it again alerts its team mates (that CAREFUL-DEFENSE is achieved). The subteam players once again execute SIMPLE-DEFENSE to attempt to position themselves close to the goal. In this way, agents coordinate their defense of the goal.

3.1 New Issues

Several issues have been brought forward due to our application of the teamwork model in RoboCup. First, RoboCup is a highly dynamic, real-time domain, where reasoning from first principles of teamwork via the teamwork model can sometimes be inefficient. Therefore, to improve efficiency, we plan to compile the teamwork model, so that the typical cases of reasoning about teamwork are speeded up. This can be achieved via machine learning methods such as chunking[11] (a form of explanation-based learning)[10]. The teamwork model itself will be retained however, since unusual cases may still arise, and require the first principles teamwork reasoning offered by the teamwork model.

Second, the teamwork model may sometimes enforce a rigid coherence constraint on the team, always requiring mutual belief to be attained. However, it is not always straightforward to attain such mutual belief. In RoboCup simulation, the shouting range (i.e., the range of the "say" message) is limited. A player may not necessarily be heard at the other end of the field. A tradeoff in the level of coherence and team performance is therefore necessary. We plan to investigate this tradeoff, and suggest corresponding modifications to teamwork models.

Finally, we are extending the capabilities of agents to deal with potential inconsistencies in their beliefs. Currently in STEAM, if an agent discovers new information relevant to the team goal, it will use decision-theoretic reasoning to select a method for attaining mutual belief. In particular, it may inform other agents by communicating the belief. However, the recipients of such communications accept the acquired belief without examination even if there are inconsistencies between this belief and a recipient's existing "certain" beliefs. This can be a problem.

For instance, in RoboCup if a defender sees the ball as being close and tells other defenders, all the defenders will establish a joint intention to approach the ball; if at the same time the player is informed by another defender who stands far away from the ball that the ball is far, an inconsistency occurs. The joint goal will be terminated, and there will be recurring processes to form and end the joint intentions, and the whole team will get stuck at this point.

To address this problem, we have taken an approach whereby agents model the beliefs of other agents in order to detect inconsistencies and to decide whether to negotiate. A key idea in our approach is that negotiation itself takes time, which can be a significant factor in RoboCup. Thus, it is often the case that an agent should decide not to argue with its disagreeing teammates and, instead, go along with the temporary inconsistency.

To detect an inconsistency, the agent receiving a message has first to compare its own beliefs with the belief conveyed in the message sent. Since the conveyed belief may not conflict explicitly with the agent's existing beliefs, the agent uses belief inference rules to figure out the relevant beliefs of the message sender. Further, since both senders and recipients may have beliefs which they are more or less certain about, the detection of inconsistency takes into account whether beliefs are "certain" (e.g., supported by direct sensory evidence as opposed to plausible default beliefs).

To decide whether to negotiate over inconsistencies, an agent uses a decision-theoretic method to select an approach to addressing the inconsistencies. Briefly, the agent always has three choices. First of all it can just keep its own belief and work on it without any argument with its teammates. Second, it can choose to accept without argument what the sender says. Finally, the third option is to expend the effort both to detect a clash in beliefs and to negotiate a resolution. In that case, the negotiation may lead to agreement with the sender's observation or alternatively may persuade the sender to accept the recipient's belief.

Under different circumstances, the cost and utility of a decision will vary a lot. In the RoboCup case proposed above, the cost of the "negotiation" may be substantial, not only because of the large amount of resources consumed, but more importantly because of the time pressure in the soccer competition. Furthermore, since the situation may change greatly (the ball may have rolled to a far-away position) after the agents are set with their arguments, the possibility and benefit of the "right" decision will decrease significantly. All these lead to a rather low expected utility of negotiation. So in this specific case, the player involved will either stick to its own belief or turn to other's observation directly, rather than bothering to argue to resolve the disagreement.

4 Lower-level skills and Learning

Inspired by previous work on machine learning in RoboCup[15, 9], we focused on techniques to improve individual players' skills to kick, pass, or intercept the ball. Fortunately, the two layer ISIS architecture helps to simplify the problem for skill learning. In particular, the lower-level in ISIS is designed to provide several recommendations (such as various kicking directions) to the higher-level, but it need not arrive at a specific decision (one specific kicking direction). Thus, an individual skill, such as a kicking direction to clear the ball, can be learned independently of other possible actions.

Learning has currently been applied to (i) selection of an intelligent direction to shoot a ball when attempting to score a goal and (ii) selection of a plan to intercept an incoming ball.

Scoring goals is clearly a critical soccer skill. However, our initial hand-coded, approaches to determining a good direction to kick the ball, based on heuristics such as "shoot at the center of the goal", or "shoot to a corner of the goal", failed drastically. In part, this was because heuristics were often foiled by the fact that small variations in the configuration of players around the opponent's goal or a small variation in the shooter's position may have dramatic effects on the right shooting direction.

To address these problems, we decided to rely on automated, *offline* learning of the shooting rules. A human expert created a set of shooting situations, and selected the optimal shooting direction for each such situation. The learning system trained on these shooting scenarios. C4.5[12] was used as the learning system, in part because it has the appropriate expressive power to express game situations and can handle both missing attributes and a large number of training cases.

In our representation, each C4.5 training case has 39 attributes, such as the shooters angles to the other visible players. The system was trained on over roughly 1400 training cases, labeled by our expert with one of UP, DOWN, and CENTER (region of the goal) kicking directions. The result was that given a game situation characterized by the 39 attributes, the decision tree selected the best of the three shooting directions. The resulting decision tree provided a 70.8%-accurate set of shooting rules.

These learned rules for selecting a shooting direction were used successfully in RoboCup'97. The higher-level typically selected this learned shooting direction when players were reasonably close to the goal, and could see the goal.

In contrast to the *offline* learning of shooting direction, we have also begun to explore *online* learning of intercept plans. In our initial implementation, ISIS players used a simple hand-coded routine to determine the plan for intercepting the ball. Our experience at RoboCup97 was that the resulting intercept plans work fine under some playing conditions, but fail under others. The result often depends on such external factors as network conditions, frequency of perceptual updates from the soccer server and the style of play of the opposing team. Unlike real soccer players, our ISIS players' intercept skills were not adapting very well to differing external factors.

To address this problem, we are exploring how players can adapt their intercept online, under actual playing conditions. Of course, an adaptive intercept has inherent risks. In the course of a game, there are not many opportunities to intercept the ball, and worse, inappropriate adaptations can have dire consequences. Therefore, it is important for adaptation to be done rapidly, reasonably and smoothly.

To assess these risks and the overall feasibility of an adaptive intercept, we have started to explore simple adaptive approaches. In keeping with the risks, our current approach uses hill-climbing search in the space of plans where evaluation of success or failure is driven by an "oracle", ISIS's higher-level, decision-making tier. In addition, we have adopted a conservative approach of using distinct searches for distinct input conditions, so for instance balls that are moving towards the player may be treated separately from balls moving away.

As a player uses the intercept assigned by some input condition, failure to meet expectations will result in a new intercept plan for this input condition if there has been a history of similar failures. ISIS's higher-level drives the evaluation since it has the necessary context to model the failure. For instance, failure due to a blocking player is treated differently from failure due to an improper turn angle.

This work is preliminary and has not been fully evaluated. However, it has been tested under fixed, ideal conditions (e.g., reliable perceptual updates). In these tests, the method exhibits consistent and rapid convergence on simple turn and dash plans that are at least as good as the manually derived plans used at RoboCup97. In addition to more extensive evaluation under varying conditions, we are now considering enhancements such as allowing the learning under one input condition to influence similar input conditions.

5 Evaluation

There are several aspects to evaluation of ISIS. As mentioned earlier, ISIS successfully participated in RoboCup'97, winning the third-place prize in the simulation league tournament, in the 29 teams that participated. Overall at RoboCup97, ISIS won six out of the seven games in which it competed, outscoring its components 37 goals to 19.

Another key aspect of evaluation is measuring the contribution of the explicit teamwork model (STEAM) to ISIS. STEAM's contribution is both in terms of improved teamwork performance and reduced development time. To measure the performance improvement due to STEAM, we experimented with two different settings of communication cost in STEAM. At "low" cost, ISIS agents communicate a significant number of messages. At "high" communication cost, ISIS agents communicate no messages. Since the portion of the teamwork model in use in ISIS is effective only with communication, a "high" setting of communication cost essentially nullifies the effect of the teamwork model.

For each setting of communication cost, ISIS played 7 games against a fixed opponent team of roughly equivalent capability. With low communication cost, ISIS won 3 out of the 7 games. It scored 18 goals against the opponents, and had 22 goals scored against it. With high communication cost, ISIS won none out of the 7 games it played. It scored only 3 goals, but had 20 goals scored against it.

The results clearly illustrate that the STEAM teamwork model does make a useful contribution to ISIS's performance. Furthermore, by providing general teamwork capabilities, it also reduces development time. For instance, without STEAM, all of the communication for jointly initiating and terminating all of the team operators (about 20 in the current system) would have had to be implemented via dozens of domain-specific coordination plans.

6 Related Work

In terms of work within RoboCup, ISIS was the only team at RoboCup'97 that investigated the use of a general, domain-independent teamwork model to guide agent's communication and coordination in teamwork. Some researchers investigating teamwork in RoboCup have used explicit team plans and roles, but they have relied on *domain-dependent* communication and coordination. Typical examples include [2, 1]. Other investigations of teamwork in RoboCup have used implicit or emergent coordination. A typical example is Yokota et al.[20].

Our application of learning in ISIS agents is similar to some of the other investigations of learning in RoboCup agents. For instance, Luke et al.[8] use genetic programming to build agents that learn to use their basic individual skills in coordination. Stone and Veloso[16] present a related approach, in which the agents learn a decision tree which enables them to select a recipient for a pass.

In terms of related work outside RoboCup, the use of a teamwork model remains a distinguishing aspect of teamwork in ISIS. The STEAM teamwork model used in ISIS, is among just a very few implemented general models of teamwork. Other models include Jennings' *joint responsibility* framework in the GRATE* system[6] (based on Joint Intentions theory), and Rich and Sidner's COLLAGEN[13] (based on the SharedPlans theory), that both operate in complex domains. STEAM significantly differs from both these frameworks, via its focus on a different (and arguably wider) set of teamwork capabilities that arise in domains with teams of more than two-three agents, with more complex team organizational hierarchies, and with practical emphasis on communication costs (see [19] for a more detailed discussion).

7 Summary

We have discussed teamwork and learning, two important research issues in multi-agent systems. The vehicle for our research is ISIS, an implemented team of soccer playing agents, that successfully participated in the simulation league of the RoboCup'97 soccer tournament. We have taken a principled approach in developing ISIS, guided by the research opportunities in RoboCup. Despite the significant risk in following such a principled approach, ISIS won the third place in the 29 teams that participated in the RoboCup'97 simulation league tournament.

There are several key issues that remain open for future work. One key issue is improved agent- or team-modeling. One immediate application of such modeling is recognition that an individual, particularly a team member, is unable to fulfill its role in the team activity. Other team members can then take over the role of this failing team member. Team modeling can also be applied to recognize opponent behaviors and counter them intelligently.

Acknowledgement

This research is supported in part by NSF grant IRI-9711665. We thank Bill Swartout, Paul Rosenbloom and Yigal Arens of USC/ISI for their support of the RoboCup activities described in this paper.

References

1. T. F. Bersano-Begey, P. G. Kenny, and E. H. Durfee. Agent teamwork, adaptive learning, and adversarial planning in robocup using a prs architecture. In *RoboCup-97: The first robot world cup soccer games and conferences.* Springer-Verlag, Heidelberg, Germany, 1998.
2. S. Ch'ng and L. Padgham. Team description: Royal merlbourne knights. In *RoboCup-97: The first robot world cup soccer games and conferences.* Springer-Verlag, Heidelberg, Germany, 1998.
3. P. R. Cohen and H. J. Levesque. Teamwork. *Nous*, 35, 1991.
4. J. Firby. An investigation into reactive planning in complex domains. In *Proceedings of the National Conference on Artificial Intelligence (AAAI)*, 1987.
5. B. Grosz and S. Kraus. Collaborative plans for complex group actions. *Artificial Intelligence*, 86:269–358, 1996.
6. N. Jennings. Controlling cooperative problem solving in industrial multi-agent systems using joint intentions. *Artificial Intelligence*, 75, 1995.
7. H. J. Levesque, P. R. Cohen, and J. Nunes. On acting together. In *Proceedings of the National Conference on Artificial Intelligence.* Menlo Park, Calif.: AAAI press, 1990.
8. S. Luke, Hohn C., J. Farris, G. Jackson, and J. Hendler. Co-evolving soccer softbot team coordination with genetic programming. In *RoboCup-97: The first robot world cup soccer games and conferences.* Springer-Verlag, Heidelberg, Germany, 1998.
9. H. Matsubara, I. Noda, and K. Hiraki. Learning of cooperative actions in multi-agent systems: a case study of pass play in soccer. In S. Sen, editor, *AAAI Spring Symposium on Adaptation, Coevolution and Learning in multi-agent systems*, March 1996.
10. T. M. Mitchell, R. M. Keller, and S. T. Kedar-Cabelli. Explanation-based generalization: A unifying view. *Machine Learning*, 1(1):47–80, 1986.
11. A. Newell. *Unified Theories of Cognition.* Harvard Univ. Press, Cambridge, Mass., 1990.
12. J. R. Quinlan. *C4.5: Programs for machine learning.* Morgan Kaufmann, San Mateo, CA, 1993.
13. C. Rich and C. Sidner. COLLAGEN: When agents collaborate with people. In *Proceedings of the International Conference on Autonomous Agents (Agents'97)*, 1997.
14. P. S. Rosenbloom, J. E. Laird, A. Newell, , and R. McCarl. A preliminary analysis of the soar architecture as a basis for general intelligence. *Artificial Intelligence*, 47(1-3):289–325, 1991.

15. P. Stone and M. Veloso. Towards collaborative and adversarial learning: a case study in robotic soccer. In S. Sen, editor, *AAAI Spring Symposium on Adaptation, Coevolution and Learning in multi-agent systems*, March 1996.
16. P. Stone and M. Veloso. Using decision tree confidence factors for multiagent control. In *RoboCup-97: The first robot world cup soccer games and conferences.* Springer-Verlag, Heidelberg, Germany, 1998.
17. M. Tambe. Teamwork in real-world, dynamic environments. In *Proceedings of the International Conference on Multi-agent Systems (ICMAS)*, December 1996.
18. M. Tambe. Agent architectures for flexible, practical teamwork. In *Proceedings of the National Conference on Artificial Intelligence (AAAI)*, August 1997.
19. M. Tambe. Towards flexible teamwork. *Journal of Artificial Intelligence Research (JAIR)*, 7:83–124, 1997.
20. K. Yokota, K. Ozako, Matsumoto A., T. Fujii, Asama H., and I. Endo. Cooperation towards team play. In *RoboCup-97: The first robot world cup soccer games and conferences.* Springer-Verlag, Heidelberg, Germany, 1998.

A Hybrid Agent Model, Mixing Short Term and Long Term Memory Abilities
An Application to RoboCup Competition

Fausto Torterolo[1] and Catherine Garbay[2]

[1] DIE, Dipartimento di Ingegneria Elettrica, Universitá di Palermo Viale delle Scienze, 90128 Palermo, Italy.
[2] TIMC/IMAG, Institut Albert Bonniot, Domaine de La Merci 38706 La Tronche Cedex, France.

Abstract. We present in this paper a novel approach for the modeling of agents able to react and reason under highly dynamic environments. A hybrid agent architecture is described, which allows to integrate the capacity to react rapidly to instantaneous changes in the environment with the capacity to reason more thoroughly about perceptions and actions. These capacities are implemented as independent processes running concurrently, and exploiting different memorizing abilities. Only a short-term memory is made available to reactive agents, whilst long-term memorizing abilities together with the possibility to reason about incomplete information is provided to cognitive agents. This model is currently experimented and tested under the framework of the RoboCup competition. An application example is provided to support the discussion.

1 Introduction

The recent development of robotic soccer competition has resulted in the designing of a variety of player models and architectures. Central to these developments is the specification of adapted, robust and efficient perception-decision-action cycle. Pure cognitive agent architectures have been proposed, like in Gaglio [3], where perception is rather approached in terms of static representations. Reactive agent architectures have been conversely designed, like the subsumption architecture of Brooks [1], where the emphasis on the contrary is on the notion of behavior. Most of these approaches, however, fail to consider the specificity of reasoning under dynamic environments, when the sensory information can be imprecise, uncertain and incomplete. Indeed, it has to be considered that the degree of veracity of sensory information decreases rapidly with time, since the environment is continuously evolving : consequently, what one agent sees at one time-instant may be not valid at the next time-instant. Most environments in the real world are of this kind, and it appears difficult to perform reasoning and interpretation with standard artificial intelligence techniques. The problem of dynamic environment was for example investigated in the path-planning domain where the changing of the environment with time has been shown to transform the correctly found plan to something totally unusable [5]. To solve this specific problem some authors have tried to model the incertitude of the information by stochastic methods. Also to be considered, and even a more complex problem, is the incompleteness of sensory information, that makes the agent uncertain about its current world model, a difficulty that is rarely tackled in front [12]. By incompleteness, we mean the fact that only some part of the world is perceived at each time by a given agent, this perception being in turn subject to incompleteness, depending on the distance and occlusions between objects. We present in this paper an approach to this problem, which has been developed in the framework of RoboCup, a competition between multi-agent soccer systems. In the soccer domain, the agents have different and contrasting goals. The agents in a team are supposed to act in a collaborative way, while the two teams compete together. They evolve in a highly dynamic environment comprising the ball, the other agents, and a few static elements like the goal-area. Different types of challenge are the objectives of this competition, as debated by Itsuiki et alt. in [10]. Our work takes place in the simulation section, more precisely in the cooperative and opponent-modeling challenge. Agents are simulations of robots in the soccer domain. They have limited sensory abilities and a small visioning cone. This application therefore appears as an excellent domain for the study and development of new reasoning models able to cope with the dynamicity and incompleteness of information. This kind of problem was already investigated In the work of Stone et al. [13], but to a limited extent. We have extended and deepened this approach by analyzing (i) how to represent and maintain knowledge about the world and (ii) how to use this knowledge in presence of incomplete information, in order to increase their completeness. Specific inference schemes have been developed for this purpose. Also specific to our approach is the definition of (i) a reference system to simplify the complexity of the computation, and (ii) a short term and a long term memory (respectively ST_memory and LT_memory). A hybrid agent model is finally proposed, which allows a mixture of reactive and cognitive behaviors [11, 9, 14]. That architecture realizes the Turing Machine concepts proposed by Ferguson [2], and has been inspired by a previous paper by B. Hayes-Roth [4].

2 Motivations

We try to clarify in this section the definition and use of basic notions like the ones of reactive and cognitive agents, or short-term and long-term memory. Some emphasis is put on the notions of world model and incompleteness of information.

Pure Reactive Player. A pure reactive player makes use of the mere direct information coming through its input sensors to react, i.e. to perform actions. There is no need to register any past sensory information, nor to perform any integration within an internally built word model. New incoming information is the only information used, and it is used instantaneously to decide for the next action. A pure reactive player "reacts" at the time of perception and all the available sensory information is used until new sensory information is available. Such agent therefore possesses a short term memory (ST_memory) that holds information valid only for the time interval between two perceptions.

Full Cognitive Player. A full cognitive player is able to elaborate an internal model of the world, based on the past and present sensory information. This dynamic model in turn may be used to maintain information about moving objects, and to estimate the certainty of this information. This knowledge is used to deliberate and decide about new actions. Such agent therefore possesses a long term memory (LT_memory) that holds information allowing the development of long term goals.

To illustrate these notions, lets take the example of a simple task like the one consisting in following another player (FP). This task will be performed differently, depending if the player is designed as a reactive or a cognitive agent.

For a pure reactive player a pseudo-code for this task can be:

1. Wait for new sensory information (copy it in ST_memory)
2. Look for the X_player position in the ST_memory
3. (a) IF {found} THEN {go into the corresponding direction for one time interval}
   (b) ELSE {move at a new position to look for X_player and go to 1}
4. go to 1

For a full cognitive player the pseudo-code to perform the FP task can be:

1. Wait for new sensory information (copy it in ST_memory)
2. Look for the X_player position in the ST_memory
3. (a) IF {not found} THEN {go to 4}
   (b) ELSE {follow the X_player for one time interval and go to 1}
4. estimate the X_player position (LT_memory)
5. (a) IF {estimation fails} THEN {go to 1}
   (b) ELSE {(go into the computed direction for one time interval and go to 1}

Conceptually speaking, the two behaviors are very similar, since they are made of the same basic actions. However, the way information is processed, and more precisely the coupling between information processing and action is different, since in the first case, deciding for a new action is the only way to react to the absence of information, whilst reasoning (a mental, or epistemic action) is used in the second case to try and recover more information. In fact, the two abilities should be seen as complementary, since the reactive player will behave in a faster way, and be able to react to small instantaneous changes in the environment, whilst the cognitive player may spend most time reasoning without having to think many decisions about action. To face this complementarity, we have developed a hybrid player model Fig.2 by mixing reactive and cognitive features in the same internal structure. Reactive and cognitive processes run concurrently in this model, under different priorities of execution : a high priority will be given to cognitive processes in case enough time is available, whilst the reactive part will drive the agent otherwise. The ST_memory is rather used by the reactive part of the agent, while the LT_memory is rather used by the cognitive part of the agent. In addition, some restricted exchange of information is allowed to take place between the two processes, when needed by the reactive part of the agent.

The Sensory Information. We have focused our attention on the visual information, because it represents what the player sees. Visual information is provided to the agent as a list of objects describing the relative distance, angle, distance variation, and angle variation, as shown in Fig. 1. To be noticed is the fact that the soccer simulator introduces noise, imprecision and incompleteness, as illustrated in Fig. 1. As shown in this figure, the information about the three "observed" players is transmitted to the "observing" player with different levels of accuracy and completion. For A1, full information about team name, player number and position is given with high precision. For A2 on the contrary, its relative distance and occluded position results in less accurate and complete information : no information about the team number is available, only its position and velocity are transmitted by the server to the "observing" agent. For an even farer player like A3, the only information is position and relative velocity, and its accuracy is not very high.

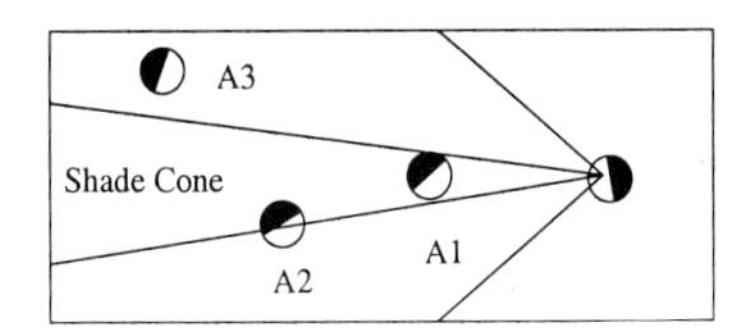

Visual information:
For A1: ((player team_name uniform_number) distance angle distance_variation anglevariation)
For A2: ((player team_name) distance angle)
For A3: ((player) distance angle)

Fig. 1. Accuracy and completeness of perception

As a consequence, the accuracy and completeness of the information available to a given player depends on its relative position and distance to others. It may happen in fact that it is impossible to guess whether another player belongs to the same team or not, depending on its distance from the observer, or whether he is moving or not. As a consequence, some processing must be included, in order to integrate the client information with the time and to build a consistent model of the world, thus allowing the estimation of lacking information and the solving of possible conflicts.

3 A Hybrid Agent Architecture

The agent architecture that we propose in this section is born from the necessity to integrate two complementary ways to process information in a single player : a cognitive one and a reactive one. A logical view of the proposed agent architecture (LA) is given in Fig.2, in terms of functional units. The functional units are designed as processes running in parallel and communicating by message passing to improve the global performances.

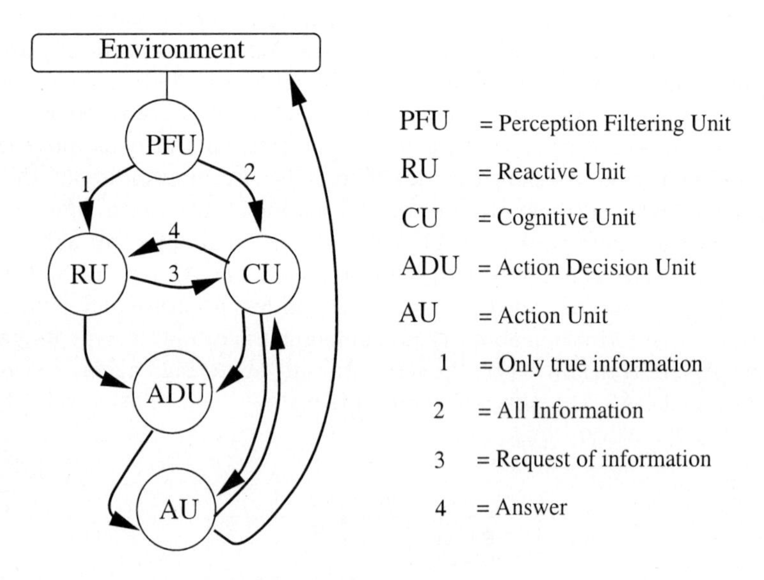

Fig. 2. Player Logic Architecture

Five main functional units may be distinguished in the proposed LA. New information is received at any time interval by the *Perception and Filtering Unit (PFU)*, from the simulator, in order to be processed. The role of the PFU is to split incomplete information from complete information, based on a straightforward analysis of the corresponding objects (see Fig. 1). The sole complete information is then transmitted to the *Reactive Unit (RU)* by the PFU, whilst all information is simultaneously sent to the *Cognitive Unit (CU)*. The RU then starts processing the transmitted information, and if a decision can be made sends a request to the *Action Decision Unit (ADU)*. The CU may concurrently infer a new action and send a request to the ADU. The role of the ADU is then to decide which action to perform from a list of actions, coming from both the RU and the CU, according to some a priori provided criteria. The decision is then transmitted to the *Action Unit (AU)* whose role is to perform the action by sending a command to the simulator. A trace of the performed action is also transmitted to the CU.

3.1 The Reactive Unit and its ST_memory

The reactive unit (see Fig.3), as told in previous sections, is a unit that gives the player the ability to react to new stimuli coming from the environment via the PFU. The RU comprises one *Short Term memory (ST_memory)*, one *Communication Unit (ComU)* and many different *Reactive Processes (RPs)*. The RPs work in parallel and share the common ST_memory. This memory is merely designed as the hard copy of the information coming from the PFU. It may also happen, as already mentioned, that one RP is lacking some information to perform its task. It is then given the possibility to access the LT_memory, which may then become a shared resource for all RPs. Such access is performed via the ComU. All decisions of action coming from the RU are sent to the ADU. The global processing of the RU is a cyclic process of the type "(i) read information (ii) decide action". The LT_memory is designed as a simple blackboard where information is maintained by the PFU, and in read-only access for the RPs. Communication via the ComU should be performed under a specific language (for example the KQML [15] language based on the KIF [8] format, developed at Stanford), but in the present implementation, it is reduced to a direct access to the LT_memory with read-only permission.

3.2 The Cognitive Unit and its LT_memory

The cognitive and reactive units work in parallel. The role of the cognitive unit is to develop more complex behaviors, like planning, or cooperation with others. It has to be noticed however, that the CU functional architecture (see Fig. 4) is more complex than the RUs one, which may increase the time spent in reasoning before reacting to changes in the environment. It may happen therefore, if the environment is changing rapidly, that the CU is not given sufficient time to decide for the next action (the decisions are then taken from the RU) or even that the decisions are inconsistent with respect to the evolution of the environment. Conversely, when the environment is changing less rapidly, no urgent decisions are taken by the RUs, and more time is given to the CU to reason, so that the overall behavior will reveal more elaborated. The proposed approach is therefore consistent with the generic idea of anytime execution, where more accurate processing is performed when more time is available. The interesting feature of this approach also lies in the fact that pure reactive behavior is performed when sufficient information about the world is available, thus allowing a flexible and dynamically determined alternation between more reactive and more cognitive behaviors.

The CU is composed of the following modules: New Information, LT_memory, World Model, Action Model, Cognitive Inference System, and Cognitive Action Decision. The New Information modules entails the information that is perceived by the agent at each time interval. It is the only interface between the external world and the Cognitive Inference System. The role of the World Model is to store knowledge about how the world is built, in terms of rules and constraints. It is composed of two parts: the Dynamic Models describe for all objects how

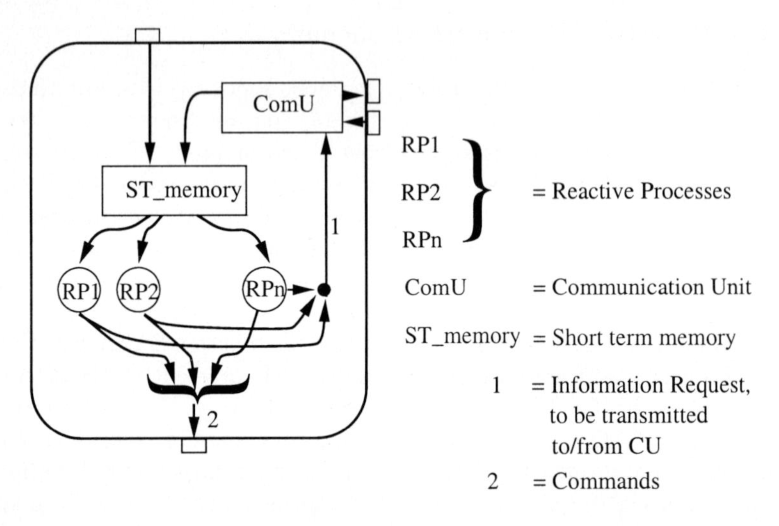

Fig. 3. Reactive Unit (RU) functional architecture

their status may change as a function of their own characteristics, while the Linguistic Models describe how to derive new object descriptions. The Action Model comprises a list of actions and models describing, for any action, its influence on the environment. The role of the Cognitive Inference System is to evaluate and modify the LT_memory at each time interval. The LT_memory is the repository of the knowledge gained by the agent about the world, either by direct external perception, or by internal reasoning. It is the central data base for the cognitive component of the agent. The refreshment cycle of the LT_memory depends on the events that can change its status. These events may be either new information or new actions. When new information is received via the inference system, the LT_memory is updated, based on the current LT_memory and World Model rules and constraints. When new actions are performed by the agent, the LT_memory is also updated, depending on the Action Model and the current LT_memory. The updating process of the LT_memory is a kind of non monotonic reasoning, with some restrictions benefiting from the peculiarities of this application (a priori knowledge of whether some information of the external world is complete or not) to allow a fast processing.

3.3 The Way Information Is Routed

The information is divided by the PFU into two categories: complete and incomplete. An information is said to be complete when each element in the object list is given a description, thus allowing the agent to behave in a perfectly (locally) known environment. A reactive behavior is launched in this case (i.e. the information is transmitted to the RU). An information is said to be incomplete when some elements in the object list are lacking. A cognitive behavior is launched in this case (i.e. the information is transmitted to the CU). In fact, some more rea-

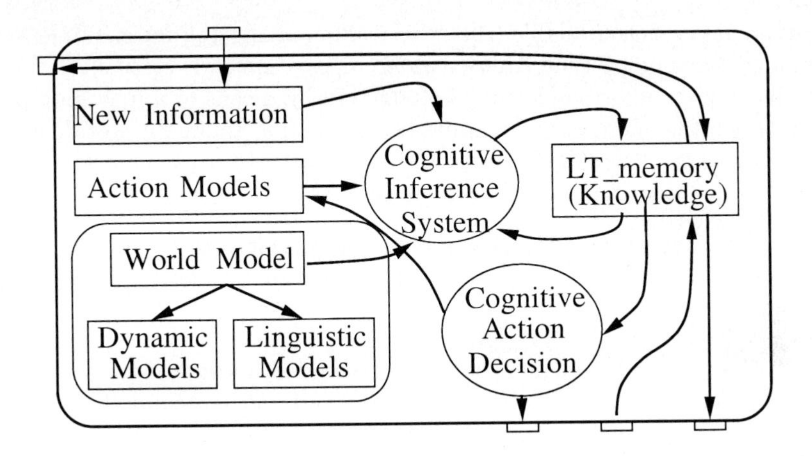

Fig. 4. Cognitive Unit (CU) functional architecture

soning is necessary in this case to ensure that the agent behaves in a consistent way.

Consider for example the two following information messages:

- Information 1: ((player team_1 uniform_10) 10 20 3 5)
- Information 2: ((player) 10 30)

In the first case all information about the player is available whilst much less is known in the second case. More information is needed in this case before deciding for any action. Due to the possibility given to the RU to communicate with the CU to obtain more information, a wide range of situations are in fact considered, ranking from pure reactive situations, where the RU works as a stand alone module, to pure cognitive situations.

3.4 Processing Incomplete Information via the CU

Three different cases have to be distinguished, as regards the completeness of the information received by the CU:

- *Case 1:* Every object is correctly and completely perceived;
- *Case 2:* Some information is lacking, because of the limited perception abilities of the player (distant objects, objects lying out of the agent vision cone);
- *Case 3:* What is perceived is incomplete and imprecise, due to environmental conditions : objects lying within the vision cone may be incompletely seen or even lacking, if occluded for example by another object.

No specific problem is encountered in the first case: the LT_memory can be directly updated with the new incoming information. Some reasoning has to be performed in the second case to retrieve lacking information : past information may be used for this purpose, and provide estimates for lacking objects, based on the dynamic models available (the new position of the ball for example may be estimated based on past information about its position and velocity). In the third case, incomplete information may be retrieved by means of a matching process under which the effective and estimated information about the world, stored in the LT_memory, is matched against the incomplete descriptions available.

The whole information management process may be summarized as follows:

- *Case 1:* Update the LT_memory based on complete object descriptions;
- *Case 2:* Try and estimate lacking information by simulating the evolution of previously seen objects, if currently out of scope;
- *Case 3:*
  1. Update the LT_memory with the estimated information.
  2. When information is incomplete, try and find matching descriptions in the available object lists, then complete this information;
  3. Update the LT_memory with the completed information.

4 The Application

The proposed agent architecture, its various processing units and modules, have been designed as general purpose components. We have tested the validity and performance of these components by building dedicated agents for the RoboCup competition.

The system is currently under implementation; it is running under UNIX and implemented in C++.

We focalize in what follows on the way reactive and cognitive players differently process information, due to their distinct memory abilities. We propose to analyze the two initial and final situations depicted in Fig.5. The opposing team is figured out in white on this display, it is based on the Humboldt University program team, winner of RoboCup 97. Our team, figured out in black, is comprising only two players, in order that each of them has maximal sensory information available.

Player 1 in our team is designed as a pure reactive player, and denoted as R_Player, whilst player 2 in our team is designed as a hybrid player, and denoted as H_Player. R_Player only uses reactive components and the ST_memory. H_Player, on the contrary, is designed according to the proposed hybrid architecture: it may use reactive as well as cognitive components, and makes use of a LT_memory. As may be observed from Fig.5, these two players are inactive, and do not move from the initial to the final situation, whilst the opponent players and the ball are actively moving : the objective of the following sections is to show how the knowledge and representations of R_Player and H_Player evolve under these conditions.

4.1 Memory Structure

Every sensory information is transmitted in the form of a list of object descriptors starting with *(see x ....)* where x is the referred time for the action of seeing. The

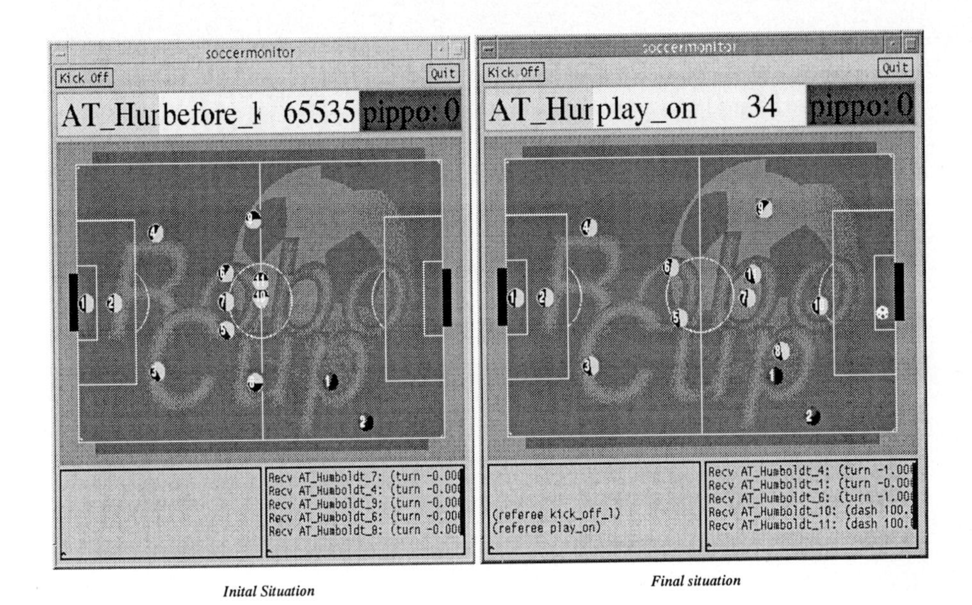

Fig. 5. An example game situation

objects are represented by their name and their features. As already exemplified, *((player) 73.7 1)* means *(object = player, distance = 73.7, angle = 1)*. A more detailed explanation for the grammar can be found in the soccer server manual [6].

Whatever the kind of memory possessed by the agent, information is stored as a list of object descriptors. For each object a list of features is reported in the corresponding object descriptor array. The meaning of each element is as follows:

obj : object identification number, used to find the object in the object descriptor array;
type : object type identification number, used to represent an object type (for example, $0 \equiv$ ball, $2 \equiv$ opponent player, etc.);
id : internal identification (for internal use);
see : number used to represent the time at which the player has seen the object;
sim : number used to represent the amount of simulations for the dynamic object (used only in the LT_memory);
ux,uy,px,py,vx,vy,ax,ay,dv : array used to represent various geometric features (position, velocity and so on.);
ϑ : flag (for internal use);

4.2 Reactive vs Cognitive Player in the Initial Situation

In this section we analyze how the two players handle the information available in the initial situation shown in Fig.5. The two players see the opponent players, the ball as well as other information. The corresponding agents are launched for the very initial time at time 0, so they have no memory of the past (if any).

Below is the information available to R_Player at time 0.

R_PLAYER sensory information at time 0. " (see 0 ((goal l) 75.2 15) ((flag c b) 24.5 -34) ((flag l t) 90 36) ((flag l b) 73.7 -10) ((flag p l t) 68.7 35) ((flag p l c) 59.7 19) ((flag p l b) 56.3 0) ((player) 73.7 15) ((player) 66.7 17) ((player) 49.4 3) ((player) 60.3 36) ((player AT_Humboldt 5) 33.1 23 -0 0) ((player AT_Humboldt) 40.4 41) ((player AT_Humboldt 7) 36.6 33 -0 0) ((player AT_Humboldt 8) 22.2 0 0 -0) ((line l) 72.2 -90)) ".

This sensory information is processed by the R_player reactive units. According to the current design, only information about the players and the ball is considered. From this set of information, only complete descriptions are used, like : *(player AT_Humboldt 5) 33.1 23 -0 0)*, where the team name *(AT_Humboldt)*, the player number *(5)*, the distance, the angle, the distance variation and the angle variation *(33.1 23 -0)* are known. Information like *((player) 60.3 36)* is discarded, because incomplete.

The processing of this information by the R_player produces a new representation in ST_memory reported in Tab.1.

Table 1. State of the R_player ST_memory at time 0

Object	obje	type	id	see	sim	ux	uy	px	py	vx	vy	ax	ay	dv	ϑ
Humboldt 5	16	2	0	0	0	0	0	10.7	−6.5	0	0	0	0	0	1
Humboldt 7	18	2	0	0	0	0	0	10.9	0.4	0	0	0	0	0	1
Humboldt 8	19	2	0	0	0	0	0	2.5	−19.4	0	0	0	0	0	1

The information reported in Tab.1 is the result of a rough filtering and processing of the sensory information transmitted by the server. Only three objects are correctly seen : the opponent player 5,7 and 8, respectively represented under identifiers 16, 18 and 19 in the ST_memory. Information about eight players in total has in fact been transmitted, the other players being out of viewing (like *opponent 11* for example). Among these eight players, five of them are too far away or partially occluded (like players 6,4 and 2), which results in the transmission of partial information which is not considered by the R_player.

Below is the information available to H_Player at time 0.

H_PLAYER sensory information at time 0. "(see 0 ((goal l) 87.4 -25) ((flag c t) 70.8 19) ((flag l t) 104.6 -7) ((flag p l t) 83.1 -8) ((flag p l c) 72.2 -20) ((flag p l b) 66.7 -36) ((ball) 40.4 0) ((player) 81.5 -24) ((player) 81.5 -22) ((player) 60.3 -33) ((player) 73.7 -7) ((player AT_Humboldt) 44.7 -15) ((player AT_Humboldt) 54.6 -2) ((player) 49.4 -8) ((player AT_Humboldt 8) 33.1 -27 -0 0) ((player)

60.3 12) ((player) 44.7 0) ((player AT_Humboldt) 44.7 4) ((player pippo 1) 13.5 0 -0 0) ((line t) 90 -45))"

The processing of this information by the H_player produces a new representation in LT_memory reported in Tab.4.

Table 2. State of the H_player LT_memory at time 0

Object	obj	type	id	see	sim	ux	*uy*	*px*	*py*	*vx*	*vy*	ax	ay	dv	ϑ
Ball	0	4	0	0	0	0	0	−1.4	−1.0	0	0	0	0	0	1
Pippo 1	1	3	0	0	0	0	0	−20.4	−20.0	0	0	0	0	0	1
Humboldt 8	19	2	0	0	0	0	0	1.4	−19.4	0	0	0	0	0	1

As may be seen from Tab.4, only three objects are correctly seen : the opponent player 8, the ball and player 1.

To be noticed is the fact that the information available to the R_Player and the H_Player is very similar, when initially launched. At this very initial time in fact, there is no way for the cognitive player to complete the available information by himself, since there is no past information available.

4.3 Reactive vs Cognitive Players in the Final Situation

The two players are now in the final situation depicted in Fig.5. They have the possibility, if given this capacity, to handle past as well as present information. Below is the information available respectively to R_Player and H_Player, together with the state of their respective memories. Note that the information has been gained at different times by the two players, due to the effective independence between the two players, which run concurrently.

R_PLAYER sensory information at time 34 "(see 34 ((goal l) 75.2 15) ((flag c b) 24.5 -34 0 0) ((flag l t) 90 36) ((flag l b) 73.7 -10) ((flag p l t) 68.7 35) ((flag p l c) 59.7 19) ((flag p l b) 56.3 0) ((player) 73.7 15) ((player) 66.7 17) ((player AT_Humboldt) 49.4 3) ((player) 60.3 36) ((player AT_Humboldt 5) 30 29 -0 0) ((player AT_Humboldt) 40.4 44) ((line l) 72.2 -90))"

Table 3. State of the R_player ST_memory at time 34

Object	obj	type	id	see	sim	ux	*uy*	*px*	*py*	*vx*	*vy*	ax	ay	dv	ϑ
Humboldt 5	16	2	0	34	34	0	0	6.5	−4.8	0	0	0	0	0	1

H_PLAYER sensory information at time 36 "Buff:(see 36 ((goal l) 87.4 -25) ((flag c t) 70.8 19) ((flag l t) 104.6 -7) ((flag p l t) 83.1 -8) ((flag p l c) 72.2 -20) ((flag p l b) 66.7 -36) ((player) 81.5 -24) ((player) 81.5 -22) ((player) 60.3 -33) ((player) 73.7 -7) ((player AT_Humboldt) 44.7 -10) ((player AT_Humboldt) 54.6 0) ((player AT_Humboldt) 33.1 14) ((player AT_Humboldt 8) 18.2 17 0.364 -0.1) ((player) 54.6 31) ((player AT_Humboldt 11) 36.6 24 -0 1) ((player pippo 1) 13.5 0 -0 0) ((line t) 90 -45))

Table 4. State of the H_player LT_memory at time 36

Object	obj	type	id	see	sim	ux	uy	px	py	vx	vy	ax	ay	dv	β
Ball	0	4	0	17	41	0	0	−46.0	−2.6	−1.3	−0.01	0	0	0	1
Pippo 1	1	3	0	36	36	0	0	−20.4	−20.0	0.0	0.0	0	0	0	1
Humboldt 7	18	2	0	33	37	0	0	−12.7	−1.3	−0.1	0.01	0	0	0	1
Humboldt 8	19	2	0	36	36	0	0	−21.4	−13.5	0.1	0.3	0	0	0	1
Humboldt 10	21	2	0	23	39	0	0	−20.1	−3.4	−0.7	−0.3	0	0	0	1
Humboldt 11	22	2	0	36	36	0	0	−16.9	4.5	−0.5	0.2	0	0	0	1

We have seen that in the initial situation (time 0), the memorizing behaviors remain substantially similar. In the final situation on the contrary, the dynamism of the environment result into more substantial changes. For the R_player in fact, the environment is formed by only one object (object 16). This player has no possibility to reason about changes in the environment, and only possesses an instantaneous representation of the world state. The H_player on the contrary takes advantage of a more complete representation of the environment state and history. To be noticed is the fact that only objects seen at time 36 are directly perceived. Information about other objects (e.g. previously seen objects 0, 18 and 21) has been estimated in the course of a simulation process. The matching process is not implemented yet and therefore not available to the H_player.

Whereas R_player is only given the possibility to react to instantaneous information about the world, H_player is given the possibility to exploit a more complete range of past and present information, thus enlarging his a priori restricted vision abilities by reasoning about non directly perceived objects.

5 Team Description

Our approach to the task of Soccer Competition have been inspired by real soccer competition.[7] In the real soccer the field is divided in three longitudinal zones (A,B,C) and three vertical zones (1,2,3) (See Fig.6). The type of game changes in function of the zone in which the ball and the players are situated. Zones A and C (lateral corridors) are extern and favorable to fast game. The central zone, (zone B), is a zone full of players and very busy. Any player in this zone has the objective of passing the ball to the external zone to support game in zone C and A. Conversely at the zones 1,2,3 are associated different game spirit. Zone 1 is the defensive zone. Players situated in this zone have the intention to avoid goal and to capture the ball. Zone 2 is an intermediate zone. In this zone players prepare and control the game. The zone 3 is an offensive zone, where all must be done to realize the goal. Player in this zone must be very reactive and opportunist.

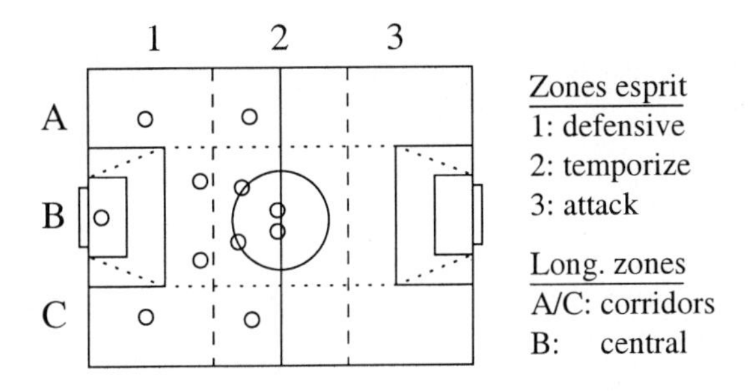

Fig. 6. The field partition, and the team distribution

A formation consists of eleven players, each with a role. The following are distinguish in: one Goalie, 2 Defenders, 4 Midfielders, 2 Wing-Forwards, 2 Attachers. The behavior and the position in the field is determined by the player's role. Players have an adaptation to the field position and changes the spirit of game from attack to defensive following the zone spirit. The role is characterized by some special behaviors. For example Goalie has the behavior of catch the ball. So, our team structure has the follows features: (i) dynamic behavior specialization, (ii) flexible positioning and game orientation, (ii) flexible and dynamic game spirit.

6 Conclusion

We have proposed a hybrid agent architecture as being built from a combination of reactive and cognitive behaviors running in a concurrent way. The core difference between these two kinds of behaviors lies in the associated memory abilities : only short-term memory abilities are provided to the reactive behaviors, whilst full long-term memorizing capacities are provided to the cognitive one. The short-term memory is the mere replication of the initial sensory information that is made available to the agent by the simulator. To be noticed is the fact that parts of this information may furthermore be discarded, if incomplete. The long-term memory on the contrary holds parts of the initial sensory information, completed with information estimated in an autonomous way, based on the reasoning abilities provided to the cognitive behavior. A varied range of decisions may therefore be taken by the ADU, from pure reactive ones in case sufficient information is available to full cognitive ones when sufficient time is available. Specific to our approach indeed is the fact that reactive and cognitive processes are considered as separate paths, each processing information in an autonomous way, rather than designed in a hierarchical way, as in traditional approaches, where the cognitive processes are meant to control the activation of reactive processes.

References

1. R. A. Brooks. Intelligence without representation. *Artificial Intelligence*, 47:139–159, 1991.
2. Innes A. Ferguson. Toward an architecture for adaptative rational mobile agents. In *Decentralized A.I. 3*, 1992.
3. S. Gaglio, A. Chella, and M. Frixione. A cognitive architecture for artificial vision. *Artificial Intelligence*, 89:73–112, 1997.
4. B. Hayes-Roth. An architecture for adaptive intelligent systems. *Artificial Intelligence*, 72:329–365, 1995.
5. Housheng Hu and Michael Brady. Dynamic global path planning with uncertainty for mobile robots in manufacturing. *IEEE Transaction on robotics and automation*, 13(5), 1997.
6. N. Itsuki. Soccer server: a simulator of robocup. In *Proceedings of AI symposium '95*, pages 29–34. Japanese Society for Artificial Intelligence, December 1995.
7. A. Laurier. *Football. Culture Tactique et Principes de Jeu.* Chiron Sport, 1985.
8. R. Fikes M. Genesereth and alt. Knowledge interchange format, version 3.0 reference manual. Technical report, Computer Since Department, Stanford University, 1992.
9. N. R. Jennings M. Wooldridge. Intelligent agents: Theory and practice. *The Knowledge Engineering Review*, 10:2:115–152, 1995.
10. Itsuki Noda and al. The robocup synthetic agent challenge 97. Available as:http://www.cs.cmu.edu/afs/cs/usr/pstone/public/papers/97syntetic-challenge/synthetic-challenge.html, 1997.
11. M.J.Wooldridge N.R.Jennings. Agent theories, architectures, and languages: A survey. In Springer-Verlag, editor, *Lecture Note in Artificial Intelligence*, volume 890. 1995.
12. M. Veloso P. Stone. Towards collaborative and adversarial learning: A case study in robotic soccer. *International Journal of Human-Computer System*, 1997.
13. Manuela Veloso Peter Stone, Mike Bowling. Predictive memory for an inaccessible environment. In *International Conference on Intelligent Robots and Systems (IROS'96)*, 1996.
14. Y. Shoham. Agent-oriented programming. *Artificial Intelligence*, 60:51–92, 1993.
15. Y. Labrou T. Finin. A semantic approach for kqml, a general purpose communication language for software agents. In *Third International Conference on Information and Knowledge Management (CIKM'94)*, November 1994.

Team-Partitioned, Opaque-Transition Reinforcement Learning *

Peter Stone and Manuela Veloso

Computer Science Department, Carnegie Mellon University
Pittsburgh, PA 15213
{pstone,veloso}@cs.cmu.edu

Abstract. We present a novel multi-agent learning paradigm called team-partitioned, opaque-transition reinforcement learning (TPOT-RL). TPOT-RL introduces the use of action-dependent features to generalize the state space. In our work, we use a *learned* action-dependent feature space to aid higher-level reinforcement learning. TPOT-RL is an effective technique to allow a team of agents to learn to cooperate towards the achievement of a specific goal. It is an adaptation of traditional RL methods that is applicable in complex, non-Markovian, multi-agent domains with large state spaces and limited training opportunities. TPOT-RL is fully implemented and has been tested in the robotic soccer domain, a complex, multi-agent framework. This paper presents the algorithmic details of TPOT-RL as well as empirical results demonstrating the effectiveness of the developed multi-agent learning approach with learned features.

1 Introduction

Reinforcement learning (RL) is an effective paradigm for training an artificial agent to act in its environment in pursuit of a goal. RL techniques rely on the premise that an agent's action policy affects its overall reward over time. As surveyed in [3], several popular RL techniques use dynamic programming to enable a single agent to learn an effective control policy as it traverses a stationary (Markovian) environment.

Dynamic programming requires that agents have or learn at least an approximate model of the state transitions resulting from its actions. Q-values encode future rewards attainable from neighboring states. A single agent can keep track of state transitions as its actions move it from state to state. Even in the POMDP model [3], in which agents must estimate their state due to hidden information, a single agent can still control its own path through the state space.

In contrast, we consider domains in which agents can control their own destiny only intermittently. In these domains, agents' actions are *chained*, i.e., a single agent's set of actions allows the agent to select which other agent will act next, or be chained after, in the pursuit of a goal. A single agent cannot control directly the full achievement of a goal, but a chain of agents will. In robotic soccer, the substrate of our work, the chaining of actions corresponds to passing

* This research is sponsored in part by the DARPA/RL Knowledge Based Planning and Scheduling Initiative under grant number F30602-97-2-0250. The views and conclusions contained in this document are those of the authors and should not be interpreted as representing the official policies or endorsements, either expressed or implied, of the U. S. Government.

a ball between the different agents. There are a variety of other such examples, such as information agents that may communicate through message passing and packet routing agents. (These domains contrast with, for example, grid world domains in which a single agent moves from some initial location to some final goal location; domains where agents take actions in parallel though also possibly in coordination — two robots executing tasks in parallel; and game domains in which the rules of the game enforce alternating actions by an agent and its opponent.) Because of our chaining of agents and the corresponding lack of control of single agents to fully achieve goals, we call these domains *team-partitioned.*

In addition, we assume agents do not know the state the world will be in after an action is selected, as another—possibly hidden—agent will *continue* the path to the goal. Adversarial agents can also intercept the chain and thwart the attempted goal achievement. The domain is therefore *opaque-transition.*

In this paper we present team-partitioned, opaque-transition reinforcement learning (TPOT-RL). TPOT-RL can learn a set of effective policies (one for each team member) with very few training examples. It relies on action-dependent dynamic features which coarsely generalize the state space. While feature selection is often a crucial issue in learning systems, our work uses a previously *learned* action-dependent feature. We empirically demonstrate the effectiveness of TPOT-RL in a multi-agent, adversarial environment, and show that the previously learned action-dependent feature can improve the performance of TPOT-RL. It does so by compressing a huge state space into a small, local feature space and is effective because the global team performance correlates with the local cues extracted by the learned feature.

The remainder of the paper is organized as follows. Section 2 formally presents the TPOT-RL algorithm. Section 3 details an implementation of TPOT-RL in the simulated robotic soccer domain with extensive empirical results presented in Section 4. Section 5 relates TPOT-RL to previous work and concludes.

2 Team-Partitioned, Opaque-Transition RL

Formally, a policy is a mapping from a state space S to an action space A such that the agent using that policy executes action a whenever in state s. At the coarsest level, when in state s, an agent compares the expected, long-term rewards for taking each action $a \in A$, choosing an action based on these expected rewards. These expected rewards are learned through experience.

Designed to work in real-world domains with far too many states to handle individually, TPOT-RL constructs a smaller feature space V using action-dependent feature functions. The expected reward $Q(v, a)$ is then computed based on the state's corresponding entry in feature space.

In short, the policy's mapping from S to A in TPOT-RL can be thought of as a 3-step process:

State generalization: the state s is generalized to a feature vector v using the state generalization function $f : S \mapsto V$.

Value function learning: the feature vector v is used to estimate the expected reward for taking each possible action using the changing (learned) value function $Q : (V, A) \mapsto \mathbb{R}$.

Action selection: an action a is chosen for execution and its real-world reward is used to further update Q.

While these steps are common in other RL paradigms, each step has unique characteristics in TPOT-RL.

2.1 State Generalization

TPOT-RL's state generalization function $f : S \mapsto V$ relies on a unique approach to constructing V. Rather than discretizing the various dimensions of S, it uses *action-dependent* features. In particular, each possible action a_i is evaluated locally based on the current state of the world using a fixed function $e : (S, A) \mapsto U$. Unlike Q, e does not produce the expected long-term reward of taking an action; rather, it classifies the likely short-term effects of the action. For example, if actions sometimes succeed and sometimes fail to achieve their intended effects, e could indicate something of the following form: if selected, action a_7 is (or is not) likely to produce its intended effects.

In the multi-agent scenario, other than one output of e for each action, the feature space V also involves one coarse component that partitions the state space S among the agents. If the size of the team is m, then the partition function is $P : S \mapsto M$ with $|M| = m$. In particular, if the set of possible actions $A = \{a_0, a_1, \ldots, a_{n-1}\}$, then

$$f(s) = \langle e(s, a_0), e(s, a_1), \ldots, e(s, a_{n-1}), P(s)\rangle, \text{ and so}$$
$$V = U^{|A|} \times M.$$

Thus, $|V| = |U|^{|A|} * m$. Since TPOT-RL has no control over $|A|$ or m, and since the goal of constructing V is to have a small feature space over which to learn, TPOT-RL will be more effective for small sets U.

This state generalization process reduces the complexity of the learning task by constructing a small feature space V which partitions S into m regions. Each agent need learn how to act only within its own partition. Nevertheless, for large sets A, the feature space can still be too large for learning, especially with limited training examples. Our particular action-dependent formulation allows us to reduce the effective size of the feature space in the value-function-learning step. Choosing features for state generalization is generally a hard problem. While TPOT-RL does not not specify the function e, our work uses a previously-learned dynamic feature function.

2.2 Value Function Learning

As we have seen, TPOT-RL uses action-dependent features. Therefore, we can assume that the expected long-term reward for taking action a_i depends only on the feature value related to action a_i. That is,

$$Q(\langle e(s, a_1), \ldots, e(s, a_{n-1}), P(s)\rangle, a_i) = Q(\langle e(s', a_1), \ldots, e(s', a_{n-1}), P(s')\rangle, a_i)$$

whenever $e(s, a_i) = e(s', a_i)$ and $P(s) = P(s')$. In other words, if $f(s) = v$, $Q(v, a_i)$ depends entirely upon $e(s, a_i)$ and is independent of $e(s, a_j)$ for all $j \neq i$.

Without this assumption, since there are $|A|$ actions possible for each feature vector, the value function Q has $|V| * |A| = |U|^{|A|} * |A| * m$ independent values.

Under this assumption, however, the Q-table has at most $|A| * |U| * m$ entries: for each action possible from each position, there is only one relevant feature value. Therefore, even with only a small number of training examples available, we can treat the value function Q as a lookup-table without the need for any complex function approximation. To be precise, Q stores one value for every possible combination of action a, $e(s, a)$, and $P(s)$.

For example, Table 1 shows the entire feature space for one agent's partition of the state space when $|U| = 3$ and $|A| = 2$. There are $|U|^{|A|} = 3^2$ different entries in feature space with 2 Q-values for each entry: one for each possible action. $|U|^{|A|} * m$ is much smaller than the original state space for any realistic problem, but it can grow large quickly, particularly as $|A|$ increases. However, notice in Table 1 that, under the assumption described above, there are only $3 * 2$ independent Q-values to learn, reducing the number of free variables in the learning problem by 67% in this case.

$e(s, a_0)$	$e(s, a_1)$	$Q(v, a_0)$	$Q(v, a_1)$
u_0	u_0	$q_{0,0}$	$q_{1,0}$
u_0	u_1	$q_{0,0}$	$q_{1,1}$
u_0	u_2	$q_{0,0}$	$q_{1,2}$
u_1	u_0	$q_{0,1}$	$q_{1,0}$
u_1	u_1	$q_{0,1}$	$q_{1,1}$
u_1	u_2	$q_{0,1}$	$q_{1,2}$
u_2	u_0	$q_{0,2}$	$q_{1,0}$
u_2	u_1	$q_{0,2}$	$q_{1,1}$
u_2	u_2	$q_{0,2}$	$q_{1,2}$

$\Longrightarrow$

$e(s, a_i)$	$Q(v, a_0)$	$Q(v, a_1)$
u_0	$q_{0,0}$	$q_{1,0}$
u_1	$q_{0,1}$	$q_{1,1}$
u_2	$q_{0,2}$	$q_{1,2}$

Table 1. A sample Q-table for a single agent when $|U| = 3$ and $|A| = 2$: $U = \{u_0, u_1, u_2\}$, $A = \{a_0, a_1\}$. $q_{i,j}$ is the estimated value of taking action a_i when $e(s, a_i) = u_j$. Since this table is for a single agent, $P(s)$ remains constant.

The Q-values learned depend on the agent's past experiences in the domain. In particular, after taking an action a while in state s with $f(s) = v$, an agent receives reward r and uses it to update $Q(v, a)$ as follows:

$$Q(v, a) = Q(v, a) + \alpha(r - Q(v, a)) \quad (1)$$

Since the agent is not able to access its teammates' internal states, future team transitions are completely opaque from the agent's perspective. Thus it cannot use dynamic programming to update its Q-table. Instead, the reward r comes directly from the observable environmental characteristics—those that are captured in S—over a maximum number of time steps t_{lim} after the action is taken. The reward function $R : S^{t_{lim}} \mapsto \mathbb{R}$ returns a value at some time no further than t_{lim} in the future. During that time, other teammates or opponents can act in the environment and affect the action's outcome, but the agent may not be able to observe these actions. For practical purposes, it is crucial that the reward function is only a function of the observable world *from the acting agent's perspective.* In practice, the range of R is $[-Q_{max}, Q_{max}]$ where Q_{max} is the reward for immediate goal achievement .

The reward function, including t_{lim} and Q_{max}, is domain-dependent. One possible type of reward function is based entirely upon reaching the ultimate goal. In this case, an agent charts the actual (long-term) results of its policy in the environment. However, it is often the case that goal achievement is very infrequent. In order to increase the feedback from actions taken, it is useful to use an internal reinforcement function, which provides feedback based on intermediate states towards the goal. We use this internal reinforcement approach.

2.3 Action Selection

Informative action-dependent features can be used to reduce the free variables in the learning task still further at the action-selection stage if the features themselves discriminate situations in which actions should not be used. For example, if whenever $e(s, a_i) = u_1$, a_i is not likely to achieve its expected reward, then the agent can decide to ignore actions with $e(s, a_i) = u_1$.

Formally, consider $W \subseteq U$ and $B(s) \subseteq A$ with $B(s) = \{a \in A | e(s, a) \in W\}$. When in state s, the agent then chooses an action from $B(s)$, either randomly when exploring or according to maximum Q-value when exploiting. Any exploration strategy, such as Boltzman exploration, can be used over the possible actions in $B(s)$. In effect, W acts in TPOT-RL as an action filter which reduces the number of options under consideration at any given time. Of course, exploration at the filter level can be achieved by dynamically adjusting W.

(a)

$e(s,a_0)$	$e(s,a_1)$	$Q(v,a_0)$	$Q(v,a_1)$
u_0	u_0	$q_{0,0}$	$q_{1,0}$
u_0	u_1	$q_{0,0}$	—
u_0	u_2	$q_{0,0}$	$q_{1,2}$
u_1	u_0	—	$q_{1,0}$
u_1	u_1	—	—
u_1	u_2	—	$q_{1,2}$
u_2	u_0	$q_{0,2}$	$q_{1,0}$
u_2	u_1	$q_{0,2}$	—
u_2	u_2	$q_{0,2}$	$q_{1,2}$

(b)

$e(s,a_0)$	$e(s,a_1)$	$Q(v,a_0)$	$Q(v,a_1)$
u_0	u_0	—	—
u_0	u_1	—	—
u_0	u_2	—	$q_{1,2}$
u_1	u_0	—	—
u_1	u_1	—	—
u_1	u_2	—	$q_{1,2}$
u_2	u_0	$q_{0,2}$	—
u_2	u_1	$q_{0,2}$	—
u_2	u_2	$q_{0,2}$	$q_{1,2}$

Table 2. The resulting Q-tables when **(a)** $W = \{u_0, u_2\}$, and **(b)** $W = \{u_2\}$.

For example, Table 2, illustrates the effect of varying $|W|$. In the rare event that $B(s) = \emptyset$, i.e. $\forall a_i \in A, e(s, a_i) \notin W$, either a random action can be chosen, or rough Q-value estimates can be stored using sparse training data. This condition becomes rarer as $|A|$ increases. For example, with $|U| = 3, |W| = 1, |A| = 2$ as in Table 2(b), $4/9 = 44.4\%$ of feature vectors have no action that passes the W filter. However, with $|A| = 8$ only $256/6561 = 3.9\%$ of feature vectors have no action that passes the W filter. If $|W| = 2$ and $|A| = 8$, only 1 of 6561 feature vectors fails to pass the filter. Thus using W to filter action selection can reduce the number of free variables in the learning problem without significantly reducing the coverage of the learned Q-table.

By using action-dependent features to create a coarse feature space, and with the help of a reward function based entirely on individual observation of the environment, TPOT-RL enables team learning in a multi-agent, adversarial environment even when agents cannot track state transitions.

3 TPOT-RL Applied to a Complex Multi-Agent Task

Our research has been focussed on multi-agent learning in complex, collaborative and adversarial environments. Our general approach, called *layered learn-*

ing, is based on the premise that realistic domains are too complex for learning mappings directly from sensor inputs to actuator outputs. Instead, intermediate domain-dependent skills should be learned in a bottom-up hierarchical fashion [9]. We implemented TPOT-RL as the current highest layer of a layered learning system in the RoboCup soccer server [7].

The soccer server used at RoboCup-97 [4] is a much more complex domain than has previously been used for studying multi-agent policy learning. With 11 players on each team controlled by separate processes; noisy, low-level, real-time sensors and actions; limited communication; and a fine-grained world state model including hidden state, the RoboCup soccer server provides a framework in which machine learning can improve performance. Newly developed multi-agent learning techniques could well apply in real-world domains.

A key feature of the layered learning approach is that learned skills at lower levels are used to train higher-level skills. For example, we used a neural network to help players learn how to intercept a moving ball. Then, with all players using the learned interception behavior, a decision tree (DT) enabled players to estimate the likelihood that a pass to a given field location would succeed. Based on almost 200 continuous-valued attributes describing teammate and opponent positions on the field, players learned to classify the pass as a likely success (ball reaches its destination or a teammate gets it) or likely failure (opponent intercepts the ball). Using the C4.5 DT algorithm [8], the classifications were learned with associated confidence factors. The learned behaviors proved effective both in controlled testing scenarios [9, 11] and against other previously-unseen opponents in an international tournament setting [4].

These two previously-learned behaviors were both trained off-line in limited, controlled training situations. They could be trained in such a manner due to the fact that they only involved a few players: ball interception only depends on the ball's and the agent's motions; passing only involves the passer, the receiver, and the agents in the immediate vicinity. On the other hand, deciding where to pass the ball during the course of a game requires training in game-situations since the value of a particular action can only be judged in terms of how well it works when playing with particular teammates against particular opponents. For example, passing backwards to a defender could be the right thing to do if the defender has a good action policy, but the wrong thing to do if the defender is likely to lose the ball to an opponent.

Although the DT predicts whether a player can execute a pass, it gives no indication of the strategic value of doing so. But the DT reduces a detailed state description to a single continuous output. It can then be used to drastically reduce the complex state and provide great generalization. In this work we use the DT as the crucial action-dependent feature function e in TPOT-RL.

3.1 State Generalization Using a Learned Feature

In the soccer example, we applied TPOT-RL to enable each teammate to simultaneously learn a high-level action policy. The policy is a function that determines what an agent should do *when it has possession of the ball.*[2] The input of the policy is the agent's perception of the current world state; the output is a target destination for the ball in terms of a location on the field, e.g. the opponent's goal. In our experiment, each agent has 8 possible actions as illustrated in Figure 1(a). Since a player may not be able to tell the results of other players' actions, or even when they can act, the domain is opaque-transition.

A team formation is divided into 11 positions ($m = 11$), as also shown in Figure 1(a) [11]. Thus, the partition function $P(s)$ returns the player's position. Using our layered learning approach, we use the previously trained DT as e. Each possible pass is classified as either a likely success or a likely failure with a confidence factor. Outputs of the DT could be clustered based on the confidence factors. In our experiments, we cluster into only two sets indicating success and failure. Therefore $|U| = 2$ and $V = U^8 \times \{PlayerPositions\}$ so $|V| = |U|^{|A|} * m = 2^8 * 11$. Even though each agent only gets about 10 training examples per 10-minute game and the reward function shifts as teammate policies improve, the learning task becomes feasible. This feature space is immensely smaller than the original state space, which has more than 22^{10^9} states.[3] Since e indicates the likely success or failure of each possible action, at action-selection time, we only consider the actions that are likely to succeed ($|W|=1$). Therefore, each player learns 8 Q-values, with a total of 88 learned by the team as a whole. Even with sparse training and shifting concepts, such a learning task is tractable.

3.2 Internal Reinforcement through Observation

As in any RL approach, the reward function plays a large role in determining what policy is learned. One possible reward function is based entirely upon reaching the ultimate goal. Although goals scored are the true rewards in this domain, such events are very sparse. In order to increase the feedback from actions taken, it is useful to use an internal reinforcement function, which provides feedback based on intermediate states towards the goal. Without exploring the space of possible such functions, we created one reward function R.

R gives rewards for goals scored. However, players also receive rewards if the ball goes out of bounds, or else after a fixed period of time t_{lim} based on the ball's average lateral position on the field. In particular, when a player takes action a_i in state s such that $e(s, a_i) = u$, the player records the time t at which the action was taken as well as the x coordinate of the ball's position at time t, x_t. The reward function R takes as input the observed ball position over time t_{lim} (a subset of $S^{t_{lim}}$) and outputs a reward r. Since the ball position over time depends also on other agents' actions, the reward is stochastic and non-stationary. Under the following conditions, the player fixes the reward r:

1. if the ball goes out of bounds (including a goal) at time $t + t_o$ ($t_o < t_{lim}$);
2. if the ball returns to the player at time $t + t_r$ ($t_r < t_{lim}$);
3. if the ball is still in bounds at time $t + t_{lim}$.

[2] In the soccer server there is no actual perception of having "possession" of the ball. Therefore we consider the agent to have possession when it is within kicking distance.

[3] Each of the 22 players can be in any of 680*1050*3600 (x, y, θ) locations, not to mention the player velocities and the ball position and velocity.

In case 1, the reward r is based on the value r_o as indicated in Figure 1(b): $r = \frac{r_o}{1+(\phi-1)*t_o/t_{lim}}$. Thus, the farther in the future the ball goes out of bounds (i.e. the larger t_o), the smaller the absolute value of r. This scaling by time is akin to the discount factor used in Q-learning. We use $t_{lim} = 30sec.$ and $\phi = 10$.

In cases 2 and 3, the reward r is based on the average x-position of the ball over the time t to the time $t+t_r$ or $t+t_{lim}$. Over that entire time span, the player samples the x-coordinate of the ball at fixed, periodic intervals and computes the average x_{avg} over the times at which the ball position is known. Then if $x_{avg} > x_t$, $r = \phi * \frac{x_{avg}-x_t}{x_{og}-x_t}$ where x_{og} is the x-coordinate of the opponent goal (the right goal in Figure 1(b)). Otherwise, if $x_{avg} \leq x_t$, $r = -\phi * \frac{x_t-x_{avg}}{x_t-x_{lg}}$ where x_{lg} is the x-coordinate of the learner's goal.[4] Thus, the reward is the fraction of the available field by which the ball was advanced, on average, over the time-period in question. Note that a backwards pass can lead to positive reward if the ball then moves forward in the near future.

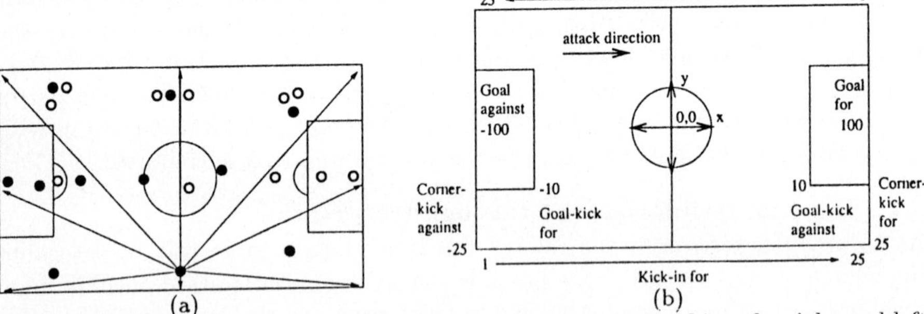

Fig. 1. (a) The black and white dots represent the players attacking the right and left goals respectively. Arrows indicate a single player's action options when in possession of the ball. The player kicks the ball towards a fixed set of markers around the field, including the corner flags and the goals. (b) The component r_o of the reward function R based on the circumstances under which the ball went out of bounds. For kick-ins, the reward varies linearly with the x position of the ball.

The reward r is based on direct environmental feedback. It is a domain-dependent internal reinforcement function based upon heuristic knowledge of progress towards the goal. Notice that it relies solely upon the player's own impression of the environment. If it fails to notice the ball's position for a period of time, the internal reward is affected. However, players can track the ball much more easily than they can deduce the internal states of other players as they would have to do were they to determine future team state transitions.

As teammates learn concurrently, the concept to be learned by each individual agent changes over time. We address this problem by gradually increasing exploitation as opposed to exploration in all teammates and by using a learning rate $\alpha = .02$ (see Equation 1). Thus, even though we are averaging several reward values for taking an action in a given state, each new example accounts for

[4] The parameter ϕ insures that intermediate rewards cannot override rewards for attaining the ultimate goal.

2% of the updated Q-value: rewards gained while teammates were acting more randomly are weighted less heavily.

4 Results

Empirical testing has demonstrated that TPOT-RL can effectively learn multi-agent control policies with few training instances in a complex, dynamic domain. Figure 2(a) plots cumulative goals scored by a learning soccer team playing against an otherwise equally-skilled team that passes to random destinations over the course of a single long run equivalent in time to 160 10-minute games. In this experiment, and in all the remaining ones, the learning agents start out acting randomly and with empty Q-tables. Over the course of the games, the probability of acting randomly as opposed to taking the action with maximum Q-value decreases linearly over periods of 40 games from 1 to .5 in game 40, to .1 in game 80, to point .01 in game 120 and thereafter. As apparent from the graph, the team using TPOT-RL learns to vastly outperform the randomly passing team. During this experiment, $|U| = 1$, thus rendering the function e irrelevant: the only relevant state feature is the player's position on the field.

A key characteristic of TPOT-RL is the ability to learn with minimal training examples. During the run graphed in Figure 2(a), the 11 players got an average of 1490 action-reinforcement pairs over 160 games. Thus, players only get reinforcement an average of 9.3 times each game, or less than once every minute. Since each player has 8 actions from which to choose, each is only tried an average of 186.3 times over 160 games, or just over once every game. Under these training circumstances, very efficient learning is clearly needed.

TPOT-RL is effective not only against random teams, but also against goal-directed, hand-coded teams. For testing purposes, we constructed an opponent team which plays with all of its players on the same side of the field, leaving the other side open as illustrated by the white team in Figure 1. The agents use a hand-coded policy which directs them to pass the ball up the side of the field to the forwards who then shoot on goal. The team periodically switches from one side of the field to the other. We call this team the "switching team."

Were the opponent team to always stay on the same side of the field, the learning team could advance the ball up the other side of the field without any regard for current player positions. Thus, TPOT-RL could be run with $|U| = 1$, which renders e inconsequential. Indeed, we verified empirically that TPOT-RL is able to learn an effective policy against such an opponent using $|U| = 1$.

Against the switching team, a player's best action depends on the current state. Thus a feature that discriminates among possible actions dynamically can help TPOT-RL. Figure 2(b) compares TPOT-RL with different functions e and different sets W when learning against the switching team.

With $|U| = 1$ (Figure 2(b.1)), the learning team is unable to capture different opponent states since each player has only one Q-value associated with each possible action, losing 139-127 (cumulative score over 40 games after 160 games of training). Recall that if $|U| = 1$ the function e cannot discriminate between different classes of states: we end up with a poor state generalization.

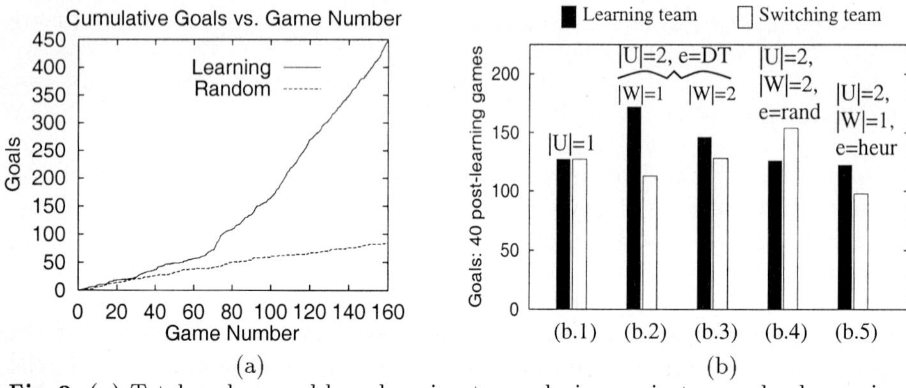

Fig. 2. **(a)** Total goals scored by a learning team playing against a randomly passing team. The independent variable is the number of 10-minute games that have elapsed. **(b)** The results after training 5 different TPOT-RL runs against the switching team.

In contrast, with the previously trained DT classifying passes as likely successes or failures (e = DT) and TPOT-RL filtering out the failures, the learning team wins 172-113 (Figure 2(b.2)). Therefore the learned pass-evaluation feature is able to usefully distinguish among possible actions and help TPOT-RL to learn a successful action policy. The DT also helps learning when $W = U$ (Figure 2(b.3)), but when $|W| = 1$ performance is better.

Figure 2(b.4) demonstrates the value of using an informative action-dependent feature function e. When a random function e = rand is used, TPOT-RL performs noticeably worse than when using the DT. For the random e we show $|W| = 2$ because it only makes sense to filter out actions when e contains useful information. Indeed, when e = rand and $|W| = 1$, the learning team performs even worse than when $|W| = 2$ (it loses 167-60). The DT even helps TPOT-RL more than a hand-coded heuristic pass-evaluation function (e = heur) based on one that we successfully used on our real robot team [13] (Figure 2(b.5)).

Final score is the ultimate performance measure. However, we examined learning more closely in the best case experiment (e = DT, $|W| = 1$ — Figure 2(b.2)). Recall that the learned feature provides no information about which actions are *strategically* good. TPOT-RL must learn that on its own. To test that it is indeed learning to advance the ball towards the opponent's goal (other than by final score), we calculated the number of times each action was predicted to succeed by e and the number of times it was actually selected by TPOT-RL after training. Throughout the entire team, the 3 of 8 actions towards the opponent's goal were selected $6437/9967 = 64.6\%$ of the times that they were available after filtering. Thus TPOT-RL learns that it is, in general, better to advance the ball towards the opponent's goal.

To test that the filter was eliminating action choices based on likelihood of failure we found that 39.6% of action options were filtered out when e = DT and $|W| = 1$. Out of 10,400 actions, it was never the case that all 8 actions were filtered out.

5 Discussion and Conclusion

Typical RL paradigms update the value of a state-action pair based upon the value of the subsequent state (or state distribution). As presented in [3], the typical update function in Q-learning is $Q(s,a) = Q(s,a) + \alpha(r + \gamma max_a Q(s',a) - Q(s,a))$ where s' is the state next reached after executing action a in state s and γ is the discount factor. While characterized as "model-free" in the sense that the agent need not know the transition function $T : (S, A) \mapsto S$, these paradigms assume that the agent can observe the subsequent state that it enters.

However, the vast amount of hidden state coupled with the multi-agent nature of this domain make such a paradigm impossible for the following reasons. Having only local world-views, agents cannot reliably discern when a teammate is able to take an action. Furthermore, even when able to notice that a teammate is within kicking distance of the ball, the agent certainly cannot tell the feature values for the teammate's possible actions. Worse than being model-free, multi-agent RL must deal with the inability to even track the team's state trajectory. Thus we use Equation 1, which doesn't rely on knowing s'.

Previous multi-agent reinforcement learning systems have typically dealt with much simpler tasks than the one presented here. Littman uses Markov games to learn stochastic policies in a very abstract version of 1-on-1 robotic soccer [5]. There have also been a number of studies of multi-agent reinforcement learning in the pursuit domain, including [12]. In this domain, four predators chase a single prey in a small grid-like world.

Another team-partitioned, opaque transition domain is network routing as considered in [2]. Each network node is considered as a separate agent which cannot see a packet's route beyond its own action. A major difference between that work and our own is that neighboring nodes send back their own value estimates whereas we assume that agents do not even know their neighboring states. Thus unlike TPOT-RL agents, the nodes are able to use dynamic programming.

In other soccer systems, there have been a number of learning techniques that have been explored. However, most have learned low-level, individual skills as opposed to team-based policies [1, 10]. Interestingly, [6] uses genetic programming to evolve team behaviors from scratch as opposed to our layered learning approach.

TPOT-RL is an adaptation of RL to non-Markovian multi-agent domains with opaque transitions, large state spaces, hidden state and limited training opportunities. The fully implemented algorithm has been successfully tested in simulated robotic soccer, such a complex multi-agent domain with opaque transitions. TPOT-RL facilitates learning by partitioning the learning task among teammates, using coarse, action-dependent features, and gathering rewards directly from environmental observations. Our work uses a learned feature within TPOT-RL.

TPOT-RL represents the third and currently highest layer within our ongoing research effort to construct a complete learning team using the layered learning paradigm [9]. As advocated by layered learning, it uses the previous learned layer—an action-dependent feature—to improve learning. TPOT-RL can learn against any opponent since the learned values capture opponent characteristics. The next learned layer could learn to choose among learned team policies based on characteristics of the current opponent. TPOT-RL represents a crucial step towards completely learned collaborative and adversarial strategic reasoning within a team of agents.

References

1. Minoru Asada, Shoichi Noda, Sukoya Tawaratumida, and Koh Hosoda. Purposive behavior acquisition for a real robot by vision-based reinforcement learning. *Machine Learning*, 23:279–303, 1996.
2. J. A. Boyan and M. L. Littman. Packet routing in dynamically changing networks: A reinforcement learning approach. In J. D. Cowan, G. Tesauro, and J. Alspector, editors, *Advances In Neural Information Processing Systems 6*. Morgan Kaufmann Publishers, 1994.
3. Leslie Pack Kaelbling, Michael L. Littman, and Andrew W. Moore. Reinforcement learning: A survey. *Journal of Artificial Intelligence Research*, 4:237–285, May 1996.
4. Hiroaki Kitano, Yasuo Kuniyoshi, Itsuki Noda, Minoru Asada, Hitoshi Matsubara, and Eiichi Osawa. RoboCup: A challenge problem for AI. *AI Magazine*, 18(1):73–85, Spring 1997.
5. Michael L. Littman. Markov games as a framework for multi-agent reinforcement learning. In *Proceedings of the Eleventh International Conference on Machine Learning*, pages 157–163, San Mateo, CA, 1994. Morgan Kaufman.
6. Sean Luke, Charles Hohn, Jonathan Farris, Gary Jackson, and James Hendler. Coevolving soccer softbot team coordination with genetic programming. In Hiroaki Kitano, editor, *RoboCup-97: Robot Soccer World Cup I*, pages 398–411, Berlin, 1998. Springer Verlag.
7. Itsuki Noda, Hitoshi Matsubara, and Kazuo Hiraki. Learning cooperative behavior in multi-agent environment: a case study of choice of play-plans in soccer. In *PRICAI'96: Topics in Artificial Intelligence (Proc. of 4th Pacific Rim International Conference on Artificial Intelligence, Cairns, Australia)*, pages 570–579, Cairns,Australia, August 1996.
8. J. Ross Quinlan. *C4.5: Programs for Machine Learning*. Morgan Kaufmann, San Mateo, CA, 1993.
9. Peter Stone and Manuela Veloso. A layered approach to learning client behaviors in the RoboCup soccer server. *Applied Artificial Intelligence*, 12:165–188, 1998.
10. Peter Stone and Manuela Veloso. Towards collaborative and adversarial learning: A case study in robotic soccer. *International Journal of Human-Computer Studies*, 48(1):83–104, January 1998.
11. Peter Stone and Manuela Veloso. Using decision tree confidence factors for multiagent control. In Hiroaki Kitano, editor, *RoboCup-97: Robot Soccer World Cup I*, pages 99–111. Springer Verlag, Berlin, 1998.
12. Ming Tan. Multi-agent reinforcement learning: Independent vs. cooperative agents. In *Proceedings of the Tenth International Conference on Machine Learning*, pages 330–337, 1993.
13. Manuela Veloso, Peter Stone, Kwun Han, and Sorin Achim. The CMUnited-97 small-robot team. In Hiroaki Kitano, editor, *RoboCup-97: Robot Soccer World Cup I*, pages 242–256. Springer Verlag, Berlin, 1998.

Cooperative Behavior Acquisition in a Multiple Mobile Robot Environment by Co-evolution

Eiji Uchibe, Masateru Nakamura, Minoru Asada

Dept. of Adaptive Machine Systems, Graduate School of Eng., Osaka University,
Suita, Osaka 565-0871, Japan

Abstract. Co-evolution has recently been receiving increased attention as a method for multi agent simultaneous learning. This paper discusses how multiple robots can emerge cooperative behaviors through co-evolutionary processes. As an example task, a simplified soccer game with three learning robots is selected and a GP (genetic programming) method is applied to individual population corresponding to each robot so as to obtain cooperative and competitive behaviors through evolutionary processes. The complexity of the problem can be explained twofold: co-evolution for cooperative behaviors needs exact synchronization of mutual evolutions, and three robot co-evolution requires well-complicated environment setups that may gradually change from simpler to more complicated situations so that they can obtain cooperative and competitive behaviors simultaneously in a wide range of search area in various kinds of aspects. Simulation results are shown, and a discussion is given.

1 Introduction

Realization of autonomous robots that organize their own internal structures to accomplish given tasks through interactions with their environments is one of the ultimate goals of Robotics and AI. Especially, emergence of cooperative behaviors between multiple robots has been receiving increased attention as a problem of multi agent simultaneous learning. Because, it seems difficult to apply conventional learning algorithms such as reinforcement learning to co-evolve cooperative agents since the environment including other agents may cause unpredictable changes in state transitions for learning agents.

Uchibe et al. proposed a reinforcement learning supported by system identification [10] and learning schedule [9] in multi agent environments. Their method estimates the relationships between learner's behaviors and other robot ones through interactions. However, in their method, only one robot may learn and other robots should have fixed policy in order for the learning to converge.

Recently, co-evolution has been receiving increased attention as a method for multi agent simultaneous learning. Existing methods have mostly focused on two competing individuals such as a prey and a predator. Cliff and Miller [2] have analyzed the relationship between a prey and a predator, and Floreano and Nolfi [3] have implemented real robot experiments which co-evolved prey and predator robots of which skills gradually leveled up under certain conditions. Luke et al.

[7] apply the co-evolution technique to the soccer game to evolve teams each of which can be regarded as an individual and attempts to beat other teams, that is, co-evolution for competition.

In the realm of nature, we can see, however, various aspects of behaviors emerged from multi agent environments, not only competition but also cooperation, ignorance, and so on. That means there could be artificial co-evolution for other than competition. This paper discusses how multiple robots can obtain cooperative behaviors through the co-evolutionary process. As an example task, a simplified soccer game with three learning robots is selected and a GP (genetic programming) method [5, 6], a kind of genetic algorithms based on tree structure with more abstracted node representation than gene coding in ordinary GAs, is applied so as to experimentally evaluate obtained behaviors in the context of cooperative and competitive tasks. Each robot has its own individual population, and attempts to acquire desired behaviors through interactions with environment that is ever changing in the co-evolutionary process. The complexity of the problem can be explained twofold: 1) co-evolution for cooperative behaviors needs exact synchronization of mutual evolutions, and 2) three robot co-evolution requires well-complicated environment setups that may contribute to providing a wide variety of searching area from simpler to more complicated situations in which they seek for better strategies so that they can emerge cooperative and competitive behaviors simultaneously.

The rest of this article is organized as follows. First, we describe our views on co-evolution in the context of cooperative and competitive tasks. Next, we explain our example task, a simplified soccer game in which cooperative and competitive tasks are involved. Then, we give a brief explanation of the GP and setting parameters. Finally, the preliminary results of computer simulation are shown, and a discussion is given.

2 Co-evolution in cooperative tasks

Generally, we have following difficult problems in multi agent simultaneous learning:

1. **Unknown Policy**
   Learning agents do not know other agents' policies in advance, therefore they need to estimate them through observations and actions. What's the worse is that the agent policies may change through a learning process.
2. **Synchronized Learning**
   Mutual learning robots have to improve their learned policies simultaneously. If the opponent learning converged much earlier than itself, one robot could not improve its strategy against the difficult environment its opponent has already fixed.
3. **Credit Assignment**
   Credit assignment to learning robots for cooperation seems difficult. If the credit involves group evaluation only, one robot may accomplish a given

task by itself and others do just actions irrelevant to the task as they do not seem to interfere the one robot's actions. Else if only individual evaluation is involved, robots may compete others each other. This trade-off should be carefully dealt.

Co-evolution is one of potential solutions for the first problem by seeking for better strategies in a wide range of searching area in parallel. The second and third ones might be solved by careful designs of environmental setups and fitness functions. Emerging patterns by co-evolution can be categorized into three ones.

1. **Cycles of switching fixed strategies**
   This pattern can be often observed in case of a prey and predator which often shift their strategies drastically to escape from or to catch the opponent. The same strategies iterates many times and no improvements on both sides seem to be seen.
2. **Trap to local maxima**
   This corresponds to the second problem stated above. Since one side overwhelmed its opponents, both sides reached to one of stable but low skill levels, and therefore no change happens after this settlement.
3. **Mutual skill development**
   In certain conditions, every one can improve its strategy against ever-changing environments due to improved strategies by other agents. This is real co-evolution by which all agents evolve effectively.

As a typical co-evolution example, a competitive task such as prey and predator has been often argued [2, 3] where heterogeneous agents often change their strategies to cope with the current opponent one. That is, the first pattern was observed. In case of homogeneous agents, Luke et al. [7] co-evolved teams consisting of eleven soccer players among which cooperative behavior could be observed. However, co-evolving cooperative agents has not been addressed as a design issue on fitness function for individual players since they applied co-evolving technique to teams.

We believe that between one to one individual competition and team competition, there could be other kinds of co-evolution than competition. Thus, we challenge to evaluate how the task complexity and fitness function affect co-evolution processes in case of multi agent simultaneous learning for not only competitive but also cooperative tasks through a series of systematic experiments. First, we show the experiments for a cooperative task, that is, shooting supported by passing between two robots in Section 4.1 where unexpected cooperative behavior that can be regarded as the second pattern was emerged. Next, we add a stationary obstacle before the goal area into the first experimental set up in Section 4.2 where the complexity is higher and expected behavior was observed after longer generation changes than the previous one. Finally, we add an active learning opponent instead of the stationary obstacle to evaluate how both cooperative and competitive behaviors are emerged in Section 4.3. We have tried several fitness functions, and we may conclude that the same level fitness function among them seems better to co-evolve cooperative and competitive agents, and other ones tend to evolve only one side, that is the second pattern. In the following, we describe them in detail.

3 Task and assumptions

3.1 Environment and robots

Before explanation of the proposed method, we show a concrete task for reader's understanding of the method. We have selected a simplified soccer game consisting of two or three robots as a testbed for the problem because both competitive and cooperative tasks are involved as stated in RoboCup Initiative [4]. We built an original soccer simulator which models real mobile robots we have been using so far in [1, 8, 9]. The environment consists of a ball and two goals, and a wall is placed around the field except the two goals. The sizes of the ball, the goals and the field are the same as those of the middle league of RoboCup.

The robots modeled have the same body (power wheeled steering system) and the same sensor (on-board TV camera), that is, homogeneous agents. In this simulator, the robot can not obtain the complete information because of limitation of its sensing capability and occlusion of the objects. **Fig.1** shows the size of the environment.

3.2 Function and terminal sets

As sets of functions, we prepare the simple conditional branching function that executes its first branch if the condition "a is b" is true, otherwise executes its second branch, where a is a kind of image features, and b is its category. **Table 1** shows the details of this function "IF_a_is_b".

Terminals in our task are actions that have effects on the environment. A terminal set consists of the following four behaviors :

1. `shoot` : the robot shoots a ball into the opponent goal based on the visual information about the ball and the opponent's goal.
2. `pass` : the robot kicks a ball to one teammate based on the visual information about the ball and other robots including the teammate.
3. `avoid` : the robot avoids collisions with other robots based on the visual information about them.
4. `search` : the robot searches the ball by turning to the left or right based on the visual information about the goal.

Although we design these behaviors by hand in this experiments, these primitive behaviors can be acquired by other learning algorithms such as ones in [1, 8, 9].

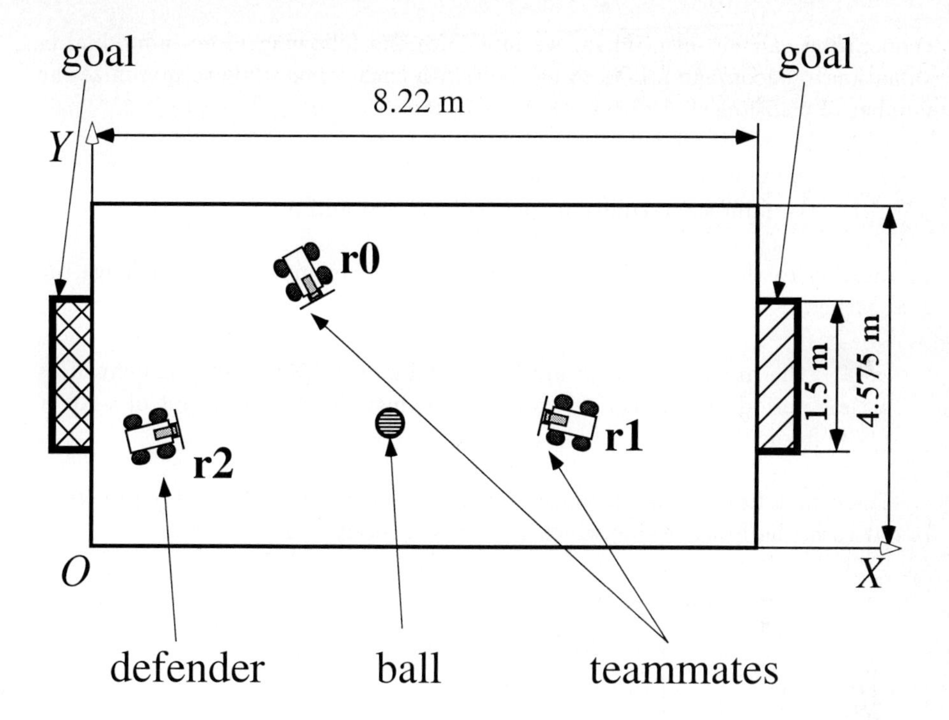

Fig. 1. Environment

Table 1. Function sets

a	ball, goal, other robot 0, other robot 1, $\cdots$
b	left, middle, right, small, medium, large, lost

3.3 Fitness measure

One of the problems to apply an evolutionary algorithm is the design of fitness function which leads robots to purposive behaviors. We utilize the standardized fitness representation, that has a positive value. The smaller is the better (0.0 is the best). We first consider the following parameters to evaluate team behaviors such as cooperation between teammates and competition with opponents:

- $G(i)$: the total number of achieved goals for the team to which robot i belongs,
- $L(i)$: the total number of lost goals for the team to which robot i belongs.

With these parameters only, most robots tends to be idle (passive cooperation) except one that attempts at achieving the goal for itself, and therefore no active cooperation can be seen. Then, we introduce the following more individual evaluation to encourage robots to interact with each other while to minimize the number of collisions:

- $K(i)$: the number of ball-kicking by robot i,
- $C(i)$: the number of collisions between robot i and other ones.

In addition to the above, the following is involved to make robots achieve the goal earlier.

- *steps* : the number of steps until one trial ends, where a step is defined as a time period for one action execution against the sensory input of a robot (1/30 [msec]).

The fitness function is calculated by linear combination of these parameters. In our case, the fitness value which the robot i receives is given by :

$$\begin{aligned} f_s(i) &= \alpha_k h(K(i), \beta) + \alpha_g h(G(i), T_{max}) + \alpha_l * L(i) \\ &\quad + \alpha_c * C(i) + \alpha_s * steps \qquad (1) \\ h(x,y) &= \begin{cases} y - x \text{ if } x < y \\ 0 \quad \text{otherwise} \end{cases}, \end{aligned}$$

where T_{max} denotes the maximum number of trials, and $\alpha_k \sim \alpha_s$ and β are constants. In the following experiments, we set $\alpha_k = \alpha_g = 1$, $\alpha_l = 0.5$, $\alpha_c = 0.05$, $\alpha_s = 0.0001$, and $\beta = 10$. If two or more individuals have the same fitness value, we prefer to one with more compact tree depth.

3.4 Other parameters in genetic programming

Other parameters in GP here are: the size of each population is 80, the number of generations for which the evolutionary process should run is 60, the maximum depth that must not be exceeded during the creation of a genetic tree is 10, and the maximum tree depth by crossing two trees is 25.

The best performing tree in the current generation will be moved unchanged to next generation. In order to select parents for crossover, we use tournament selection with size 10. The crossover probability is set to 95 %, reproduction probability is set to 5 % mutation probability is set to 10 %.

After each population selects one individual separately, the selected individuals participate in the game. We perform 20 games to evaluate them. One trial is terminated if the robot shoots a ball into the goal or *steps* exceed 1000. As a result, it needs 1600 trials to alter a new generation. The hardware used for the simulation is Sun SPARC Station Ultra2, which takes about one day to evaluate one experiment.

4 Simulation results

4.1 Two learners

At first, we demonstrate the experiments to acquire cooperative behaviors between two robots. Both robots belong to the same team, and they obtain the score if they succeed in shooting a ball into the goal. The number of function sets is $28(= 7(\text{ball}) + 2 \times 7(\text{two goals}) + 7(\text{teammate}))$.

Figs.2 (a) and (b) show the results of evolution process in the case of two robots. The fitness values of the best individuals converged in generation 20 (See (a)). The tree depths and the numbers of nodes of the best **r0** and **r1** are (29, 637) and (21,611), respectively. In this case, **r0** does not kick the ball by itself but shakes its body by repeating the behaviors `search` and `avoid`. On the other hand, **r1** approaches the ball and passes the ball to **r0**. After **r0** receives the ball, it executes `shoot` behavior to shoot the ball into the goal. However, **r1** approaches the ball faster than **r0**. As a result, **r0** shoots the ball into the goal while **r1** avoid collisions with **r0**. The successful behaviors are shown in **Fig.3**.

Although we tested several fitness functions, the resultant behaviors are similar to the behavior shown in Fig.3. In this task, **r0** does not kick the ball toward **r1** through all the generation. We suppose that the reasons why they acquire the cooperative behaviors as shown in Fig.3 are as follows:

- In order for **r0** to pass the ball to **r1**, **r1** has to shoot the ball which is passed back from **r0**. This means that in this situation the development of both robots needs to be exactly synchronized. It seems very difficult for such a synchronization to be found.
- **r1** may shoot the ball by itself whichever **r0** kicks the ball or not. In other words, **r1** does not need the help by **r0**.

In this task, **r0** and **r1** do not have even complexity of the tasks. As a result, the behavior of **r1** dominates this task while **r0** does not improve its own behavior. This is the second pattern explained in Section 2

4.2 Two learners and one stationary robot

Next, we add one robot as an stationary obstacle to the environment described in Section 4.1. The number of function sets is $35(= 7(\text{ball}) + 2 \times 7(\text{two goals}) + 2 \times 7(\text{teammate and opponent}))$.

Fig.4 (a) and (b) show the results of evolutionary process where a good synchronization between the best individuals of **r0** and **r1** can be seen (See (b)). The tree depths and the numbers of nodes of the best **r0** and **r1** are (11,63) and (19, 577), respectively. Although both learning robots are placed in the same way as in the previous experiments, the acquired cooperative behaviors are quite different because of the one stationary opponent. Since it becomes more difficult for **r1** to shoot the ball for itself because of the existence of **r2**, **r1** has to evolve behaviors with **r1** synchronously. In other words, the complexity of the task for **r0** increased around the same level of **r1**.

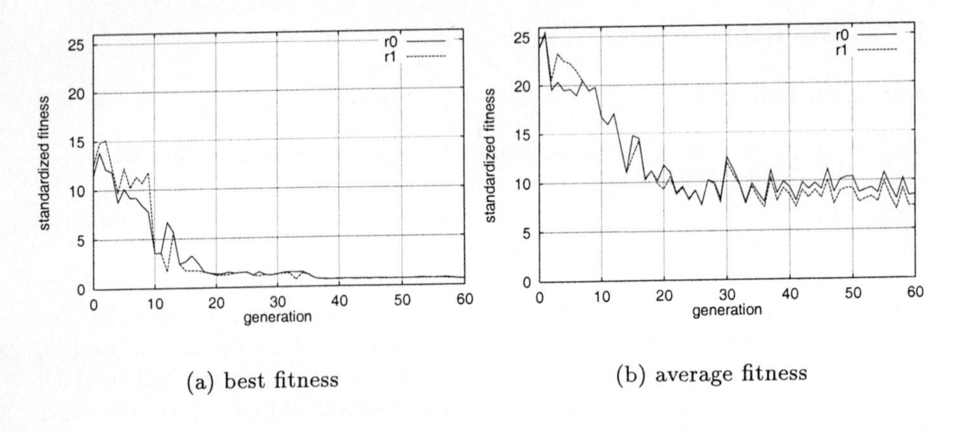

Fig. 2. fitness in case of two learners

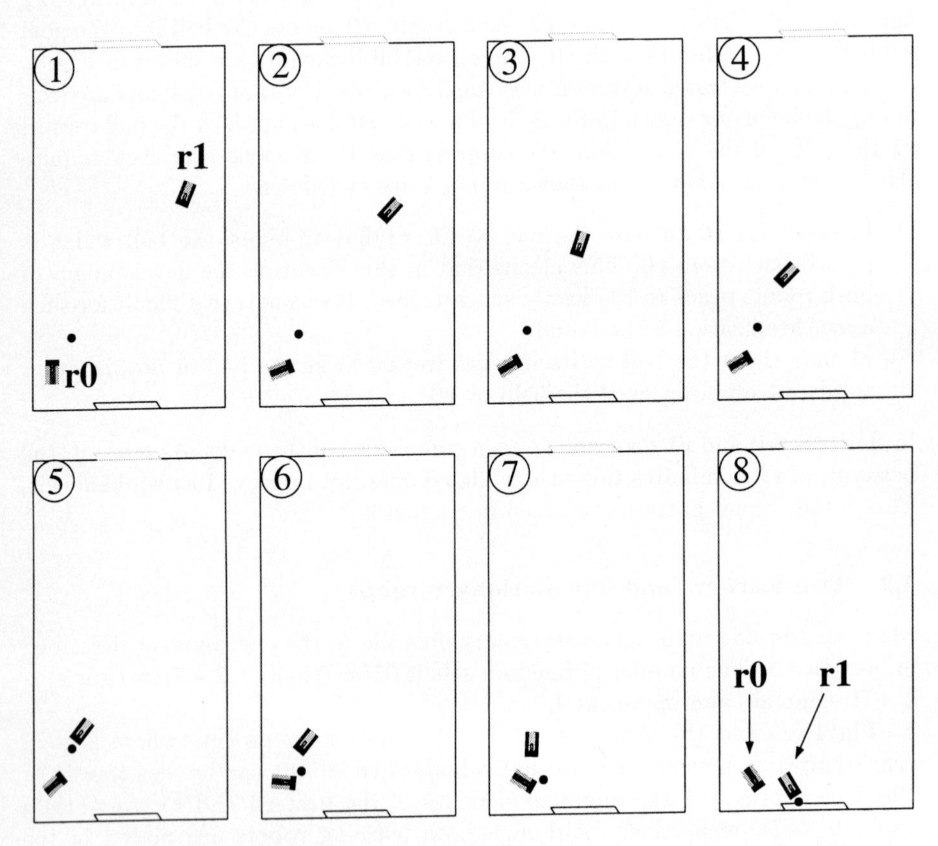

Fig. 3. Two robots (**r0** and **r1**) succeed in shooting a ball into the goal

A history of evolution is as follows. Although both **r0** and **r1** chase after the ball and kick the ball until generation 4, **r0** begins to kick the ball towards **r1**. However **r1** can not shoot the ball from the **r0** directly because **r0** can not pass the ball to the **r1** precisely. Therefore, **r1** kicks the ball to the wall and continues to kick the ball to the opponent's goal along the wall until generation 15. After a number of generations, both robots improve their own behaviors and acquire cooperative behaviors shown in **Fig.5** in generation 61, where **r0** kicks the ball to the front of **r1**, then **r1** shoots the ball into the opponent's goal. Although it intends to shoot the ball for itself, **r0** makes a way for **r1** to avoid collisions with other robots. As a result, both robots improve the cooperative behaviors synchronously. This is a kind of the third pattern described in Section 2

4.3 Three learners

Finally, we test the co-evolution among three robots. That is, **r2** added in Section 4.2 evolves its behavior with **r0** and **r1** simultaneously. The difference from Sections 4.1 and 4.2 is involvement of competition between **r2** and **r0** and **r1**. The number of function sets is as many as the case of Section 4.2.

We setup the same fitness function (Eqn. (1)) so that we make **r0**, **r1** and **r2** equal. The results are shown in **Fig.6**. As compared with the only cooperative tasks in Section 4.2, fitness values rather oscillate than stay stable. The tree depths and the numbers of nodes of the best **r0**, **r1**, and **r2** are (24,1143), (15, 1093) and (21, 749), respectively.

We can see two typical settlements in this three-robot soccer game. One is the same behaviors described in Section 4.2 : **r0** kicks the ball toward **r1**, then **r1** shoots the ball into the goal avoiding collisions with **r2** (See **Fig.7**). The other one is that **r2** intercepts the ball and shoots the ball into the goal (See **Fig.8**). The ratio between the former and the latter is about 25 % : 75 %. The aim of **r0** is to pass the ball to **r1** while the aim of **r2** is going to intercept the ball. It depends on each other for **r0** and **r2** to achieve each goal. However, **r2** can observe the ball and the opponent's goal at the same time and it may shoot the ball by itself while **r0** needs to pass the ball to **r1**. As a result, we suppose that the predominance of **r2** may be caused by the different complexity of the given tasks, that is, task complexity for **r0** and **r1** is higher than that for **r2**.

5 Concluding remarks

This paper showed how co-evolution technique could emerge not only competitive behaviors but also cooperative ones through a series of experiments in which two or three robots play a simplified soccer game. In order to co-evolve cooperative agents, it should be noted that robots must synchronize their evolutionary processes. Otherwise, there are many traps to local maxima (suboptimal strategies) as we can see in Section 4.1.

In case of more complicated situation (three agents and both cooperation and cooperation are involved), the task complexity should be equal to all agents

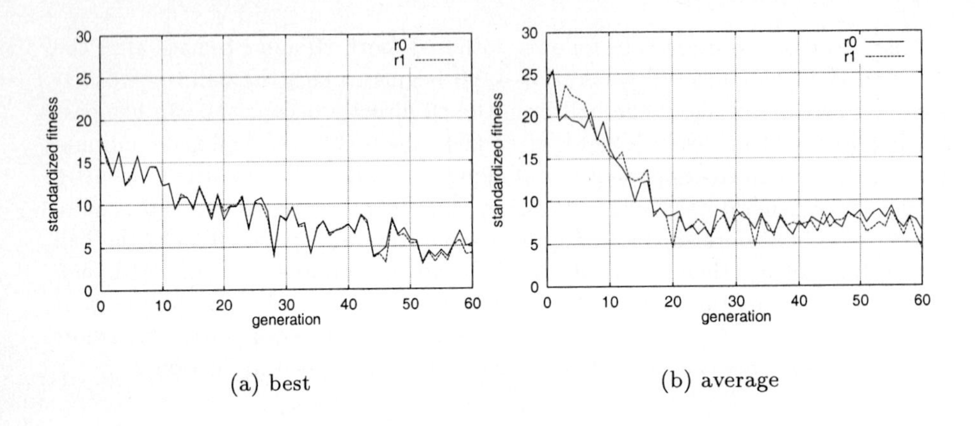

(a) best (b) average

Fig. 4. fitness in case of the two learners and one stationary robot

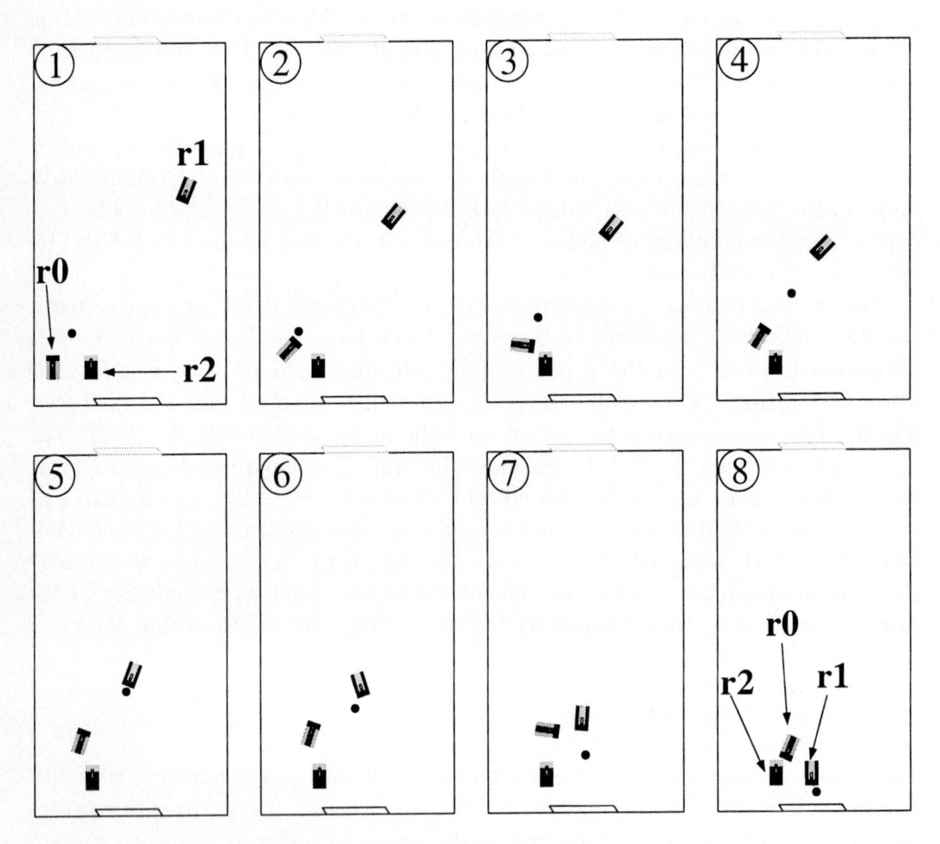

Fig. 5. Two robots (**r0** and **r1**) succeed in shooting a ball into the goal against the stationary keeper (**r2**)

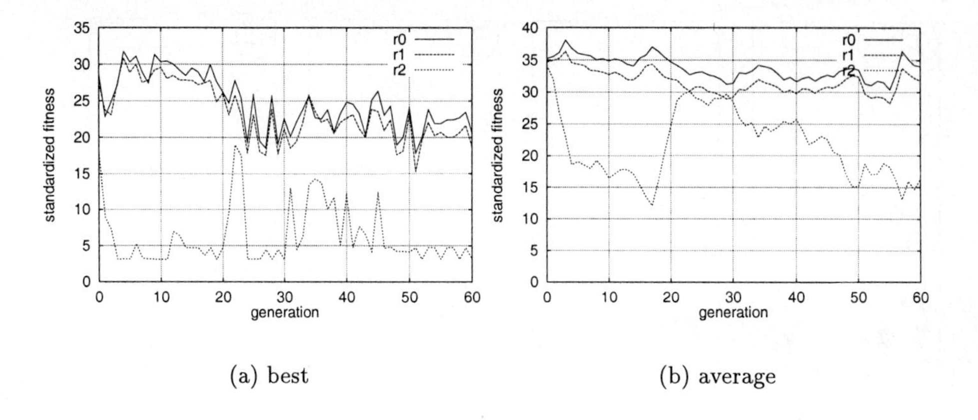

(a) best

(b) average

Fig. 6. fitness in case of three learners

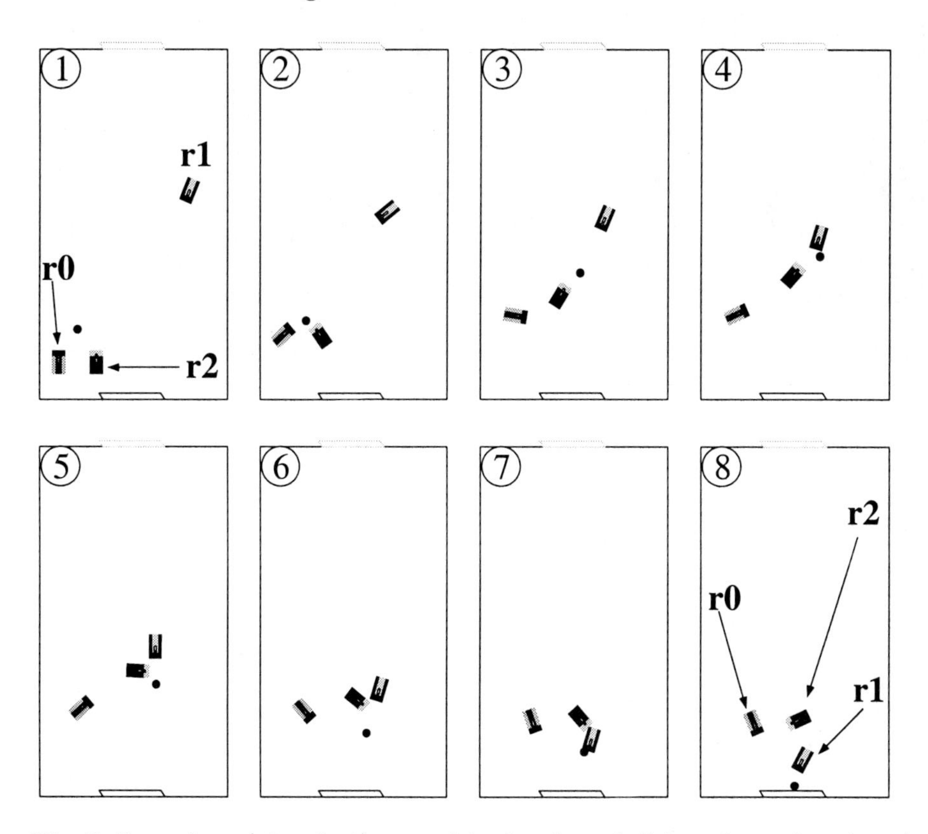

Fig. 7. Two robots (**r0** and **r1**) succeed in shooting a ball into the goal against the keeper (**r2**)

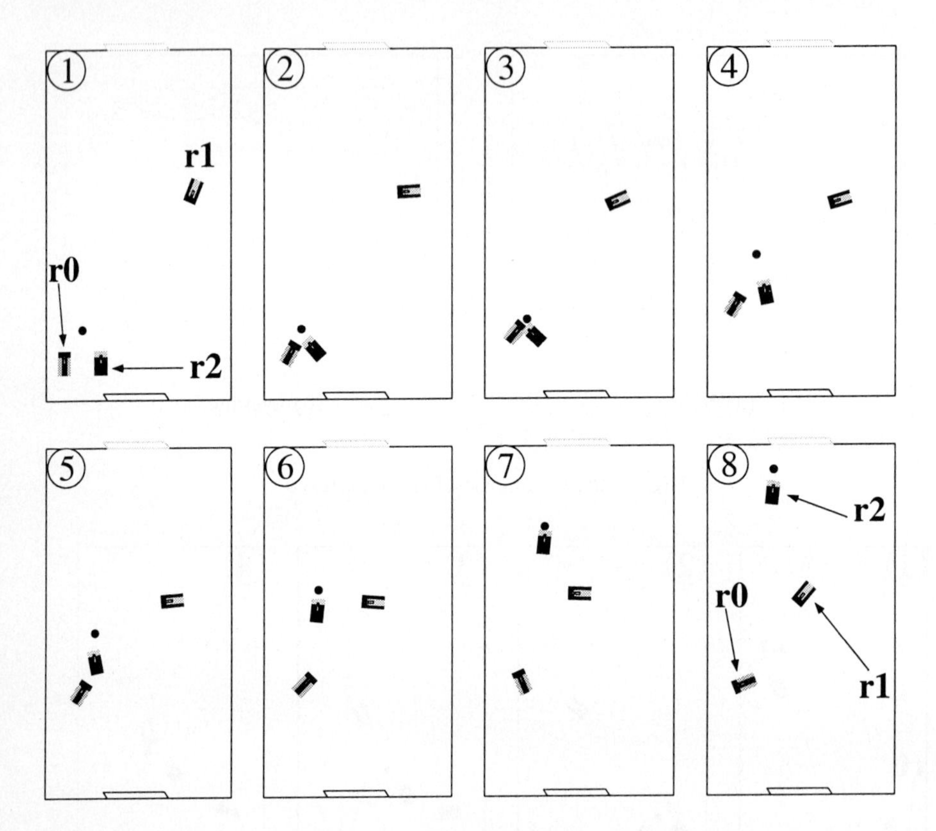

Fig. 8. The keeper (**r2**) succeeds in shoot a ball into the goal against the two robots (**r0** and **r1**)

so as to co-evolve cooperative and competitive agents simultaneously. This also suggests that the environment itself should co-evolve from simpler to more complicated situations to assist the development of desired skills of cooperations and competitions. Otherwise, co-evolution is prone to be settled into suboptimal strategies as shown in Section 4.3.

More systematic understanding is, however, needed to make clear what are necessary and sufficient conditions to lead co-evolutionary processes to successful situations. Design issues of environments including agents, tasks, and fitness functions are our future work. Also, we are planning to implement real experiments to check the validity of the proposed method and the obtained behaviors.

Acknowledgement

This research was supported by the Japan Society for the Promotion of Science, in Research for the Future Program titled Cooperative Distributed Vision for Dynamic Three Dimensional Scene Understanding (JSPS-RFTF96P00501).

References

1. M. Asada, S. Noda, S. Tawaratumida, and K. Hosoda. Purposive Behavior Acquisition for a Real Robot by Vision-Based Reinforcement Learning. *Machine Learning*, 23:279–303, 1996.
2. D. Cliff and G. F. Miller. Co-evolution of Pursuit and Evasion II : Simulation Methods and Results. In *Proc. of the 4th International Conference on Simulation of Adaptive Behavior: From Animals to Animats 4.*, pages 506–515, 1996.
3. D. Floreano and S. Nolfi. Adaptive Behavior in Competeing Co-Evolving Species. In *Fourth European Conference on Artificial Life (ECAL97)*, pages 378–387, 1997.
4. H. Kitano, M. Asada, Y. Kuniyoshi, I. Noda, E. Osawa, and H. Matsubara. RoboCup A Challenge Problem for AI. *AI Magazine*, 18(1):73–85, 1997.
5. J. R. Koza. *Genetic Programming I : On the Programming of Computers by Means of Natural Selection.* MIT Press, 1992.
6. J. R. Koza. *Genetic Programming II : Automatic Discovery of Reusable SubPrograms.* MIT Press, 1994.
7. S. Luke, C. Hohn, J. Farris, G. Jackson, and J. Hendler. Co-Evolving Soccer Softbot Team Coordination with Genetic Programming. In *Proc. of the RoboCup-97 Workshop at the 15th International Joint Conference on Artificial Intelligence (I-JCAI97)*, pages 115–118, 1997.
8. E. Uchibe, M. Asada, and K. Hosoda. Behavior Coordination for a Mobile Robot Using Modular Reinforcement Learning. In *Proc. of the 1996 IEEE/RSJ International Conference on Intelligent Robots and Systems*, pages 1329–1336, 1996.
9. E. Uchibe, M. Asada, and K. Hosoda. Cooperative Behavior Acquisition in Multi Mobile Robots Environment by Reinforcement Learning Based on State Vector Estimation. In *Proc. of IEEE International Conference on Robotics and Automation*, pages 1558–1563, 1998.
10. E. Uchibe, M. Asada, and K. Hosoda. State Space Construction for Behavior Acquisition in Multi Agent Environments with Vision and Action. In *Proc. of International Conference on Computer Vision*, pages 870–875, 1998.

Integrated Reactive Soccer Agents

Wei-Min Shen, Jafar Adibi, Rogelio Adobbati, Srini Lanksham, Hadi Moradi, Behnam Salemi, and Sheila Tejada

Computer Science Department / Information Sciences Institute
University of Southern California
4676 Admiralty Way, Marina del Rey, CA 90292-6695
{shen, dreamteam}@isi.edu

Abstract. Robot soccer competition provides an excellent opportunity for robotics research. In particular, robot players in a soccer game must perform real-time visual recognition, navigate in a dynamic field, track moving objects, collaborate with teammates, and hit the ball in the correct direction. All these tasks demand robots that are autonomous (sensing, thinking, and acting as independent creatures), efficient (functioning under time and resource constraints), cooperative (collaborating with each other to accomplish tasks that are beyond individual's capabilities), and intelligent (reasoning and planing actions and perhaps learning from experience). To build such integrated robots, we should use different approaches from those employed in separate research disciplines. In the 1997 RoboCup competition, the USC/ISI robot team, called Dreamteam, fought hard and won the world championship in the middle-sized robot league. These robots all share the same general architecture and basic hardware, but they have integrated abilities to play different roles (goal-keeper, defender or forward) and utilize different strategies in their behavior. Our philosophy in building these robots is to use the least possible sophistication to make them as robust as possible. This paper describes our experiences during the competition as well as our new improvements to the team.

1 Introduction

Since individual skills and teamwork are fundamental factors in the performance of a soccer team, Robocup is an excellent test-bed for integrated robots [RoboCup]. Each soccer robot (or agent) must have the basic soccer skills— dribbling, shooting, passing, and recovering the ball from an opponent, and must use these skills to make complex plays according to the team strategy and the current situation on the field. For example, depending on the role it is playing, an agent must evaluate its position with respect to its teammates and opponents, and then decide whether to wait for a pass, run for the ball, cover an opponent's attack, or go to help a teammate.

Figure 1: Integrated Soccer Robots

To build agents with soccer-playing capabilities, there are a number of tasks that must be addressed. First, we must design an architecture to balance the system's performance, flexibility and resource consumption (such as power and computing cycles). This architecture, integrating hardware and software must work in real-time. Second, we must have a fast and reliable vision system to detect various static and dynamic objects in the field. Such a system must be easy to adjust to different lighting conditions and color schema (since no two soccer fields are the same, and even in the same field, conditions may vary with time). Third, we must have an effective and accurate motor system and must deal with uncertainties (discrepancy between the motor control signals and the actual movements) in the system. Finally, we must develop a set of software strategy for robots to play different roles. This can add considerable amount of flexibility to our robots.

Several previous works have considered these problems. For example, before the publication of [RoboCup96], layered-controlled robots [Brooks86] and behavior-based robots [Arbib81 and Arkin87] already began to address the problem of integrated robots. In a 1991 AI Spring symposium, the entire discussion [AISS91] was centered around integrated cognitive architectures. We will have more detailed discussion on related work later.

Since building integrated robots for soccer competition requires integration of several distinct research fields, such as robotics, AI, vision, etc., we have to address some of the problems that have not been attacked before. For example, different from the small-sized league and most other teams in the middle-sized league, our robots perceive and process all visual images on-board. This will give much higher noise-ratio if one is not careful about how the pictures are taken. Furthermore, since the environment is highly dynamic, uncertainties associated with the motor system will vary with different actions and with the changes of power supply. This posts additional challenges on real-time reasoning about action than systems that are not integrated as complete and independent physical entities.

Our approach to built the robots is to use the least possible sophistication to make them as robust as possible. It is like teaching a kid to slowly improve his/her ability. Instead of using sophisticated equipment, programming very complicated algorithms, we use simple but fairly robust hardware and software (e.g., a vision system without any edge detection). This proved to be a good approach and showed its strength during the competition.

In the following sections of this paper, we will address the above tasks and problems in detail. The discussion will be organized as descriptions of component in our systems, with highlights on key issues and challenges. The related work will be discussed at the end.

2. The System Architecture

Our design philosophy for the system architecture is that we view each robot as a complete and active physical entity, who can intelligently maneuver and perform in realistic and challenging surroundings. In order to survive the rapidly changing environment in a soccer game each robot must be physically strong, computationally fast, and behaviorally accurate. Considerable importance is given to an individual robot's ability to perform on its own without any off-board resources such as global, birds-eye view cameras or remote computing processors. Each robot's behavior must base on its own sensor data, decision-making software, and eventually communication with teammates.

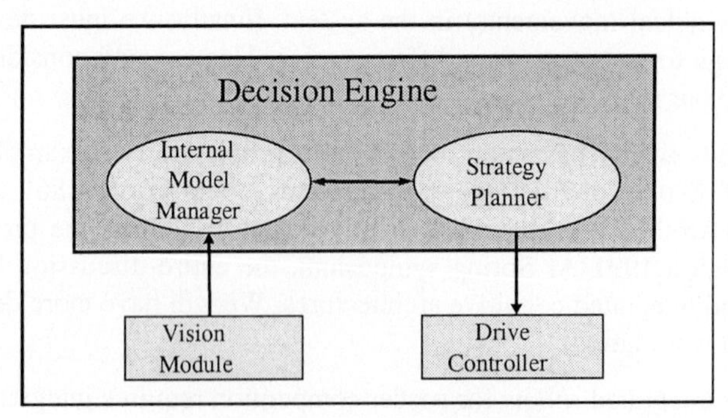

Figure 2: The System Architecture

The hardware configuration of our robot is as follows (see examples in Figure 1). The basis of each robot is a 30x50cm, 4-wheel, 2x4 drive, DC model car. The wheels on each side can be controlled independently to make the car spin fast and maneuver easily. The two motors are controlled by the on-board computer through two serial ports. We designed built the hardware interface between the serial ports and the motor control circuits on the vehicle. The robot can be controlled to move forward and backward, and turn left and right. The "eyes" of the robot are commercial digital color cameras called QuickCam made by Connectix Corp. The images from both cameras are sent into the on-board computer through a parallel port. The on-board computer is an all-in-one 133MHz 586 CPU board extensible to connect various I/O devices. There are two batteries on board, one for the motor and the other for the computer and camera.

We upgraded the robots based on our previous experiences from Robocup97. We replaced the robot drivers with faster and more reliable drivers. Each single board computer has equipped with 8MB flash disk that holds the minimum Linux kernel and the robots executable program. The flash disk provides automatic boot-up capability that is necessary in case of a sudden power failure. The flash disk is connected through PC104 extension that gives us further flexibility in adding new hardware to our robots, such as a second camera.

The software architecture of our robot is illustrated in Figure 2. The three main software components of a robot agent are the vision module, the decision engine, and the drive controller. The task of the vision module is to drive the camera to take pictures, and to extract information from the current picture. Such information contains an object's type, direction, and distance. This information is then processed by the decision engine, which is composed of two processing units - the internal model manager and the strategy planner. The model manager takes the vision module's output and maintains an internal representation of the key objects in the soccer field. The strategy planner combines the internal model with its own strategy knowledge, and decides the robot's next action. Once the action has been decided, a command is sent to the drive controller which properly executes the action. Notice that in this architecture, the functionality is designed in a modular way, so that we can easily add new software or hardware to extend its working capabilities. Our design is mainly driven by two factors: feasibility and robustness.

3. The Vision Module

Human vision is fundamental in playing real soccer, and so is computer vision for robotic soccer agents: to develop a useful strategy in the field, robots have to determine the direction and distance of objects in the visual field. These objects include the ball, the goals, other players, and the lines in the field (sidelines, end of field, and penalty area). All this information is extracted from images of 658x496 RGB pixels, received from the two on-board cameras (front and back) via a set of basic routines from a free package called CQCAM, provided by Patrick Reynolds from the University of Virginia. The two color digital cameras are fitted with wide-angle lenses for extended field of view coverage (Figure 3).

Due to the very limited on-board computing resources in an integrated robot, it is a challenge to design and implement a vision system that is fast and reliable. To achieve a fast processing of visual information, we incorporate two main strategies to our system. First, we take pictures with our "frontal" (with respect to the robot's current direction) camera, and only resort to getting images from the back camera if the object being searched cannot be detected in the front image. Second, we use a sample-based

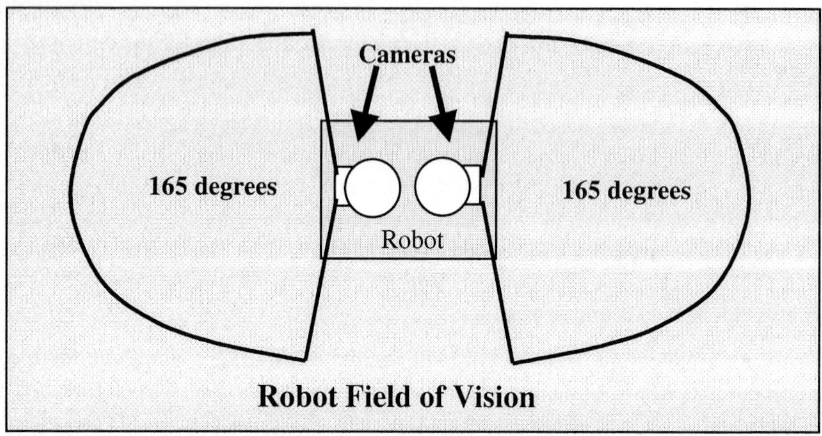

Figure 3: Robot Field of Vision

method that can quickly focus attention on certain objects. Depending on the object that needs to be identified, this method will automatically select certain number of rows or columns in an area of the frame where the object is most likely to be located. For example, to search for a ball in a frame, this method will selectively search only a few horizontal rows in the lower part of the frame. If some of these rows contain segments that are red (the color of the ball), then the program will report the existence of the ball. Domain knowledge about soccer is useful here to determine where and how the sample pixels should be searched: for example, since the ball is often on the floor, only the lower part of the image needs to be searched when we are looking for the ball. Using this strategy, the speed to reliably detect and identify relevant objects is greatly improved.

To increase the reliability of object recognition, two additional real time checks are embedded in the vision system. One is the conversion of RGB to HSV, and the other is "neighborhood checking" to determine the color of pixels. The reason we convert RGB to HSV is that HSV is much more stable to identify colors than RGB in variable light conditions. Neighborhood checking is an effective way to deal with noisy pixels when determining colors. The basic idea is that pixels are not examined individually for their colors, but rather grouped together into segment windows and using a majority-vote scheme to determine the color of a window. For example, if the window size for red is 5 and the voting threshold is 3/5, then a line segment of "rrgrr" (where r is red and g is not red) will still be judged as red.

Object's direction and distance are calculated based on their relative position and size in the image. This is possible because the size of ball, goal, wall and others are known to the robot at the outset. Furthermore, our vision system compensates for peripheral object shape and size distortion (this is caused by the inherent fisheye distortion of the wide-angle lenses on our cameras), so object information can be extracted from an effective field of view of 150 degrees. Also, position information can be used to detect object motion by comparing an object's position in two consecutive image frames. To make this vision approach more easily adjustable when the environment is changed, we have kept the parameters for all objects in a table, in a separate file. This table contains the values of camera parameters such as brightness and contrast, as well as window size, voting threshold, average HSV values, and search fashion (direction, steps, and area). When the environment is changed, only this file needs to be changed and the vision program will function properly. We are currently working on a parameter learning algorithm to automatically determine the values for each object in a new environment.

Vision modules such as the one described here also face problems that are unique for integrated robots. For example, images will have much higher noise-ratio if the robot is not careful about when and how the pictures are taken. It took us quite a long time to realize this problem. At first, we were very puzzled by the fact that although the vision system is tested well statically, our robot would sometimes behave very strangely as if it is blind. After many trials and errors, we noticed that pictures that are taken while the robot is still moving have very low quality. Such pictures are not useful at all in decision-making. Since then, special care has been given to the entire software system; furthermore, the robot takes pictures only when it is not moving.

4. Drive Controller

As specified in the system architecture, the drive controller takes commands from the decision engine, and sends the control signals to the two motors in parallel via two serial ports and a special-purpose hardware interface board. The interface provides a bridge between the two systems (the computer and the robot body) that have different power supplies.

Since the two motors (one for each side of the robot) can be controlled separately, the robot can respond to a large set of flexible commands. The basic ones include turning left and right, moving forward and backward. Others include making a big circle in the forward-left, forward-right, back-left and back-right direction. This is done by giving different amounts of drive force to the different sides. In the competition, however, we only used the basic actions for reliability reasons.

One challenge for building this simple drive controller is how to make the measured movements, such as moving forward 10 inches or turning left 35 degree. We solve this problem first by building a software mapping from the measurements of movement to the time duration of the motor running. For example, a command turning left for 30 degree would be translated by this mapping to forwarding the right-motor and backwarding the left-motor for 300ms. This solution works well when all components in the system, especially the batteries, are in perfect condition and floor material is good for wheel movement. But the accuracy of this open-loop control "deteriorates" when the power decreases or as the environment changes. Once this happens, the whole robot will behave strangely because the motor movements are no longer agreeing with the control signals.

To solve this problem, we have made all motor controls closed-loop in the entire system. Instead of saying "turning 75 degree," we also specify the termination criteria for such a turn command. For example, if the purpose of this turning is to find a goal, then the program will repeat issue smaller turnings until the goal is found. With these closed-loop control commands the reliability of motor control has increased considerably and become more robust with respect to power fluctuation. This approach proved robust during the competition

5. The Decision Engine

Our model-driven integrated robot architecture is based on the existing theories of autonomous agents (see for example [ShenBook]). The model-driven principle has guided our design and implementation of the brain of our robots, namely the Decision Engine. Compared to other model-less and pure-reactive approaches, our approach could in principle demonstrate more intelligent behaviors without sacrificing the ability to quickly react to different situations.

As one can see in Figure 2, the Decision Engine receives input from the vision module and sends move commands to the drive controller. The decision engine bases its decisions on a combination of the received sensor input, the agent's internal model of its environment, and knowledge about the agent's strategies and goals. The agent's internal model and strategies are influenced by the role the agent plays on the soccer field. There are three types of agent roles or playing positions: goal-keeper, defender, and forward. The team strategy is distributed into the role strategies of each individual agent. Depending on the role type, an agent can be more concerned about a particular area or object on the soccer field, e.g. a goal keeper is more concerned about its own goal, while the forward is interested in the opponent's goal. These differences are encoded into the two modules that deal with the internal model and the agent's strategies.

The decision engine consists of two sub-modules: the internal model manager and the strategy planner. These sub-modules communicate with each other to formulate the best decision for the agent's next action. The model manager converts the vision module's output into a "map" of the agent's current environment, as well as generating a set of object movement predictions. It calculates the salient features in the field and then communicates them to the strategy planner. To calculate the best action, the strategy planner uses both the information from the model manager and the strategy knowledge that it has about the agent's role on the field. It then sends this information to the drive controller and back to the model manager, so that the internal model can be properly updated.

5.1 Model Manager

For robots to know about their environment and themselves, the model manager uses the information detected by the vision module to construct or update an internal model. This model contains a map of the soccer field and location vectors for nearby objects.

A location vector consists of four basic elements; distance and direction to the object and the change in distance and direction for the object. The changes in distance and direction are used to predict a dynamic object's movement; these are irrelevant for objects that are static. Depending on the role a robot is playing, the model manager actively calls the vision module to get the information that is important to the robot and updates the internal model. For example, if the robot is playing goal keeper, then it needs to know constantly about the ball, the goal, and its current location relative to the goal.

An internal model is necessary for several reasons. First, since a robot can see only the objects within its current visual frame, a model is needed to keep information that is perceived previously. For example, a forward robot may not able to see the goal all the time. But when it sees the ball, it must decide quickly in which direction to kick. The information in the model can facilitate such decision readily. Second, the internal model adds robustness for a robot. If the camera fails for a few cycles (e.g. due to a hit or being blocked, etc.), the robot can still operate using its internal model of the environment. Third, the model is necessary for predicting the environment. For example, a robot needs to predict the movement of the ball in order to intercept it. This prediction can be computed by comparing the ball's current direction with its previous one. Fourth, the internal model can be used to provide feedback to the strategy planner to enhance and correct its actions. For example, in order to perform a turn-to-find-the-ball using the closed-loop control discussed above, the internal model provides the determination criteria to be checked with the current visual information.

5.2 Strategy Planner

In order to play a successfully soccer game, each robot must react appropriately to different situations in the field. This is accomplished by the strategy planner, which resides as a part of the decision engine on each robot. Internally, a situation is represented as a vector of visual clues such as the relative direction and distance to the ball, goals, and other players. A strategy is then a set of mappings from situations to actions. For example, if a forward player is facing the opponent's goal and sees the ball, then there is a mapping to tell it to perform the kick action.

For our robots, there are five basic actions: forward, backward, stop, turn-left and turn-right. These actions can be composed to form macro actions such as kick, line-up, intercept, homing, and detour. For example, a detour action is basically a sequence of actions to turn away from the ball, move forward to pass the ball, turn back to find the ball again, and then forward to push the ball. These compound actions represent a form of simple planning. This simple reasoning and planning of actions is very effective to create an illusion that the robots are "intelligent."

5.3 Role Specifications

There are five roles that a robot can play for its team: left-forward, right-forward, left-defender, right- defender, and goal-keeper. Each role is actually implemented as a set of mappings from situations to actions, as described above. Each role has its own territory or zone and a home position. For example, the left-forward has the territory of the left-forward quarter of the field, and its home position is near the center line and roughly 1.5 meter from the left board line. Similarly, the left-defender is in charge of the left-back quarter of the field and its home position is at the left front of the base goal. The mappings for each role, that is forward, defender, and goal-keeper, are defined briefly as follows:

5.3.1 Forward

The strategy for the forward role is relatively simple compared to the defense. Its task is to push the ball towards the opponent's goal whenever possible. A forward must look for the ball, decide which direction to kick when the ball is found, and perform the kick or detour action appropriately. This strategy proved to be fast and effective in the competition.

5.3.2 Defender

The defender's strategy is very similar to that of the forward, except that the distance to the opponent goal is substantially larger compared to the position of the forward. Similar to the goal keeper, it tries to position itself between the ball and its own goal. The most difficult action for a defender is to reliably come back to its position after it chases the ball away.

5.3.3 Goal Keeper

The goal-keeper is the most complex role to play for humans as wells as robots. We have concentrated most of our effort on improving the goal-keeper by adding new sensors (a second camera) and developing a decision-making framework. The framework for the goal-keeper is to model many of its tasks as a tree of decision-making steps. We want to show that many tasks can be naturally modeled as an effort not to decide about a situation or postpone the decision for later stages. But in some cases there is an extremely need to make the decision quickly. The main idea is that decision support can maximally influence a decision if it is delivered at the time when the agent needs it most. An agent should make the decision as efficient as possible. How ever the question is when to make the decision to move?

In our particular example the question is: when should the goal-keeper move for the ball and which direction? A simple overview on the Robocup97 competition shows there was only 1 or 2 goals saved by the goal keepers of all teams. However, there were only 9 goals made by attackers. This observation shows although there were not many attacks during competition, goal-keepers missed most of the them. The main problem in this environment is the uncertainty about the ball position and about the internal model of each robot that tries to predict the next position of the ball and make the best decision.

Active / Passive

An agent in general and a soccer player in particular could be viewed as a passive agent or active agent. By passive behavior for a soccer player we mean an agent that remains in its position and does not make decisions to move most of the time. We may call this a conservative approach. The reliability of the sensed information could change the level of activity. An agent prefers to stay in the same position if it can not rely on the vision, as the error of any action is equal or greater than its current position. Active behavior implies that agent can predict the next state of the ball (or environment) and tries to go to the best position in the next state. In this approach the probability of error increase especially if the input data comes with the noise.

Essentially, the goal keeper flip between two types of states -- known states in which the goal keeper knows about its position and unknown states in which the goal keeper has lost its position. In each state agent tries to gets back to the known state which it might have several sub-states. The transition from a known state to an un-known state is an important issue, which we would like to address in our current research.

Goal keeper Optimization

In this section we attempt to simplify the goal keeper behavior in a way that we can build a simple model for this agent. Figure 4 shows the goal-keeper in its normal position. We may define the following probabilities:

$P(center) = \beta$ ball approaches the goal from the center of the field

$P(side) = \alpha_1$ ball approaches the goal from the right side of the field

$P(side) = \alpha_2$ ball approaches the goal from the left side of the field

If we consider an uniform distribution for both side of the field:

$$\alpha_1 = \alpha_2 = \alpha$$

We may also consider the following probability for agent behavior:

P (g) Get to a certain position

P (i) The probability of the truth of sensed information

P (r) Probability of return to the normal position after a given action

A goal keeper's action is a combination of *Action + Return.* Hence to optimize the goal keeper's action to be as accurate as possible we will have:

$$P\,(Save) = \sum_{field} P(save) \times P(Ball) =$$

$$P\,(save\ in\ side)\ P\,(side) + P\,(save\ in\ center)\ P\,(center)$$

In which *P (Save)* could be calculated form the *P (g), p (I) and P (r).*

If *P (center) >> P (side)* means goal keeper prefers to stay in the center and do not any movement, and

If *P (center) << P (side)* means goal keeper may save more ball if goal keeper moves to the sides for the ball.

A simple observation from the Robocup97 competition shows that the *P (center) << P (side),* which supports the idea of an active agent rather than a passive agent. The whole idea of goal keeper optimization is to minimize the *Error* of the goal keeper behavior which may compute of the following:

$$Error(Golaie) = \sum_{field} save - \sum_{field} miss$$

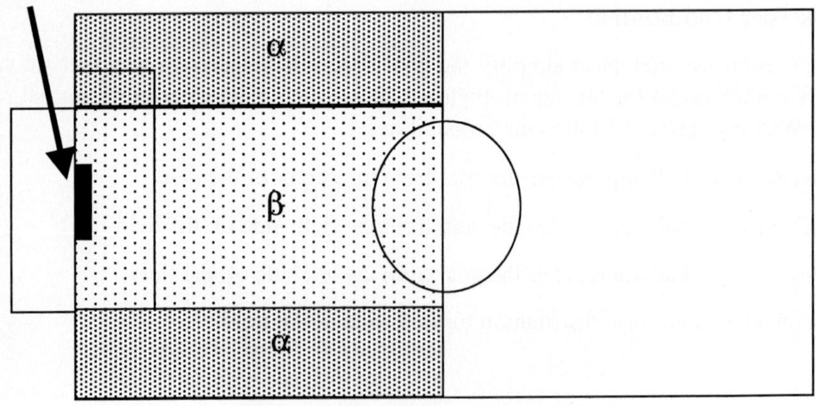

Figure 4: GoalKeeper Optimization

To address all above issues the Dream Team goal-keeper uses two cameras, one in the front and one in the back. The goal-keeper uses the back camera mostly to adjust its position the center of the goal and to find the original position. The front camera, which is equipped with a wide-angle lens, is the critical sensor for saving a ball. However as the back camera also is also equipped with a similar lens it can see the ball in the corner of the field. The power of new vision system, which gives the robots about 320 degrees of view, helps the goal keeper to switch to an active mode rather than being in a passive mode.

6. Related Work

Our current approach follows an earlier, integrated system called LIVE [Shen91] for prediction, learning, planning and action, and a theory of autonomous learning in general [Shen94]. This work also shares ideas with much architecture in [AISS91] that integrate planning and reacting for mobile robots, as well as recent progress in Agent research such as [GR97]. In our current model, however, a robot uses the internal model and the closed-loop control to guide its actions based on visual feedback. Our earlier work includes a silver medal winner robot called YODA in the 1996 AAAI Robot competition [YODA97].

Our approach is also closely related to Arkin's behavior-based robots [Arkin87]. They differ, however, in the fact that a soccer team is inherently a multi-agent problem. Our current approach is to collaborate without any explicit communication. This is possible because each robot's internal model is kept consistent with the environment and the models of its teammates.

Finally, although most of this work is experimental, the team's performance during the competition indeed demonstrated the merits of this approach. The Dreamteam has scored 8 out of the 9 goals made in the entire middle-sized RoboCup tournament (including the 2 goals against our own, as we described earlier). At the current stage, it seems the most effective approach for soccer robots is to build integrated robots using the least-sophistication to achieve the most robustness.

7. Future Work and Conclusions

In building integrated robots that are autonomous, efficient, collaborative, and intelligent, we have demonstrated a simple but effective approach. This is, however, not the end of the story. In the future, we will continue following our design strategy but improving our robots to make them truly integrated. We plan to add communication and passing capacities to increase their ability to collaborate, provide better sensors to increase awareness, and allow them to learn from their own experience.

References

[Arbib81] Arbib,M. 1981. Perceptual Structures and Distributed Motor Control. I Handbook of Physiology- The Nervous System, II, ed. V. B. Brooks, 1449-1465. American Physiological Society.

[Arkin87] Arkin, R.C. 1987. Motor Schema-Based Mobile Robot Navigation. International Journal of Robotics Research, 92-112

[Brooks86] Brooks, R. A. 1986. A Robust Layered Control System for a Mobile Robot. IEEE Journal of Robotics and Automation 2(1).

[GR97] Garcia-Alegre M. C., Recio F. Basic Agents for Visual/Motor Coordination of a Mobile Robot, Proceeding of the first International Conference on Autonomous Agents, Marina del Rey, CA, 1997, 429:434.

[RoboCup] Kitano H., Asada M. , Kuniyoshi Y., Noda I., Osawa E. Robocup: The Robot World Cup Initiative, Proceeding of the first International Conference on Autonomous Agents, Marina del Rey, CA, 1997, 340-347.

[AISS91] Laird, J.E. (ed) Special Issue on Integrated Cognitive Architectures. ACM SIGART Bulletin 2(4).

[YODA96] Shen, W.H., J. Adibi, B. Cho, G. Kaminka, J. Kim, B. Salemi, and S. Tejada. YODA—The Young Observant Discovery Agent. AI Magzine, Spring 1997. 37-45.

[Shen94] Shen, W.M. 1994. Autonomous Learning From Environment. W. H. Freeman, Computer Science Press. New York.

[Shen91] Shen, W. M. 1991. LIVE: An Architecture for Autonomous Learning from the Environment. ACM SIGART Bulletin 2(4): 151-155.

An Innovative Approach to Vision, Localization and Orientation Using Omnidirectional Radial Signature Analysis

Andrew R. Price Dr. Trevor Jones

School of Science and Technology
Deakin University, Geelong, Victoria Australia
arprice@deakin.edu.au trevj@deakin.edu.au

The greatest risk of innovative design is that it may not prove successful. That does not mean, by any account, that it should not be tried. Nor does it make the idea irrelevant. What is important is that those that follow are made aware of the problems encountered. The Omnidirectional Radial Signature Analysis Network (ORSAN) was an attempt to overcome the problems associated with the Robocup environment that were evident at RoboCup 97. In particular, lighting inconsistencies and a steering problem with the Omnidirectional Ball Based Driving Mechanism developed by this team and presented at Robocup 97. Through a series of difficulties and setbacks following the successful Nagoya event, only 16 weeks were available to produce an entire team of robots for Paris. In the end, only the prototype was ready, and so the Deakin Black Knights attempt at Robocup-98 was really over before it began. This paper details the development of the Omnidirectional Radial Signature Analysis Network, the problems it was designed to solve and the eventual conclusions that were drawn about this innovative approach to Robocup 98.

Introduction

The Omnidirectional Radial Signature Analysis Network, or ORSAN, began as an idea to solve three major problems that were evident in our previous team that competed in the first Robocup competition in Nagoya (Price 1997). The omnidirectional ball based driving mechanism exhibited at Robocup 97 had a steering problem that was exacerbated by the soft carpet of the arena. At the time, the mechanism was not equipped with any form of orientation stabilizer such as a compass. Compass modules were available, but trials before Japan revealed that these modules were severely affected by the magnetic fields of nearby motors as well as the presence of ferrous metals anywhere in the vicinity of the robot. As a result they were of little value in the Robocup environment. During Robocup 97 the importance of directional control on the omnidirectional chassis was demonstrated, and it was clear that an alternative to magnetic compasses had to be found.

The most notable problem that was experienced by all teams in Nagoya was the variation in lighting across the field. At the time, we had only a relatively simple RGB based color recognition system that was able to distinguish only a few very distinct colors with any great assurance. The most significant problem however was caused by the fact that we were using an overhead vision system that was configured to observe the entire field at one time.

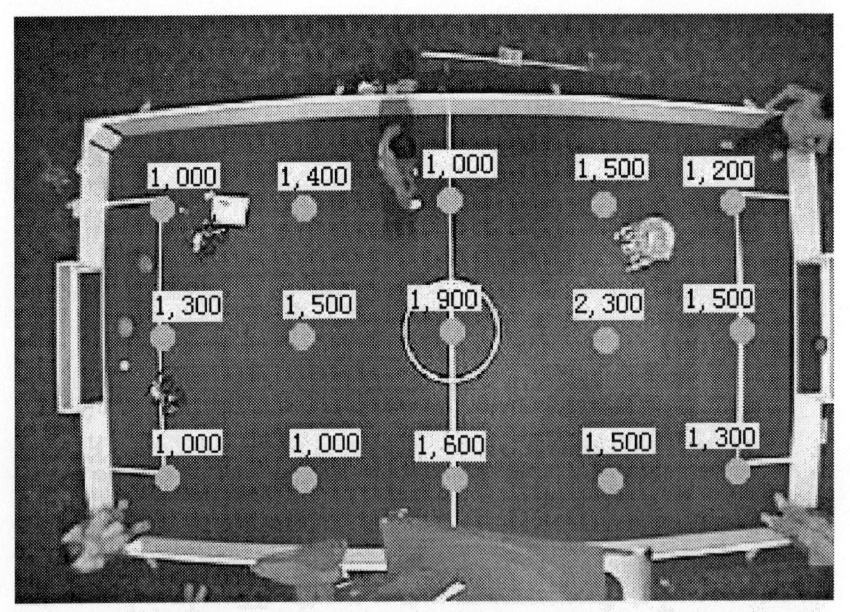

Fig. 1. Variation in lighting intensity at RoboCup 97

Because of the automatic image enhancement of the camera that was used, it was impossible to adjust the vision system to cope with the significant change in lighting intensity across the surface. As a result, when the camera was adjusted for the darker areas (the goals), the brighter areas appeared white. When the camera was adjusted for the lighter areas (down the centre) the darker areas appeared black. Although the steering problem could be corrected to some degree due to the start stop nature of the competition, the change in lighting meant that only 30% of the field was observable at any one time. This created a serious problem. For these reasons, ORSAN was developed. In keeping with this team's ideals on innovative design, we were aware that ORSAN might not be competitive. The significance of examining an alternative approach to vision, that could not only identify surrounding objects without the use of color but also perform self orientation and localization and do it affordably far outweighed the risk.

Vision in an Adversarial Environment

A significant amount of research effort has gone into producing vision systems that can cope with complex environments and difficult conditions (Shoji Suzuki 1997). Many approaches rely on existing camera technology and 'off the shelf' hardware (Shen 1997). While many of these systems are very highly competitive, the abundance of design and construction skills available at Deakin University prompted the notion that some improvement in vision might be possible through new and innovative vision hardware.

Examination of the RoboCup environment leads to an interesting set of characteristics that make an alternative approach to vision and orientation possible.

1. The shape of the environment (walls and boundaries, center markings) is known and constant
2. The shape of the ball is known and constant (and also constant in three dimensions)
3. The shape and position of the goals are constant
4. The shape and position of corners are constant.
5. The shape of one team (our own) is known
6. The shape of the opposition team is unknown (other than basic dimensions as per the rules)

There is only one unknown in the environment, the shape of the opposition robots. Opposition robots may thus be deduced, since any object that is not identified by its shape as being one of the known signatures must be the opposition. Actual physical shape is unaffected by color or various effects due to lighting differences, and no artificial markings are required.

The premise of our new vision system is that each component of the RoboCup environment possesses a unique physical characteristic or signature.

ORSAN: Omnidirectional Radial Signature Analysis Network

ORSAN is a shape based vision system that detects and identifies the unique signatures generated by objects within the RoboCup environment when a pattern of concentric circles of laser light is deflected by the object

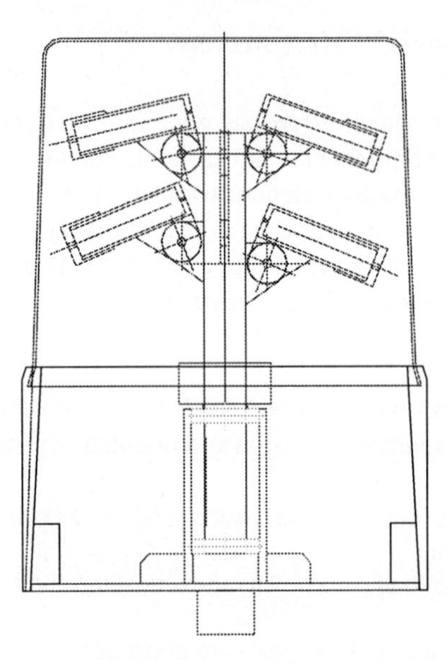

Fig. 2. The ORSAN module

Figure 2 shows the fundamental component of ORSAN: a compact module that generates concentric circles of laser light. This module is mounted centrally on top of the robot so that the circle pattern is generated on the surrounding floor to a distance up to one metre. In practice however the greater the radius, the more powerful the light source required. A distance of 1m is achievable using low cost 1mW laser modules.

Though linear laser striping is fairly common (Liu 1997), ORSAN is a truly omnidirectional image system. At present, speeds of up to 25 frames per second may be analyzed. The limiting factor is the speed of the detector system.

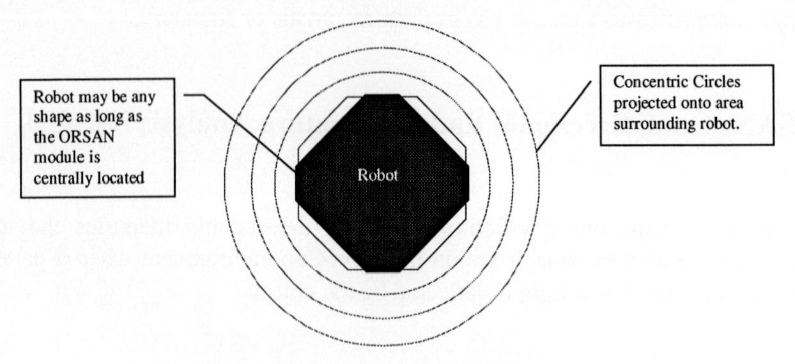

Fig. 3. Robot surrounded by concentric circles of light

The second component of ORSAN is the detector system that is located centrally, directly above the laser module as shown below. At present, the detector is a monochrome CCD camera that has been modified with a wide-angle lens and filtered to accept the red spectrum above 520nm.

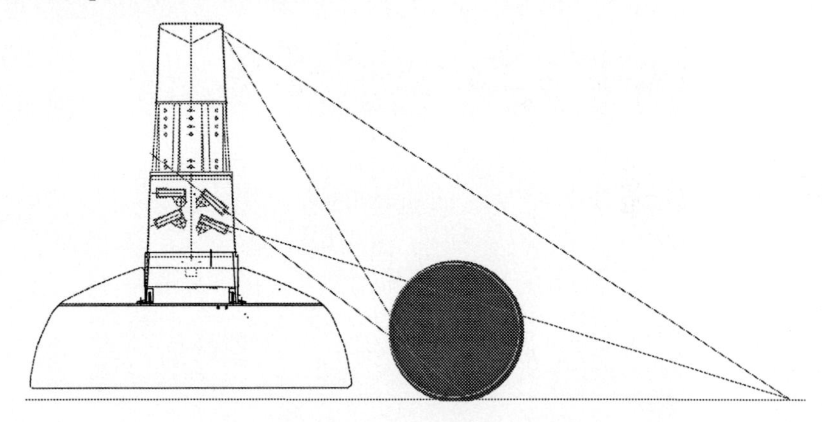

Fig. 4. Robot Chassis complete with ORSAN

Spinning the lasers using a specially designed rotating turret (Figure 2) generates circles. The diameter of each circle may be adjusted such that any combination of circles can be generated within the physical limitations of the light source.
Under normal conditions, when the robot is completely alone on a flat space, the detector records the presence of concentric circles and hence no obstacles present. However when an obstacle, such as the ball, enters the range of the vision system, part of the circle pattern is deflected or obscured. In either case the result is the same. When the detector system scans the circles, it notes that part of the circle is effectively missing. This missing part forms a chord. Multiple chords missing from adjacent circles form a signature. In the RoboCup environment the signatures are unique for each entity. By analysis of the chords an accurate identification and position of each object is obtained.

Figure 5 shows the information that is returned to the sensor. When no obstacles are present unbroken circles are returned. However when the ball is within the circle pattern the circles are deflected or obscured by the curvature. Detection of the difference is achieved by scanning points that make up the regular circles. Only points along the circle are scanned and the ends of each missing cord are identified.

Fig. 5. The chords produced by the RoboCup ball.

Starting with chords from the inner most circle, the following algorithm is applied:

1. Obtain tangent to the chord, (i.e. a line from the centre of the circle perpendicular to the chord)
2. Record the tangent angle.
3. Apply rules for each of the known entities within the environment.
4. If the chord pattern matches, identify object and location based on further rules
5. If the chord pattern does not match, assign contact as hostile robot (given no other unknowns)

Each circle is independent, and does not necessarily have to be equally spaced. This does not affect the complexity of analysis, since the mathematics behind signature recognition is independent of the relationship between two circles. In order to recognize a signature however the object must deflect a minimum of two consecutive concentric circles. One circle does not provide enough information to reliably distinguish objects although it can provide a crude measure of range and baring. The general algorithm of signature analysis is as follows:

If a cord exists on a Circle C_1 of Radius R_1 at an angle of Φ_1, and a Cord exists on a Circle C_2 of Radius R_2 at angle Φ_2 then an object exists at X, Y.

Figure 6 shows the relationships between cords and entities within the RoboCup environment.

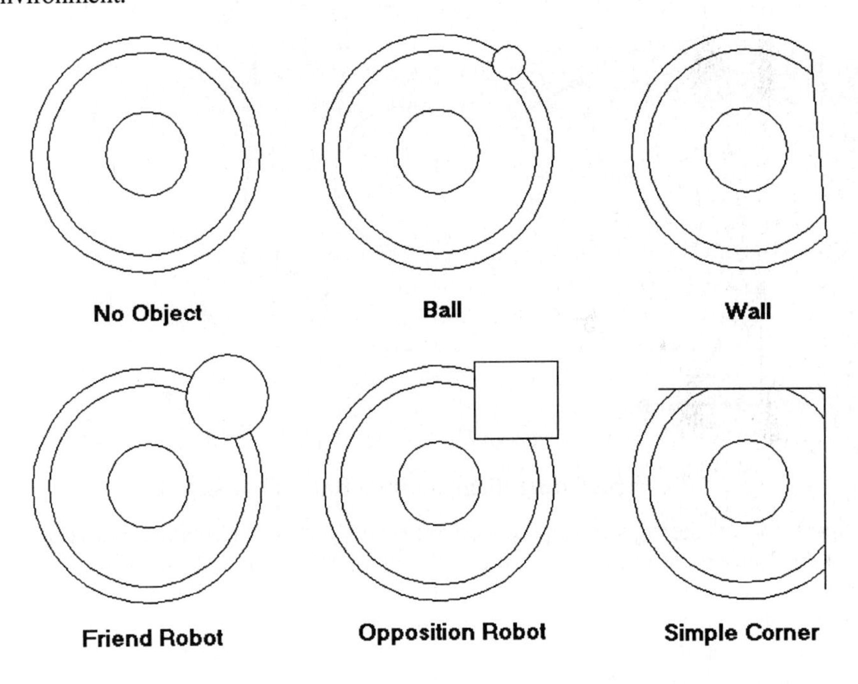

Fig. 6. Characteristic signatures generated by ORSAN

Processing the signatures

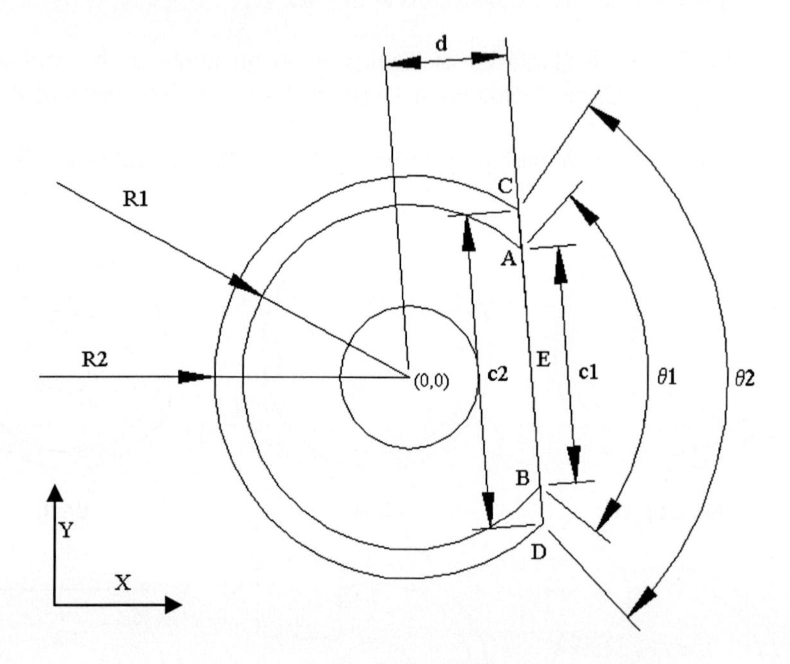

Fig. 7. **Signature diagram for a flat wall or surface**

Since each of the signatures produced are both dynamic and unique it is necessary to establish the relationship between the chords of the circle and the object for each type of object. The simplest example is the flat surface or wall. In this case, two chords should exist whose distance d, from the origin is equal and whose normals O,E are at equivalent angles. For two cords to exist in two independent circles in such a manner, it is probable that they were both deflected by a flat surface

The distance d, is expressed by:

$$d = R\cos(\frac{1}{2}\theta) \quad \textbf{(1)}$$

if the distance d, is the same for chords of circles with radii R_1 and R_2 then:

$$R_1 \cos(\frac{1}{2}\theta_1) = R_2 \cos(\frac{1}{2}\theta_2) \quad \textbf{(2)}$$

d is the distance OE in the right angle triangle OEA. The length of the chord C_1 is the distance AB. In Cartesian form this is the distance between the points (x_A,y_A) and (x_B,y_B) which are the coordinates of the endpoints of the chord AB relative to the Origin O of the circle, hence:

$$C_1 = \sqrt{\left(\left|x_A - x_B\right|\right)^2 + \left(\left|y_A - y_B\right|\right)^2} \quad \textbf{(3)}$$

the distance d may be expressed:

$$d = \sqrt{R_1^{\,2} - \left(\frac{\sqrt{\left(\left|x_A - x_B\right|\right)^2 + \left(\left|y_A - y_B\right|\right)^2}}{2}\right)^2} \quad \textbf{(4)}$$

Equation 2 may therefore be expressed in terms of the coordinates of the endpoints of the chords.

$$\sqrt{R_1^{\,2} - \left(\frac{\sqrt{\left(\left|x_A - x_B\right|\right)^2 + \left(\left|y_A - y_B\right|\right)^2}}{2}\right)^2} = \sqrt{R_2^{\,2} - \left(\frac{\sqrt{\left(\left|x_C - x_D\right|\right)^2 + \left(\left|y_C - y_D\right|\right)^2}}{2}\right)^2} \quad \textbf{(5)}$$

From the captured image, data pertaining to the end points of the chords is acquired in Cartesian form, therefore Equation 5 does not require conversion of the data from its natural form.

If the condition of Equation 5 is met, then two chords of acceptable length exist in circles of radius R_1 and R_2. It is possible, however unlikely that the two chords are unrelated by angle. Therefore it is necessary to test the angular relationship of two chords also.

The normal to the chord OE, may be expressed as the line between the points *(0,0)* and (x_E, y_E). Therefore:

$$x_E = \frac{x_A + x_B}{2} \qquad\qquad y_E = \frac{y_A + y_B}{2} \quad \textbf{(6)}$$

If the angles of the normals are the same then the angle of the line OE and OE' will be identical, hence:

$$\tan^{-1}\left(\frac{y_A + y_B}{x_A + x_B}\right) = \tan^{-1}\left(\frac{y_C + y_D}{x_C + x_D}\right) \quad \textbf{(7)}$$

The angles of equation 7 are in the first quadrant and must be compensated for other quadrants based on the sign of each denominator and numerator.

Thus for two concentric circles a relationship exists such that:

If the distance d, from the centre of the circles to the midpoint of each chord is the same and the angle to the normal of each chord, ϕ is the same, then there is a high probability that the circles are being deflected by a flat wall at distance d, angle ϕ.

A more complex and dynamic example is that of a cylinder, or cylindrical shaped robot, such as the omnidirectional ball based driving mechanism (Price 1998).

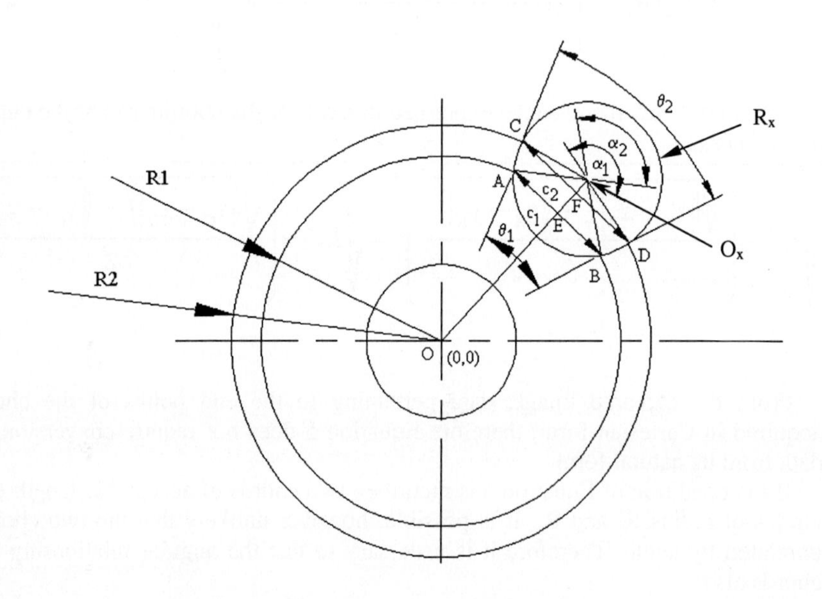

Fig. 8. Signature diagram for a cylindrical object

When the cylinder deflects the circles, two chords are generated as before. In the case of a cylindrical object the distance d, from the origin O, to the midpoint of each chord E, and F respectively, is different, however both chords remain symmetrical about a common radius.

Given that a chord of length C_1 exists in the inner circle, it is necessary to examine all the chords that may exist in the outer circle to see if one of length C_2 exists at the same angle as C_1. This necessitates predicting the length C_2 based on data obtained from C_1, and then comparing this prediction with all the chords in the outer circle to find a potential match.

From Figure 8 it can be seen that:

$$OE + EO_x = OF + O_x F \tag{8}$$

from the equation for the length of a chord it can be shown that:

$$C_2 = 2\sqrt{R_2^{\ 2} - OF^2} = 2\sqrt{R_x^{\ 2} - O_x F^2} \qquad \textbf{(9)}$$

$$R_2^{\ 2} - OF^2 = R_x^{\ 2} - O_x F^2 \qquad \textbf{(10)}$$

$$O_x F \ = \sqrt{R_x^{\ 2} - R_2^{\ 2} + OF^2} \qquad \textbf{(11)}$$

Combining equations 8 and 11 yields:

$$OE + EO_x = OF + \sqrt{R_x^{\ 2} - R_2^{\ 2} + OF^2} \qquad \textbf{(12)}$$

$$(OE + EO_x - OF)^2 = R_x^{\ 2} - R_2^{\ 2} + OF^2 \qquad \textbf{(13)}$$

Let $X^2 = R_x^{\ 2} - R_2^{\ 2}$ and $Y = OE + EO_x$

Then:

$$(Y - OF)^2 = X^2 + OF^2 \qquad \textbf{(14)}$$

$$Y^2 - 2YOF + OF^2 = X^2 + OF^2 \qquad \textbf{(15)}$$

$$\frac{Y^2 - X^2}{2Y} = OF \qquad \textbf{(16)}$$

Substituting into Equation 9 yields:

$$C_2 = 2\sqrt{R_2^{\ 2} - OF^2} = 2\sqrt{R_2^{\ 2} - \left(\frac{Y^2 - X^2}{2Y}\right)^2} \qquad \textbf{(17)}$$

Since X and Y are both in terms of constants or data obtainable from the chord in the inner circle the length of the chord C_2 can be predicted from C_1 if the object impinging on the circles is a cylinder of radius R_x.

Since $Y = OE + EO_x$

$$Y = \sqrt{R_1^{\ 2} - \left(\frac{\sqrt{\left(|x_A - x_B|\right)^2 + \left(|y_A - y_B|\right)^2}}{2}\right)^2} + \sqrt{R_x^{\ 2} - \left(\frac{\sqrt{\left(|x_A - x_B|\right)^2 + \left(|y_A - y_B|\right)^2}}{2}\right)^2}$$

(18)

Given that a chord of length C_2 exists, its angle must match that of C_1. This may be obtained as per equations 6 and 7 when a chord of suitable length has been found.

Thus a relationship exists such that

If a chord of length C_1, exists on the inner circle and a chord of length C_2 exists on the outer circle and the angle to the normal of each chord, ϕ is the same, then there is a high probability that the circles are being deflected by a cylindrical object at distance Y, angle ϕ.

Signatures are developed from the inner most laser circle outwards. There are several reasons for this. The most critical region is the space immediately surrounding the robot. The objects that can be determined in this area are of more immediate significance than any other. While more distant objects play a part in planning future motion, knowing if you are holding the ball is of greater value.

Secondly two factors influence the precise shape of the deflections and the visible perception of the deflections. The angle of the laser beam striking the object, and the angle of the camera with respect to both the reflected beam and the object is of significant importance. Signatures are developed from the point of view of the leading surface of the object interfering with the circles. It is quite feasible that the object itself will obscure the view of the camera and the lasers themselves. Signatures may only be detected up to the largest dimension of the object. Beyond that point the actual length of the chords is likely to be obscured. Hence, chords in the smallest diameter circle are examined first and then each circle increasing outwards.

An object such as a sphere creates special problems. Unlike a cylinder a sphere has non-uniform cross section in the vertical plane. Thus a laser beam striking low down on the sphere will actually be obscured from the sensor by the sphere itself.

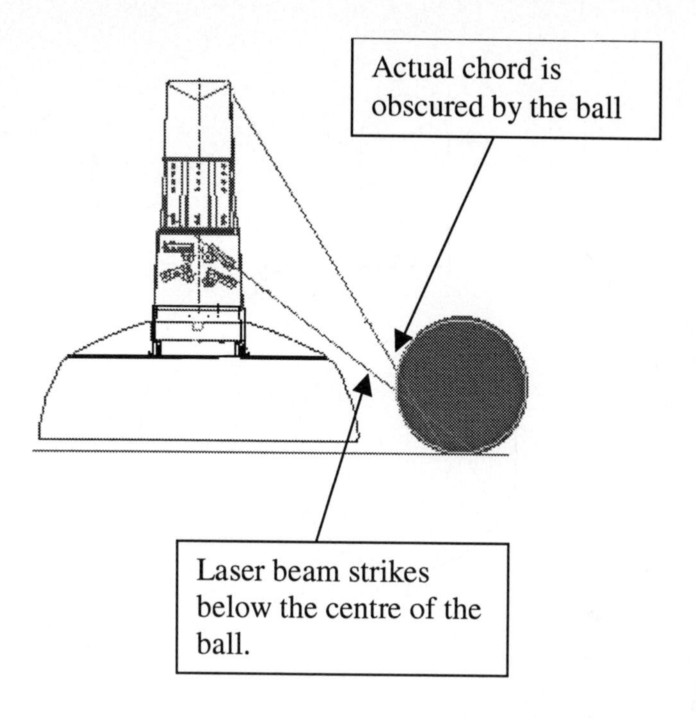

Fig. 9. The curvature of the ball Obscures the true deflection of the laser

The generated chord as perceived by the sensor will not be with respect to the actual point at which the laser strikes the sphere, but to the largest dimension of the sphere. As it happens, this dimension is the diameter of the sphere, and represents a circle in the vertical plane. A sphere may therefore be approximated as a cylinder of radius R_x. Signatures of spheres are thus approximated by cylinders. The important characteristic however is that signatures are significantly unique such that a high degree of certainty is obtained when identifying patterns. Since a sphere, or ball is being approximated by a cylinder the mathematics behind the prediction of each signature is the same. The omnidirectional chassis of the test environment is 400mm in diameter. The standard Robocup ball is 200mm in diameter. While the modelling technique is the same, the results for each object are unique and substantially different:

Suppose a chord of length 100mm is located along the inner circle as shown in figure 10.

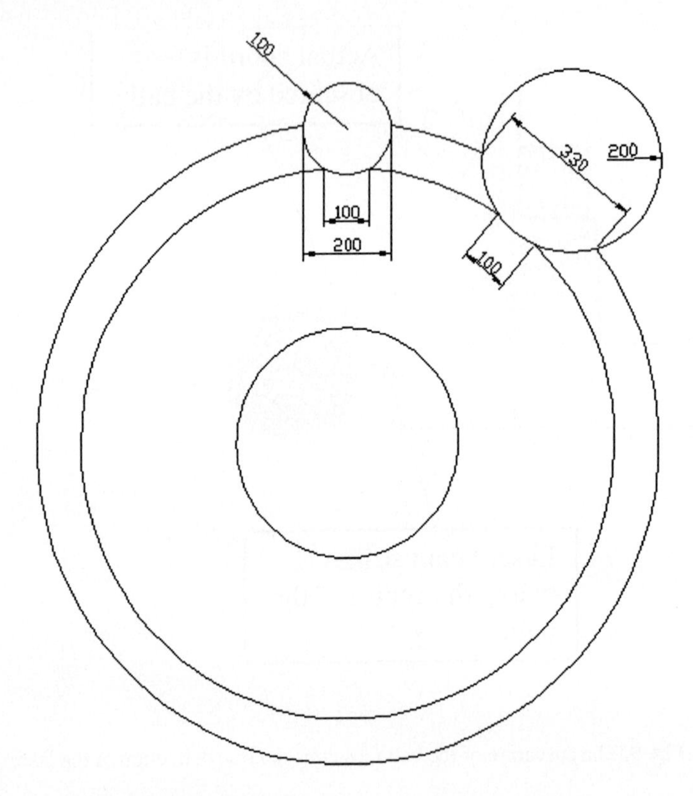

Fig. 10. Comparision of two cylindrical signatures

If the ball caused the chord, a companion chord of approximately 200mm would be expected in the outer circle. If a cylindrical robot caused the chord, a companion chord of approximately 330mm would be expected. The mathematics used in both predictions varies only in the diameters of the impinging objects, however a substantial and measurable difference in the expected signatures exists.

It is evident, that the more complex the shape, the more complex the signature becomes. However as long as there is a relationship between the lengths of chords in two circles and the object, then the object may be classified to some degree. For the test environment of Robot Soccer it has been demonstrated that the ball, friend robots and the walls may all be identified. It is significant however that the shape of the opposition robot is unknown. Given that in the playing area the opposition robots are the only unknown shape, it is possible to deduce their location and identity.

There is a real and limiting factor to the process of signature analysis as evidenced by figure 11. A cube is a shape whose symmetry changes with its orientation. A cube whose face is normal to and bisected by the radius of the circles is symmetrical, as is a cube at 45 degrees.

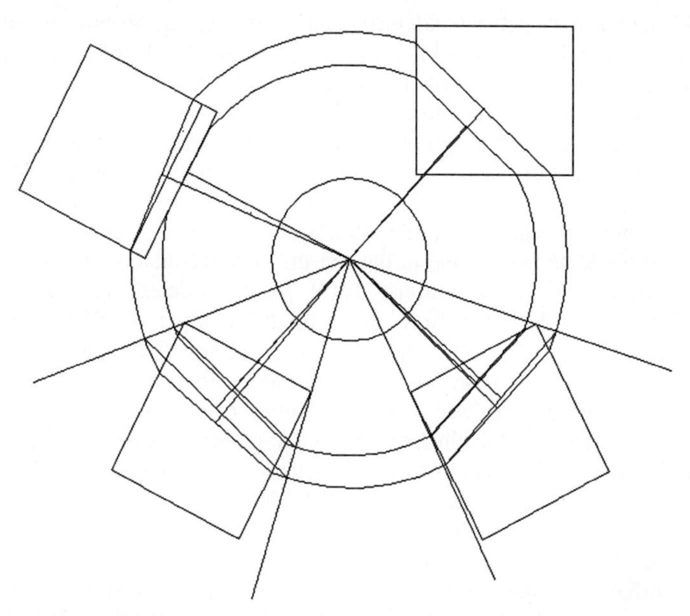

Fig. 11. The multiple identical signature dilemma of a cube

Unfortunately as shown in figure 11, other orientations are not symmetrical about the radius. This leads to an unpredictable length in both chords. To compound this problem, it is highly possible that the cube in many different orientations may generate the same length chord in the inner circle. Unlike the sphere, or ball, where the occlusion effect is predictable due to the nature of the shape, occlusions caused by cubes cannot be reliably predicted using limited information. This severely interferes with the relationship between the object and the chords. It is therefore extremely difficult to make reliable predictions as to the length of the second chord and hence the exact orientation of a cube. While it is not feasible to determine the orientation of a cube or more complex shape reliably using signature analysis, its identity may be inferred from the fact that in a limited test environment such as Robocup, the only unknown shape or object is that of the opposition robots. Hence a secondary relationship exists whereby:

If the signature caused by the impingement of an object on two or more circles does not correlate with any predictable signature, then there is a reasonable probability that the object is an opposition robot of unknown shape.

Evaluation of ORSAN

ORSAN makes it possible to identify objects within the RoboCup environment without relying on color or special markings. The uniquely identifiable shape of each object deflects the concentric laser circles in a predictable and identifiable pattern. The angle of the normal to the chords in a signature allows the robot to orient itself with respect to objects within the environment or to fixed features such as walls and corners. In general, the problems encountered at RoboCup 97 for which ORSAN was designed to overcome have been achieved although success was limited.

Primarily, ORSAN requires that the circles of concentric laser light generated by the ORSAN module be perceptible to the sensor. Several factors hinder this process. For cost reasons and legal reasons, 1mW laser modules were used. Though economically viable, the light from these modules when distributed around the circumference of a 1m diameter circle greatly decreases. This was overcome to a workable level by aligning more than one laser along the same circular path so that the light from one module was only distributed over a quarter of the circle's circumference. In this way the circles were visible on the dark carpet of the RoboCup environment, however the level of ambient light significantly swamped the red of the lasers, limiting the effectiveness of the ORSAN system within the environment. Curiously enough, in low ambient light environments ORSAN works with rather spectacular results. For wall following applications where orientation and range are desirable, ORSAN is able to provide precise baring and distance references to surrounding objects in the range of +/- 2mm using very economical hardware, in all directions surrounding the robot at once.

In a competitive environment such as RoboCup the amount of additional hardware required to carry the ORSAN system borders on excessive. With kicking mechanisms of greatly increased force, damage to the module is a genuine possibility.

The range of the ORSAN hardware was limited by the fact that the CCD sensor was located only 500mm above the ground. This limited the effective range to a radius of 1m. While it is easy enough to increase the height, the value of a compact lightweight robot with high maneuverability is lost.

ORSAN can identify objects if their shape is symmetrical about any given radius from the centre of the concentric circles. Balls, cylinders, walls and corners (a composite of two walls) are all symmetrical. In other words, the signature generated by these objects does not depend on the orientation of the object. Cubes, rectangles and complex shapes may provide a range and baring, but identification is not possible unless it is the only unknown signature in the environment.

Conclusion

The ORSAN system was an experiment to try and overcome problems with the RoboCup environment that were identified at Nagoya. Objects within the environment may be identified using only their shape. In addition, permanent features of the environment may be used to allow the robot to calibrate its orientation, without the need for ferro-magnetically sensitive compass modules. While ORSAN is capable of achieving the objectives for which it was designed, its success was limited due to the high ambient light, and by the amount of equipment that must be carried into a hostile environment. Ideally ORSAN is more suited to omnidirectional range finding in environments with extremely low ambient light levels, or using lasers of higher output. Signature analysis is effective in limited environments, where no more than one unknown signature exists, and objects are symmetrical about the radius of the concentric circles.

References

Liu, C. (1997). “A Gopher Robot.” Proceedings of Field And Service Robotics Conference, December 97 Canberra Australia.

Price, A. (1997). “Robocup 97: An Omnidirectional Perspective.” Robocup-97: Robot Soccer World Cup I **Springer Verlag Lecture Notes In Artificial Intelligence 1395**.

Price, A. (1998). “Omnidirectional Robots: New Challenges In Mobility, Vision and Tactics for Robot Soccer.” Field and Service Robotics, Springer Verlag Lecture Notes In Computer Science.

Shen, W. M. (1997). “Autonomous Soccer Robots.” Robocup-97: Robot Soccer World Cup I **Springer Verlag Lecture Notes In Artificial Intelligence 1395**.

Shoji Suzuki, Y. T., E Uchibe, M Nakamura, C Mishima, H Ishizuka, T Kato, M Asada (1997). “Vision-Based Robot Soccer Learning Towards RoboCup: Osaka University "Trackies".” Robocup-97: Robot Soccer World Cup I **Springer Verlag Lecture Notes In Artificial Intelligence 1395**.

An Application of Vision-Based Learning in RoboCup for a Real Robot with an Omnidirectional Vision System and the Team Description of Osaka University "Trackies"

Sho'ji Suzuki[1], Tatsunori Kato[1], Hiroshi Ishizuka[1],
Yasutake Takahashi[1], Eiji Uchibe[1], and Minoru Asada[1]

Dept. of Adaptive Machine Systems, Graduate School of Engineering,
Osaka University, Suita, Osaka 565-0871, Japan

Abstract. This paper gives a team description of Osaka University "Trackies" for RoboCup-98, and related research issues. We focus on behavior learning of our goalie robot which has an omnidirectional vision system. A Q-learning method is applied by defining substates from visual information of the ball and the goal. To reduce the learning time, we propose an attention control method for an omnidirectional vision by means of an active zoom mechanism. We perform computer simulation and real robot experiments to show the validity of the proposed method.

1 Introduction and the Team Description of Osaka University "Trackies-98"

One of major issues in robotics is to make a robot adapt itself to changes in dynamic environments. We are interesting in how a robot acquires a behavior in dynamic environments and how robots cooperate without explicit communication in the context of cooperative distributed vision [1]. For the first step of an application of cooperative distributed vision, we have build a goalie robot with an omnidirectional vision system and applied a Q-learning method. In order to reduce search space for learning, we propose an attention control method into the omnidirectional vision by means of an active zoom mechanism. In this paper, we summarize our method and experimental results.

Followed the team description of Osaka University for RoboCup-98, the rest of the paper is organized as follows. First, we examine the relationship between the target position in the omnidirectional view in terms of the focal length and the distance between the robot and the target. Then, we set up a control low to realize a zoom servoing, Finally, we design the state space for the robot to acquire the desired behavior based on the reinforcement learning scheme.

1.1 The Team Description of Osaka University "Trackies-98"

The team of Osaka University "Trackies-98" consists of four heterogeneous attackers and a goalie (see Figure 1). Three attackers have been replaced from the team "Trackies-97" which has following features;

1. The team consists of four homogeneous attackers and a goalie.
2. Every robot is controlled by its remote host computer via radio link.
3. An attacker has a CCD camera fixed on its body without any active mechanism and its shooting behavior is acquired by a Q-learning method.
4. The goalie has an omnidirectional vision and its behavior is hand-coded.

Details are given in [8].

(a) attacker with no active camera

(b) attacker with a panning camera

(c) self-contained attacker

(d) goalie

Fig. 1. Robots of Osaka University "Trackies" team

Attackers of "Trackies-97" have three major problems;

1. the cooperative behavior has not been realized.

2. the robot is easy to lose the ball and difficult to find it since the camera has a narrow view angle and no pan or tilt mechanism.
3. the control of the robot is not reliable because of noises on radio links.

Therefore, we have build three types of attackers shown in Figure 1 (a), (b), and (c) to cope with these problems. Since we are interest in cooperation without explicit communication, we do not use a global vision system and an inter-robot communication system to share/exchange information between robots.

The robot in Figure 1 (a) is the one used as an attacker of "Trackies-97". We apply a genetic programming method so that robots acquire a cooperative behavior. Since robots have no inter-robot communication system, they need recognize other robot's behavior through its vision system. We perform a computer simulation for a pass behavior by two robots. Details are discussed in [9] and [10]. The robot in Figure 1 (b) has a pan mechanism to extend a view angle. A Q-learning method is applied to acquire a shooting behavior with panning motion of the camera. The robot in Figure 1 (c) is a completely self-contained type which includes a CPU board, an image capture board, and motor drivers. The behavior of the robot is hand-coded. The robot shown in Figure 1 (d) is the goalie whose behavior is acquired by a Q-learning method which is described in the rest of this paper.

2 The Task and the Robot System for a Goal Keeping Behavior

Several applications of the omnidirectional vision have been proposed, such as autonomous navigation [2], visual surveillance and guidance [3], video conference, virtual reality, and site modeling [4]. These methods have focused on its opto-geometric features to reconstruct 3-D scene structure. Our approach differs from their applications in two fold: we do not reconstruct any geometric structure from the omnidirectional views. Rather, we use it as a sensory system for a goal defending mobile robot. We apply Q-learning method [7] with a state space consisting of ball and goal images.

We introduce an active zoom mechanism into the omnidirectional vision in order to accelerate the learning. We implement a zoom servoing [5] so that the target image can be captured at the constant position when the target moves on the ground plane. This servoing is realized by controlling focal length of the camera. Due to the active zoom mechanism, the target motion in the radius direction can be canceled, and only circular motions around the image center can be observed. This simplifies the image processing and target tracking.

Our robot is shown in Figure 2(a) where an omnidirectional vision system is installed onto the 2-DOFs non-holonomic vehicle such that its optical axis can be coincident with the axis of vehicle rotation. The robot is controlled by a remote host computer via radio link. Figure 2(b) shows actions of the robot. The remote computer sends motor commands to control the robot motion.

An omnidirectional vision system consists of a conic mirror and a TV camera [2] of which optical axis is aligned with the vertical axis of the mirror as

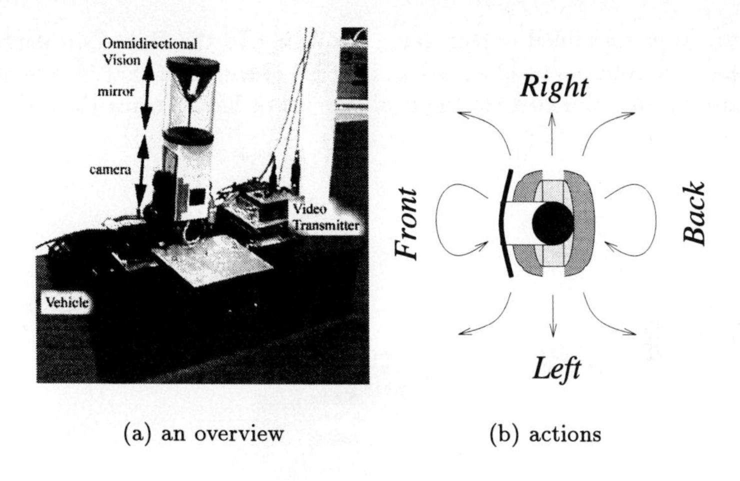

(a) an overview (b) actions

Fig. 2. The robot

shown Figure 3(a). The projection onto the image plane is determined by the the shape of the mirror and the camera configuration parameters (height of the camera, distance between the mirror and the lens, and focal length) which are designed according to individual purposes. Our omnidirectional vision system has a hyperbolic mirror and a sample of its image is shown in Figure 3(b). The omnidirectional image is transmitted to the remote computer via video transmitter and processed on it.

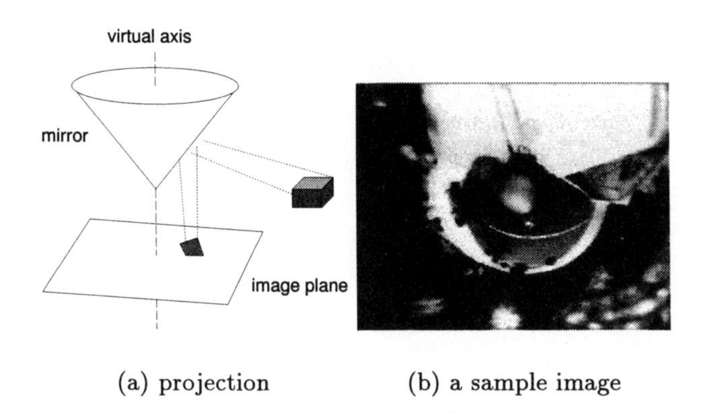

(a) projection (b) a sample image

Fig. 3. Projection and a sample image by an omnidirectional vision

We set up a simplified soccer game according to the RoboCup context [6]. The task of the robot is to block a ball in front of the goal, that is, a goalie task (see Figure 4). In order to keep the goal, the robot has to track the moving ball and move to appropriate position.

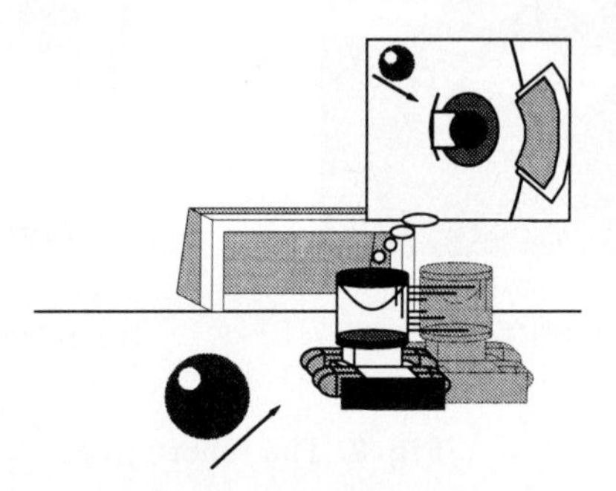

Fig. 4. The task of the robot

3 Learning by a Robot with an Omnidirectional Vision System and an Embedded Zoom Control

3.1 Active Zoom Control on the Omnidirectional Vision

The coordinate system and parameters are shown in Figure 5(a). Let $P(R, \theta, Z)$ and $p(r, \theta)$ denote a point in the environment and a projected point of P in the image plane, respectively. We assume that the object is on the ground plane, therefore Z becomes a constant and P is uniquely projected onto p. The relation between P and p is given by,

$$Z = R \tan \alpha + c + h,$$

$$\tan \gamma = \frac{b^2 + c^2}{b^2 - c^2} \tan \alpha + \frac{2bc}{c^2 - b^2} \frac{1}{\cos \alpha}, and \tag{1}$$

$$r = \frac{f}{\tan \gamma},$$

where a and b are the parameters of the hyperbolic mirror, $\frac{R^2}{a^2} - \frac{Z^2}{b^2} = -1$, and $c = \sqrt{a^2 + b^2}$. h is the height of the sensor. In our system these parameters are $a^2 = 233.3, b^2 = 1135.7$ and $h = 250[mm]$.

We add an attention control on an omnidirectional vision by controlling focal length of the camera in order to reduce the search space of the learning for

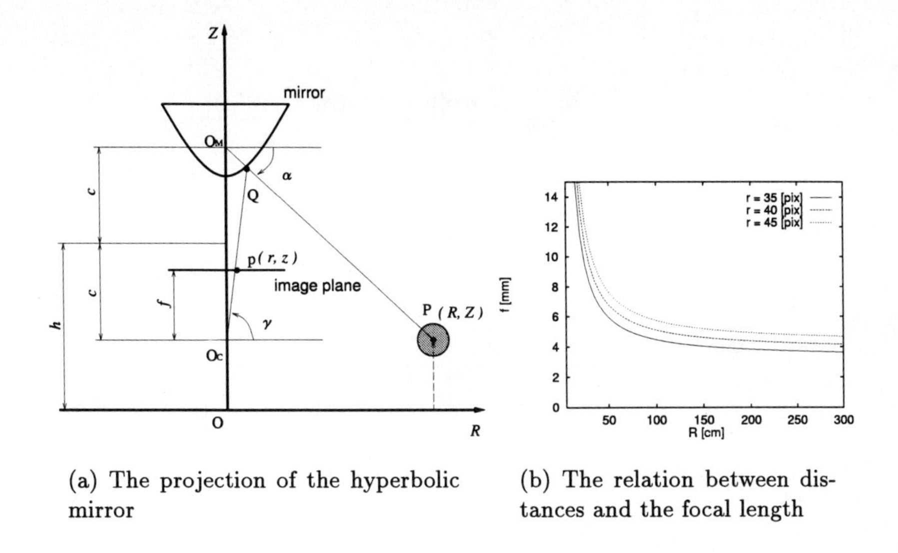

(a) The projection of the hyperbolic mirror

(b) The relation between distances and the focal length

Fig. 5. The basics of the hyperbolic mirror

behavior acquisition. In general, an attention control is realized by tracking an object in the image plane, which is implemented by controlling pan and tilt angles of the camera. However, in an omnidirectional vision system, matching of the object with the target image is not simple. Therefore, we propose an attention control by observing the object in a certain distance from the center in the image plane. The change of the distance of the target in the image is tracked by changing focal length of the camera as shown in Figure 6.

Figure 5(b) shows the relation between the distance of the object in the image from the center r, the distance of the object in the environment R, and the focal length of the camera f. We control the focal length of the camera with a following equation;

$$u_f = K({}^I r_d - {}^I r), \tag{2}$$

where u_f is the change of the focal length of the camera, ${}^I r_d$ is the desired distance in the image, ${}^I r$ is the current distance in the image. For example, if ${}^I r$ is smaller than ${}^I r_d$ it comes closer by increasing f as shown in Figure 6.

3.2 Learning of a Goal Keeping Behavior

We apply Q-learning, one of major reinforcement learning methods, to acquire a goal defending behavior. The state space needs to be defined from the image observed from the robot [7]. We define the substates as shown in the first column in Table 1. The second column shows the numbers of quantization for each

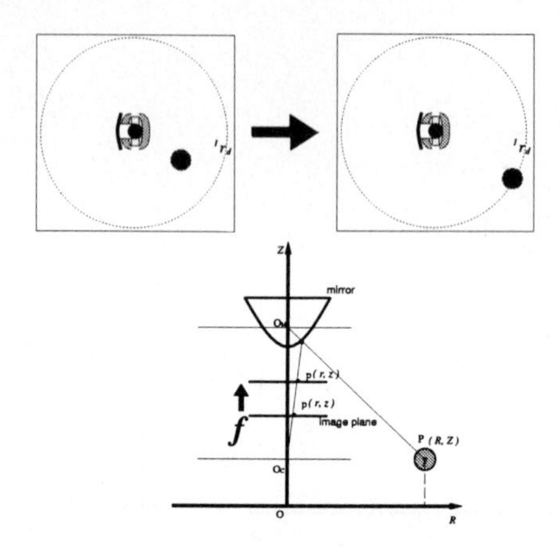

Fig. 6. Attention control on the omnidirectional vision

substates. In addition, the numbers of the quantization of the substates without the zoom servoing are shown.

zoom servoing	with	without
direction of the ball in the image	8	8
change of the direction of the ball	3	3
distance of the ball in the image	–	3
change of the distance of the ball	–	3
direction of the goal in the image	8	8
distance of the goal in the image	2	2
total number of the state	3456	320

Table 1. Substates

In the case of no zoom servoing, the states are defined in terms of the direction and the distance of the ball and the goal in the image. We define 8 substates for the direction of the ball as shown in Figure 7(a) and 3 substates (far, medium and near) for the distance as shown in Figure 7(b). In addition we define temporal changes of the direction and the distance, (clock wise, counter clock wise, no change) and (farther, nearer, no change), respectively. The numbers of the substates for the direction of the goal is 8, the same quantization for the ball, and substates of the distance is 2 (far and near). The total number of states is $3456 (= 8 \times 3 \times 3 \times 3 \times 8 \times 2)$.

In the case of active zoom servoing, substates for the distance of the ball and its temporal change are not necessary since the distance in the image is constant.

The direction of the ball and the goal are defined in the same manner as above. The distance of the goal is far and near, however, the observed image of the goal changes when the attention control is used. The distance of the goal can be represented by a monotonic function in terms of the actual distance between the robot and the ball. The total number of states is $320 (= 8 \times 3 \times 2 \times 8)$.

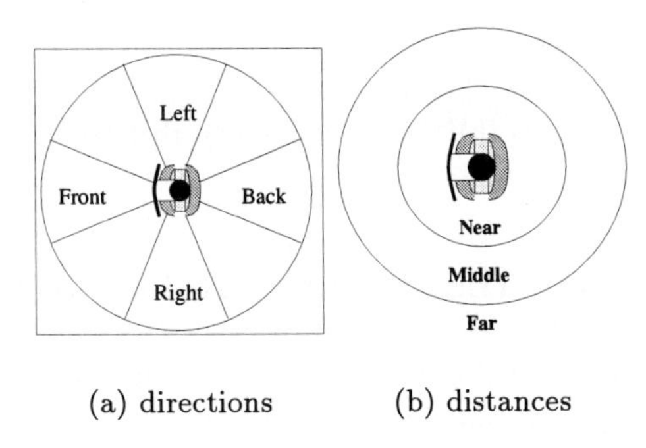

(a) directions (b) distances

Fig. 7. Substates

4 Experiments

First, we performed a computer simulation to acquire a goal defending behavior. Figure 8(a) shows the environment and the initial positions of the robot and the ball. The environment is built according to the RoboCup middle league regulations. The size of the field is 4575[mm] in width and 4110[mm] in length which is equivalent to the half size of the regulations. The goal size is 1500[mm] in width and 600[mm] in height and the diameter of the ball is 200[mm]. The goal and the ball are painted in blue and red respectively for easy detection.

The ball is located on a half circle defined by the center of the goal and two corners, and the robot is located inside the circle randomly. The ball rolls toward the goal at a constant velocity. One trial terminates when the ball comes into the goal or goes out from the field. Figure 8(b) shows the task success rate with the learned behavior. When the attention control is used the robot learn quicker than the case without the attention control.

After learning in the simulation, the acquired behavior is implemented on the real robot. Figures 9(a)-(f) show a sequence of the real behavior, when the robot succeeded in blocking the ball in front of the goal.

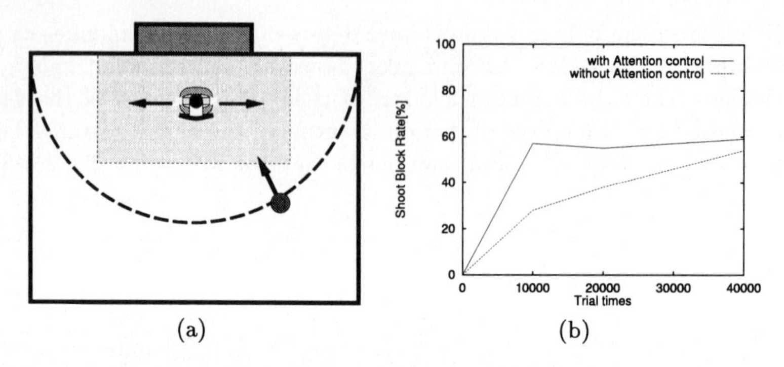

Fig. 8. (a)Initial position and (b)Result

5 Conclusions and the Result of RoboCup-98

We have proposed an attention control for an omnidirectional vision by controlling focal length of the camera and implemented it on a mobile robot. We have applied Q-learning method for acquisition of a goal defending behavior and shown that the attention control effectively worked to reduce the learning time. In this paper, we have shown a case that an embedded servo worked effectively for learning of the robot. However, we have not considered on a trade-off between installing a servoing mechanism and reduction of the learning time. This is the future work.

Though the team "Trackies-98" is third place in the middle size robot league, we are not satisfied the performance of our robots. From the view point of learning, we have not treated a behavior in an environment with opponents. This is the next issue.

This work was supported by the Cooperative Distributed Vision project in the Research for the Future Program of the Japan Society for the Promotion of Science (JSPS-RFTF96P00501).

References

1. T. Matsuyama. Cooperative Distributed Vision – Dynamic Integration of Visual Perception, Action, and Communication –. In *Proc. of Image Understanding Workshop*, 1998.
2. Y. Yagi and S. Kawato. Panoramic scene analysis with conic projection. In *Proc. of IEEE/RSJ International Conference on Intelligent Robots and Systems 1990 (IROS'90)*, pages 181–187, 1990.

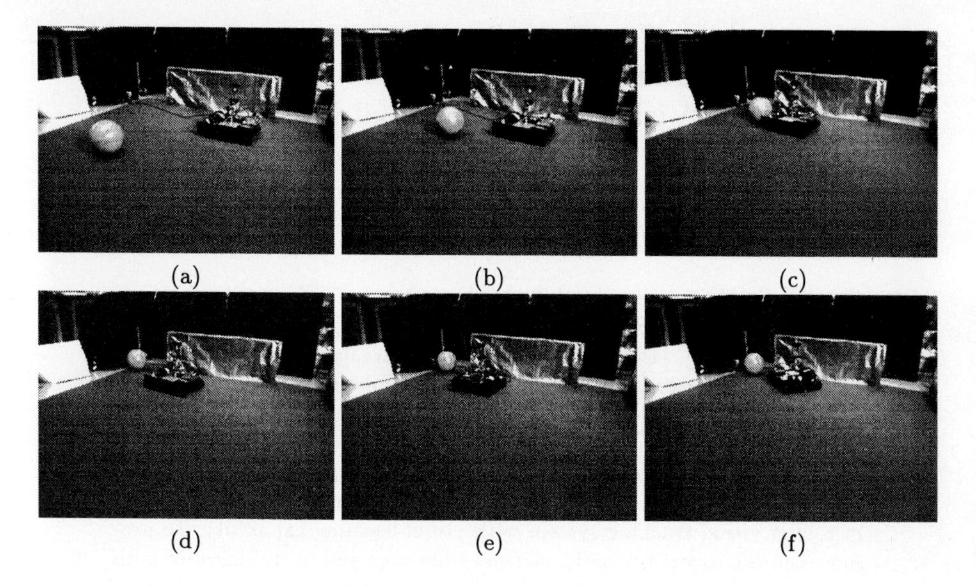

Fig. 9. A sequence of behavior

3. H. Ishiguro. Distributed vision systems: A perceptual information infrastructure for robot navigation. In *Proc. of IJCAI-97*, pages 36–41, 1997.
4. V. N. Peri and S. Nayar. Generation of perspective and panoramic video from omnidirectional video. In *Proc. of 1997 Image Understanding Workshop*, pages 243–245, 1997.
5. K. Hosoda, H. Moriyama, and M. Asada. Visual servoing utilizing zoom mechanism. In *Proc. of IEEE Int. Conf. on Robotics and Automation*, pages 178–183, 1995.
6. H. Kitano, M. Asada, Y. Kuniyoshi, I. Noda, E. Osawa, and H. Matsubara. "RoboCup: A challenge problem of Al". *AI Magazine*, 18:73–85, 1997.
7. M. Asada, S. Noda, S. Tawaratsumida, and K. Hosoda. "Purposive Behavior Acquisition for a Real Robot by Vision-Based Reinforcement Learning". *Machine Learning*, 23:279–303, 1996.
8. S. Suzuki, Y. Takahashi, E. Uchibe, M. Nakamura, C. Mishima, and M. Asada. "Vision-Based Learning Towards RoboCup: Osaka University 'Trackies' ". *RoboCup-97: Robot Soccer World Cup I*, Springer, pp.305–319, 1997.
9. E. Uchibe, M. Nakamura, and M. Asada. "Cooperative Behavior Acquisition in A Multiple Mobile Robot Environment by Co-evolution". *Proc. of the second RoboCup Workshop*, pp.237-250, 1998.
10. E. Uchibe, M. Nakamura, and M. Asada. "Cooperative Behavior Acquisition in A Multiple Mobile Robot Environment by Co-evolution". *RoboCup-98: Robot Soccer World Cup II*, Springer.

The RoboCup-NAIST: A Cheap Multisensor-Based Mobile Robot with Visual Learning Capability

T. Nakamura* K. Terada H. Takeda A. Shibata

Nara Institute of Science and Technology Dept. of Information Systems
8916-5, Takayama-cho, Ikoma, Nara 630-0101, Japan
*E-mail:takayuki@is.aist-nara.ac.jp
URL: http://cairo.aist-nara.ac.jp/~ takayuki/robocup-naist.html

Abstract. Our contribution is composed of two parts: one is development of a cheap multisensor-based mobile robot, the other is development of robust visual tracking system with visual learning capability. To promote robotic soccer research, we need a low cost and portable robot with some sensors and a communication device. This paper describes how to construct a robot system which includes a lightweight and low cost mobile robot with visual, tactile sensors, TCP/IP communication device, and portable PC where Linux is running. In real world, robust color segmentation is a tough problem because color signals are very sensitive to the slight changes of lighting conditions. In order to keep visual tracking systems with color segmentation technique running in real environment, a learning method for acquiring models for image segmentation should be developed. In this paper, we also describe a visual learning method for color image segmentation and object tracking in dynamic environment. An example of the developed soccer robot system and preliminary experimental results are also shown.
Keywords: Multisensor-Based, Portable PC, Linux, Visual Learning, Color Image Segmentation and Tracking

1 Introduction

Robotic soccer is a new common task for artificial intelligence (AI) and robotics research[1, 2]. The robotic soccer provides a good testbed for evaluation of various theories, algorithms, and agent architectures. Through the research for accomplishing this task, a number of technical breakthroughs for AI and robotics are expected to be discovered. We focus on two points among RoboCup physical agent challenges [2]: one is **platform** and the other is **perception**.

So far, many researchers have been studying robotic soccer and have proposed a variety of theories and methods for controlling, planning and so on. They built a team of robotic platforms for playing soccer by themselves, or purchased robotic platforms (for example, [3]). There is no standard robotic platform design for robot soccer. Generally, contemporary robotic systems involve large amounts of expensive, special purpose hardware for motor control and image processing. In this paper, we describe how to construct a cheap multisensor-based mobile robot and its control system mainly made from a state-of-the-art portable PC, a battery-powered R/C model car, a CCD camera and a set of tactile sensors. Since recent portable PC is affordable and powerful, such a PC is used as a central controller which manages processing sensor information, controlling motor and communication between robots. As a chassis of the mobile robot, a 4-wheel drive R/C model car is utilized. The important feature of our robot is that this platform has all its essential capabilities on board. Our platform consists of driving, visual sensing, tactile sensing, motor control, communication and decision-making system. Since each system is made of devices commercially obtainable, we can reduce both of the cost and complexity of the system. According to our design principle for soccer robot system, those who are interested in the robotic soccer would easily utilize or build this robotic platform by themselves.

In real world, robust color segmentation is a tough problem because color signals are very sensitive to the slight changes of lighting conditions. Currently, human programmer adjusts parameters used in discriminating colored objects in response to the changes of surroundings. In order to keep visual tracking systems with color segmentation technique running in real environment, a learning method for acquiring models for image segmentation should be developed. In this paper, we apply a visual learning method to the problem of color image segmentation and object tracking in dynamic environment. To realize a visual learning, our method utilizes the competitive learning algorithm called *rival penalized competitive learning* (RPCL) [4] which can automatically find out the number of classes in the sample data that a perceived color image consists of. After this learning, our method uses discovered classes as color models for objects. Using this color models, a color image is segmented into several regions which correspond to some objects. Then, based on segmented regions, our method performs visual tracking.

To evaluate the developed system, we have implemented some behaviors for playing soccer and a visual learning method which can perform color image segmentation and object tracking. Preliminary experimental results are also shown.

2 Our Hardware Architecture

In order that our soccer robots are used by not only roboticists but also researchers in other research communities, our soccer robots should be manageable. Furthermore, in order that our robot system satisfies the requirements of a standard platforms, it is important to reduce the cost and time for building our robot system. To address this issue, we use a portable PC as a central controller of robot system which is recently affordable and powerful.

2.1 Driving System

As a chassis of the mobile robot, we utilize a 4-wheel drive R/C model car which is commercially available. Actually, we utilize a chassis of "BLACK BEAST" (NIKKOH [1]) (See **Fig. 1**). This chassis is composed of a PWS (Power Wheeled Steering) system with two independent motors. Because of this mechanical structure, our robot can rotate at the same place. This system is useful for avoiding the situation that its body gets stuck into corners. Existing motors provided by NIKKOH are comparatively powerful. However, if we put something whose weight is more than 1 Kg on the existing chassis, the body can't move around

[1] NIKKOH is a Japanese toy company. BLACK BEAST is also commercially available outside of Japan.

by those motors. In order to make the motor more powerful, a planetary gear box is attached to the existing motor. As a result, the chassis is able to carry something that weighs about 4 Kg. As the planetary gear box, we utilize the gear box [2] for a toy model car.

2.2 Tactile Sensing System

A tactile sensing system is used for detecting contact with the other objects such as a ball, teammates, opponents and a wall. It is also important for soccer robots to have tactile sensing capability, because soccer robots frequently collide with each other, walls or a ball in a soccer field. Furthermore, tactile sensing system can compensates for limitation of visual sensing. Since the field of view of the camera mounted on the robot is limited, if collision between the robots or between the robot and the wall or the ball occurs on the outside of the field of view, it is difficult to detect these happenings based on the image information. Tactile sensing system where tactile sensors are set around the body of soccer robot is very useful for solving this problem. Since the cost of producing a tactile sensing system is generally high, this prevents it being used widely.

Here, we construct a cheap tactile sensing system (See **Fig. 2**) by remodeling a keyboard which is usually used as an input device for PC. A keyboard consists of a set of tactile sensors each of which is a ON/OFF switch called a key. If a key is pressed, the switch is ON. If not, the switch is OFF. Since we can get a keyboard at a low price, it is possible to construct this tactile sensing system for soccer robots at a low cost.

If a tactile sensor (key) hits an object such as a ball or an opponent, the sensor outputs an ASCII code corresponding to the key. In case several sensors have contact with the other object, an output of this sensing system is a sequence of ASCII codes.

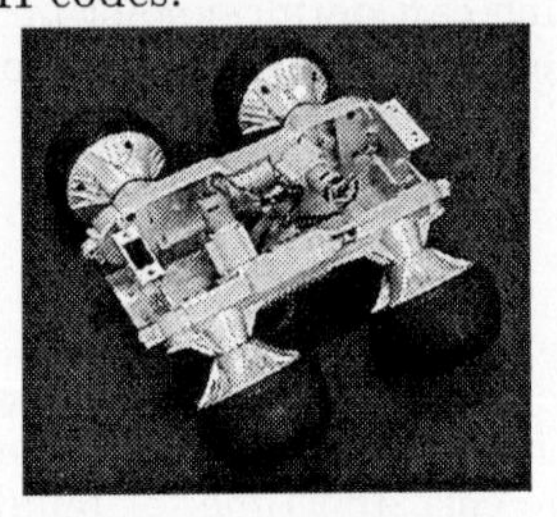

Fig. 1. Our driving system.

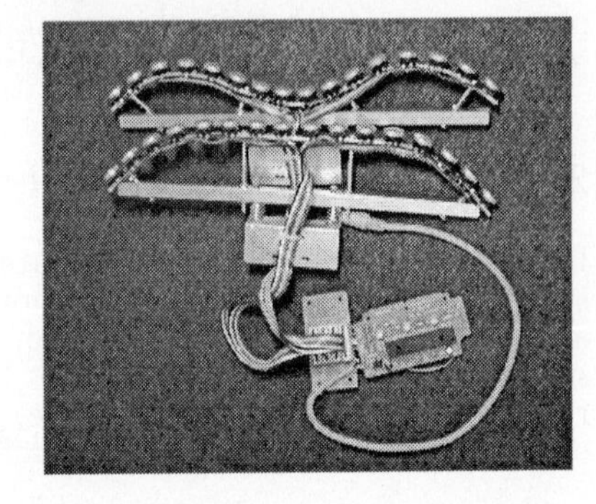

Fig. 2. Our tactile sensors made of a key board.

2.3 Visual Sensing System

Our robotic soccer project aims the development of robotic soccer players with on-board visual sensor like human soccer players. So, a visual sensing system in our soccer robot plays a fundamental role in acquiring visual information and recognizing it. Our soccer robots make a pass or tackle and shoot a ball into a goal based on the images taken by the on-board camera. In order to build such visual sensing system, we have chosen to use a commercial video capture

[2] The gear box is commercially available from a Japanese toy company TAMIYA

PCMCIA card (IBM Smart Capture Card II, hereafter SCCII) which can be easily plugged into a portable PC and a color CCD camera (SONY EVI D30, hereafter EVI-D30) which has a motorized pan-tilt unit.

SCCII is a PCMCIA type-II video capture card which can capture at 30 frame-per-second at maximum resolution 320-by-240 in 16-bit RGB formats. We can feed video to SCCII in NTSC or PAL format, and the card provides jacks for both composite-video and S-Video input. A device driver for the use of SCCII on Linux OS[5] is distributed as a free software. We utilize this device driver in order to capture images on Linux OS.

EVI-D30 is a high-performance color CCD camera, because it has auto target tracking function based on color information and motion detection function. We can control eyes of EVI-D30 with a motorized pan-tilt device which can be managed by a portable PC through RS232C. The pan and tilt angle of this device ranges from -100 to $+100$ and from -25 to $+25$, respectively. In this way, this camera can cover wide field of view. Since our soccer robot has such sensing capability, our robot can find a ball by moving its camera head without moving its body.

2.4 Motor Control System

A motor control system is used for driving two DC motors and is actually an interface board between a portable PC and motors on the chassis of our soccer robot (see **Fig. 3**). This control board is plugged into a parallel port on the portable PC. Our motor control system manages only the direction of current to a DC motor. The control circuit in this board consists of mainly 4 relays in terms of one motor (see **Fig. 3**). These relays are used as just like an ON/OFF switch and for controlling the direction of current. This board is powered by a 7.2 V battery for a R/C model car. As a result, this board can sends three control commands to right and left motors such as "(Forward, Stop, Backward)". The motor control command is actually 2 bits binary commands for one motor. Therefore, totally 4 bits (D0!&D1 or D2!&D3 in **Fig. 3**) in the parallel port are used for transmitting motor commands to the control board. Since we can send the motor control command to each of the two motors separately, our soccer robot has 3 sub-action primitives, forward, stop and backward in term of one motor. All together, our soccer robot can take 9 action primitives.

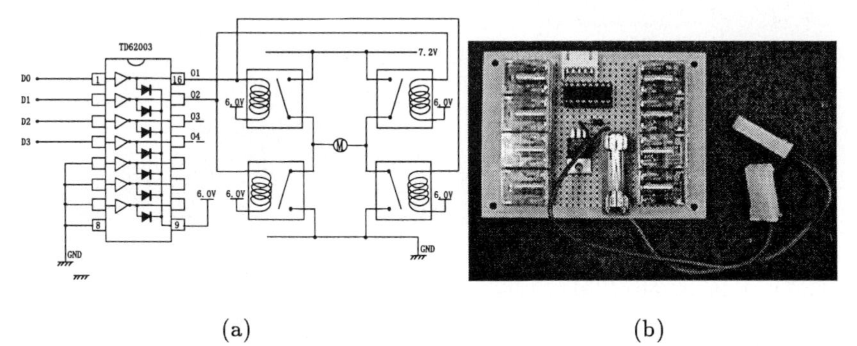

(a) (b)

Fig. 3. Our motor drive board.

2.5 Communication System

In the soccer game, teammates need to communicate each other for accomplishing a given task in cooperative manner. So, we set a wireless LAN device for communication on our soccer robot. The wireless LAN device is actually WaveLAN(AT&T) which can be plugged into a portable PC. The system operates in 2.4GHz frequency band. The rate of transmitting data is 2Mbps. The maximum transmission range will reach several hundred meters when there is a clear line of sight between the transmitter and receiver.

2.6 Intelligent Control System

We call a central controller for processing sensor information and controlling the body of mobile robot and camera "intelligent control system". The intelligent control system consists of software, programming environment and OS. In order to adopt an OS as the central manager of robotic system, the OS should have some characteristics as follows:(1)It is possible to run multiple independent processes. (2)It is possible to make a process abort or wait for running again. (3)The system provides mechanisms for simple and high-speed process synchronization and communication.

In this work, we have chosen to use Linux OS as an OS for intelligent control system. Linux is a freely distributable, independent UNIX-like OS. Much of the software available for Linux is developed by the Free Software Foundation's GNU project. It supports a wide range of software, including X Windows, Emacs, TCP/IP networking (including SLIP/PPP/ISDN).

We cannot guarantee user-mode processes to have exact control of timing because of the multi-tasking nature of Linux. Our process might be scheduled out at any time for anything from about 10 milliseconds to a few seconds (on a system with very high load). However, for most applications in RoboCup competition so far, this does not seem to really matter. If we want more precise timing than normal user-mode processes, there is a special kernel RT-Linux that supports hard real time(See [6] for more information on this.).

3 System Configuration of Our Soccer Robot

Currently, we have developed a vision-based mobile robot for robotics soccer as shown in **Fig. 4**. As a portable PC, we have chosen to use a Libretto 60 (Toshiba) which is small and light-weight PC. The total cost of this soccer robot is about $ 4,800.

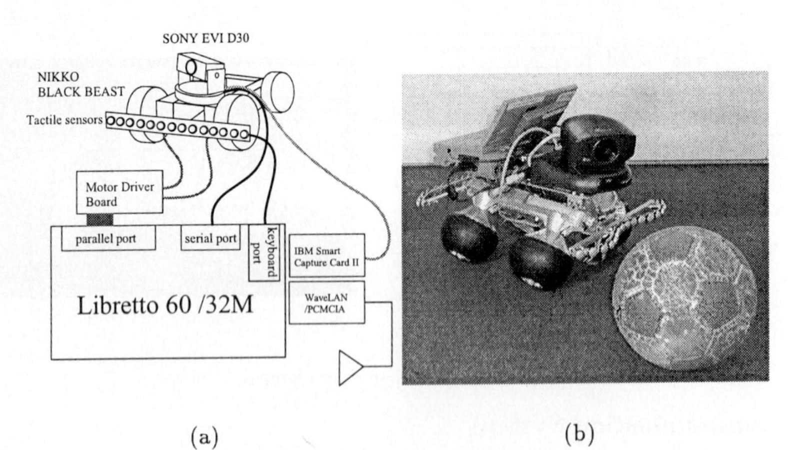

(a) (b)

Fig. 4. Our soccer robot.

4 Our Software Architecture

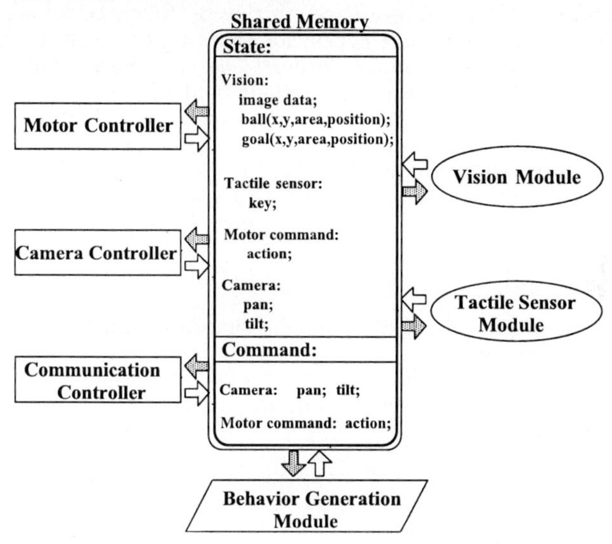

Fig. 5. Software architecture.

In order to control our hardware systems and coordinate between them, we use a shared memory [7] and 5 software components which are the motor controller, camera controller, tactile sensor module, vision module and behavior generator. **Fig. 5** shows an interactions between these software components. Note that this figure shows the software architecture of our current robotic soccer system. All software components read and write the same shared memory. Using this shared memory, they can communicates each other unsynchronously. As shown in **Fig. 5**, we define the structure of the shared memory. For example, the behavior generator takes the state of camera, vision, tactile and motor in the shared memory as input vectors. Then, it combines these information with programmer's knowledge and decides the robot's action at next time step. Finally, it writes the motor command for the motor controller on the shared memory. In the same way, other software components read states and write commands in each timing.

4.1 Motor Controller

We assume that a motor command is defined by a action primitive and its duration. In our robotic system, an action consists of a combination of 4 action primitives (move forward, backward, turn left, and turn right) and 4 kinds of the duration ($100msec$, $150msec$, $200msec$, $300msec$). Furthermore, we add one action for kicking a ball strongly to the actions. This action is produced by a combination of "move forward" and $500msec$ duration. Totally, our mobile robots can take 17 actions. Motor controller module reads the command from the shared memory every $100msec$. If there is a command, it executes the command and rewrites the executed command as the state of motor.

4.2 Camera Controller

We can control the onboard camera (SONY EVI-D30) through RS232C with the VISCA protocol provided by Sony Corp. Using VISCA, we can control the pan, tilt angle and the focal length of the camera and take its focus. Furthermore, we can turn on and off the camera through this protocol. In robotic soccer task, panning the camera is important action for tracking the objects such as a ball, a goal, teammates and opponents. In order to detect where a ball is in the field, our robots always try to track a ball in their field of view using the motion of their camera head. Actually, pan and tilt angles are controlled so that the center of the ball image may coincide with the center of the captured image.

Since the soccer robots frequently lose the ball in the field, they must find the ball again as soon as possible. In order to realize the procedure for finding a ball, it is considered that panning the camera is very useful. We called such behavior of the camera "finding behavior". We implement the finding behavior as follows:

repeat
  Read the area of ball region from the shared memory.
  Make the camera **pan**.
  if the camera rotates to the limit,
    make the direction of its panning **opposite**.
until the ball is in view.

4.3 Tactile Sensor Module

Since our tactile sensor system is actually a keyboard, an output of the sensor system is an ASCII code corresponding to the key. We can get this ASCII code via X Event [8] which is a library function of X11 for detecting all events in X Window system. Our tactile sensor module maintains a table of ASCII codes and the configuration of tactile sensors. All tactile sensors are numbered from 1 to 32 so that the left front of tactile sensor unit might be numbered 1 and the right back 32. In case a sensor has contact with an object, the sensor module can detect which sensor has contact with the object using the table that shows corresponding between ASCII codes and the index number of tactile sensors. Then, the tactile sensor module rewrites the index number of the detected sensor as the state of the tactile sensor system on the shared memory.

4.4 Vision Module

The vision module provides some information about the ball and goal in the image.

To date, in RoboCup competition, each soccer robot tried to discriminate such objects based on color information. In our study, we also use color information for segmenting and tracking objects (a ball, goals, white lines, teammates and opponents). Furthermore, in order that such color image segmentation and object tracking should be correct even if surroundings such as lighting condition changes, our vision module has visual learning capability based on RPCL [4]. After the color segmentation, we calculate the coordinates of the center of ball and goal position, and the both maximum and minimum horizontal coordinates of the goal and so on. (See **Fig. 5**.) Then, based on segmented regions, our robots perform visual tracking. Our vision module also discriminates in which position center of ball or goal appears among three positions (right, center and left of an image).

Construction of Color Model for Segmentation In order to make initial color models for some obects, we use the competive learning alogirhtm called *rival penalized competitive learning* (RPCL) [4] which can automatically find out the number of classes in the sample data. After this learning, we use discovered classes as color models for objects. We briefly explain the procedure of RPCL according to [4]. RPCL algorithm repeats the following two steps until the prototype vectors converge on constant vectors.

STEP1: Randomly take a sample $\boldsymbol{x} = (x_1, x_2, \cdots, x_k)$ from a data set. Let $\boldsymbol{w}_i = (w_{i1}, w_{i2}, \cdots, w_{ik})$ be a prototype vector ($i = 1 \sim, N$). Then, calculate a parameter u_i defined as follows:

$$u_i = \begin{cases} 1, & \text{if } i = c \text{ such that} \\ & \gamma_c d(\boldsymbol{x}, \boldsymbol{w}_c) = \min_j \gamma_j d(\boldsymbol{x}, \boldsymbol{w}_j) \\ -1, & \text{if } i = r \text{ such that} \\ & \gamma_r d(\boldsymbol{x}, \boldsymbol{w}_r) = \min_{j \neq c} \gamma_j d(\boldsymbol{x}, \boldsymbol{w}_j) \\ 0, & \text{otherwise.} \end{cases}$$

where $\gamma_j = n_j / \sum_{i=1}^{k} n_i$ and n_i is the cumulative number of the occurrences of $u_i = 1$. $d(\boldsymbol{x}, \boldsymbol{w})$ denotes a distance between $\boldsymbol{x}$ and $\boldsymbol{w}$. Generally, $d(\boldsymbol{x}, \boldsymbol{w}_i) = ||\boldsymbol{x} - \boldsymbol{w}_i||^2 = \sum_{j=1}^{k} |x_j - \boldsymbol{w}_{ij}|^2$. Moreover, $\boldsymbol{w}_c$ and $\boldsymbol{w}_r$ denote the winner vector which wins the competition for adapting to the input vector and the second winner vector called "rival", respectively.

STEP2:Update the prototype vector $\boldsymbol{w}_i$ by

$$\Delta \boldsymbol{w}_i = \begin{cases} \alpha_c(\boldsymbol{x} - \boldsymbol{w}_i) & \text{if} u_i = 1 \\ -\alpha_r(\boldsymbol{x} - \boldsymbol{w}_i) & \text{if} u_i = -1 \\ 0 & \text{otherwise.} \end{cases}$$

where $0 \leq \alpha_c, \alpha_r \leq 1$ are the learning rates for the winner and rival vector, respectively.

Fig. 6 shows examples of an image captured by SCCII and segmented image based on our method. Although the size of a captured image is 320×240 pixels, we shrink it so as to reduce computational cost of CPU on a portable PC. Actually, the size of a processed image is 80×60 pixels. As shown in **Fig. 6** (b), our color segmentation algorithm succeeds in extracting a red ball, a yellow goal, green field (ground), white line, and white wall. Currently, it takes about 230 $msec$ (about $4Hz$) for one cycle of this procedure in case $N = 57$.

Simple Color-Based Tracking Our simple tracking method is based on tracking regions with similar color information from frame to frame. We define a fitness function $\Phi_{target}(x, y)$ at a pixel (x, y) as a criterion for extracting a target region in the image,

$$\Phi_{target}(x, y) = \begin{cases} 1 & \boldsymbol{C}(x, y) \in \boldsymbol{CM}_{target} \\ 0 & \text{Otherwise} \end{cases}$$

,where $\boldsymbol{C}(x, y)$ and $\boldsymbol{CM}_{target}$ show a $Yr\theta$ value at (x, y) and a color model for a target represented by a cuboid, respectively. Based on $\Phi_{target}(x, y)$, the best estimate $(\hat{x}_{target}, \hat{y}_{target})$ for the target's location is calculated as follows:

$$\hat{x}_{target} = \frac{\sum_{(x_i, y_i) \in R} x_i \Phi_{target}(x_i, y_i)}{\sum_{(x_i, y_i) \in R} \Phi_{target}(x_i, y_i)}, \quad \hat{y}_{target} = \frac{\sum_{(x_i, y_i) \in R} y_i \Phi_{target}(x_i, y_i)}{\sum_{(x_i, y_i) \in R} \Phi_{target}(x_i, y_i)},$$

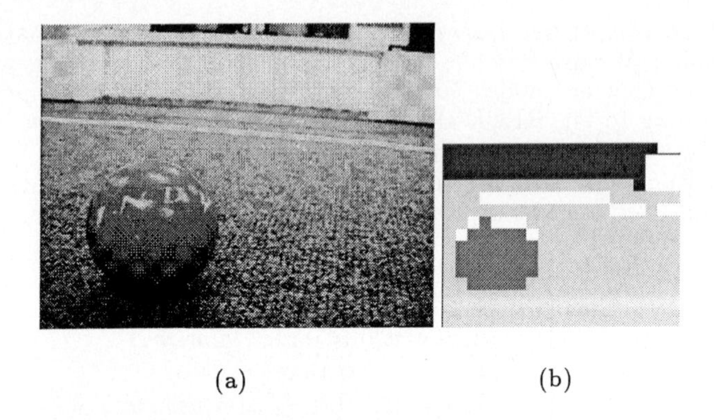

(a) (b)

Fig. 6. A example of processed images taken by the robots

where R shows the search area. Initially, R implies an entire image plane. After initial estimation for the location of the target, we can know the standard deviations $\sigma(\hat{x}_{target})$ and $\sigma(\hat{y}_{target})$ regarding $(\hat{x}_{target}, \hat{y}_{target})$. Therefore, based on the deviations, R is restricted to a local region during the tracking process as follows:

$$\begin{aligned} R : \{(x, y)| \\ \hat{x}_{target} - 2.5\sigma(\hat{x}_{target}) \leq x \leq \hat{x}_{target} + 2.5\sigma(\hat{x}_{target}), \\ \hat{y}_{target} - 2.5\sigma(\hat{y}_{target}) \leq y \leq \hat{y}_{target} + 2.5\sigma(\hat{y}_{target})\}. \end{aligned}$$

$\sum_{(x_i,y_i)\in R} \Phi_{target}(x_i, y_i)$ shows the area of the target in the image. Based on this value, we judge the appearance of the target. If this value is lower than the pre-defined threshold, the target is considered to be lost, then R is set to be the entire image plane for estimation at next time step. We set this threshold for the target area $= 0.05 * S$, where S shows the area of the entire image. This process helps to reduce the computational cost for extracting regions with similar color.

4.5 Behavior Generator

The behavior generator decides the robot's behavior such as avoiding a wall (called avoiding behavior), shooting a ball into a goal (called shooting behavior) and defending a goal from opponent's attack (called goalie behavior).

Avoiding Behavior We implemented avoiding behavior so that the robot may avoid a wall using tactile sensors. We divided 32 tactile sensors into 4 groups (G1,G2,G3,and G4);

$G1(1 \cdots 8)$: left front, $G2(9 \cdots 16)$: right front,
$G3(17 \cdots 24)$:left back, $G4(25 \cdots 32)$:light back

Avoiding behavior is implemented as follows:

Read the state of tactile sensor from the shared memory.
position ← **the state**.
switch(position)
  G1: move backward and turn right.

G2: move backward and turn left.
G3: move forward and turn right.
G4: move forward and turn left.

While the area of the ball region is less than a threshold, this behavior has top priority over all other behaviors. As a result, whenever the robot collides with an object in case that that condition is valid, it always avoids it. On the contrary, when the area of the ball is more that the threshold, this behavior is suppressed by the other behavior, for example, shooting behavior that is explained in the following.

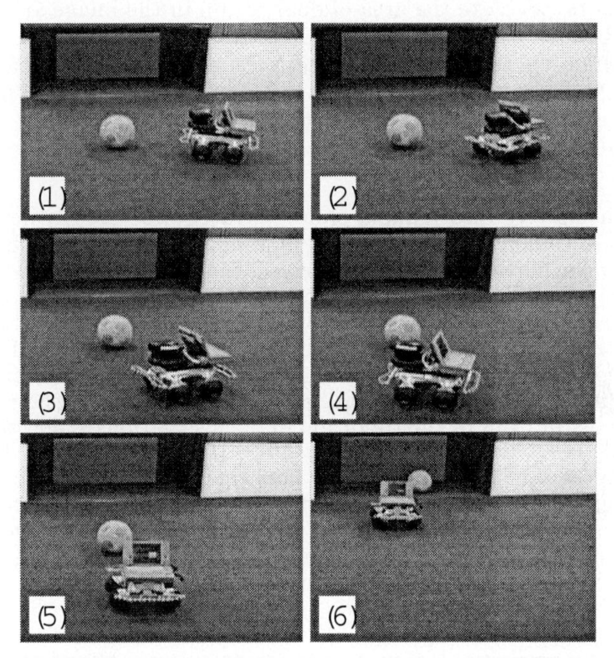

Fig. 7. Shooting behavior. (1):Approach the ball. (2),(3),(4):Round the ball. (5),(6):Kick the ball.

Shooting Behavior We make a simple strategy for shooting the ball into the goal. To shoot the ball to the goal, it is important that the robot can see both ball and goal. Therefore, the robot must round the ball until the robot can see both ball and goal with the camera toward the ball. Finally, the robot kicks the ball strongly. **Fig. 7** shows the shooting behavior.

The concrete procedure of shooting behavior is follows:

1)Find the ball
2)Approach the ball
  While approaching the ball
  Read the area of ball from the shared memory.
  if the area > 20 **then** stop
3)Round the ball
  Detect the direction of goal **d** $\leftarrow$ **the direction**
  switch(d)
    right: clockwise round the ball

with the camera toward the ball
left: counterclockwise round the ball
with the camera toward the ball
if the robot can see both ball and goal **then** stop
4)**Make** the body of the robot **turn** toward the ball
5)**Kick** the ball strongly

Here, we explain how to detect the direction of the goal which plays important role in shooting behavior. To find out the opponent's goal, the robot turns its camera head from -100 ° to +100 °. During the motion of the camera head, the robot continue to calculate the area of goal region in the image and find out the angle where the area of the goal is maximum. If the angle ranges in the right/left hand of the robot, the robot recognizes that the direction of the goal is right/left on the basis of the forward direction of the robot. In this way, the motion of the camera head enables the relative configuration between the robot and the goal to be detected.

Goalie Behavior We also make a simple strategy for preventing a ball from entering a goal. To defend the goal from opponent's attack, it is important that the robot always moves with the center of the robot body toward the ball. Therefore, the robot must coordinate the motion of camera head and its body so as to find a ball in the penalty area. When the ball is far from the robot, the robot turns its camera head to track a ball without the motion of its body. As long as the rotational angle of the camera head is within a certain range, the robot doesn't need to change its position to defend the goal. The reason is that the ball is almost in front of the robot under such situation. When the rotational angle for tracking a ball is over a certain range, the robot moves to the right/left side of penalty area in parallel with the goal line. In this way, according to the rotational angle of the camera head for tacking a ball, the goalie robot changes a strategy for defending the goal. This strategy comparatively worked well at the RoboCup-98 competition. **Fig. 8** shows an example sequence of the goalie behavior that is recorded at the RoboCup-98 competition.

Fig. 8. Goalie behavior

5 Discussion

In this paper, we described how to construct a cheap multisensor-based mobile robot system which consists of mainly made from state-of-the-art portable PC, a battery-powered R/C model car, a CCD camera and a set of tactile sensors by remodeling a keyboard. Since these components are commercially available, we can construct the total system at comparative low cost. Our robot system might be used as a personal robot which can be used at home since its price would be low and its performance would be high. Now, we use this multisensor-based mobile robot as a standard platform for robotics soccer research. In the future, we plan to realize

- robust behavior based on sensor fusion between visual and tactile information, and
- cooperative behavior with other robots.

We also describe the software architecture of our robots and how to implement some kinds of behaviors for playing soccer. In respect of the implementation of soccer behavior, it is found that the motion of camera head on the robot gives cue for identifying the relative position between the robot and the goal or the ball. Currently, all soccer behaviors are programmed by the designers of the robot. This is a problem because these behaviors are not so adaptive to the unexpected change of the environment. Therefore, online learning method is required for adapting the robot to the unexpected events in its environment. Such learning method should operate on a mechanism that creates the adaptive behaviors in addition to the behavior programmed by the designer. In the future, we plan to realize

- online learning mechanism based on the evolution of the pre-programmed behaviors.

In this paper, we also describe a method for visual learning for color image segmentation on the basis of RPCL. Through the RoboCup-98 competition, it is found that the performance of the visual tracking based on our method works well. However, when the lighting condition of the environment suddenly changes, our robots have failed to track a ball. The reason is that the our current visual learning method is basically off-line. So, in order to make our visual tracking capability more robust, we should develop a mechanism of on-line visual learning for tracking the objects in the unexpected environment.

References

1. H. Kitano, M. Tambe, Peter Stone, and et.al. "The RoboCup Synthetic Agent Challenge 97". In *Proc. of The First International Workshop on RoboCup*, pages 45–50, 1997.
2. M. Asada, Y. Kuniyoshi, A. Drogoul, and et.al. "The RoboCup Physical Agent Challenge:Phase I(Draft)". In *Proc. of The First International Workshop on RoboCup*, pages 51–56, 1997.
3. Inc. Nomadic Technologies. `http://www.robots.com/robotdiv.html`.
4. L. Xu, A. Krzyzak, and E. Oja. "Rival Penalized Competitive Learning for Clustering Analysis, RBF Net, and Curve Detection". *IEEE Trans. on Neural Networks*, 4:4:636–649, 1993.
5. Linux. `http://www.linux.org`.
6. RT-Linux. `http://luz.cs.nmt.edu/~rtlinux`.
7. W. Richard Stevens. *UNIX NETWORK PROGRAMMING.* Prentice Hall, Inc., 1990.
8. Adrian Nye. *Xlib programming manual : for version X11 of the X Window System, 3rd ed.* O'Reilly, 1992.

Andhill-98: A RoboCup Team which Reinforces Positioning with Observation *

Tomohito Andou

C&C Media Research Laboratories, NEC Corporation
Miyazaki 4-1-1, Miyamae-ku, Kawasaki, 216-8555, Japan
E-mail: tandou@ccm.cl.nec.co.jp

Abstract. On reinforcement learning with limited exploration, an agent's policy tends to fall into a worthless local optimum. This paper proposes Observational Reinforcement Learning method with which the learning agent evaluates inexperienced policies and reinforces it. This method provides the agent more chances to escape from a local optimum without exploration. Moreover, this paper shows the effectiveness of the method from experiments in the RoboCup positioning problem. They are advanced experiments described in our RoboCup-97 paper[1].

1 Introduction

Andhill [2] won the second prize of RoboCup-97 simulator league and the championship of RoboCup Japan Open 98 simulator league. In these competitions, Andhill's on-line learning mechanism was only used in a few games because of the insufficiency of the advantage from fixed-positioning strategies. In RoboCup-98, the offside rule was introduced and it was expected that positioning strategies of many teams would be changed. A learning mechanism would be effective if strategies of the opponent were unimaginable. Therefore, we used an on-line learning mechanism in many games at RoboCup-98.

The on-line learning mechanism of Andhill-98 was implemented with consideration of the following three points. First, the agents should behave on their best policies during a whole game because the game is not a practice game. So agents are set to explore very little policies. Second, the learning results must be reflected within a very short game. The learning agent's policy can be started from a randomly initialized policy, a zero initialized policy, or an already learned policy. We chose a zero initialized policy because it is the most sensitive way for short-term learning. Third, learning from zero is too dangerous especially at the first period of the game. To avoid the danger, Andhill-98 started with the fixed-positioning strategy while learning, and switched to the learned-positioning strategy when the team scores.

---

* This work was mainly done when the author was in Dept. of Mathematical and Computing Sciences, Tokyo Institute of Technology.

[2] The team name Andhill is a parody of "anthill" in which ants live. This comes from its character of a multi-agent system.

The first point above, namely, limited exploration is an essential difficulty on reinforcement learning. To deal with this difficulty, Andhill-98 used Observational Reinforcement Learning method. This paper describes the RoboCup positioning problem in section 2, Observational Reinforcement Learning method in section 3, experiments in section 4, and conclusions in section 5.

2 The RoboCup Positioning Problem

A RoboCup agent has mainly the following three routines:

1. If the ball is within kickable range: Where will it kick the ball to?
2. If the ball is close from it: How will it catch the ball?
3. If the ball is far from it: Where will it run to?

Andhill-98 is designed as (1) and (2) are already set by human, and (3) can be obtained by learning.

About (3), most of the RoboCup teams applied fixed-positioning strategies while dynamic positioning strategies are general in real-world soccer games. Dynamic positioning can be determined by something like a positioning function which inputs an agent's environment and outputs its suitable position. We attempted to learn such a dynamic positioning function by on-line reinforcement learning. We designed the function as a three layer neural network $(5-6-1)$ which inputs the following elements for each place and outputs the suitability of the place for positioning.

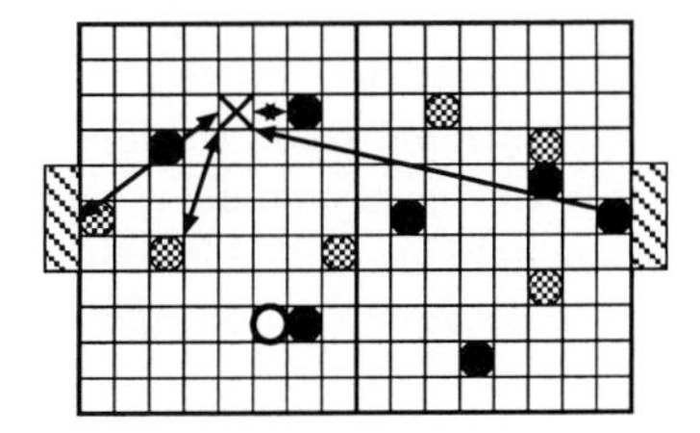

1. Distance to the goal of the agent's side
2. Distance to the goal of the opponents' side
3. Distance to the closest team-mate
4. Distance to the closest opponent
5. Which team possesses the ball

Fig. 1. Field information

3 Observational Reinforcement Learning

3.1 Difficulties in an ordinary reinforcement learning method

In an ordinary reinforcement learning method, an agent can only reinforce its experienced policies[2]. This restriction causes the following two difficulties in the RoboCup positioning problem. The first difficulty is the fact that positioning is a combined action. While previous reinforcement learning techniques had only dealt with primitive actions, we had to also apply reinforcement learning to combined actions. In RoboCup problems, positioning consists of two kinds of primitive actions, dash and turn. A combined action usually costs much time and/or something else. The RoboCup positioning costs much time and great stamina. This means exploration can not be executed sufficiently in on-line learning.

The second difficulty is the fact that positioning is a cooperative action. In the case of on-line learning, all agents have to learn simultaneously in a cooperative multi-agent environment. The learning of an agent requires the other agents behave stably in their best policies. It means that in order to learn better, they can not explore. This dilemma limits trials and errors. Both of the difficulties limit exploration of an agent. In an ordinary reinforcement learning method with limited exploration, an agent's policy tends to fall into a worthless local optimum. We proposed Observational Reinforcement Learning method which offers more chances to escape from worthless local optimal policies without exploration.

3.2 Observational Reinforcement Learning method

Observational Reinforcement Learning method is a reinforcement learning method in which an agent can also reinforce an inexperienced policy which is evaluated as good from its observation. In the RoboCup positioning problem, an agent can evaluate some positions as good just only from its observation. One evaluation is like: A place where the ball comes frequently will suit for positioning. In this method, an agent can reinforce suitable places immediately by this evaluation. The agent needs no actual experience of positioning to reinforce the policy. Therefore, an agent can reinforce various positioning independently of the cost of the positioning or the agent's actual behaviors. Observational Reinforcement Learning method enables an agent to reinforce not only low-costed policies but various policies. Moreover, an agent can reinforce various policies while behaving in its best policies. Consequently it offers more chances to escape from worthless local optimal policies.

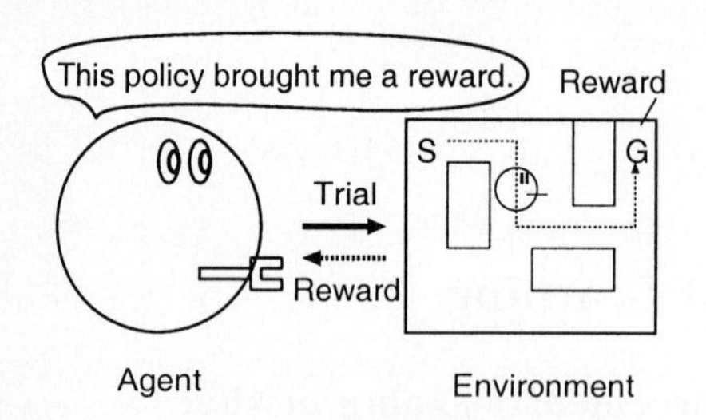

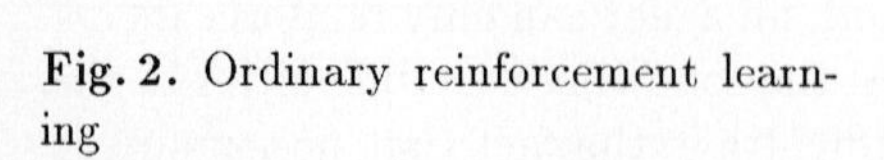

Fig. 2. Ordinary reinforcement learning

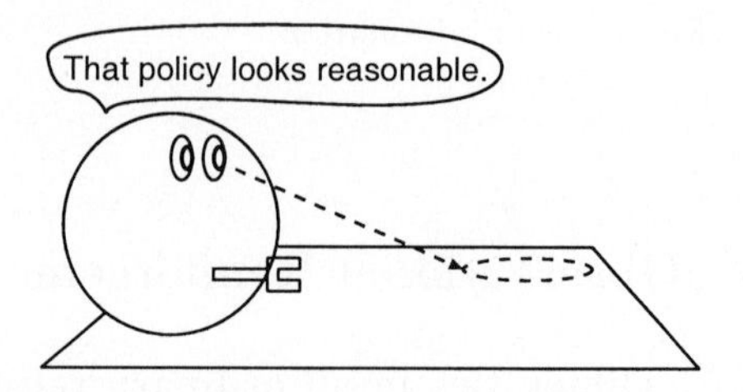

Fig. 3. Observational Reinforcement Learning

Observational Reinforcement Learning method could be regarded as a combination method of reinforcement learning and supervised learning. Therefore, supervised learning techniques like a complementary error back propagation algorithm can be applied in this method.

3.3 Comparison with ordinary reinforcement learning

Ordinary reinforcement learning can be regarded as an imitation of the process of gaining "confidence", because the learning agent gets rewards when it acts something good and becomes sure that the action is good. Observational Reinforcement Learning is, however, an imitation of "regret", because the learning agent judges its recent action was worse than another action and becomes sure that the unexecuted action would be good.

As mentioned above, Observational Reinforcement Learning offers more chance to escape from a local optimal policy and this advantage is remarkable in online learning. However, it also has a weakness especially in multi-agent learning. Generally, a multi-agent system works efficiently by diversity among the agents. Diversity among the agents is important on a cooperative behavior. Ordinary reinforcement learning agents have some diversity because experience is unique for each agent, but Observational Reinforcement Learning agents have little diversity because observation tends to similar among all agents. Thus both can make up the weakness of the other.

	Type of Learning	On-line Learning	Diversity in a Multi-Agent System
Ordinary RL	Confidence	Difficult	Various
Observational RL	Regret	Easier	Monotonous

4 Experiments

We attempted the RoboCup positioning problem with Observational Reinforcement Learning method. We compared the following three experiments:

- **Experiment 1 – Ordinary reinforcement learning**: An agent gets a reward when it kicks the ball, and then it reinforces its recent policies.
- **Experiment 2 – Observational Reinforcement Learning**: An agent reinforces the ball location when all team-mates are far from the ball. See Fig. 4.
- **Experiment 3 – Combination of ordinary reinforcement learning and Observational Reinforcement Learning**: Both reinforcements in Experiment 1 and Experiment 2 are to be done.

Experiments were executed in an unending on-line game between a learning team and a fixed position team on the RoboCup-97 rule. Ten players of the learning team excepting a goal-keeper were learning simultaneously. Exploration is not used explicitly. That is, all learning agents choose policies which are they

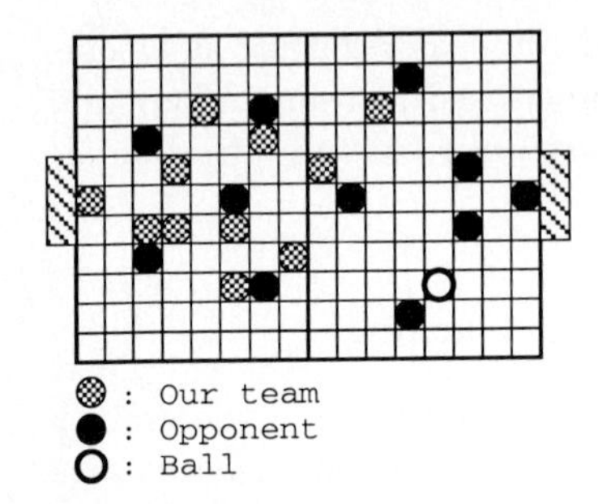

The ball location is far from all team-mates. Then, they will reinforce a policy of staying at the place.

Fig. 4. A situation in which agents regret

think the best. In the experiments, the fixed position team was Andhill-97, and the learning team was a position learning team who has the same faculties of Andhill-97 excepting the positioning. The learning was started with randomly initialized policies. Each experiment needed about 20 hours until policies of all the learning agents would converge.

The next table shows the experimental results. Values of the table represent averaged scores per a game of over 60 games. In the case that both of the teams are the same fixed position teams, the game will end in score 6.0 to 6.0 on the average. The learning team won against Andhill-97 only in Experiment 3.

	Experiment 1	Experiment 2	Experiment 3
Learned team score	2.6	4.5	5.4
Fixed position team score	3.0	9.3	5.0
Margin	−0.4	−4.8	0.4

Details of each experimental results are described in Section 4.1, Section 4.2, and Section 4.3.

4.1 Results of Experiment 1 (Ordinary RL)

In Experiment 1, we attempted the RoboCup positioning problem with an ordinary reinforcement learning method. In this method, an agent will get a reward when the agent is doing something good, and will reinforce the recent behaviors. The reward was defined as: When an agent kicks the ball, a reward is given to it.

Fig. 5. shows the policy transitions of 10 agents of the learning team. The horizontal axis means the game time count, and the vertical axis means the distance from their goal. The values of the vertical axis are regularized into $(-1.0, 1.0)$ and averaged from the behaviors of 10 games. This figure shows that the agents' policies branched into two types. There were seven agents who has defensive policies of positioning, and four agents who has offensive policies. This

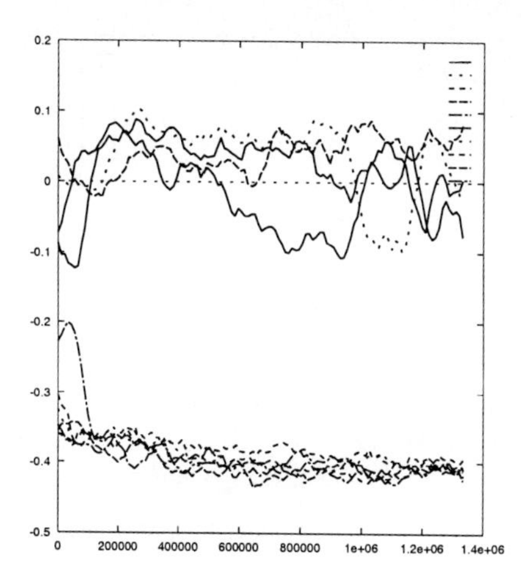

Fig. 5. Policy transitions of 10 agents in Experiment 1: The vertical axis means the distances from their goal.

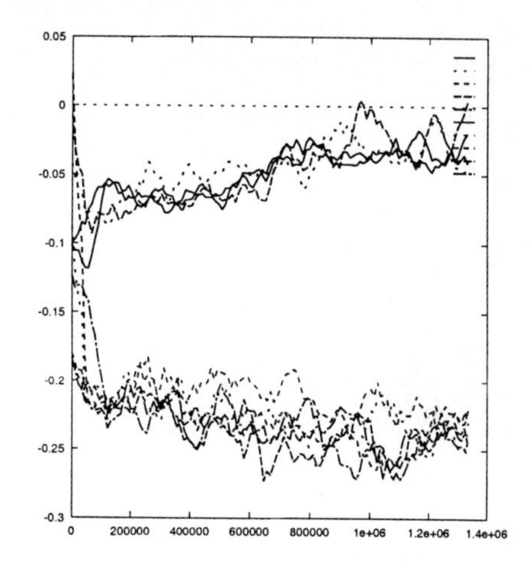

Fig. 6. Policy transitions of 10 agents in Experiment 1: The vertical axis means the distances from their closest team-mates.

strategy is a little too defensive, so the score rate was slightly less than that of the fixed positioning.

The vertical axis of Fig. 6. means the distance from a learning agent to the closest team-mate. The values are also regularized into $(-1.0, 1.0)$ and averaged from the behaviors of 10 games. There were seven agents which gather into a lump, and four agents which keep away from other agents. Two types of policies never crossed nor joined. This means that it had little possibilities of escaping from local optimal policies. This was the most important difference from the other two experiments.

4.2 Results of Experiment 2 (Observational RL)

In Experiment 2, Observational Reinforcement Learning was used independently. A learning agent would reinforce good looking policies. The good looking policy was defined as: When all team-mates are far from the ball, the ball location would be a good place for positioning.

Fig. 7. corresponds to Fig. 5. of Experiment 1. The horizontal axis means the game count, and the vertical axis means the distance from their goal. This figure shows that the all agents' policies were similar. As mentioned in section 3, Observational Reinforcement Learning tended to make the diversity monotonous. Fig. 8. corresponds to Fig. 6. of Experiment 1. This figure shows the same tendency that the diversity was monotonous. In this strategy, all the agents were keeping around the ball. This looked good positioning, but not so good in practice because of little diversity of agents' policies and factors of stamina and so on.

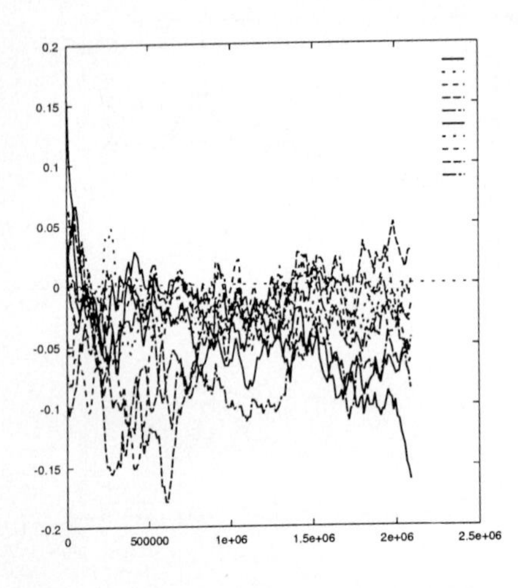

Fig. 7. Policy transitions of 10 agents in Experiment 2: The vertical axis means the distances from their goal.

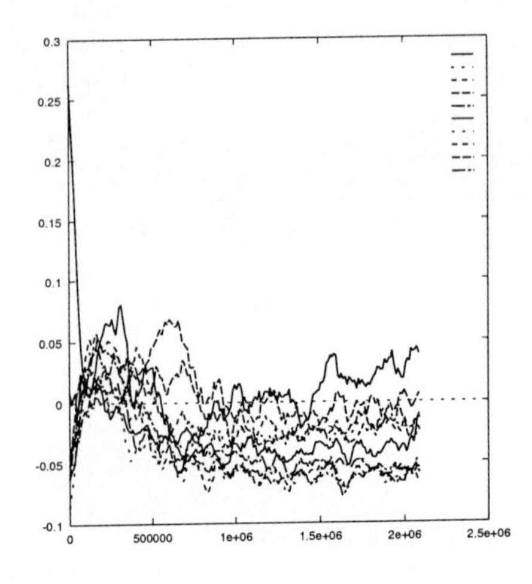

Fig. 8. Policy transitions of 10 agents in Experiment 2: The vertical axis means the distances from their closest team-mates.

4.3 Results of Experiment 3 (Ordinary RL and Observational RL)

In Experiment 3, a combination method of ordinary reinforcement learning and Observational Reinforcement Learning was used. A learning agent would get a reward when the agent kicked the ball. This was the same way of Experiment 1. A learning agent would also reinforce good looking positioning which was the ball location when all team-mates were far from it. This was the same way of Experiment 2. It was expected that there are more possibilities of escaping from local optimal policies than those of Experiment 1, and that there is more diversity of agents' policies than that of Experiment 2.

Fig. 9. corresponds to Fig. 5. of Experiment 1 and Fig. 7. of Experiment 2. There were two types of policies. Three agents were defensive and eight agents were offensive. This means that the learning agents had more diversity than that of Experiment 2. It is the same tendency of Experiment 1, but this has an important difference. The agents' policies crossed or joined occasionally. This means that there were more possibilities of escaping from local optimal policies than those of Experiment 1.

Fig. 10. corresponds to Fig. 6. of Experiment 1 and Fig. 8. of Experiment 2. This figure shows the most complicated process of learning. There is a great change of policies before time count 600000. We believe that this is just an important process of getting cooperative policies in multi-agent learning.

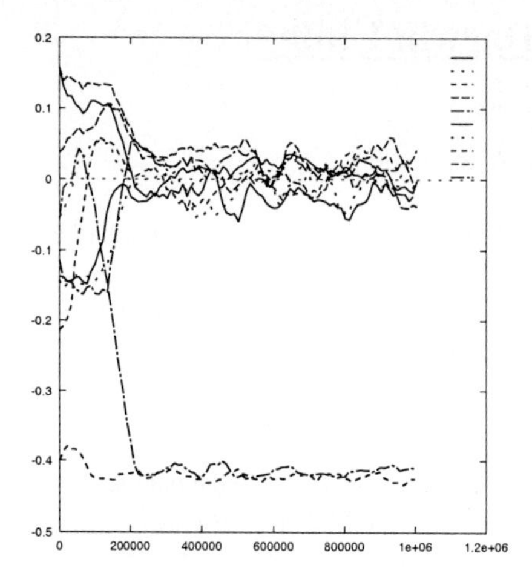

Fig. 9. Policy transitions of 10 agents in Experiment 3: The vertical axis means the distances from their goal.

Fig. 10. Policy transitions of 10 agents in Experiment 3: The vertical axis means the distances from their closest team-mates.

5 Conclusions

This paper proposed the Observational Reinforcement Learning method and compared it with an ordinary reinforcement learning method. Experiments showed that both of the methods have important roles of learning. The followings are the comparison among the three methods.

- **Ordinary reinforcement learning independently**: This is effective on developing diversity of the learning agents. However, it tends to fall into a worthless local optimal policy.
- **Observational Reinforcement Learning independently**: This is effective on escaping from worthless local optimal policies. However, it tends to lose diversity of the learning agents.
- **Combined method of ordinary reinforcement learning and Observational Reinforcement Learning**: This can fill up the each other method's weak point.

References

1. Andou, T.: "Refinement of Soccer Agents' Positions Using Reinforcement Learning", In *RoboCup-97: Robot Soccer World Cup I, pp.373 – 388* (1998).
2. Kaelbling, L. P.: "Reinforcement Learning: A Survey", *Journal of Artificial Intelligence Research 4, pp.237 – 285* (1996).

Evolving Team Darwin United

David Andre[1] and Astro Teller[2]

[1] Computer Science Division, U.C. Berkeley, 387 Soda Hall , Berkeley, CA 94702-1776
dandre@cs.berkeley.edu
[2] Computer Science Department, Carnegie Mellon University, Pittsburg PA 15213
astro@cs.cmu.edu

Abstract. The RoboCup simulator competition is one of the most challenging international proving grounds for contemporary AI research. Exactly because of the high level of complexity and a lack of reliable strategic guidelines, the pervasive attitude has been that the problem can most successfully be attacked by human expertise, possibly assisted by some level of machine learning. This led, in RoboCup'97, to a field of simulator teams all of whose level and style of play were heavily influenced by the human designers of those teams. In contrast, our 1998 team was "designed" entirely by the process of genetic programming. Our evolved team placed in the middle of the pack at Robocup98, despite the fact that it was largely machine learned rather than hand coded. This paper presents our motivation, our approach, and the specific construction of our team that created itself from scratch.

1 Introduction

Imagine a group of human programmers attempting to create a team for RoboCup-98. Not only do they have to think on the level of strategies, but also on the level of complex behaviors to achieve simple tasks. They have to learn the details of the simulator, not just as it ought to work, but as it actually works when these two are not the same. These human programmers have to try to not only impart their own soccer experiences to these procedures, but also to imagine ways in which the simulator differs from the real world and so too should the individual player and team strategies. On top of all this, the human programmers of such a RoboCup-98 team must design all this under the changing conditions of new rules, new simulator characteristics, and all of the noise and sensory limitations built into the fundamental model of the simulated world.

If a technique existed that allowed all of these issues to be solved automatically, such a technique would have certain advantages in the competition. In addition, such a technique would also be of immediate interest outside the soccer simulator domain, a claim that is more difficult to make for hand-coded solutions to the soccer simulator domain. It has been repeatedly been asserted that this problem is just too difficult for such a technique to exist. "Because of the complexity of the [soccer server] domain, it is futile to try to learn intelligent behaviors straight from the primitives provided by the server [Stone and Veloso., 1998].

The motivation for team Darwin United was to demonstrate that this claim is not true and that, in particular, genetic programming can be used successfully as a technique for training a team using the basic percepts and actions of the simulator. Luke's impressive results (1997a) at RoboCup-97 showed that genetic programming could tackle the task of generating competitive strategies, given a suite of hand-coded complex, low-level behaviors. Obviously, it is desirable that a technique for program induction work on multiple levels when applied to difficult problems. Thus, our goal has been to evolve a team to compete in the simulator league at RoboCup-98 that attacks the problem "from the ground up." This paper discusses our research, the techniques that we are using to achieve the stated goals, and our results at Robocup98.

It seemed highly unlikely that we would get superlative behavior on a known-to-be difficult problem without some modification to the standard genetic programming (GP) paradigm. We replaced the standard suite of hand-coded behaviors with a complex fitness function that provides reward for good play even when no goals are scored. As Koza, Andre, Bennett, and Keane have noted in their research on evolving analog circuits [Koza et al., 1997], it is significantly easier to write specifications for complex behavior than to write the programs to achieve the behavior. As part of the specifications, we introduced a graduated fitness function that tests each individual for increasing levels of skill, described in more detail in section 2. Additionally, although we would like GP to solve the entire problem of team soccer play, it turns out that GP is remarkably slow to learn generalizable routines to reliably run to and kick the ball when given only the most basic of primitives. Thus, we give each of the teams in the initial generation a set of automatically defined functions (ADFs) [Koza, 1994] that encode some simple functionality such as running to and kicking the ball. These subroutines are very simple and highly non-optimal. At first, these subroutines seem to violate the basic premise of our work, that we want Darwin United to learn to play soccer in the simulator in a style all its own. Two factors when taken together, show that this is still the case. The first is that a total of perhaps 2 hours of human time was spent creating these ADFs. More importantly, these ADFs are subject to evolution as are each of the players. The hand-created solutions are simply seeds from which the learning process begins.

2 The Coach: Specifications vs. Programming

To avoid the problem of attempting to learn only from games won or points scored, we wrote a list of specifications that describe the desired behavior.

There are two ways in which the specifications are ordered. First, we specify that each team is to play in a graduated series of games, where if the team's performance in any game is too poor, it does not proceed to later games and gets a maximal value of fitness for those games that it does not play (in GP, lower fitness is better). This technique limits the amount of time spent evaluating poor teams and is reminiscent of many previous techniques for focusing computation on individuals with reasonable chances of success (e.g., [Teller and Andre, 1997]). Second, within each game, there is a ordered list of scores where each score is a kind of success in the game, and each successive element in the list is more important than all the previous elements combined. Scoring more than the opponent is the last, and therefore most important element of the list. The lexical ordering is achieved by scaling each element to a single place value in a decimal score. In other words, the smallest score is scaled to be less than a thousandth, the next is scaled to be less than a hundredth, and so on. It is important to repeat that lower fitness is better.

Below are the elements of the scoring list from least important to most important.

- **Getting Near the Ball.** Each player gets a value of 1.0 to begin with (given that its initial distance from the ball is X) and can reduce this value only by moving at least once and seeing the ball at distance X' (X' < X). *Scale: 1/1000*
- **Kicking the Ball.** This score factor ranges from 0.0 to 1.0 and encourages kicking the ball and penalizes not kicking it when it is kickable. *Scale 1/100.*
- **Sides.** Each player on a team gets a value that expresses the amount of time that the ball was on their half. *Scale 1/10.*
- **Being "Alive".** Each player receives a penalty of 1.0 unless it turns at least once and runs forward at least once during the game. *Scale 0.1 to 1.0.*
- **Scoring a Goal.** Each player on a team receives a negative bonus fitness point for each goal scored. *Scale 1 to 10.*
- **Winning The Game.** Each player on a team receives 10 fitness points for a win, 30 points for a tie, and 40 points for a loss.

The following point can not be overstated. Because of this lexical dominance (each successive element in the list being more important than all previous elements combined), **once a team gets the hang of scoring goals against an opponent, none of these other factors has any appreciable effect on fitness**. Since winning is the only real metric for success, it should be (and is here) the final and dominant measure of fitness. This technique allowed us to help Darwin United get up to speed in the domain, but then allowed it to create its own unbiased solution, and all this without a human ever writing detailed code to teach Darwin United how to play.

Thus we can obtain a score for each game that a team plays in. If the team does well enough in a given game, it can move on to play a more difficult game (or set of games). A team's total fitness is an average over these separate game's measures. Each team in the population follows the following schedule. First, the team is tested against an empty field. It passes this test if it scores within 30 seconds, and fails otherwise. Although this sounds easy, it requires a bit of evolution to obtain a team that reliably dribbles and shoots the ball into the opposing goal, rather than shooting out of bounds, towards one's own goal, etc. Second, the team plays against a hand-coded team of "kicking posts" - players that simply stay in one spot, turn to face the ball, and kick it towards the opposite side of the field whenever it is close enough. This promotes teams that can either dribble or pass around obstacles. When a team scores against the kicking posts, it then plays the winning team from the 1997 RoboCup championships, the team from Humboldt University, Germany. Then, only if the team does well enough against the German team (i.e., scores at least one goal), is the team allowed to play three games in a tournament with other winning teams.

3 Team Structure and Team Transformations

This section provides an overview of how each team is represented in the evolutionary process and how teams are changed in the exploration phase of the machine learning (search) process. Each team is composed of eleven distinct members. Each member is represented by an evolved program written entirely using the primitives shown in Table 1. Notice that all eleven team members in a team share the same set of 8 ADFs. Each team, however, has its own set of evolving subroutines. In this way each team can develop and share a certain style of play, but there is still diversity of these styles across the population of teams. This structure is shown in Figure 1.

In evolution, the two dominant forms of search operators are crossover and mutation. Crossover takes multiple (usually two) individuals, and exchanges "genetic material" between them. Mutation typically selects some aspect of the evolving individual to be changed and replaces that aspect with a new, randomly generated piece of "genetic material." In this case, "genetic material" is the lisp-like code written using the primitives shown in Table 1.

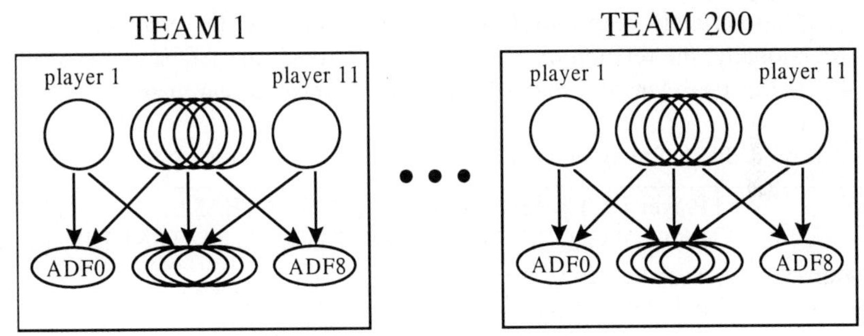

Fig. 1. The population and team structures

There are no looping constructs in the primitives list for the evolving programs, because the evolved program for each team member is run every 100ms and returns one of three primitives actions (turn, dash, kick). Notice also that two of the primitives listed in Table 1 are READ and WRITE, which give the evolving programs access to an indexed memory. This memory is not cleared between executions of the program, which allows the evolving programs to learn to act in ways more complex than as purely reactive agents. Now when a team is selected for "recombination" (a search step), either crossover or mutation is selected. If crossover is selected then a second team is also included with which genetic material can be exchanged. The details are not appropriate for this paper, but the key insight is that most of the time, when genetic material is exchanged between teams, genetic material from player 6 is exchanged with player 6 from the other team. This creates what is referred to (in both biology and in evolutionary computation) as a set of "niches." Niches tend to foster diversity of behavior, which is exactly what we desire of such a system. The goalie and the center forward should not act in similar ways, so it is appropriate that those two "types" of players rarely exchange piece of their code.

Notice though that because Darwin United is evolved from scratch, there are not notions of "positions" (aside from the goalie). That is to say that if one player wants to (evolves to) play on the left side of the field, it will do this independent of whether a teammate is already in this space. It has been our experience however that positioning is not something that tends to evolve in the way that humans enforce it. However, whether this is a deficiency in how evolution makes soccer players or whether it is a deficiency of how human's think about soccer is still an open question.

It would initially seem reasonable to do "All Star" teams, that is, taking the "best" player from each "position" and making a team out of those individuals. First, this is to imply that the best team is always composed of the best individuals. This is known not to be the case among human soccer teams. Second, as has just been pointed out, there is no enforced relation between player number and position, so there is some danger that this process would selected "too many" forwards or "too many" defenders. Furthermore, each of a team's players share certain behaviors (or parts of behaviors) through the common ADFs. This is a form of implicit communication through coordination of activity. There are a number of reasons to think that this will be useful in this domain and has certainly been useful in related domains (e.g. the predator- prey domain). Additionally, the relation of team members through the ADFs makes it much harder to separate them in a way that is likely to produce an all star team.

Table 1. The list of primitives for the evolving soccer players

Inputs:	Player.{X,Y}.Pos, {DIST,DIR}.To.{BALL, GOAL}, Ball.{DIST,DIR}.Delta, T-{DIST,DIR}(a), O-{DIST,DIR}(a)
Constants:	Real Valued Constants
Memory:	READ(x), WRITE(x,y)
Calculations:	ADD, SUB, MULT, DIV, SIN, COS, IFLTE
Actions:	KICK(a,b), TURN(a), DASH(a), GRAB (for goalie)
Team:	ADF(1-8)

4 Program Primitives for Evolution

To minimize the design time (which is, after all, part of the point of automatic programming) we use libsclient and give GP some of the inputs provided by libsclient. Specifically, each player has access to both the X and Y components of its position (XPOS and YPOS), its direction (DIR), the distance to the ball (BDIST), the angle to the ball (BDIR), the change in distance to the ball from the last time step (BCHNG), and the change in angle to the ball from the last time step (BDIRCHNG). Additionally, the players have access to a function that can provide them information about the distances and angles to each of the other players that can be seen. T-DIST(a) is a one argument function that returns the distance to the player that is the 'a'th far away. If the argument is larger than or equal to the number of teammates seen, then undefined is returned. If the argument is below zero, the value of T-DIST(0) is returned. O-DIST(a) is similarly defined for the opponents. T-DIR(a) and O-DIR(a) give the directions to the visible players. For any value that cannot be determined (such as the ball direction if the ball is out of view), a value of -999 is returned. The function set for the players consists of a variety of functions for computation, memory, and setting the command to be executed. The programs can use the four argument IFLTE function (if x ! y then u else v) as a conditional. Random constants are also included. Each player has access to its own array of indexed memory with 10 cells of memory. The one argument READ function allows the player to access the memory, and the 2 argument WRITE function allows the player to set the values of the memory cells.

The player also has access to three functions to set the command that will be executed on the next time step. KICK(a,b) sets the command to be a kick, with power a and direction b. DASH(a) sets the command to be a running movement, with power a. TURN(a) sets the command to be a turning movement with power a. The last of these functions to be executed is set to be the actual command issued to the server. There are also eight two argument functions ADF1 through ADF 8. For the ADFs, the terminal set additionally consists of the 2 zero argument functions ARG0 and ARG1 (i.e., "first parameter" and "second parameter"). All these primitives for the evolution of the soccer playing programs are shown in Table 1.

5 Results

At Robocup98, only 17 other teams placed above Team Darwin. We had one win, one loss, and one draw. The draw was particularly interesting. Against the team that won our initial group, Team Darwin forced a draw by utilizing an offsides trap that had evolved when playing against the 1997 champion team. This non-human style program was able to force a draw against a much better team by not following the standard human-like strategies. Overall, we found our performance respectable given that we focused on machine learning and not on robocup specific strategies.

Acknowledgments

The authors gratefully thank Peter Stone and Sean Luke for advice and assistance throughout this project. David is supported by a National Defense Science and Engineering Grant, and Astro is supported through the generosity of the Fannie and John Hertz Foundation.

References

[Koza et al., 1997] John R. Koza, Forrest H Bennett III, David Andre, Martin A. Keane, and Frank Dunlap. Automated synthesis of analog electrical circuits by means of genetic programming. IEEE Transactions on Evolutionary Computation, 1(2):109-128, July 1997.

[Koza, 1994] J. Koza. Genetic Programming 2. MIT Press, 1994.

[Stone and Veloso., 1998] Peter Stone and Manuela Veloso. A layered approch to learning client behaviors in the robocup soccer server. Applied Artificial Intelligence, 12, 1998.

[Teller and Andre, 1997] Astro Teller and David Andre. Automatically choosing the number of fitness cases: The rational allocation of trials. In Koza et al. editors, Genetic Programming 1997: Proceedings of the Second Annual Conference, pages 321-328, Stanford University, CA, USA, 13-16 July 1997. Morgan Kaufmann.

UBU: Utility-Based Uncertainty Handling in Synthetic Soccer

Magnus Boman, Helena Åberg, Åsa Åhman, Jens Andreasen, Mats Danielson, Carl-Gustaf Jansson, Johan Kummeneje, Harko Verhagen, Johan Walter

DECIDE/K2LAB (www.dsv.su.se/DECIDE)
Department of Computer & Systems Sciences
Stockholm University & The Royal Institute of Technology
Electrum 230
SE-164 40 KISTA, SWEDEN
mab@dsv.su.se

The UBU RoboCup team is described. Intuitive ideas and general objects for the participating researchers and their students are presented. The UBU team participates in the simulation league. The key idea is to repeatedly use the advice provided by a normative pronouncer (or decision module) when choosing what to do next. The pronouncer acts on input from each individual player, the basis of which is stored in a local information base. The team is under continuous development: At the time of writing, no version of UBU makes extensive use of pronouncer calls.

Team Description Format

We begin by describing the methodology of the UBU project in Section 2. The pivotal concept of pronouncer is briefly presented in Section 3, and principles of normative artificial decision making are discussed in Section 4. We then present a very coarse program architecture, together with some implementation details, in Section 5. The final section on future research is important, since we are currently in the final stages of completing the basic functionalities of the team.

Project Methodology

The DECIDE research group has since its inception focussed on normative decision analysis, and on tools for evaluation in particular (EDB96, EDB97, DE98). In recent years, some of the attention has been given to *artificial decision making* (BE95, B97, B98). As the term indicates, not only the decision makers, but the entire procedure of reaching a decision is artificial. The concept of autonomous artificial agents deciding what to do without human intervention is currently studied in several on-going projects with participants from DECIDE. These include agent assistance in securing energy contracts on a de-regulated market, agent-based intelligent building control, machine learning driven pollution control, and decision making agents in telecommunications. The projects are academic, but all have industrial participants. Some of the projects use customised simulation testbeds. The RoboCup domain is a good complement to these testbeds, since it offers a relatively simple dynamic real-time environment. Hence, we have chosen to test some our hypotheses formulated within the mentioned projects on a RoboCup team.

The UBU research team is basically a group of students and their supervisors. Magnus Boman (team captain), Mats Danielson, Carl-Gustaf Jansson, and Harko Verhagen constitute the latter category. Johan Kummeneje has recently become a graduate student in the DECIDE group with the dissertation topic directed towards RoboCup.

Pronouncers

A requirement that must be met for artificial agents to behave intelligently is that they can ask for advice. The base case is when the agent asks itself what it should do next. Nearly all AI research, as well as more than half of all agent research, can be placed in this simpler category. Many of the classical AI problems, such as the frame problem and the knowledge representation problem appear immediately, and must be addressed. The even more difficult case is when the idle agent asks someone (or something) else. This case can in turn be analysed by considering two sub-cases. Firstly, the agent may ask other agents, belonging to the same multi-agent system (MAS). Second, the agent may ask an entity that is not part of the MAS and that may in fact not be an agent at all. This entity may come in different guises, usually a blackboard, a control panel, a pronouncer, a decision module, an ontology, a knowledge repository, a daemon, or an oracle. Each of the guises just mentioned have too many variations to allow for them to be studied in precise terms: a blackboard, for instance, does not entail the same agent architecture or model to all researchers that claim to use them.

The availability of data runs from full to zero. In the former case, the entity is in some sense omniscient. Put simply, if each agent represents all its known or believed information in a knowledge base, the entity has access to a database containing the union of all such knowledge bases, with each entry typed to the agent in whose knowledge base the entry appeared. In the latter case, one must first define zero data availability. In the strictest possible sense, it means the entity accepts no input, since each input consists of data. Hence, it is solipsistic. Recalling that its sole purpose was to give advice, it is also useless. In a slightly less strict variant, zero data availability means that input may be given to the entity, but that this input reveals no information about any of the agents in the MAS. A syntactic realisation could be that input cannot contain typed or modal formulae, i.e. the entity knows neither the identity of the agent with which it communicates, nor the identity of any of the agents that might be mentioned in the input message it receives. In this case, the entity may be used as a random procedure; i.e. it can give advice of a quality equal to casting the dice.

The quality of data basically runs from precise and certain to imprecise and uncertain. The reason for saying 'basically' is that under some circumstances, the quality of imprecise data is equal to that of precise data. For instance, when solving a system of linear equations representing constraints on agent behaviour in order to find out what an agent should do next, imprecise data generally speaking yields a solution set of a cardinality bounded only by the number of variables in the system, while precise data yields a unique solution. A well-studied example is situations in games giving rise to multiple equilibria: it is the assumptions about the quality of data available to the players that determine whether the game has a unique equilibrium point or not.

Next, we name the entity giving advice. We study pronouncements, and therefore a candidate for naming the entity is *pronouncer*. This term is new to the computer science literature. It implies that the advice given is formal and authoritative, giving the entity a normative status, so it should be used with care, but would be appropriate for the purpose of the UBU team. Another candidate is *decider*. It is neutral, being simply the nominative form of 'decide'. It is, however, common to used decider as short for 'decision procedure' in recursion theory. Moreover, it is the name of at least one commercial product in the area of risk analysis. A third possibility is *decision module*. The word 'module' suggests something internalised, i.e. that we are studying one module among others, intrinsic to an intelligent agent. This is less appropriate for how advice is provided in the case of UBU. The term has been used extensively in the MAS literature before, usually in connection with planning, but also for bases of heuristics in expert systems, and for software providing normative advice to an inquirer. By contrast, *oracle* suggests something extrinsic. Oracle too has been used in the MAS literature. Unfortunately, oracle already has a well-defined meaning in complexity theory. It also implies high quality of the advice given, if not omniscience. In view of the above, we will choose the decision module term.

There are two possibilities for situating the entity. One is to define a decision module as local to each agent. Just as each agent might have its own list of goals, the decision module is treated as a customised tool for decision support. Hence, the entity is not merely copied into each agent, but is adapted to the agent to which it belongs from the outset, and increasingly so during its life-span. The alternative is to have a global entity that querying agents call upon repeatedly. The entity is then a resource to be shared among the agents, and will then amount to a function, the input of which will have to carry all information about the decision situation, and the output of which will be a recommended action. This function would be centralised in the same way as a facilitator in a federated architecture (GK94). We choose the second possibility.

If our sole concern was individually rational agents, and we also relied only on the principle of maximising the expected utility, the input could be a decision tree, possibly weighted with probabilities and utilities. The function would then amount to a calculator recommending (one of) the action(s) with the highest expected value. However, we attempt to produce socially rational agents, and must therefore add group constraints, or use similar means to qualifying individually rational behaviour to achieve social intelligence (B98). This cannot be achieved by merely modifying the weights in the decision tree. Instead, such constraints are part of a local information base, with respect to which each evaluation is carried out by the decision module. The necessity of such local bases was previously realised in the context of risk constraints (EBL98): Not all risk attitudes can be modelled using decision trees.

Naturally, one can imagine simple MAS in which each agent has the same responsibility towards the group, and even in such systems non-trivial problems arise (KJ98), but the realistic and interesting case is where each agent has unique obligations towards the other agents. For instance, a MAS might consist of 200 agents in which a particular agent has obligations towards the entire population (including itself), but also towards two overlapping strict subsets of, say, 20 and 25 agents that constitute coalitions. These coalitions might be dynamically construed, something which will affect the nature of obligations heavily over time.

Interestingly enough, procedures for updating the local information base can be viewed as learning procedures. In particular, the adaptation to particular coalitions, i.e. to group constraints, can be viewed as learning how to function socially. These issues cannot be pursued in this brief team description, but are currently under investigation (BV98).

Artificial Decision Making

We make the following two provisos, more concise motivations for which are available in (B97) and (E96), respectively.

Proviso 1: Agents act in accordance with advice obtained from their individual decision module, with which they can communicate.

Proviso 2: The pronouncer contains algorithms for efficiently evaluating supersoft decision data concerning probability, utility, credibility, and reliability.

Every change of preference (or belief revision, or assessment adjustment) of the agent is thought of as adequately represented in the local information base. This gives us freedom from analysing the entire spectrum of reasoning capabilities that an agent might have, and its importance to the use of the decision module. The communication requirement presents only a lower bound on the level of sophistication of agent reasoning, by stating that the agent must be able to present its decision situation to the pronouncer, and that the agent can represent this information in the form of an ordinary decision tree, extended by general risk constraints (EBL98).

The second proviso also requires some explanations. Supersoft decision theory is a variant of classical decision theory in which assessments are represented by vague and imprecise statements, such as "The outcome o is quite probable" and "The outcome o is most undesirable" (M95). Supersoft agents need not know the true state of affairs, but can describe their uncertainty by a set of probability distributions. In such decisions with risk, the agent typically wants a formal evaluation to result in a presentation of the action (in some sense) optimal with respect to its assessments, together with measures indicating whether the optimal action is much better than any other action, using a distance measure. The basic requirement for normative use of such measures is that (at least) probability and utility assessments have been made, and that these can be evaluated.

In the local information base are non-linear systems of equations representing supersoft data about

- probabilities of the occurrence of different consequences of actions
- utilities of outcomes of different consequences of actions
- credibilities of the reports of other agents on different assessments
- reliabilities of other agents on different agent ability assessments

The preferences of the agents can be stated as intervals of quantitative measures or by partial orderings. Credibility values are used for weighting the importance of relevant assessments made by other agents in the MAS. Reliability values mirror the reliability of another agent as it in turn assesses the credibility of a third agent (E96). All bases except the utility base are normalised. Note that a MAS without norms is treated is this paper as a social structure where group utility is irrelevant to the individual agent. The presence of norms can manifest itself in various ways (B98), but unfortunately we cannot discuss this matter further here.

Notes on the Implementation

The goalkeeper of the UBU team is implemented in Java. The code was written in accordance with several key concepts of Java, e.g., threads, encapsulation, and communication. We have designed the goalkeeper agent to consist of three subsystems for communication, memory, and deliberation (or reasoning). The latter is the only role-specific code module of the goalkeeper agent. The goalkeeper is a mixed-behaviour agent in that it is reactive, e.g., in its use of the `catch` command (applied whenever the ball is within the `catchable_area`), while more deliberative in its positioning.

The field players are written in C with a separate module for basic interaction with the server including navigation, and a communications module letting each of the agents provide hints about the positions of other players and the ball. On top of these two modules, local information bases are being implemented as a 'magnets' that attract agents to areas of strategic importance. 'Negative magnetism' is used to reject agents from danger-zones (such as other players). The positions of the magnets are dynamically re-calculated, using the estimated positions of all players.

The decision module to be used is basically DELTA (D97), with certain modifications made to it. DELTA was written in C.

Future Research

The UBU team is unfinished. In Paris, UBU won one game and lost three. That version of the team had no pronouncer calls, and not even all the basic functionality was implemented. At PRICAI, a much-improved version will compete. This version will have almost full basic functionality, e.g., a reasonable treatment of offside strategies. Hopefully, it will also contain pronouncer calls. In any case, the long-term goal is the Stockholm competition in 1999: playing home has some great advantages! For that competition, a version of UBU depending highly on pronouncer calls is planned.

References

(Åb98) H. Åberg: "Agent Roles in RoboCup Teams", M. Sc. Thesis, DSV, 1998.

(Åh98) Å. Åhman: "Decision Control in RoboCup Teams", M. Sc. Thesis, DSV, 1998.

(BE95) M. Boman & L. Ekenberg: "Decision Making Agents with Relatively Unbounded Rationality", *Proc DIMAS'95*: I/28-I/35, AGH, Krakow, 1995.

(B97) M. Boman: "Norms as Constraints on Real-Time Autonomous Agent Action", Boman & Van de Velde (eds.) *Multi-Agent Rationality (Proc MAAMAW'97)*: 36-44, LNAI 1237, Springer-Verlag, 1997.

(B98) M. Boman: "Norms in Artificial Decision Making", to appear in *Artificial Intelligence and Law*.

(BV98) M. Boman & H. Verhagen: "Social Intelligence as Norm Adaptation", Dautenhahn & Edmonds (eds.) *Proc WS on Socially Situated Intelligence*, SAB'98, Zuerich, 1998.

(D97) M. Danielson: Computational Decision Analysis, Ph.D. thesis, DSV, 1997.

(DE98) M. Danielson & L. Ekenberg: "A Framework for Analysing Decisions Under Risk", *European Journal of Operations Research* **104**: 474-484, 1998.

(E96) L. Ekenberg: "Modelling Decentralised Decision Making", Tokoro (ed.), *Proc ICMAS'96*: 64-71, AAAI Press, 1996.

(EDB96) L. Ekenberg, M. Danielson & M. Boman: "From Local Assessments to Global Rationality", *Intl Journal of Intelligent Cooperative Information Systems* **5**(2&3): 315-331, 1996.

(EDB97) L. Ekenberg, M. Danielson & M. Boman: "Imposing Security Constraints on Agent-Based Decision Support", *Decision Support Systems* **20**(1): 3-15, 1997.

(EBL98) L. Ekenberg, M. Boman, and J. Linnerooth-Bayer: "General Risk Constraints", Presented at the DAS98 WS at the Intl Institute for Applied Systems Analysis (IIASA), Laxenburg, Austria. Submitted.

(EK97) H. Engström & J. Kummeneje: "DR ABBility: Agent Technology and Process Control", M. Sc. Thesis, DSV, 1997.

(GK94) M. R. Genesereth & S. P. Ketchpel: "Software Agents", *Communications of the ACM* **37**(7): 48-53, 1994.

(KJ98) S. Kalenka & N. R. Jennings: "Socially Responsible Decision Making by Autonomous Agents", *Proc Fifth Intl Colloq on Cognitive Science*, San Sebastian, 1998.

(M95) P-E. Malmnäs: "Methods of Evaluations in Supersoft Decision Theory", unpublished manuscript, 1995. Available on URL: `http://www.dsv.su.se/~mab/DECIDE`.

AT Humboldt in RoboCup-98 (Team description)

Pascal Gugenberger, Jan Wendler, Kay Schröter, and Hans-Dieter Burkhard

Institute of Informatics
Humboldt University Berlin, D-10099 Berlin, Germany **
email: muellerg/wendler/hdb/kschroet@informatik.hu-berlin.de
WWW: http://www.ki.informatik.hu-berlin.de

Abstract. The paper describes the scientific goals of the virtual soccer team "AT Humboldt 98", which became vice champion in RoboCup-98 in Paris. It is the successor of the world champion "AT Humboldt" from RoboCup-97 in Nagoya.

1 Introduction

The virtual soccer teams "AT Humboldt 97" and "AT Humboldt 98" (AT stands for "Agent Team") are implemented by our AI group at the Institute of Informatics at the Humboldt University Berlin. The work is done by groups of students as practical exercises for the advanced course "Modern methods in AI" during summer semester. A core group of three students maintains the coordination and the programs.

The new program "AT Humboldt 98" [1] is based on the architectural concepts of the successful program from RoboCup-97 in Nagoya. We decided to make a re-implementation for more rigid structuring. The new team has more skills, more complex deliberation processes, and new facilities for on-line learning.

The team from Nagoya took part in RoboCup-98 under the name "AT Humboldt 97". The idea behind its nomination was the possibility to compare the development in virtual soccer from Nagoya to Paris. We therefore didn't want to change the program, but this was not possible because of the new rules (changes in the soccer server). We tried to make only as few changes as possible. Necessary changes concerned the new parameters. At the end, "AT Humboldt 97" still was one of the top 16 teams in Paris. The main handicap arose from the new offside rule: There was no feature in "AT Humboldt 97" to avoid offside. Teams exploiting the offside rule could easily stop the offense of AT Humboldt 97.

2 Scientific Goals

We are interested in virtual soccer for the development and the evaluation of our research topics in artificial intelligence which concern the fields of

** The work was partly sponsored by infopark online service

[1] the sources are available from our web pages: http://www.ki.informatik.hu-berlin.de/RoboCup/RoboCup98/index_e.html

- Agent oriented techniques (AOT),
- Multi-Agent Systems (MAS),
- Case Based Reasoning (CBR).

Thus many aspects of our soccer program are heavily influenced by these fields, but it is important not to consider these fields in isolation: to create our soccer agents, we also needed a lot of contributions from other fields of computer science (e.g. programming techniques, synchronization, concurrency) and from mathematics. Thereby we gain deeper insights for integration of AI techniques in software development. This aspect is especially important for the education of our students.

2.1 Agent oriented techniques (AOT)

Our understanding of AOT is closely related to new developments and new requirements in software technologies, which are driven by new expectations to programs and intelligent systems by a broad audience (not limited to computer scientists: this distinguishes AOT from e.g. object oriented techniques). Characteristic aspects of related agent-programs (we do not want to give one more definition of "agents") are e.g. autonomy, cooperation, rational behavior and mental qualities (the programs use "knowledge" for their "decisions" and they can deal with "orders" by their users), etc. AOT should support related functionalities. Up to now there is no common understanding of what agent oriented techniques may be (but remember that object oriented programming needed 20 years of development).

RoboCup is an ideal environment for testing appropriate structures and programming techniques. Our agent architecture uses a mental deliberation structure which is best described by a belief-desire-intention architecture (BDI) [3]. Distinct from other (e.g. logically motivated) approaches our approach is closely related to procedural thinking, and we use object oriented programming (C++, Java) for the implementation.

2.2 Multi Agent Systems (MAS)

Our interest in RoboCup for MAS concerns the cooperation between agents in the presence of opponents. Special emphasis is given to emergent cooperation: How can agents cooperate only by observing each other (or better to say: how can we implement cooperative behavior by using our knowledge about the programmed behavior of our agents). Social behavior results from common individual rules. In the future we will try to compare emergent cooperation with cooperation by explicit communication of intentions. Experiments with communication of world state information did not lead to significant improvements.

2.3 Case Based Reasoning (CBR)

Our understanding of CBR [2] means learning from former experiences (cases) especially for situations where we have not enough information to induce rules.

Successful CBR needs efficient case memories which permit the retrieval ("reminding") of old cases in short time.

RoboCup offers a lot of scenarios for learning from experiences. We distinguish between "off-line learning " (training), where we can make a lot of experiments for collecting cases, and "on-line learning " during the matches where we can collect only few cases in order to learn the opponents tactics and skills.

Cases from "off-line learning " can be used to extract rules for behavior and to tune parameters.

3 The Architecture of "AT Humboldt 98"

Figure 1 depicts the overall structure of "AT Humboldt 98". The arrows indicate the data-flow. The sensors parse the information coming from the Soccer Server and cause their integration into the internal world model. The deliberation component decides what to do based on the data in the world model. As a result of its deliberation it creates a plan of atomic actions using the available skills, which also make use of the world model. The deliberator hands the plan over to the effectors that manage its execution and inform the world model about the actions the agent has sent to the server.

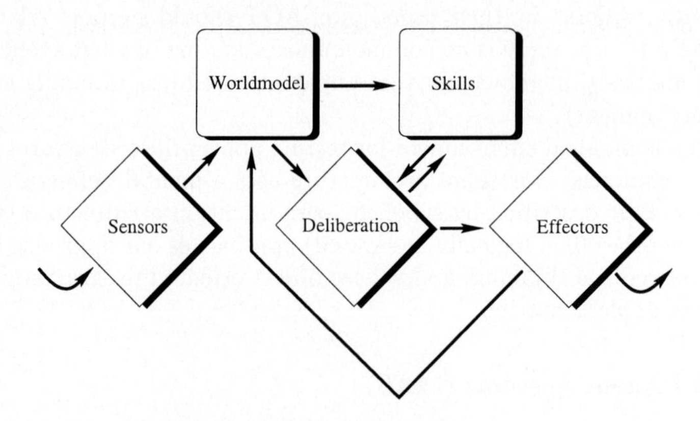

Fig. 1. Overall agent architecture

3.1 Skills

Skills enable the deliberator to work on a more abstract level by encapsulating "subconscious" tasks like running, kicking the ball or dribbling. They decompose intentions like "Move to position x, y" into a series of atomic actions. Often it is not necessary to compute the complete series, since new incoming sensor information frequently changes world model so significantly that the part of the plan that hasn't been executed yet has to be adapted to this information or even completely recalculated.

One of the most important skills is the kicking of the ball. The implemented skill accelerates the ball in a given direction, trying to achieve a given final speed.

If this speed cannot be reached, the skill tries to maximize it. If the player is in the way of the ball, the skill moves the ball around the player [2].

Another skill is running to a given position. It tries to compute a plan that enables the player to reach this position as fast as possible. At the same time it ensures that the stamina of the player does not decrease below a certain threshold depending on the current intention. Obstacles on the path to the target position are avoided. The player can also run backwards with this skill.

The third important skill is dribbling. It allows the player to move forward while keeping the ball within the player's control radius. The skill tries to keep the players body between the ball and the closest opponent. Just like the previous skill dribbling doesn't let the player's stamina drop below a given threshold.

3.2 World Model (belief)

All information concerning the outside world is a part of the belief. New sensor information updates the world model. Parameters of objects outside of the visual range are estimated by simulation. The world model can perform simulations into the future and estimate e.g. shortest paths for intercepting the ball.

Data that belongs to the same simulation step is stored together in an object called situation. It contains the representation of the players and the ball. Speed and position values are stored together with reliability values that indicate how old the underlying sensor information is. The world model also contains internal knowledge about the base position of the player, which usually changes during the course of a game, and the role of the player, which remains constant.

3.3 Deliberation (desire, intention, plan)

Deliberation starts whenever an update of the world model is completed or the current plan is finished. First it looks for an existing plan and evaluates the conditions for continuation. If it does not decide to continue this plan, then it evaluates all options by calculating a rough estimate of their expected utilities. Options with a utility above a certain limit are chosen as desires, i.e. as candidates for a new intention.

During the second phase an intention is chosen out of the current desires. Starting with the highest scored desire we check if the desire is feasible, i.e. if a related plan can be computed. If so, this desire becomes the agent's intention. Otherwise the remaining desires are examined.

Currently we are using only single intentions. Further conditions are modelled as *constraints*. The utilities of obeying the constraints are regarded while computing the utility estimates for the options.

After the commitment to an intention, a new plan is computed based on related skills. As a result, a sequence of actions is given to the effector module which sends them to the soccer server while maintaining synchronization. If

[2] actually, our kick implementation was not completely finished in AT Humboldt 98

an intention resp. plan is later dropped by the deliberator, then a new plan overwrites the old one in the effector.

The stability of intentions/plans is always in conflict with the re-deliberation and adaptation to new situations. We use the following procedures: There is a specification for the current intention under which conditions it must not be cancelled. This condition is checked whenever the deliberator starts. Furthermore, we compare a new chosen intention with the old one at the end of the deliberation process.

4 Learning techniques

We distinguish between off-line learning ("training") and on-line learning (adaptation during the matches). Our approaches are in an experimental stage.

The success of skills depends on appropriate choices of the consecutive actions. Thereby, learning should not concern a single action, but the whole sequence of actions. Learning for skills can be performed as off-line learning. We have made some experiments for learning, but up to now the skills have been hand coded according to an analysis of situations.

Our experiments concern the training of the ball-shooting skill which is performed by several kick commands. Data is collected using an automatized coach mode. This data is analyzed in order to find optimal parameters and hopefully to find rules for computing the optimal parameters.

Another experiment concerns the choice of good base positions using on-line learning. Good positions depend strongly on the opponents. We have recorded positions during a game in a raw grid of the field and then adapted the player positions to that knowledge. We have used this strategy in some of our matches in Paris, but we have not really been satisfied by the results.

Choices in the deliberator depend on several parameters (especially for utility calculations). These parameters may be tuned in a general way (off-line learning) and regarding the behavior of opponents during a match (on-line learning). We have experimented with CBR for the deliberation concerning dashing to a good position in a concrete situation. Cases contain information about positions, expected behavior, and stamina. Our strategy could be used for both off-line learning and on-line learning, respectively. First experiences are reported in [4].

5 Implementation Issues

The re-implementation of our programs for "AT Humboldt 98" follows a consequent object oriented design methodology. We use C++ for the implementation. Java would have been an alternate choice under the aspect of software technologies, but was ruled out because of of the slow implementation of the Java Virtual Machine on our machines. AOT is used for the structure (architecture) of our programs as described above.

To support the concurrent development we use the freely available source code management system CVS [5] and the documentation system doc++ [6].

6 The Development of the "AT Humboldt" Programs

We started in early 1997 with the implementation of a first prototype (March 1997). The design and implementation of a soccer agent was the topic of practical exercises for the advanced course "Modern methods in AI" during summer semester 1997 (April – July). Different architectures and learning concepts have been discussed and partially implemented in C++ and Java. The best concepts were chosen for the final implementation of the program of "AT Humboldt" for RoboCup 97 in Nagoya. Because of performance we decided to use C++.

The first running version was built in the beginning of August by a group of 8 students. The structuring according to AOT allowed significant improvements in the remaining short period of time. The usage of learning (especially CBR) could not be realized. The reasons for success in Nagoya could be seen in the efficient skills and the emergent cooperation based on simple principles.

The re-implementation for "AT Humboldt 98" was the work of a core group of three students. The extensions and new features were again the topic of the practical exercises for the course "Modern methods in AI" during summer semester, although we had serious timing problems because RoboCup-98 was scheduled more than a month earlier than in 1997.

Our experiences with RoboCup under educational aspects are very promising: Students work in a larger project which they have to organize by themselves. The project includes the development of own concepts (at the beginning, it was completely open, which concepts would be useful). Successful implementation needs the integration of a lot of different concepts not restricted to AI.

Acknowledgment

The authors want to give their special thanks to all students involved in our RoboCup projects and to the sponsor infopark online service.

References

1. Burkhard, H. D., Hannebauer ,M., Wendler, J. : AT Humboldt — Development, Practice and Theory. RoboCup-97: Robot Soccer World Cup I, Springer 1998, 357–372.
2. Lenz, M., Bartsch-Spörl, B., Burkhard, H. D., Wess, S.: Case Based Reasoning Technology. From Foundations to Applications. LNAI 1400, Springer 1998.
3. Rao, A. S., Georgeff, M. P. : BDI agents: From theory to practice. In V. Lesser, editor, *Proc. of the First Int. Conf. on Multi-Agent Systems (ICMAS-95))*, pages 312–319. MIT-Press, 1995.
4. Wendler, J., Lenz, M.: CBR for Dynamic Situation Assessment in an Agent-Oriented Setting. In D. Aha and J. J Daniels, editors, *Proc. AAAI-98 Workshop on Case-Based Reasoning Integrations, Madison, USA, 1998.*
5. CVS: http://www.loria.fr/ molli/cvs-index.html
6. DOC++: http://www.zib.de/Visual/software/doc++/index.html

Individual Tactical Play and Pass with Communication between Players

~ Team descriptions of Team Miya2 ~

Harukazu Igarashi, Shougo Kosue, Masatoshi Miyahara
Kinki University, Higashi-Hiroshima City, 739-21, Japan

Abstract. In this paper we describe our team, Miya2, which is to participate in the simulator league of RoboCup 98. Miya2 is characterized by soccer agents that make individual tactical plays and passes by using communication between players. In our experiments, 76.6% of passes made by Miya2 soccer agents were passes made with communication. More specifically, 53.6% of passes are quick passes made with communication in which the passers only used auditory information without looking around for receivers.

1 Introduction

What is the next challenging problem to be solved by a computer after defeating a human chess champion? Robot soccer is one of the relevant candidates for the standard challenging problems in Artificial Intelligence.

At the Faculty of Engineering at Kinki University, we constructed a team, *Miya2*, of synthetic soccer agents and will participate in the simulator league of RoboCup 98 (The World Cup Robot Soccer 98)[1]. In this short paper, we give a technical description of our team. Team Miya2 is an improved version of team Miya, which participated in RoboCup97 Nagoya[2]. Team Miya was characterized by individual tactical plays[3]. Individual tactical plays do not require communication between players, so the speed of passing was rapidly increased in RoboCup 97 games, and the team sometimes behaved as if it had been taught some tactical plays. Team Miya proceeded to the quarterfinal match and was one of the best eight teams in the simulator league.

In Team Miya2, a kind of communication between players is realized by using a "say" command so that a passer can make a pass to a receiver without looking around for a receiver. This paper describes Team Miya2 and the experimental results for frequency of passes that use this communication in actual simulation games.

2 Objective

Many researchers have emphasized the importance of communication between multiple agents. In a dynamically changing environment like a soccer game, however, there is not enough time for agents to communicate with each other and confirm their teammates' explicit intentions. Furthermore, hostile agents interfere with the communication by "jamming" the other team's agents. Thus, we tried to invent agent-control algorithms that do not require communication or any exhausting calculations in Team Miya for RoboCup97 Nagoya[2][3]. The main objective of our research on

Team Miya was to develop an agent model that satisfies the following two requirements. The first requirement is that the cooperative actions of the agents be expressed in a simple form that can be modified easily. The second requirement is that the actions of the agents be quick and smooth in a real-time environment. Team Miya proceeded to the quarterfinal match and was one of the best eight teams in RoboCup97 Nagoya. This shows that the main objective was realized fairly well.

On Team Miya2, we implemented a kind of team play as a next step toward an intelligent multi-agent system. We invented a pass with communication between a passer and a receiver as a typical example of team play. Team Miya2 is expected to show the importance of communication between agents to reach a team goal in an multi-agent system.

3 Architecture

Figure 1 shows the architecture of a soccer client program controlling an agent on Team Miya and Team Miya2. A client program receives visual and auditory information, decides on an action, and compiles the action into a series of basic commands prescribed by RoboCup 98 regulations[1].

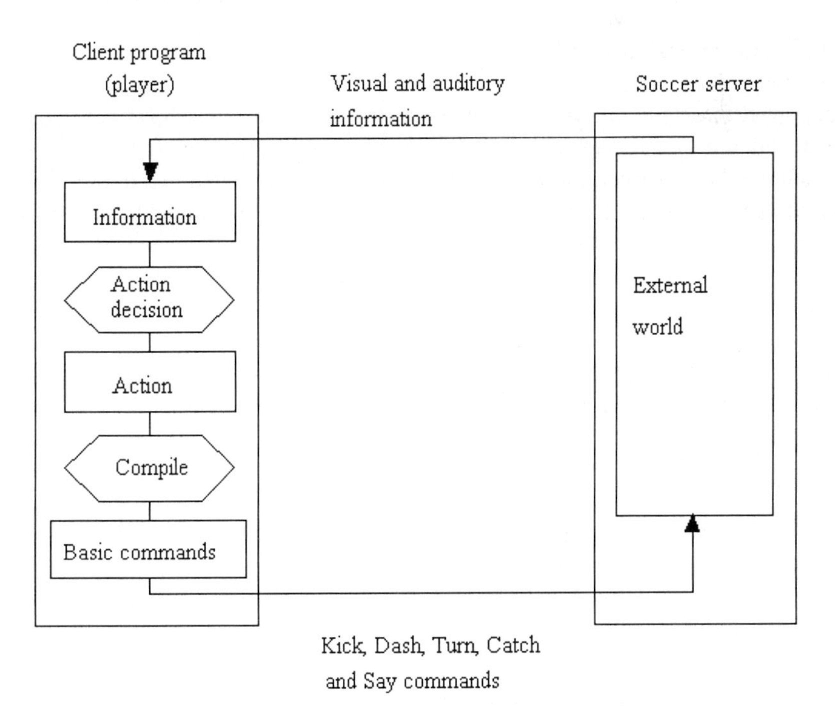

Fig. 1. Architecture of a client program controlling a soccer agent on Team Miya and Team Miya2

4 Hierarchy of Actions

The second feature of Team Miya and Team Miya2 is a hierarchy of actions. Actions are generally classified into four levels: strategy, tactics, individual play and basic commands (Table 1). A higher-level action includes more players and requires information in a wider range of time and space than a lower-level action. Coradeschi et al.[4] and Tambe[5] expressed the relationship between actions as a decision tree. We call such a decision tree an *action tree.* A soccer agent selects an action from the action tree at each action cycle by analyzing visual and auditory information and by considering the agent's current state. The action is then *compiled* into a series of basic commands: kick, turn, dash, catch and say.

Table 1. Hierarchy of actions

	Action	Definition	Examples
Level 4	Strategy	Cooperative team action	Rapid attack, Zone defense
Level 3	Tactics	Cooperative action by a few players for a specific local situation	Centering pass, Post play, Triangle pass
Level 2	Individual play	Individual player skill	Pass, Shoot, Dribble, Clear
Level 1	Basic command	Basic commands directly controlling soccer agents	Kick, Turn, Dash, Catch, Say

5 Individual Tactical Play

The action tree contains information on compiling. Tactic actions, however, require communication between players and often slow the reactions of players in a real-time environment like a soccer game. Therefore, we removed tactics and introduced individual tactical plays into the action tree on Team Miya. In contrast to this, a pass with communication between players is added on Level 3 for Team Miya2. The hierarchy of the modified action tree used for Miya2 is shown in Table 2.

Table 2. Modified hierarchy of actions

	Action	Definition	Examples
Level 4	Strategy	Cooperative team action	Rapid attack, Zone defense
Level 3	Tactics	Cooperative action by a few players for a specific local situation	Pass with communication
Level 2	Individual tactical play	Action of an individual player for a specific local situation without communication, but expecting cooperation from a teammate	Safety pass, Post play, Centering pass
	Individual play	Individual player skill	Pass, Shoot, Dribble, Clear
Level 1	Basic command	Basic commands directly controlling soccer agents	Kick, Turn, Dash, Catch, Say

Individual tactical play was introduced to reduce the delay time between decisions and actions. The *individual tactical play* is defined as an action that an individual plays in a specific local situation without communication from a teammate. However, an agent expects some cooperation from a teammate in an individual tactical play. For Team Miya and Team Miya2, we implemented three actions as individual tactical plays: the safety pass, the centering pass and the post play. These three plays speed up the tactical actions of the safety pass between two players, the centering pass from a wing player, and the post play of a forward player.

6 Action Tree

According to the role given to the agent, each agent has its own action tree based on the modified hierarchy shown in Table 2. An agent's next action is specified by prioritized rules organized into its own action tree. An example of an action tree is shown in Fig. 2. Here, if the node offense is selected, the firing conditions of action nodes at levels 2 and 3 are checked. The knowledge of selecting actions at levels 2 and 3 are expressed as if-then rules in a C program. Examples of the firing conditions include whether there are opponent players nearby, whether the player can kick the ball, whether the ball is moving, whether the player can estimate his position correctly, and whether the player can see the ball.

In addition to the if-then rules, some actions at levels 2 and 3 are prioritized. For example, pass with communication, shoot, centering pass, safety pass and safety kick have higher priority according to this order. The action selected by this is compiled into a series of basic commands at level 1. The process of deciding an action at level 2 and the compiling procedure were described as *action decision* and *compile* in Fig. 1.

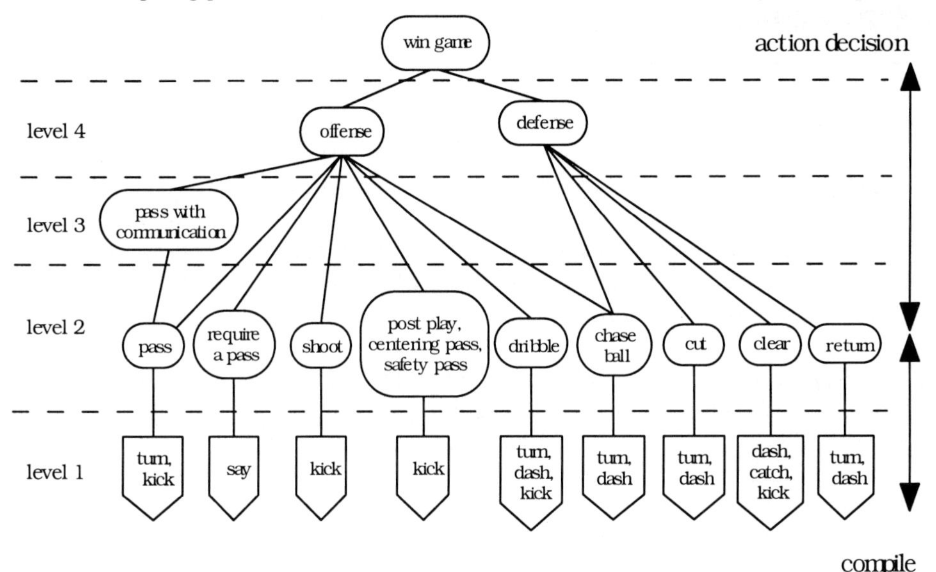

Fig. 2. Action tree used on Team Miya2

7 Safety Pass and Safety Kick

The actions of level 2 are not unrelated to one another. The actions--shoot, centering pass, post play, dribble and clear-- consist of two basic skills: the safety pass and the safety kick. The *safety pass* is a skillful pass to a receiver so that it is not easily intercepted by the opponents. Pass direction determined by machine learning techniques is quite laborious[6]. However, the effectiveness of a learning system may depend on its learning data, in this case the opponents' behavior. For this reason, we use the following rules to determine which teammate to pass to. First, the distance between passer and receiver should be smaller than 30. Second, a receiver should be forward of the passer and in the visual field of the passer. Third, every angle between a receiver and an opponent nearer than a receiver, should be more than 7.5 degrees. These three conditions on Team Miya[3] are slightly modified on Team Miya2.

The *safety kick* is a skillful kick, which eludes interception by the opponents, in the direction of the objective. An agent usually has a visual field with an angle of 90 degrees in its forward direction. Let us assume that an agent would like to kick a ball in the forward direction. We divide a region with an angle of –45 degrees to +45 degrees in the forward direction into seven equal fans. An agent searches the seven fan regions for opponents and selects the fan that has no opponents and is closest in the forward direction. The agent kicks the ball into the center of the selected fan region. In the case of Team Miya[3], we checked a region with an angle of –35 degrees to +35 degrees. Accordingly, a wider region is checked on Team Miya2 and a kick is safer than on Team Miya.

8 Pass with communication

Let us assume a situation of a passer and a receiver as shown in Fig.3. Here, there is a *pass-course line*, which is a straight line connecting the ball and a receiver. The distance from an opponent player i to the pass-course line is denoted by h_i . If all h_i's are larger than a certain value, which is set at 4 in this case, then the receiver requires the passer to pass the ball by a say command. The distance h_i can be easily calculated as $h_i = |d_i \sin q_i|$, where the distance d_i between the receiver and an opponent player i and the angle q_i between the ball and an opponent player i are included in the visual information that the receiver gets from the soccer server.

After receiving the pass request from the receiver, the passer can pass the ball to the receiver without searching for the receiver. This is because every piece of message data that a player gets from the soccer server as auditory information includes a direction to the player who sent the message. Consequently, a passer can sometimes make a pass to a receiver who is not in the visual field of the passer on Team Miya2. We call this kind of pass a *quick pass with communication*.

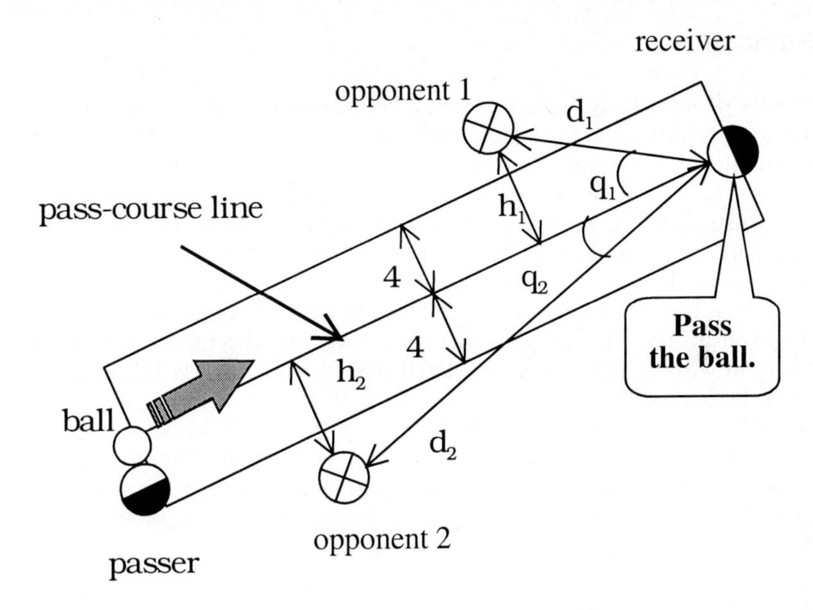

Fig. 3. Pass with communication

9 Experiments

In this section, we describe two experiments on counting the number of kicks, safety passes and two kinds of passes with communication between players. In each experiment, the numbers of the four plays were counted in ten actual simulation games.

First, ten games between Miya2 and Miya2 were played. In Experiment 1, the function of passes with communication were taken off. For each game, a ratio $r_1(i)$ (i=1,...,10), which is defined as $r_1(i)=N_{pass0}/N_{kick}*100$ was calculated. In this definition, N_{pass0} is the number of safety passes and N_{kick} is the number of kicks made in one game. The average of $r_1(i)$ was 18.2%. This means that only 18.2% of kicks are done for passes on Team Miya2 if any communication is not used for passes.

Second, another set of ten games were played between Miya2 and Miya2. In Experiment 2, the following three values were calculated for the ten games: $r_2(i)=N_{pass}/N_{kick}*100$, $r_3(i)=N_{pass1}/N_{pass}*100$ and $r_4(i)=N_{pass2}/N_{pass}*100$. In this definition, N_{pass1} and N_{pass2} are the number of passes with communication and quick passes with communication made in one game, respectively. The variable N_{pass} is the sum of N_{pass0}, N_{pass1} and N_{pass2}. As a result of Experiment 2, we obtained the values: 36.7%, 76.6% and 53.6% for the averages of the variables $r_2(i)$, $r_3(i)$ and $r_4(i)$, respectively. From this, it follows that the ratio of passes to kicks are almost doubled by introducing communication between players because most of the passes, 76.6%, are passes with communication. Moreover, more than 50% of passes are quick passes with communication on Team Miya2.

10 Summary

Team Miya2 was designed to quicken reactions in a dynamically changing environment. For this purpose, individual tactical plays and a pass with communication between players are introduced on Team Miya2. In our experiments, 76.6% of passes made by Miya2 soccer agents are passes with communication. In more than 50% of the passes, passers used only the auditory communication with receivers who are not in the visual fields of the passers. These frequent quick passes without searching for receivers will realize a long chain of rapid passes such as a triangle pass.

Consequently, Team Miya2 was one of the best six teams in RoboCup98 Japan Open(Tokyo, April 1998) and one of the best sixteen teams in RoboCup98 Paris(France, July 1998).

However, we believe that our next goal should be the creation of a team of client programs equipped with Artificial Intelligence techniques such as machine learning, inference and coordination in a multi-agent system. This goal will be achieved by building on the basic techniques invented for soccer agents on Team Miya2.

References

[1] http://www.robocup.v.kinotrope.co.jp/games/98paris/312.html

[2] http://www.robocup.v.kinotrope.co.jp/games/97nagoya/311.html

[3] Igarashi, H., Kosue, S., Miyahara, M., Umaba, T.: Individual Tactical Play and Action Decision Based on a Short-Term Goal -Team descriptions of Team Miya and Team Niken-. In: Kitano, H.(ed.): RoboCup-97: Robot Soccer World Cup I, Springer-Verlag(1998)420-427

[4] Coradeschi, S., Karlsson, L.: A decision-mechanism for reactive and cooperating soccer-playing agents. Workshop Notes of RoboCup Workshop, ICMAS 1996

[5] Tambe, M.: Towards Flexible Teamwork in RoboCup. Workshop Notes of RoboCup Workshop, ICMAS 1996

[6] Stone, P., Veloso, M.: Using machine learning in the soccer server. Proc. IROS-96 Workshop on RoboCup(1996)19-27

UFSC-Team: A Cognitive Multi-agent Approach to the RoboCup'98 Simulator League

Augusto Cesar Pinto Loureiro da Costa
Guilherme Bittencourt

Departamento de Automação e Sistemas
Universidade Federal de Santa Catarina
88040-900 - Florianópolis - SC - Brazil
{ loureiro | gb }@lcmi.ufsc.br,
WWW home page: http://www.das.ufsc.br/ufsc-team

Abstract The RoboCup Simulator League competition is a very interesting laboratory for open cognitive multi-agent systems. It presents an environment where two robot teams play soccer, with all the challenges that this task brings-up. In this environment each player has its own vision system, there is no single agent that has a global view of the field. Another restriction is that a very narrow communication band-width is allowed to inter-agent communication. The UFSC-Team adopts a cognitive multi-agent approach where most part of the cooperation process is based on visual information and where the exchanged messages are just used to decide the role each agent should play in a predefined strategy.

Content Areas: *RoboCup, Multi-Agent Systems, Cognitive Agents.*

1 Introduction

The *World Robot Cup (RoboCup)* [KTS+97] [KAK+97] Simulator league competition is a very interesting laboratory for *Open Cognitive Multi-Agent Systems* [SDB92]. In this competition a robot soccer match is played by two team with 11 players in a simulator. In this simulator each robot is completely autonomous. Each one has its own vison system and it is controlled by one or more agents.

According to the RoboCup Simulator League rules, the inter-agent communication is permitted only through the Soccer Server and limited to a very narrow band-width. Because of this, the cooperation process should be based mainly on perceptual information, rather then on inter-agent communication.

This paper describes the *UFSC-Team*, a cognitive multi-agent system approach to the RoboCup'98 Simulator League. This team uses a cooperation process based mainly on the agent visual information but that also makes use of a small amount of auditive information, i.e., message exchange according to the RoboCup Simulator terminology. The UFSC-Team agent architecture is described in Section 2. In Sections 3 to 5, the team structure, team behavior and agent behavior are described. The cooperation process is presented in section 6. Finally, in Section 7, the conclusions and perspectives are presented.

2 The Agent Architecture

The UFSC-Team agent architecture is based on the *Expert-Coop* system [BC97]. It presents a concurrent architecture, and is implemented using the C++ programming language. The agent architecture consists of the following processes: the *Interface* process; the *Coordinator* process, with its respective mail box process; and the *Expert* process, that also has its own mail box process.

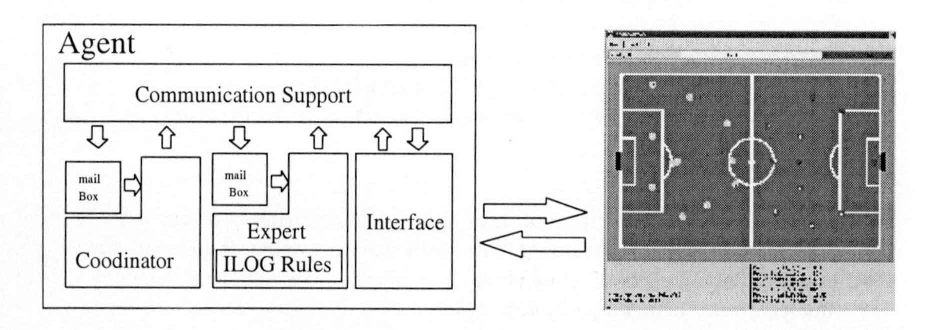

Figure1. UFSC-Team Agent Architecture

The Interface process is performed by a reactive agent responsible by the communication with the Soccer Server. It receives, from the Soccer Server, both, visual and auditive, information and transmits the action chosen by the cognitive agent.

The Coordinator process is responsible by the agent cooperation and by the communication among the team agents. It also provides the agent group management functionalities.

The Expert process supports the cognitive behavior of the agent. This process has an embodied knowledge-based system, built from the Ilog Rules 5.0 system [ILO95]. This knowledge-based system contains a set of rule bases, each one associated with a specific role, that should be performed by the agent, such as mid-fielder player role, attacker player role, etc. These rule bases codify the practiced plays associated with the respective role.

The Coordinator mail box process receives all the messages addressed to the agent, either by other agent's Coordinator processes, or by its own Interface process. The Expert mail box process just receives the messages sent to the Expert process by its associated Coordinator process.

3 Team Structure

The proposed multi-agent soccer team is composed by eleven *players*, implemented through software agents based on the Expert-Coop. The team has a *tactical formation*, like in real human soccer. Each tactical formation is represented by three numbers that specify how many players are, by default, in the *defense, misfield* and *attack* areas (e.g., 4-3-3, 4-4-2, etc.). This tactical formation can be modified dynamically according to the team goals, the match score, the time left, etc. Also like in human soccer, each player has a *tactical positioning* that defines its default position and the respective movement inside the soccer field. The tactical formation, associated with the tactical positioning, specifies a *role* and the respective *behaviors* for each player (e.g. left side player, center forward player, etc). Again analogously to the human soccer, the players are joined into four groups – goalkeeper, defender, mid-fielder and attacker – according to their respective *roles* and *behaviors.* These groups are open groups, but for one player to join a given group, it is necessary that it has the required behaviors to play the group roles. Finally, a player can have more than one role and take part in more than one group.

4 Team Behavior

The team behavior is given by a *tactical formation*, or a *tactical formation set*, [SV98] and the respective tactical positioning. In our approach, one *basic tactical formation* was chosen, (4-3-3), see figure 2. This basic tactical formation defines the general movement of the team.

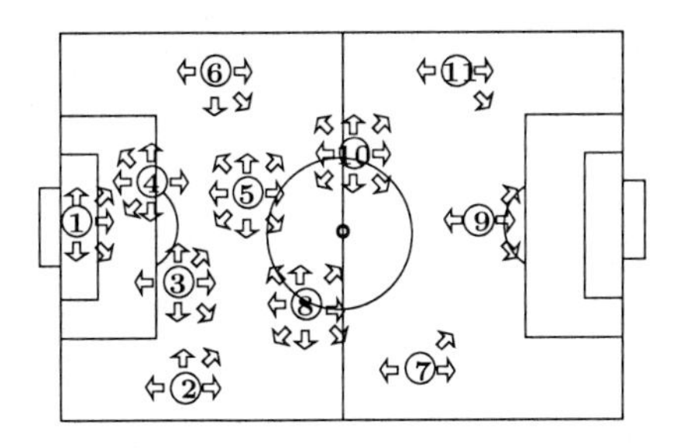

Figure2. The team home formation

Two other tactical formations, the *defensive formation* and the *attack formation*, derived from the *basic formation*, were chosen to be implemented. The first one, the *attack formation*, should be performed when the team has the control of the ball and goes forward trying to make a goal. The other one, the *defensive formation*, is adopted when the team looses the ball control and goes back either to avoid that the other team make a goal or to get the ball control back, see figure 3.

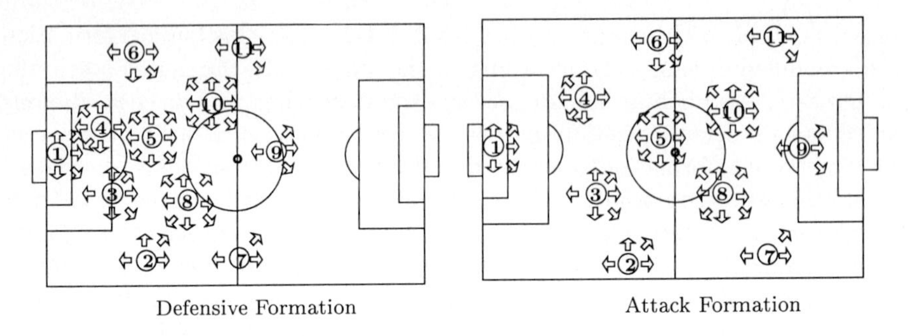

Figure3. The team defensive formation and the attack formation

5 Agent Behavior

In the adopted MAS approach, each player has its own *knowledge domain* about the environment. This knowledge domain can be divided into perceptual knowledge and social knowledge. Perceptual knowledge consists of auditive information, i.e., the messages sent by the other agents of the team; and visual information, i.e., the positions of the neighbor players, of the ball, and of the field limits. Perceptual knowledge is derived from the information provided by the Soccer Server and it is limited by the robot's perceptual range. The social knowledge consists of a set of rule bases that codify, in the form of typical reactive behaviors, the different roles that an agent should assume when involved in different practiced plays. This knowledge is determined by the agent tactical positioning. The social knowledge also contains miscellaneous information, such as the score, the time left, etc., and a meta rule base, common to all agents that belong to the same tactical group. This meta rule base is used when the agent has the ball control in order to choose the most appropriate practiced play.

6 Cooperation Process

The cooperation process among the UFSC-Team agents happens basically under two situation: when the team has the ball under control and when the team tries to get the ball control back.

In the first situation, when the team has the ball under its own control, the agent who keeps the ball will choose, using its tactical group meta rule base, one of the practiced plays, according to the acquired visual information, the agent position in the field, game score, time left, etc. Once the practiced play is chosen, the agent will inform the other agents involved in the specific practiced play, through an auditive message, which one was chosen to be performed and what are their roles in the play. Once the information received, these other agents, that do not have the ball control, will load the rule bases associated with their specific roles in the play. These rules will drive these agents to the best positioning, according to the chosen practiced play, but taking into account the current tactical formation.

In the second situation, when the team tries to get the ball control back, the agents will assume the defensive tactical formation. In this case, each agent load its default defensive rule base. These rules, according to the perceptual information, will drive the agent to perform some defensive behaviors, such as the area defensive behavior. In this case, the rules associated with the agents in the attacker and mid-fielder groups drive them back to their field and keep their auditive capability ready to receive a help request from an agent of the defender group to join in a group that is performing a specific defensive behavior.

7 Results on RoboCup'98

In RoboCup'98, the UFSC-Team played four matches and lost all of them. The best result was against the UBU team, UBU 1 x 0 UFSC-Team. RoboCup'98 was the first opportunity to evaluate in practice our research work and to compare it with other teams. The main problem presented by the UFSC-Team in RoboCup'98 was the real-time response. The team was not able to attend the soccerserver real-time restrictions, resulting in a very slow reaction time. This problem happened because one of the agent processes, the *expert* process, was executed in an interpreted mode. Because of software compatibility ploblems, during the RoboCup'98, we didn't had enough time to fix this problem before the competition.

8 Conclusion

The UFSC-Team presents an open cognitive multi-agent approach to the RoboCup Simulator League.It is based on the Expert-Coop environment and it presents a concurrent architecture to cognitive multi-agent systems that improves the agent performance and allows it to attend real-time restrictions, on best effort approach. Its cognitive behavior is based on a powerful knowledge-based system shell, the ILOG Rules 5.0 [ILO95], that allows the agents to dynamically choose an adequate rule base, according to the different roles and behaviors to be performed.

9 Future Works

Some effort to improve the UFSC-Team real-time response is being undertaken. A Case-Based Reasoning approach has been applied to the expert process knowledge base in order to improve its knowledge management capabilities. We intend also to incorporate a new inference engine into the expert process. Beside this other agent cooperation strategies are being implemented and tested to be integrated into the Coordinator process.

In the near future, a real robot soccer team should be implemented to participate in the RoboCup Small Size League and RoboCup Full Set Small Size League, where inter-agent communication has no restriction. The unrestricted communication will allow the agent team to use more sophisticated cooperations strategies that use both, agent perception and inter-agents communication, such as the *Dynamic Social Knowledge Cooperation Strategy* [CB98], or some cooperation strategies based on the *Negotiation Behavior* [Pru81] [SPJ97].

Acknowledgment

The authors express their thanks to the Ilog Inc. and to its representant in Brazil, Choose Technologies, for their support to this project in the form of a license to use the Ilog Rules software and all its associated documentation. The authors also acknowledge the RoboCup Federation and the ICMAS'98 Organizing Committee for their support that made possible the participation of UFSC-Team in the RoboCup'98.

References

[BC97] G. Bittencourt and A. C. P. L. da Costa. Expert-coop: An environment for cognitive multi-agent systems. *in pre-printers IFAC/IFIP MCPL'97, Conference on Management and Control of Production and Logistics*, 2:492–497, October 1997.

[CB98] A. C. P. L. da Costa and G. Bittencourt. Dynamic social knowledge: A cognitive multi-agent systems cooperation strategy. *Submitted to ICMAS'98 Third Conference on Multi-Agent Systems*, July 1998.

[ILO95] ILOG. *ILOG Rules 5.0 User Manual/Reference Manual.* ILOG, 1995.

[KAK+97] H. Kitano, M. Asada, Y. Kuniyoshi, I. Noda, and E. Osawa. Robocup: The robot world cup initiative. In *On-line Agent'97 proceeding http://sigart.acm.org:80/proceedings/agents97/*, August 1997.

[KTS+97] H. Kitano, M. Tambe, P. Stone, M. Veloso, S. Coradeschi, E. Osawa, H. Matsubara, I. Noda, and A. Minoru. The robocup synthetic agent challenge 97. In *Fifteenth International Joint Conference on Artificial Intelligence (IJCAI-97)*, volume 1, pages 24–29, 1997.

[Pru81] D. G. Pruitt. *Negociation Behavior.* Academic Press, 1981.

[SDB92] J.S. Sichman, Y. DEMAZEAU, and O. BOISSER. When can knowledge-based systems be called agents? *Anais do IX Seminario Brasileiro de Inteligencia Artificial*, pages 172–185, Outubro 1992. ISSN 0104-6500.

[SPJ97] C. Sierra, Faratin. P., and N. R. Jennings. A service-oriented negociation model between autonomous agents. In *8th European Workshop on Modeling Autonomous Agents in a Multi-Agent World (MAAMAW-97) International Conference on Computational Linguistic (COLIG)*, pages 15–35, 1997. Ronneby, Sweden.

[SV98] P. Stone and M. Veloso. Task decomposition and dynamic role assignment for real-time strategic teamwork. In *Submitted to Third International Conference on Multi-Agent Systems (ICMAS'98)*, 1998. http://www.cs.cmu.edu/afs/cs.cmu.edu/user/pstone/mosaic/pstone-papers.html.

Description of Team Erika

Takeshi Matsumura
(e-mail:matsu@futamura.info.waseda.ac.jp)

Graduate School of Science and Engineering,
Waseda University, Tokyo, Japan

Abstract. This paper introduces the learning of basic actions by Genetic Algorithm and the effective method for design of agents' cooperation based on role and scenario model.

1 Learning of Basic Actions

To realize cooperations of agents that are the reseach purpose of RoboCup, few actions are provided from soccer server. For example, to realize the pass play that is typical cooperation in soccer, a client would be required the ability to kick a ball in any direction.

Therefore, I made a client learn some basic actions by Genetic Algorithm (GA).

1.1 Genetic Algorithm

Genetic Algorithm(GA) is a kind of random searching technique based on random extraction of samples.[5] GA has a special feature which keeps a set of candidancies of solution and operate it. The candidancy of solution is called a chromosome or individual. An individual is consisted of gens. The set of individual is called population.

GA has a characteristic operator which is called crossover. Crossover operator takes two individuals in population and make new ones by exchange parts of them.

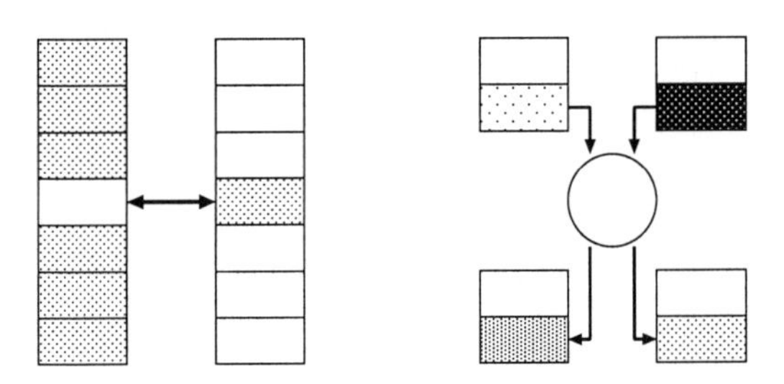

Fig. 1. Two type of crossover operator. Left one changes genes simply. The other takes a pair of genes, makes new pair and puts them back to the individuals.

Each individual has a fitness to the solution. Size of populations is a constant value so some individuals whose fitness is lower are thrown away from the new population. This is called Natural Selection. GA obtains a set of individuals that have better fitness to the solution.

The followings are required for GA to leave a set of adaptive individuals [1]:

1. Generating individuals as various as possible in the first generation.
2. Keeping diversity of generation.
3. Preserving characters of individuals that have better fitness.

To keep diversity of population I adopted MGG generation exchange method.[1] To preserve characters of individuals I adopted BLX-α crossover operator.[1]

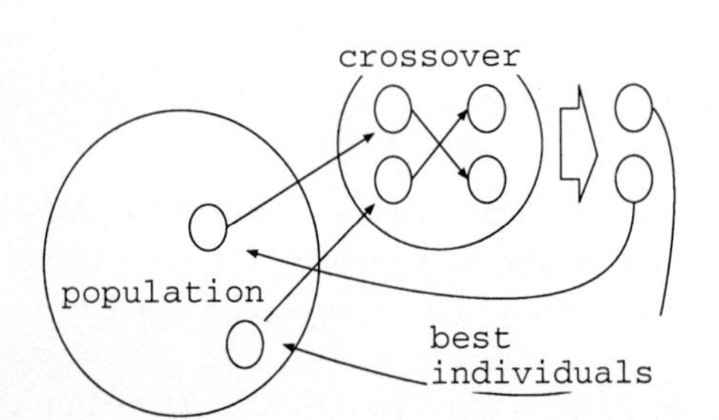

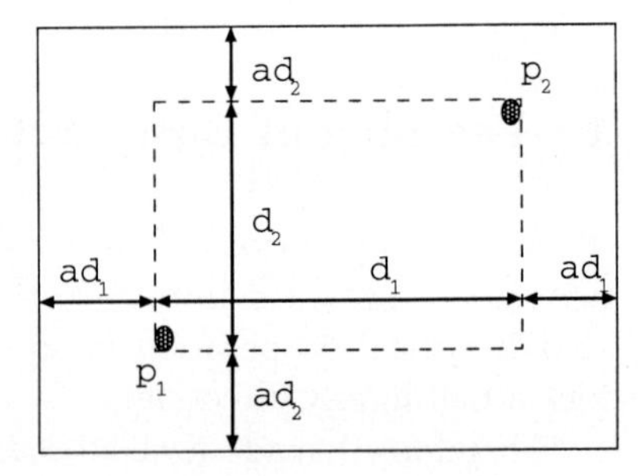

Fig. 2. Left figure showes MGG generation exchange method. A pair of individuals is taken from a population and they produce some individuals using crossover. Only a pair of individuals that have the best fitnesses in them are put back to the generation. The other figure showes BLX-α crossover operator. P_1 and P_2 are parents and a is a constant. Operator takes a pair of individuals in the area enclosed by solid line.

1.2 example of the design : kick-ball-around action

When a player kick a ball forward, we can use the kick command directory. But to kick a ball backwards, a client should kick the ball continuously at an angle tangent to his body in order to turn the ball to the disired direction and then kick it. I assumed the following expressions to calcurate the kick direction θ

$$\theta = a_b + \alpha \cdot vdist \cdot \delta t + \theta'$$
$$\theta' = cos^{-1}\frac{r_p + r_b}{d'} + 90 - tan^{-1}\frac{\epsilon}{d' sin\theta_2}$$
$$d' = d + \alpha \cdot vdist \cdot \delta t$$

where d is a distance between the player and ball, δt is one simulator cycle, a_0 is absolute angle of player, a_b is relative angle from a_0, r_p and r_b are respectively

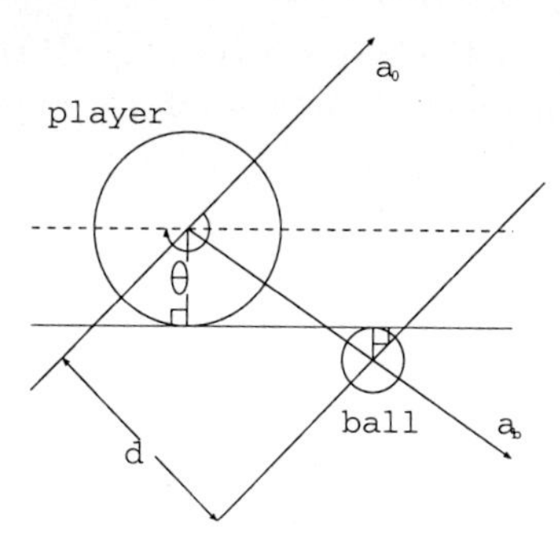

Fig. 3. kick direction

radius of player and ball. *vdist* and *vdir* are respectively DistChng and DirChng of ball which soccer-server takes. The term of $\tan^{-1}$ was added to avoid the colligion of player and ball.

Constants α and ϵ in the previous expressions, and another constant Pk that is the power player kick the ball make up an individual used by GA.

1.3 Result of Learning

I could get some result of the learning of kick-ball-around action.

α	ϵ	Pk	fitness
1.68	0.44	17	1403
1.71	0.60	17	1397
1.68	0.76	18	1396
1.68	0.23	17	1394
1.65	0.66	18	1372

Table 1. Best five individuals in the last population. α,ϵ and Pk consist an individual and the fitness is the total angle that player could turn the ball continuously.

I did this simulation with 1000 population for 10 generations. The best individual was got at the 7th generation, and the client which has these parameters could turn the ball around him continuously more than 3 times.

2 Cooperation based on Scenario and Role

2.1 Introduction

Since there may be a sudden change of situation which is caused by other player in the multi-agent soccer an intention control program is needed. Such a program will be very complex and redundant it describing all the conditions to deal with such sudden changes. Therefore, I introduced two ideas, role and scenario.[3]

2.2 Behavior of Role and Scenario

```
Scenario     : Ball interception by DF(Defender)

Role         : 1. intercept and cut shoot line
             : 2. cut pass line
             : 3. cut another shoot line

Terminate    :  * Ball has gone after the DF area.
condition        --> Back-up-of-Goalie
   and
  next          * Ball has gone before the DF area.
scenario         --> Return-his-position

                * Ball is out of side line.
                 --> Kick-in-action

                * One of defender obtains Ball.
                 --> Pass-forward-or-Clear

Condition    :  * Nearest to Ball           --> role 1
 to get         * Be on the offside line --> role 2
 a role         * Otherwise                 --> role 2,3
```

Fig. 4. Example of scenario. If a player of the Defender group is the nearest one to the ball,he gets role 1 to play and if he is able to get the ball, the terminal condition of this scenario is satisfied so he gets Pass-forward-or-Clear as the next scenario. But if there is another teammate who becomes closer to the ball while he is getting the ball, he changes his role to role 2 or 3.

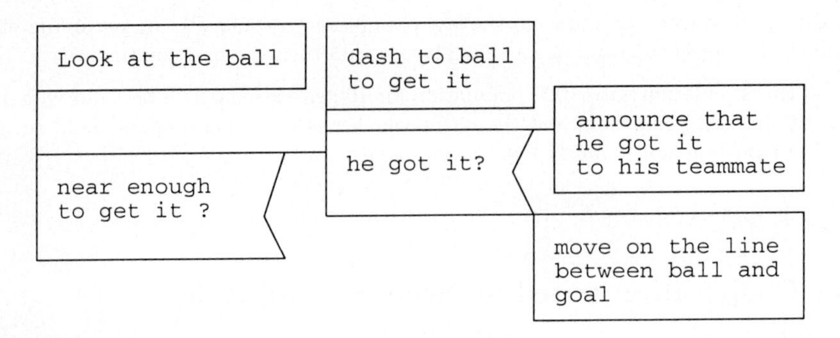

Fig. 5. Example of role. This PAD represents role 1 which is used in the example of scenario in figure 5.Noted that the role does not need any conditions to terminate the scenario.

A role is a function that an agent executes. Scenario is a structure consisted of two functions. Role-selector is a function for an agent to get his role and Scenario-terminator is terminal condition for the scenario. If Scenario-terminator is satisfied, it terminates current role, gives up the scenario and then take another one.

Every agent in the Team belongs to a group and they alway works under the same scenario. If a group gets a scenario, every agent in the group evaluates Role-selector of the scenario to get his own role to play.

Scenario-terminator is checked continuously as a back ground process. We do not take care of the terminal condition when describing a role.

After finished his role, an agent evaluates Role-selector of the scenario to get his role again.

2.3 Moving Agents between Neibouring Groups

There are four groups in my team, Forward(FW), Midfielder(MF), Defender(DF) and Goalie(GK). Every agent is in his initial group at first, but he can move to another group. If a group got a scenario that requires more agents than the group has, it can move some agents from neibouring groups.

```
Scenario  : Goal Kick
Role      : 1. Kicker
          : 2. Receiver

Terminate:  * Receiver got the ball
condition     and he returned to DF group
   and       --> Return-home-position
  next      * Opponent got the ball
Scenario
             --> Intercept

Condition:  * Goalie --> Kicker
 to get     * Otherwise --> Receiver
 a role
```

Fig. 6. Scenario example which needs to get another agent. GK group is usually consisted of only goalie, but this scenario requires 2 roles so the goalie makes another agent move to GK group to safisfy this scenario.

A group can get forcely an agent from another group without permittance of the agent or the group. It can reduce the negotiations but it may cause the struggle of the agent between the groups. Such conflictions will be solved by designing scenarios which need less roles.

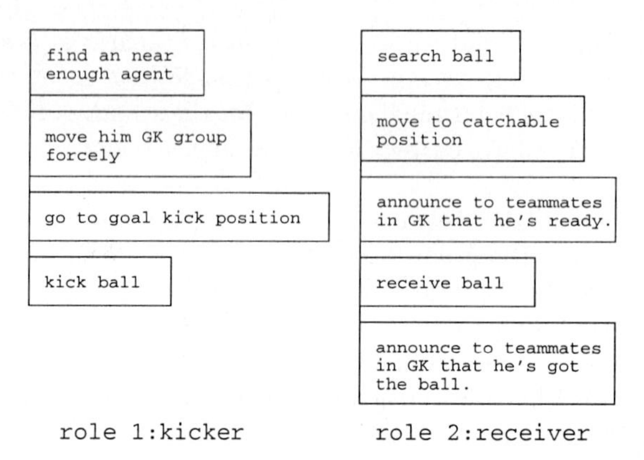

Fig. 7. Roles of kicker and receiver. Receiver is already in GK group when he gets this scenario. He do not know that he was in DF group. And he will be continuously in the GK group until another group orders him to move to.

3 Future work

3.1 Expansion of role-scenario model

At present, every agent always belongs to a group, and every group always has a scenario. There are some problems as follows:

1. It's difficult to move some agents between two groups.
2. The number of scenarios and that of groups are always same.
3. When there are scenario A and B, the scenario A may not require as many agents as the scenario B.

Here it is an improved model.

1. There are no groups.
2. Any agent can belong to any scenario and can have no scenario.
3. Any agent can suggest a scenario any time.
4. Any agent who hears that suggestion and doesn't belong to a scenario tries to take part in that suggested scenario.
5. If there is a role in the scenario which is fit to an agent, the agent takes part in the scenario and play the role.
6. In the case the agent who is belonging to an scenario A suggests another scenario B, the scenario A is terminated and every agent who was taking part in the scenario A is free.

This model gives flexibility to the relation of agent and scenario. In RoboCup soccer agents tend to depend on their places, but they should make a cooperation with other agents who are unexpectedly near of them in the more complex problems like Disaster Relief. This model is effective to make such a multi-agent system.

References

1. Masayuki Yamamura: The theory and practice of Genetic Algorithm. Tutorial of Japan Society for Software Science and Technology (1998)
2. Eiichi Osawa, Makoto Yokoo: MultiAgent. Tutorial of Japan Society for Software Science and Technology (1998)
3. Simon Ch'ng and Lin Padgham: Team description: Royal Melbourne Knights. The First International Workshop on RoboCup (1997) 125–128
4. Sean Luke, Charles hohn, Jonathan Farris, Gary Jackson, James Hendler: Co-Evolving Soccer Softbot Team Coordination with Genetic Programming. The First International Workshop on RoboCup (1997) 115–118
5. Lawrence Davis: Handbook of Genetic Algorithms. Van nostrand Reinhold, A Division of Wadsworth, Inc (1990)

Getting Global Performance through Local Information in PaSo-Team'98

E. Pagello[1,2], F. Montesello[1], F. Garelli[1], F. Candon[3], P. Chioetto[1], S. Griggio[1]

[1] Dept. of Electronics and Informatics, Padua University, Italy
[2] Inst. LADSEB of CNR, Padua, Italy
[3] University of Venice, Italy

Abstract. We illustrate new improvements in PaSo-Team (The University of Padua Simulated Robot Soccer Team), a Multi-Agent System able to play soccer game for participating to the Simulator League of RoboCup competition. PaSo-Team looks like a partially reactive system built upon a number of specialized behaviors, designed for playing a soccer game that generate actions accordingly with environmental changes. Major improvements have been obtained in the design of individual skills so that the general architecture of PaSo-Team and its coordination model are exploited in a better way.

1 Introduction

After the experiences acquired in developing PaSo-Team for RoboCup'97, we have improved the performance of the proposed architecture by trimmering individual skills. In such way we were able to better exploit the underlying ideas contained in the proposed architecture, namely the quest for a global optimization through local information. This approach is justified observing that during the game the relevant portion of the game field is mostly centered around the ball and sometimes it is stretched towards the goal. This area is assumed to be better observable than the whole field because of its reduced size, allowing the players to behave optimally in a local area but in the same time supplying the whole team with a big amount of global information.

In this perspective, the architecture's assumption that each player is doing its best (local optimum), gives a good justification for the passing technique described in PaSo-Team'97 description paper [11]. Exploiting this architectural feature as mechanism for managing cooperative tasks, any improvement in the individual ability to increase the offensive potential by correct local positioning automatically allows to increase the performance of the whole system. For this reason in designing the PaSo-Team'98 we keep unchanged the overall reactive structure while we improved mainly the individual skills of the players.

2 Architecture and Coordination Model

As it has been pointed out in literature [1] a sound arbitration mechanism is the base for an appropriate performance of a behavior-based autonomous system. In our case a further difficulty arises, due to the simultaneous presence of several playing agents in the same environment. Starting from the pioneristic subsumption architecture originally devised by Brooks [3], a number of innovative behavior-based systems have been proposed in literature (Connell [4], Maes [7], Anderson [1], Kaelbling [6].

Their proposals are dominated by the concept of arbitration which results in an either spatial or temporal ordering of behaviors. The former causes the concurrent activation of a set of primitive reflexive behaviors, also referred to as static arbitration, the latter brings about a sequential activation of different sets of primitive reflexive behaviors, also referred to as dynamic arbitration.

However, because the inclusion of temporal ordering appears too problematic when it is devised within a general multiagent framework, we have implemented a static arbitration as a special purpose behavioral module where pre-processed sensor data are always channeled to discriminate a candidate skill to be enabled as a response to typical perceived patterns. Every time sensor data are directly channeled between the perception block and the selected behavior, this behavior is activated whereas the remaining ones are inhibited.

The resulting architecture, shown in fig. 1, resembles partly the proposal of Anderson and Donath [1] and partly that of D'Angelo [5] in what the collection of boolean values (flags), updated using information supplied from sensor data pre-processing, defines a coarse-grained global state of an agent which controls behavior switches as a rough inhibitor/activation mechanism.

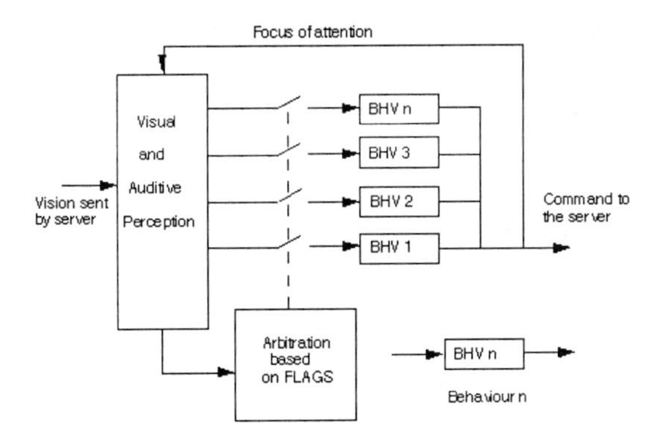

Fig. 1. agent arbitration

At each time of the game, every team player will be enabled with a behavior which depends on both its current position and orientation in the soccer game field and the pattern of the objects the agent is aware of. So, though every agent has the same cloned structure, it does not need to be activated with the same behavior. This means that is the world that makes differences among the agents.

The basic coordination mechanisms induced by the arbitration involve usually two behaviors at a time. In the attack playmode, such behaviors characterize two players belonging to the same team, that is the ball-holder and a potential receiver of the ball. We want to realize the simultaneous activation of the behavior *playball* for the player with the ball and the behavior *smarcamento* for the next ball-holder candidate. The coordination arises through the simultaneous activation of the pair {bhv_1, bhv_2} for any pair of players, where, during attacking, bhv_1 = *playball*, bhv_2 = *smarcamento*, whereas, during defending, bhv_1 = *bhv_X*, bhv_2 = *interdict*.The behavior *bhv_X* is not a real implemented behavior, but it is referred to the apparent behavior of the opponent. Actually, the opponent estimated behavior is *chasing the ball*, so our player must compete against it.

To realize this kind of interaction we have built a rigid arbitration which choose a candidate behavior looking at the *global flags* representing particular states of the whole team of agents, and at some *local flags* related to states of a single agent. In this way the emerging cooperative behavior appearing during the game may be considered as an *eusocial behavior*, a collective behavior due to the interaction of "genetically" determined individual behavior, as discussed in McFarland [8]. The proposed arbitration is an hint for the emerging of a social behavior, like an ant colony that seems to be a relatively smart being even if formed by a finite number of pure instinctive individuals. The emergence of some sort of intelligent behavior like triangulations or non explicit pass arise mainly from the interactions between the single players and the environment, namely in our case between our players and the opponents, exploiting advantageously their dynamics. For an example of *emerging collective behaviour* refer to [10], where the coordination model is accurately described. It is in this context that the improvements in individual skills support the coordination model.

3 Visual Fields and Maps

We developed an "ad-hoc" algorithm who takes care of positioning the potential receiving team-mates in the locally best position to get a ball, in the same way we introduced visual fields and maps to improve individual skills from the point of view of the ball's owner. Both these solutions are the logical consequence of the assumption that in soccer the portion of field relevant to the actions is usually centered around the ball, causing an increase of relevance for the players near the ball and their own perceptions. So, as the unmarking action performed by team-mates without ball is effectively relevant (as shown by the good performances of PaSo-Team in Nagoya), in the same way the abilities of the ball's owner are relevants. To improve the individual abilities of the player who is handling the ball we introduced a new representation for the visual information that the player is acquiring from the environment: the Visual Fields and the Visual Maps. The Visual Field is a data structure built to contain objects representing relevant features of the environment and it's istantiated once for each player of our team. The typical objects that should be inserted in this structure are *team-mates*, *opponents*, *ball* and *the goal area*. Other geometric shapes should be inserted in the structure are *the whole game field*, *both the half sides of the field*, *the penalty areas*, *the circle in the middle* and *the offside areas*.

All these objects must be dynamically inserted in the Visual Field structure, following the evolution of the environmental dynamics. This means that during the game the Visual Field structure is costantly updated inserting, updating and removing objects representing the vision frame of the considered player. For geometrical shapes, as the whole field or the offside area, the instantiation is straightforward, involving only the insertion of the object in the structure. The same operation is different in case of object as the players, team-mates or opponents. In fact they have not a predefined geometrical shape and so this operation must be done explicitely. In our case we assigned to each player a circle-shaped area, but with a parametrized diameter, depending on the team and the distance. The assumption of a circle-shaped area for the players must be considered arbitrary, because using a differently shaped area should be possible to exploit other features as the face-dir information. To deal with middle-complexity actions as *obstacle avoidance, pass to mate, dribbling,* we extended the presented Visual Fields with another structure, the Visual Maps. A map is a simple array of 360 floats built looking at the Visual Field istantiated for a given player at a given time. It is composed by 360 directions (degrees) toward which a player can decide to dash or kick; an high value in the map indicates a direction to avoid, while a zero or negative value suggests a very good way to take.

For example, suppose that our player had to dribble some opponents. At the beginning the client has a void map (all values are zero); than he asks to the objects in the Visual Field (dynamically updated during the perception phase) to create their mark in the map: in this case an opponent add some positive values in the elements of the map corresponding to bad directions. After every objects in the field have modified the map, the client applies some filters to promote a particular direction (the enemy goal port for example) or to create other useful effects. Finally the player looks for the minimum in the map and goes in the suggested direction.

It's very important notice that the parameters used in the creation of map depend by the particular action you have to do; for example the enemy radius is 0.8 m (the real radius of a player) when you must do obstacle avoidance, and is 3 m (the kickable area radius) when you must dribble your opponents. Starting from the obstacle avoidance problem, we have created a very powerful tool, easily adaptable to other actions. In the future we plan to improve the structure of Visual Map and to add reinforcement learning for the dynamic parameters.

4 Niche and Pruning of the Arbitration Module

To better exploit the potential moods of the proposed architecture we finally proceeded with a strong simplification of the arbitration module. This was made by pruning the decision tree, cutting off a lot of exception handled in it, like the action schemas and timeout-based decision. Other simplifications in the arbitration module has arisen from exploiting some features of the *environmental niche* inside whom the players live, like the "dynamical environment hypotesis". Such

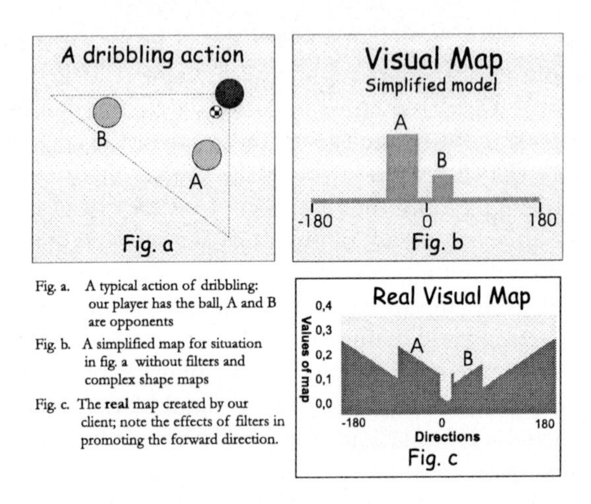

Fig. 2. Visual Field and Visual Maps

assumption is based on the idea that in such environmental niche (the soccer field) the players are assumed as objects with non-negligible motion, even if not always active during the game. This has allowed a reduction in complexity in the arbitration module, avoiding for example an explicit *monitoring the ball* action, assuming that the environmental dynamics will be sufficient to keep informed the players on the ball position. All these simplifications has allowed a more clear understanding of the decision tree functioning, allowing in the same way a better understanding of the architecture's performance.

5 Conclusions

In this paper a further improvement in exploiting the features of our architecture for multi-agent systems playing simulated soccer has been presented. The performances of PaSo-Team'98, compared with those ones of PaSo-Team'97, have shown an improvement in global coordination, allowed by a more efficient exploitation of the architecture's features, among which the "implicit coordination", already described in [9]. The assumption of the relevance of ball-centered area has shown its validity, explicitating the role of the *ecological niche* in designing autonomous agents, introduced by Pfeifer [12]. A full description of the implementation of PaSo-Team'97 Clients (The University of Padua Simulated Robot Soccer Team) is given in [2]. The coordination model implicitly contained in the architecture has been implemented with care by paying attention to the individual skills of the player who is handling the ball. An improved global performance through local interactions is achieved. The relationship between the problem of optimum in multi-agent systems and the "ecological niche" of the systems themselves has been explicited.

Acknowledgements

We thank Carlo Ferrari and Antonio D'Angelo, as well as all other undergraduate students of Electronics and Computer Engineering School of Padua University, who have cooperated to developing PaSo-Team. A particular thank is due to the Industrial Firm Calearo S.R.L., a car aerials and cables manufacturer, located in Isola, Vicenza (Italy), that have provided coverage of all expenses for participating at IJCAI Conference. We like to thank also Padua Branch (Italy) of Sun Microsystems Italia, that provided us freely a SUN ULTRA1 for developing PaSo-Team.

References

1. T.L. Anderson and M. Donath. Animal behaviour as a paradigm for developing robot autonomy. In Pattie Maes, editor, *Designing Autonomous Agents*, pages 145–168. The MIT Press, Cambridge (MA), 1990.
2. F. Bidinotto, A. Bissacco, M. Dal Santo, W. Frasson, S. Griggio, A. F. Grisotto, S. Marzolla, F. Montesello, and E. Pagello. Implementing a soccer client team for robocup '97 competition. Technical report, LADSEB-CNR, Padua (I), Nov. 1997.
3. R. Brooks. A layered intelligent control system for a mobile robot. *IEEE J. on Rob. and Aut.*, RA-2:14–23, Apr. 1986.
4. J. H. Connell. *Minimalist Mobile Robotics.* Number 5 in Perspective in Artificial Intelligence. Academic Press, 1990.
5. A. D'Angelo. Using a chemical metaphor to implement autonomous systems. In M. Gori and G. Soda, editors, *Topics in Artificial Intelligence*, volume 992 of *Lecture Notes in A.I.*, pages 315–322. Springer-Verlag, Florence (I), 1995.
6. L. P. Kaelbling and S. J. Rosenschein. Action and planning in embedded agents. In Pattie Maes, editor, *Designing Autonomous Agents*, pages 35–48. The MIT Press, Cambridge (MA), 1990.
7. Pattie Maes. Situated agents can have goals. In Pattie Maes, editor, *Designing Autonomous Agents*, pages 49–70. The MIT Press, Cambridge (MA), 1990.
8. D. McFarland. Towards robot cooperation. In *From Animals to Animats 4, Int. Conf. on Simulation of Adaptive Behavior (SAB-94)*, Brighton, 1994.
9. F. Montesello, A. D'Angelo, E. Pagello, and C. Ferrari. Implicit coordination in a multi-agent system using a behavior-based approach. In P. Dario T. Lueth, R. Dillmann and H. Worn, editors, *Distributed Autonomous Robotic Systems 3*, pages 351–360. Springer-Verlag, Karlsruhe (Germany), May 1998.
10. E. Pagello, A. D'Angelo, F. Montesello, and C. Ferrari. Emergent cooperative behavior in multi-robot systems. In M. Wada Y. Kakazu and T. Sato, editors, *Intelligent Autonomous Systems (IAS5)*, pages 45–52. IAS Press, Sapporo (Japan), May 1998.
11. E. Pagello, F. Montesello, A. D'Angelo, and C. Ferrari. A reactive architecture for robocup competition. In H. Kitano Ed., editor, *Robocup-97: Robot Soccer World Cup I*, Lecture notes in Artificial Intelligence Series. Springer-Verlag, 1998.
12. R. Pfeifer. Building fungus eaters: Design principles of autonomous agents. In *Forth Conference on Simulation of Adaptive Behavior (SAB-96)*, pages 3–12. Sept. 1996.

A Direct Approach to Robot Soccer Agents: Description for the Team MAINZ ROLLING BRAINS Simulation League of RoboCup '98

Daniel Polani, Stefan Weber, and Thomas Uthmann

Institut für Informatik, Johannes Gutenberg-Universität,
D-55099 Mainz, Germany
{polani,stefanw,uthmann}@informatik.uni-mainz.de
http://www.informatik.uni-mainz.de/PERSONEN/{Polani,Uthmann}.html

Abstract. In the team described in this paper we realize a direct approach to soccer agents for the simulation league of the RoboCup '98-tournament. Its backbone is formed by a detailed *world model.* Based on information which is reconstructed on the world model level, the rule-based decision levels chose a relevant action. The architecture for the goalie is different from the regular players, introducing heterogeneousness into the team, which combines the advantages of the different control strategies.

1 Introduction

The challenge of constructing intelligent agents capable of coordinated operation, real-time response, handling uncertain information and more has been a main motivation for setting up the RoboCup challenge. A tournament element as propagated in [1–5] introduces the character of a "co-evolutionary" development process on the level of concepts. With a tournament in mind, concepts being developed are subject to a "selection pressure" towards exploitation of "concept space" in contrast to the prevalent exploration character of pure research. The ability to measure the performance of a given concept in objective numbers (i.e. goals) introduces a tradeoff requirement between generality of the approach (allowing flexibility and extensibility) and utilization of specialized knowledge (allowing exploitation of model properties for performance).

To be able to evaluate the importance of accurate modeling versus higher generalizability attempts of creating agents must include both directions of research. In our approach we choose to place more emphasis on a *direct* model which tries to utilize accurately the knowledge available and to restrict tunable (learnable) parameters to domains where precise modeling is not possible or practical.

Furthermore we concentrate on having a world model which is as precise as possible. The world model level factorizes out the inconsistencies of the sensor data and merges them as well as possible, thereby serving as a well-founded basis

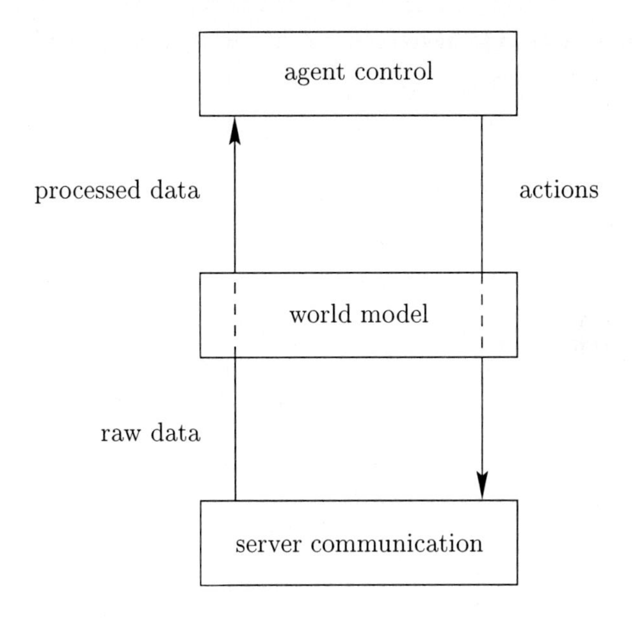

Fig. 1. General architecture of the soccer clients

for higher-level decisions and actions. For instance, it serves to reconstruct the time synchronization with the server. The particular importance of developing techniques for world model design is to be seen in conjunction with sensor fusion techniques for sensor data in real robots. These considerations warranted the particular emphasis which we put on the world model. This document describes some aspects of our MAINZ ROLLING BRAINS team, which participated at RoboCup '98 in the simulation league.

2 Agent Architecture

2.1 General Architecture and Communication Interface

The general architecture of our soccer agents is depicted in Fig. 1. The agents are realized in an object-oriented approach using C++ as language. Communication between agent and soccer server takes place via the *server communication interface*. The data are preprocessed by the *world model* level before they can be utilized by the control level. The actions performed by the control level are not sent directly to the server; instead their expected influence on the world is tracked by the world model and only then they are forwarded to the server.

The communication interface consists of several classes which are responsible for opening the communication to the server and maintaining it. They parse the information sent by the server and convert it into data structures that can be handled by the higher level classes. These raw data are never seen in this form by the *agent control*. Instead they are filtered and processed by the *world model* level which provides the agent control information not directly obtained from the server.

2.2 World Model

World Model Data Structures World model information can be e.g. position and orientation of the agent itself, estimated speed of the agent and of other agents as well as of the ball, the relative and absolute positions of objects seen, estimated positions of objects not seen and more. The position of the current agent is calculated in a straightforward way from the flags and lines seen, other objects are computed relative to it. Object movement is currently calculated from object positions in two successive time steps if they are to far away to use the velocities given by the server. Further quantities modeled are stamina and effort.

Every data in the world model also possesses a separate validity, which is dropping continuously while the object is not directly detected via the server, and reset to full, whenever the data is being refreshed from the server.

By thus processing the data the world model class provides a framework, in which the more complex decision levels can transparently access the relevant information about the world. The world models construction always assumes that there are "real" quantities of position and velocities behind the values provided by the server and that just their accuracy to which they are known to the agent is not perfect, instead of assuming a fuzzy concept of positions and other quantities as a priori structure. The control level is then free to interpret those quantities as fuzzy ones. In particular, it is a deliberate design choice not to introduce interpretation of higher quality already on the world model level.

In the current implementation the actions dispatched by the control level are only logged by the world model to be able to simulate future developments. They are not modified further, therefore, in a way, the path of the actions sent through the world model level by the control level is much "shorter" than that of the raw data arriving from the communication level.

Consistency of World Model Information An important feature of our world model is the fact that it keeps track how consistent its data are with the data received from the server. Due to the connectionless protocol used in server communication it is not possible for the world model to have a precise and consistent picture of the actions that reached the server and were performed at a certain time step. In other words, it is not known at exactly which time step an action dispatched by the agent control is going to be performed.

To perform an action at the right time step can be vital: imagine a goalie that has to intercept an incoming ball. If for instance, client or network load is high, the last observed sensor data can be several time steps older than the state currently processed by the server. To be able to handle such situations, our world model keeps a time slice of (in our case typically 20) time steps around the estimated "current" server time in memory. On request of the control level the world model provides a snapshot for a given time step inside this time slice.

The data in this snapshot can be generated by different ways. At fixed time intervals the sensor update thread (realized by timer interrupts) fills in the sensor data for the time step given by the data time stamp. If a snapshot is requested for a time step no later than that of the most recent sensor data, the original sensor data are used if present, or, if not, the missing data are interpolated from neighboring snapshots. If, however, the time step requested for the snapshot lies in the future of the most recent server data, the estimated future world state is calculated, taking into account the actions assumed to be still in the action queue. In the presence of high network or client load it can prove an advantage instead of the snapshot containing the most recent data to use the snapshot of a time step delayed to the future for the control decision.

The world model further provides a measure determining how well the world state as calculated matches the world state as obtained by the sensor update. By this the reliability of the world model forecasts is quantified and can be used as basis for decision. Using these features of the world model our agents attained a high degree of robustness. During informal games with other teams at RoboCup '98 we found that the ROLLING BRAINS agents usually were less sensitive to network or workstation load than agents from other groups. We attribute this strength in particular to the elaborate world model management described above.

3 Agent Control

Agent control structure is represented in Fig. 2. We have a large submodule which is responsible for handling general game situations as occurring during regular play. For special situations like free kick, goal kick and similar we have other submodules which take responsibility in such a case. Delegation of responsibility is not performed by syntactic separation, i.e. separation into structural different modules, but are handled on a semantic level, i.e. in principle in the same context as handling two different regular game situations. The advantage is that it is possible to handle special situations using the same level of deliberative models as regular game, allowing non-"local" strategic and tactic considerations (e.g. delaying a free kick to gain stamina or to reduce the opportunity for the opponent to obtain a goal in the remaining time). An structurally separate consideration of the different situations would have had the advantage of easier design handling, but we were interested to evaluate the merits of the unified approach.

Our current implementation realizes direct situation-based actions, i.e. the current situation as represented in the world model is mapped directly to an action, including actions which include estimates of future development. Our agents therefore act essentially as reactive ones.

The control level strategies we were and still are working on include a fuzzy-rule based and a hierarchy-rule based mechanism. The fuzzy rule approach is currently using a flat set of rules which activate in parallel the different possible actions of the agent (like *kick*, *dash* and *turn*). The degree to which an action is activated is determined by the fuzzy inference. All actions activated beyond a certain threshold are performed (according to a predefined order).

The hierarchical rule concept operates using rules consisting of two parts: the first part is the *condition* for firing the rule, the second part is an action to be performed for a *terminal rule* or subrules to be evaluated recursively for a *meta-rule* (Fig. 3).

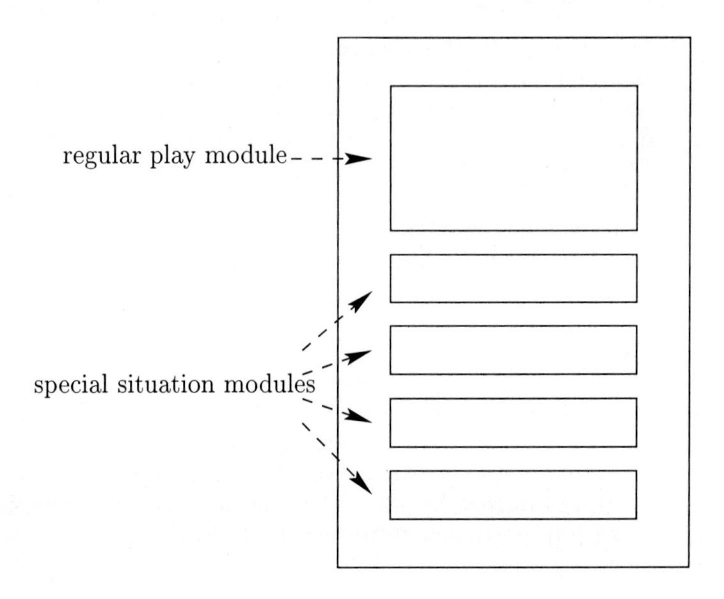

Fig. 2. Structure of agent control level

One always evaluates a set of rules, beginning from an initial starting set of rules. The condition for each rule is evaluated giving a priority for the given rule. The rule with the highest priority from the set is selected. If a terminal rule, the corresponding action (which can be an action sequence) is performed, if a meta-rule, the same rule activation algorithm is recursively applied to the set of sub-rules. Thus it is possible to implement conditions triggering higher level behavior and lower level reactive behavior in the same framework.

The selection of relevant rules and conditions in both approaches is part of the knowledge engineering task involved in the creation of the agents. Tuning the priorities for the different rules is partly expert knowledge, but mainly parameters that have to undergo adaptation for the rules to be of any use. We are attempting to approach their optimization also using Evolutionary Algorithms and related methods.

The two different approaches (fuzzy and hierarchical) have both been realized in the team. The regular players have been designed using the fuzzy representation. The goalie was a completely independent development using the hierarchical approach. It was introduced into the team in a late stage when it became clear that the original version of the goalie using the fuzzy approach was not adequate for its task. Thus the MAINZ ROLLING BRAINS indeed realized a heterogeneous team and did not only consist of clones of the same agent, whose actions only would differ by initial conditions or its position in the game. The combination of the strengths of the two different approaches is a good paradigm for a true multi-agent system where the different agent architectures complement each other to attain a common goal.

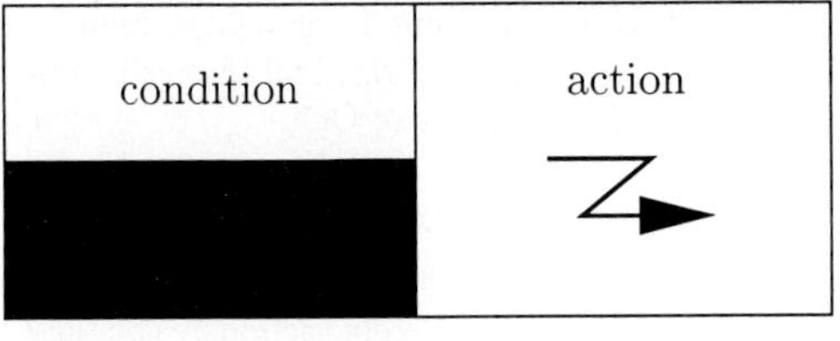

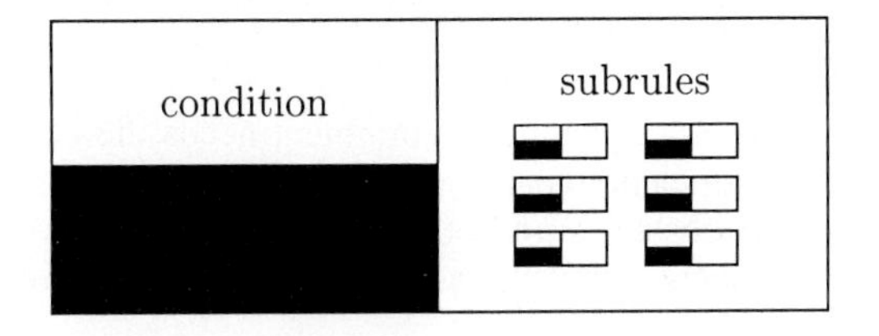

Fig. 3. Rule concept

Acknowledgements

We wish to mention here all the members of the team participating at the development of the MAINZ ROLLING BRAINS agents, namely (in alphabetical order) Christian Bauer, Marc Hellwig, Michael Junges, Oliver Labs, Achim Liese, Christian Meyer, Ralf Schmitt, and Frank Schulz. Furthermore we would like to thank Hans-Dieter Burkhard and the members of his team for their encouragement and helpful discussions, Hans-Jürgen Schröder for finding a sponsor for the project. We also gratefully acknowledge the support by the Herdt-Verlag.

References

1. Hiroaki Kitano. RoboCup: The robot world cup initiative. In *IJCAI-95 Workshop on Entertainment and AI/Alife*, August 1995.
2. Hitoshi Matsubara, Itsuki Noda, and Kazuo Hiraki. Learning of cooperative actions in multi-agent systems: a case study of pass in soccer. In *AAAI-96 Spring Symposium on Adaptation, Coevolution and Learning in Multi-Agent Systems*, pages 63–67, Mar 1996.
3. Peter Mössinger, Daniel Polani, René Spalt, and Thomas Uthmann. XRAPTOR – A synthetic Multi-Agent Environment for Evaluation of Adaptive Control Mechanisms. In F. Breitenecker and I. Husinsky, editors, *EUROSIM '95*. Elsevier, 1995.
4. Peter Mössinger, Daniel Polani, René Spalt, and Thomas Uthmann. A virtual testbed for analysis and design of sensorimotoric aspects of agent control. *Simulation Practice and Theory*, 5(7-8):671–687, 1997.
5. Itsuki Noda. Soccer server: a simulator for RoboCup. In *JSAI AI-Symposium 95*, Dec 1995.

CAT Finland: Executing Primitive Tasks in Parallel

Jukka Riekki, Jussi Pajala, Antti Tikanmäki & Juha Röning
Dept. of Electrical Engineering and Infotech Oulu
University of Oulu
FIN-90570 Oulu
{jpr,jussip,sunday,jjr}@ee.oulu.fi

Abstract. We present a novel representation for agent actions. An action map specifies the preferences for all the possible different actions. The key advantage of this representation is that it facilitates executing in parallel primitive tasks. We have utilized the action map representation in playing soccer. We describe here the system that controlled our players in the second RoboCup competition.

1 Introduction

A mobile agent operating in a dynamic environment needs the capability to react to several objects in the environment in parallel. Some reactions progress the task at hand, while some help to avoid obstacles and other mobile agents. We call these tasks of reaching and avoiding objects primitive tasks and the corresponding goals primitive goals.

In this work, we studied a situated approach, where the actions of the agent are calculated directly based on the information collected by the sensors from the environment [Brooks, 1991; Agre and Chapman, 1990; Kaelbling and Rosenschein, 1990; Arkin, 1990]. For each primitive task, there is a module producing commands that progress the task. Executing tasks in parallel by combining these commands is difficult, because the commands do not contain enough information to allow selecting a compromise that would satisfy several primitive goals.

Parallel task execution is possible to some extent in the architecture suggested by Rosenblatt and Payton [1989]. In this architecture, the modules produce votes for different commands. The votes are then summed and the command with the maximum sum is selected. However, this approach has limitations in a dynamic environment. As the heading and speed commands are processed separately, it is difficult to guarantee that the agent will reach the right location at the right time.

In this paper, we present a method for executing primitive tasks in parallel. The commands produced by independent modules are combined by a simple operation. This method is based on a novel representation of agent actions – the action map representation. The method is applied to playing soccer.

2 Action Maps

An action map specifies for each possible action how preferable the action is from the perspective of a task. The preferences are shown by assigning a weight to each action. A separate action map is calculated for each primitive task of reaching or avoiding an object. More complex tasks are performed by combining these primitive action maps.

The action maps controlling the way the agent reaches and avoids objects are called velocity maps, as they contain weights for different agent velocities. The weights make up a two-dimensional surface in a polar coordinate system (direction, speed). A weight for a velocity is calculated based on a global optimality criterion, the time to collision. The shorter the time needed to reach an object, the heavier the weight for that action. For actions that do not result in a collision, the weights are calculated by propagating, i.e., by adjusting the weights towards the weights of the neighboring actions on the map.

To preserve the weights of the actions that do reach the object, the weights are adjusted only upwards.

The resulting map contains only positive weights. A map of this kind is called a Goto map, because the performance of the action with the currently heaviest weight on such a map causes the agent to reach the corresponding object. An example of a Goto map is shown in Fig. 1.

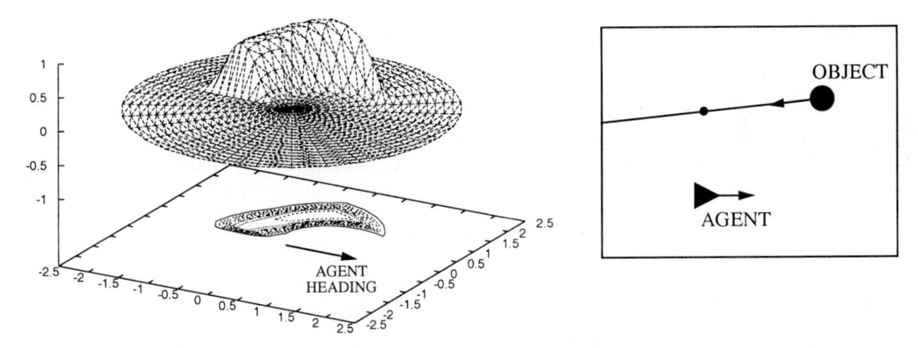

Fig. 1. Left: A Goto map. **Right:** The situation. The agent reaches the object at the location marked with the small dot, if it executes the action with the heaviest weight on the Goto map.

An action map containing negative weights is called an Avoid map. It is calculated by negating the corresponding Goto map. As only positive weights cause actions to be performed, an Avoid map does not alone trigger any motion, but only prevents some actions. In other words, an Avoid map prevents collision with an obstacle when a target is being reached with the help of a Goto map. Fig. 2 shows an Avoid map.

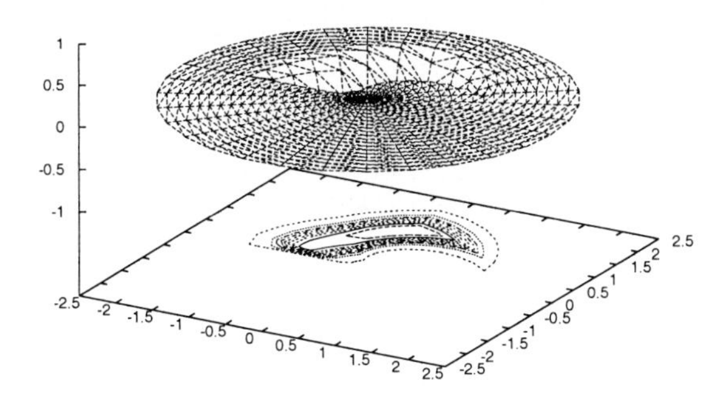

Fig. 2. An Avoid map corresponding to the Goto map shown in Fig. 1.

Tasks are executed in parallel by combining primitive Goto and Avoid maps into composite maps by the Maximum of Absolute Values (*MAV)* method. In this method, the weight with the maximum absolute value is selected for each action. In other words, the shortest time it takes to reach an object is selected for each action and the corresponding weight is stored on the composite map. Thus, the sign of a weight on a composite map specifies whether an obstacle or a target would be reached first if a certain action were performed. The absolute value of the weight specifies the time to collision with the object that would be reached first. The global optimality criterion guarantees that this method provides correct results. Fig. 3 illustrates the MAV method for compiling action maps.

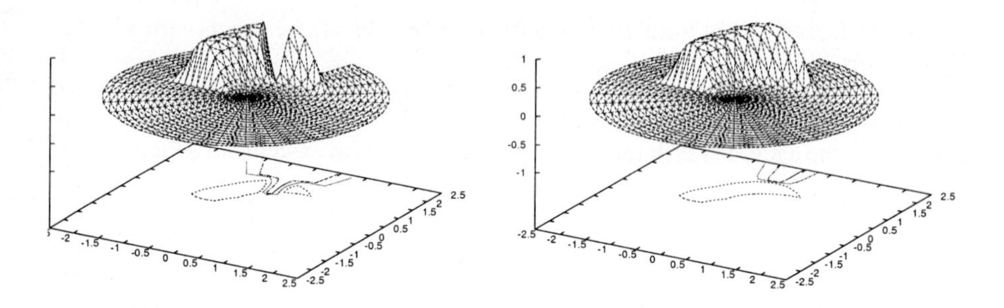

Fig. 3. A Goto map for a moving target and an Avoid map for a stationary obstacle compiled with the *MAV* method. **Left:** The obstacle is in front of the trajectory of the target, and actions in that direction are therefore forbidden. **Right:** The obstacle is behind the trajectory, and no actions are hence forbidden.

Sometimes we need to modify maps before combining them. For example, we might want to consider only actions that reach the target fast enough. Or, when avoiding an obstacle, we might want to allow those actions that cause a collision only after a considerable length of time. Modifications of this kind can be produced by filtering. Fig. 4 shows the result of filtering actions that require too long time to reach a target. Filtering can also be performed by utilizing a filter mask containing a zero for each action to be filtered and one for the rest of the actions.

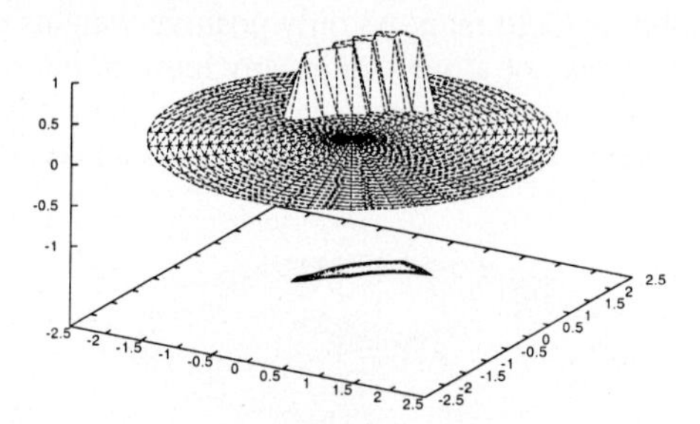

Fig. 4. Effect of filtering. The actions requiring more than 1.1 seconds to reach a target have been filtered from the Goto map shown in Fig. 1.

3 Soccer

Action maps have been utilized in playing soccer. As the control system controls the movements of both the player and the ball, the system produces two types of action maps. A velocity map specifies preferences for the different actions for reaching an object. An impulse map presents weights for all the possible different impulses (direction, magnitude) the player can give to the ball. It specifies the preferences for the different actions for kicking the ball to reach an object. Table 1 presents the primitive maps.

The opponent and team-mate maps are calculated for all opponents and team-mates whose locations are known. The GotoLoc map is utilized when a player is to kick a free kick, a goal kick, a corner kick, or a side kick. The GotoDefloc map controls a player to its default location. The KickNear and KickFar maps contain heavy weights for impulses that cause the ball to move nearer the opponents goal. The KickNear map favors small impulses, whereas the KickFar map favors big impulses.

Table 1: The primitive maps.

MAP TYPE	PRIMITIVE VELOCITY MAPS	PRIMITIVE IMPULSE MAPS
Goto	GotoBall, GotoTeam, GotoOpp, GotoLoc, GotoDefloc	KicktoGoal, KicktoTeam, KicktoOpp, KickNear, KickFar
Avoid	AvoidBall, AvoidTeam, AvoidOpp, AvoidLoc, AvoidDefloc	DontKicktoGoal, DontKicktoTeam, DontKicktoOpp, DontKickNear, DontKickFar

The primitive maps are filtered, as pointed out in Chapter 2: when avoiding opponents, the actions that require a considerable amount of time to reach the opponent are filtered from the Avoid maps. The ForbiddenSector mask is an example of a filter mask. This mask contains a zero for each impulse that would cause the ball to collide with the player kicking the ball. Fig. 5 shows an example of a forbidden sector mask.

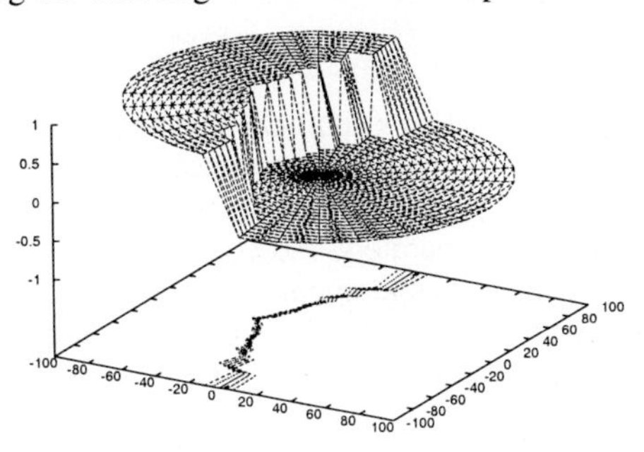

Fig. 5. A forbidden sector mask for a situation in which the ball is behind the agent and moving towards the agent. To prevent a collision, the agent should kick the ball backwards hard enough.

The composite maps are listed in Table 2. The MovetoSpecloc map controls the agent towards a good location (special location) when a goal kick, a free kick, a side kick, or a corner kick is to be performed. In this map, the AvoidBall map prevents only actions that would result in a collision with the ball in less than 1.5 seconds. Only one teammate and opponent map of a type is listed for each composite map, although all teammate and opponent maps are utilized. Each of the composite impulse maps is filtered with the ForbiddenSector mask.

Table 2: The composite maps.

MovetoSpecloc = MAV(GotoLoc, AvoidOpp, AvoidTeam, filter(AvoidBall, 1.5))
MakeGoal = filter(MAV(KicktoGoal, DontKicktoOpp, DontKicktoTeam)), ForbiddenSector)
Pass = filter(MAV(KicktoTeam, DontKicktoOpp), ForbiddenSector)
Dribble = filter(MAV(KicktoNear, DontKicktoTeam, DontKicktoOpp), ForbiddenSector)
CatchBall = MAV(GotoBall, AvoidOpp, AvoidTeam)
MovetoDefloc = MAV(GotoDefloc, AvoidOpp, AvoidTeam)

A task is performed by sending the action with the heaviest weight in the appropriate composite map to the actuators. The task to be executed at each moment is selected by an arbiter, which goes through the tasks in the following order: SpecialSituation, LookAround, MakeGoal, Pass, Dribble, CatchBall, MovetoDefloc, and TurntoBall. The arbiter selects the first task whose valid conditions are fulfilled. When a task has been selected, it remains active as long as its valid conditions are fulfilled. Among the tasks listed above, LookAround and TurntoBall are executed without utilizing action maps. The SpecialSituation task expands into a prioritized task set similar to the list above. This task set contains the MovetoSpecloc task listed in Table 2.

These task sets and the valid conditions of the tasks produce the following behavior: As long as a special situation (a free kick, goal kick, etc.) is observed, the agent performs the SpecialSituation task. Otherwise, if the location of the ball or the location of the agent is not certain the agent looks around until the certainty increases to an acceptable level. Otherwise, if the ball is near enough, the agent kicks it. It favors scoring a goal. If the weight of the best action to score a goal is too low, it considers passing the ball to a team-mate. If this cannot be done, either, the agent dribbles the ball. If the ball is not near enough to be kicked, but no team-mate is nearer the ball, the agent tries to catch the ball. Otherwise, if the agent is too far from its default location, it runs towards it. Otherwise, the agent turns towards the ball.

4 Discussion

When reactions to objects in the environment are calculated as action maps, the resulting system is situated. All reactions are grounded to the situation around the agent. Furthermore, action maps allow primitive tasks to be executed in parallel. Specifically, they allow tasks to be assessed separately and combined by a simple operation. This advantage facilitates incremental development.

The difference between this and the related work is in the amount of information encoded in a command. A larger amount of information encoded in action maps allows more versatile combination operations than the command representations in the related work. Specifically, by combining action maps, it is possible to meet primitive goals in parallel in a dynamic environment. Furthermore, action maps allow time provisions. Such rules as "Go after the ball if you can reach it in less than 2 seconds", or "Try to score a goal if the ball would reach the goal in less than 3 seconds" are easily implemented by operations on action maps.

We have tested action maps in the Soccer Server's simulated world at the RoboCup'98 in Paris. The experiments show that the action maps are a potential alternative for mobile agents operating in dynamic environments. Even when playing against the finalist, AT Humboldt, the game was quite even for long periods. This was due to our players' skill of moving to a good location and skill of selecting were to kick. However, our team did not win AT Humboldt, as our players were not good enough in dribbling, had only some simple strategies, and did not always behave in a stable way. Our goalkeeper also made some bad misses.

Soccer as an application has had a major impact on our research. The Soccer Server has facilitated performing a large amount of experiments in a highly dynamic environment. Soccer is also an interesting application, because it shows how difficult it is to decide on paper which situations are important in an application and how the agent should operate in those situations. For example, it seems to be important for a soccer player to avoid other players. Actually, the worst thing to do is to avoid an opponent approaching you. Furthermore, the skill of avoiding opponents has no noticeable correlation with the results. Due to modifications in the Soccer Server, we even run some experiments in which the opponent could not see our team. This was not obvious when observing the game. The opposing team played quite an effective game when they were not aware of our players. Further, another important skill in soccer seems to be the ability to stay on the field. However, our team plays a better game when they do not consider the boundary lines at all. As all the other players and the ball are on the field, reactions to them tend to keep also the reacting player on the field.

Maybe the most important lesson we have learnt for soccer is the improved insight about building situated agents. We have learnt that it is not necessary to be able to find in the problem space the optimal path (or even any path at all) to the goal state from every possible state. Instead, it suffices to find a feasible action that decreases the distance to the goal state. Action maps are one example: if the optimal action is prevented, an action guiding the agent closer to the goal location is selected. Another example is the behavior of our player when it is otherwise in a good location to score a goal, but the ball would bounce from the player itself if kicked towards the goal. One solution would be to calculate a sequence of kicks that would first move the ball around the player and then to the goal. Our player selects some other action in such a case. For example, if the player chooses to dribble the ball instead, it will probably be possible to score a goal a few moments later.

In other words, in a complex and dynamic environment the behavior satisfying the goals of the agent can emerge from a carefully selected set of primitive tasks. In addition to being situated, the resulting system is also considerably simpler compared to one planning optimal actions. Our players never plan ahead, they just calculate the next (turn, dash) action pair. An example of an emerging behavior is an attack performed by our team: there is no mechanism controlling the players towards the opponent's goal just because the team is attacking. Instead, the players favor the direction of the opponent's goal when kicking the ball. The location of the ball, in turn, changes the default locations of the players. As a result, when our team has the ball, they kick it towards the opponent's goal and move themselves to the same direction. In other words, they attack.

5 Conclusions

In this paper, we suggested a novel representation for agent actions. An action map specifies preferences for all the possible different actions. The central advantage of this representation is that it facilitates executing in parallel primitive tasks, i.e., tasks of reaching and avoiding objects. We have applied the action map representation to playing soccer and carried out experiments in the second RoboCup competition.

Acknowledgments

The research presented in this paper was supported by the Tauno Tönning Foundation.

References

[Agre and Chapman, 1990] Agre PE & Chapman D (1990) What are plans for? In: Maes (ed) Designing Autonomous Agents: Theory and Practice from Biology to Engineering and Back. MIT Press, Cambridge, MA, p 105-122.

[Arkin, 1990] Arkin RC (1990) Integrating behavioral, perceptual, and world knowledge in reactive navigation. In: Maes (ed) Designing Autonomous Agents: Theory and Practice from Biology to Engineering and Back. MIT Press, Cambridge, MA, p 17-34.

[Brooks, 1991] Brooks RA (1991) Intelligence without representation. Artificial Intelligence (47): 139-159.

[Kaelbling and Rosenschein] Kaelbling LP & Rosenschein SJ (1990) Action and planning in embedded agents. In: Maes (ed) Designing Autonomous Agents: Theory and Practice from Biology to Engineering and Back. MIT Press, Cambridge, MA, p 35-48.

[Rosenblatt and Payton, 1989] Rosenblatt JK & Payton DW (1989) A fine-grained alternative to the subsumption architecture for mobile robot control. Proc IEEE/INNS International Joint Conference on Neural Networks, Washington DC, 2: 317-324.

A Multi-level Constraint-Based Controller for the Dynamo98 Robot Soccer Team

Yu Zhang and Alan K. Mackworth

Laboratory for Computational Intelligence, Department of Computer Science,
University of British Columbia, Vancouver B.C. V6T 1Z4, Canada,
yzhang@cs.ubc.ca, mack@cs.ubc.ca

Abstract. Constraint Nets provide a semantic model for modeling hybrid dynamic systems. Controllers are embedded constraint solvers that solve constraints in real-time. A controller for our new softbot soccer team, UBC Dynamo98, has been modeled in Constraint Nets, and implemented in Java, using the Java Beans architecture. An evolutionary algorithm is designed and implemented to adjust the weights of constraints in the controller. The paper demonstrates that the formal Constraint Net approach is a practical tool for designing and implementing controllers for robots in multi-agent real-time environments.

1 Background and Introduction

Soccer as a task domain is sufficiently rich to support research integrating many branches of robotics and AI [3, 6]. To satisfy the need for a common environment, the Soccer Server was developed by Noda Itsuki [1] to make it possible to compare various algorithms for multi-agent systems. Because the physical abilities of the players are all identical in the server, individual and team strategies are the focus of comparison. The Soccer Server is used by many researchers and has been chosen as the official simulator for the RoboCup Simulation League [2].

Constraint Nets (CN), a semantic model for hybrid dynamic systems, can be used to develop a robotic system, analyze its behavior and understand its underlying physics [8–10]. CN is an abstraction and generalization of dataflow networks. Any (causal) system with discrete/continuous time, discrete/continuous (state) variables, and asynchronous/synchronous event structures can be modeled. Furthermore, a system can be modeled hierarchically using aggregation operators; the dynamics of the environment as well as the dynamics of the plant and the controller can be modeled individually and then integrated [7]. A controller for our new softbot soccer team, *UBC Dynamo98*, has been developed using CN.

The rest of the paper describes CN and how we use it to model and build the controller for our soccer-playing softbot *UBC Dynamo98*. Section 2 introduces the CN model of the controller for our soccer-playing softbot. Section 3 discusses constraint-based control and shows how the controller satisfies the constraints in the soccer domain. Section 4 shows our team's performance in RoboCup98. Section 5 concludes the paper.

2 The CN Architecture of the Controller for a Soccer-playing Softbot

2.1 Modeling in Constraint Nets

A constraint net consists of a finite set of locations, a finite set of transductions and a finite set of connections. Formally, a *constraint net* is a triple $CN = \langle Lc, Td, Cn \rangle$, where Lc is a finite set of *locations*, Td is a finite set of labels of *transductions*, each with an *output port* and a set of *input ports*, Cn is a set of *connections* between locations and ports. A location can be regarded as a wire, a channel, a variable, or a memory cell. Each transduction is a causal mapping from inputs to outputs over time, operating according to a certain reference time or activated by external events.

Semantically, a constraint net represents a set of equations, with locations as variables and transductions as functions. The *semantics* of the constraint net, with each location denoting a trace, is the least solution of the set of equations. For *trace* and some other basic concepts of dynamic systems, the reader is referred to [10].

Given CN, a constraint net model of a dynamic system, the abstract behavior of the system is the semantics of CN, denoted $[\![CN]\!]$, i.e., the set of input/output traces satisfying the model.

A complex system is generally composed of multiple components. A *module* is a constraint net with a set of locations as its interface. A constraint net can be composed hierarchically using modular and aggregation operators on modules. The semantics of a system can be obtained hierarchically from the semantics of its subsystems and their connections.

A control system is modeled as a module that may be further decomposed into a hierarchy of interactive modules. The higher levels are typically composed of event-driven transductions and the lower levels are typically analog control components. The bottom level sends control signals to various effectors, and at the same time, senses the state of sensors. Control signals flow down and state signals flow up. Sensing signals from the environment are distributed over levels. Each level is a grey box that represents the causal relationship between the inputs and the outputs. The inputs consist of the control signals from the higher level, the sensing signals from the environment and the current states from the lower level. The outputs consist of the control signals to the lower level and the current states to the higher level.

2.2 The CN Architecture of the Controller

The soccer-playing softbot system is modeled as an integration of the soccer server and the controller (Fig. 1). The soccer server provides 22 soccer-playing softbots' plants and the ball. Each softbot can be controlled by setting its throttle and steering. When the softbot is near the ball (within 2 meters), it can use the kick command to control the ball's movement. For the controller for one of the soccer-playing softbots, the rest of the players on the field and the ball are considered as its environment. The sensor of the controller determines the state of the plant (position and direction) by inference from a set of landmarks it 'sees'. The rest of the controller computes the desired control inputs (throttle and steering) and sends them to the soccer server to actuate the plant to move around on the field or kick the ball.

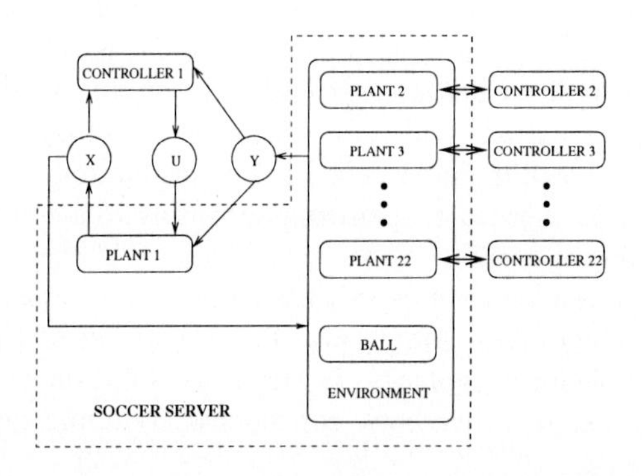

Fig. 1. The soccer-playing softbot system

For the soccer-playing softbot, we have designed the three-level controller shown in Fig. 2. The lowest level is the Effector&Sensor. It receives ASCII sensor information from the soccer server and translates it into the World model. It also passes commands from the upper level down to the soccer server. The middle level is the Executor. It tries to translate the action which comes from the upper level into a sequence of commands and sends them to the lowest level. The Executor also evaluates the situation and sends its evaluation up to the Planner. The highest level is the Planner. It decides which action to take based on the current situation and it may also consider the next action assuming the current action will be correctly finished on schedule.

The controller is composed of four CN modules. The Effector module combines with the Sensor module to form the lowest level Effector&Sensor. The Executor module forms the middle level and the Planner module forms the highest level (Fig. 2).

The controller is written in Java [4]. The Java Beans component architecture [5] is used here to implement the CN modules. Events are one of the core features of the Java Beans architecture. Conceptually, events are a mechanism for propagating state notifications between a *source* object and one or more target *listener* objects. Under the new AWT event model, an event listener object can be registered with an event source. When the event source detects that

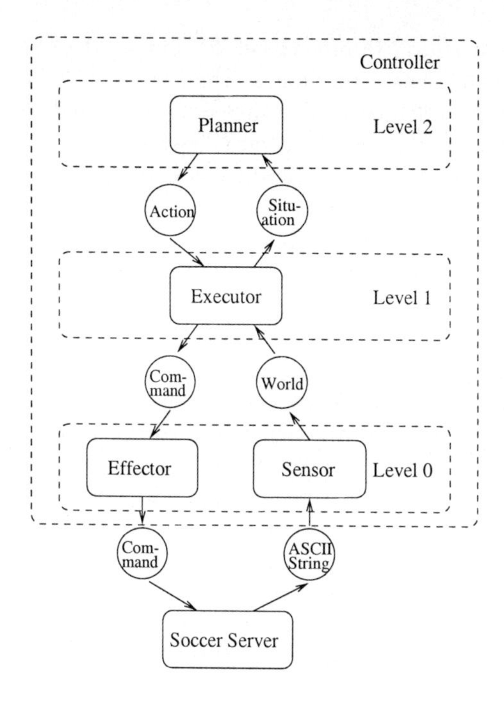

Fig. 2. The soccer-playing controller hierarchy

something interesting has happened it calls an appropriate method in the event listener object.

CN model is a data-flow model; each CN module can be run concurrently on different processors to improve the speed of the controller. Since these modules are event-driven and fixed-sample-time-driven, they are best implemented as Java threads to improve efficiency on a single CPU too. If no event arrives, they go to sleep so the CPU can deal with other softbots. In such a multi-threaded environment where several different threads may be simultaneously delivering events and/or calling methods and/or processing event objects and/or setting properties, special considerations are needed to make sure these beans properly coordinate their behaviour, using wait/notify and synchronization mechanisms.

The Sensor module wakes up when new information arrives. It then processes the ASCII information from the soccer server, updates the world model, and sends an event to the Executor. The Sensor goes to sleep when there is no information waiting on its socket.

The Executor module receives the event from the Sensor, then it processes the world model and updates the situation states. These situation states tell the Planner if it can kick the ball, if the ball is in its sight, if it is the nearest player to the ball, if there are obstacles on its way, whether the action from the Planner has finished or not, and so on. Any change of situation creates an event and triggers the higher level Planner module. This part of the Executor runs in the same thread as the Sensor module.

The main part of the Executor executes actions passed down from the Planner. It wakes up when it receives an action event from the Planner module. It produces a sequence of commands which are supposed to achieve goals (actions) when they are performed. Some of these commands are sent to the Effector's *Movement_command* buffer. Other commands are sent to the Effector's *Sensing_command* buffers, they are *Say_message* buffer, *Change_view* buffer, and *Sense_body* buffer. The Executor goes to sleep when there is no action waiting for its processing.

The Planner module wakes up when triggered by a situation-changed event from the Executor. It then produces actions and pushes them into Executor's action buffer and sends an event to trigger the Executor to execute actions. Then it goes to sleep until a new event comes.

The Effector module is a fixed-sample-time-driven module. Every 100ms, it gets one command from each non-empty buffer and sends them to the soccer server.

3 Constraint-Based Control for Soccer-playing Softbot

Constraints are considered to be relations on a set of state variables; the solution set of the constraints consists of the state variable tuples that satisfy all the constraints. The behavior of a dynamic system is constraint-based if the system is asymptotically stable at the solution set of the given constraints, i.e., whenever the system diverges because of some disturbance, it will eventually return to the set satisfying the constraints. Most robotic systems are constraint-based, where the constraints may include physical limitations, environmental restrictions, and safety and goal requirements. Most learning and adaptive dynamic systems exhibit some forms of constraint-based behaviors as well [8].

A controller is an *embedded constraint solver* if the controller, together with the plant and the environment, satisfies the given constraint-based specification. In the CN framework for control synthesis, constraints are specified at different levels on different domains, with the higher levels more abstract and the lower levels more plant-dependent. A control system can also be synthesized as a hierarchy of interactive embedded constraint solvers. Each abstraction level solves constraints on its state space and produces the input to the lower level. Typically the higher levels are composed of digital/symbolic event-driven control derived from discrete constraint methods and the lower levels embody analog control based on continuous constraint methods [7].

The Executor module can be seen as an embedded constraint solver on its world state space. It solves the constraint-based requirements passed down from the higher layer Planner module. For example, if the action from the Planner is to intercept the ball at (x_b, y_b, vx_b, vy_b), and the state variables of the robot soccer player are (x_p, y_p, vx_p, vy_p), the constraints are $x_p + vx_p * t = x_b + vx_b * t$ and $y_p + vy_p * t = y_b + vy_b * t$.

The Planner module can be seen as an embedded constraint solver on its situation state space. The ultimate constraint here is: the number of goals scored by its team should be more than its opponent's. To satisfy this ultimate constraint, the robot has to satisfy a series of other constraints first.

These constraints have their priorities. The constraints with higher priority must be solved earlier. The constraint of knowing its position and the ball's should be solved first. Then the robot will try to solve the constraints of collision and offside. In order to win, the robot will consider some other constraints, such as, its own team's time in possession of the ball should be longer than its opponent's team, the ball should be near enough to the opponent's goal, the ball should be as far away as possible from its own goal, and the ball should be kicked into opponent's goal instead of its own goal.

It chooses actions to satisfy the constraints at this level. When the robot loses its own position or the ball's position for a certain amount of time, it sends $find\_me$ or $find\_ball$ actions down to the Executor. When the robot senses that it will collide with other players, it sends *avoid_collision* action down to the Executor. It also sends down $avoid\_offside$ down to the Executor if it finds itself is at offside position. The robot tries to *intercept* the ball if it senses that it is nearer to the ball than its teammates, if not, it goes to a certain position to *assist* its teammate's interception. If the robot gets the ball, it has to choose where to kick it. The action here should best satisfy the constraints listed above. The problem is that sometimes the robot can't find a kick direction that satisfy all the constraints. For example, if the robot chooses the kick direction which can make sure that its teammates can get the ball, the ball might be kicked away from its opponent's goal and near its own goal. We solve this by combining these constraints into one utility constraint. This combined utility constraint is to maximize the *utility function*:

$$U(o) = \sum_i k_i * P_i(o) \tag{1}$$

$U(o)$ is the action o's utility. $P_i(o)$ is the probability of satisfying the constraint i when taking the action o. k_i is the weight for the constraint i. The constraint solver for this combined utility constraint will output the action o with the highest utility. These weights can be set by hand. They can also be tuned by a learning method, such as reinforcement learning. Also the utility function $U(a)$ need not be *linear*; it might be obtained by using neural network learning.

We also designed a coach program using an evolutionary algorithm to adjust the weights of constraints and other parameters in the controller. The coach maintains a *population* of *individuals*. Each *individual* consists of a pair of *chromosomes*. A *chromosome* is an array of parameters, which we call *genes*. Thus, each *individual* has two copies of each *gene* as a consequence of *biparental inheritance*. The sum of each pair of *genes* determines one parameter in the robot. The coach selects the fittest *individuals* as parents via a tournament, performs *crossover* and *mutation* on parent's *chromosomes*, then passes them down to their children. We believe this kind of simulation of *natural selection* will evolve a very good robot team if enough time and supervision are given.

The robots also communicate with each other to share information and to coordinate their actions among them. For example, if one robot comes near the ball, it says "my ball" to its teammates, the teammate who gets the message will send back "kick here" if it is in a good receiving position or go away from the ball if it is also near the ball.

4 Results

To compare our approach with other teams' that differ in models, architectures and control methods, we took part in the World RoboCup98 which was held on July 4-8, 1998 in Paris, France. The first game we played against **NIT Stones 98**. The opponent team had an interesting strategy with many of its players swarming around the ball and kicking the ball forward. We won this game, with a score of 4:1. We played against **Mainz Rolling Brains** in the second game. This team's strategy was to move the *full-backs* up in an offside trap to push the opponents' *forwards* back. But its *forwards* didn't try to avoid offside positions, they just kept their positions near the opponent's goal. This strategy was used by many teams in World RoboCup98. We drew this game, the score was 0:0. Our team's advantage is that our players can sense if they are at offside positions, and if they are, they can try to avoid that situation by moving towards their own side. Our players' low level skills like kicking backwards were not as good as those of the opponent's team. Lots of shots by the opponents were saved because their *forwards* were offside. Our players advanced near the opponent's goal many times, but their shots lacked adequate strength to score. We played against **CAT-Finland** in the third game. This team's original strategy was to keep its *full-backs* near its own goal and its *forwards* near the opponent's goal. It's a fixed position strategy and it was also used by many teams in World RoboCup98. When **CAT-Finland** competed with **Mainz Rolling Brains**, the disadvantage of their strategy was shown in the score 0:4. When **CAT-Finland** played against our team, they changed their strategy to that used by **Mainz Rolling Brains**. Some teams belonging to this category also changed their strategy later as **CAT-Finland** did. We lost this game; the score was 0:1. Lots of shots by **CAT-Finland** were also saved because their *forwards* were offside. At one point, one of our *full-backs* slowed down to keep energy, so **CAT-Finland**'s *forwards* got an chance to shoot. Our goalie missed the ball.

So our team won one game, drew one game and lost one game in World RoboCup98. Although we lost the game, we don't think our team is worse than **CAT-Finland**. We know there are many random factors in the soccer server and network communication between the server and clients is not stable either. Winning was not our purpose. Our team was successful in the World RoboCup98 from a research point of view. It shows that constraint-based control and evolutionary algorithms are effective methods in multi-agent real-time robot design. It also shows that Java is fast enough to compete in a traditional C++ world.

5 Summary and Conclusions

Constraint Nets (CN), a semantic model for hybrid dynamic systems, can be used to develop a robotic system, analyze its behavior and understand its underlying physics.

The soccer-playing softbot system is modeled as an integration of the soccer server and the controller. The three-level controller is composed of four modules. The Effector module combines with the Sensor module to form the lowest level Effector&Sensor. The Executor module forms the middle level and the Planner module forms the highest level. The controller is written in Java. The Java Beans component architecture is used here to implement the CN modules and we use the Java event mechanism to implement communication among these CN modules. They are implemented in Java threads to improve efficiency.

The controller for soccer-playing softbot is synthesized as a hierarchy of interactive embedded constraint solvers. Each level solves constraints on its state space and produces the input to the lower level. We have also designed a coach program using an evolutionary algorithm to adjust the weights of constraints and other parameters in the controller.

In short, we have demonstrated that the CN model is a formal and practical tool for designing and implementing, in Java, constraint-based controllers for robots in multi-agent, real-time environments.

References

1. Noda Itsuki. Soccer Server System. Available at http: //ci.etl.go.jp/ noda /soccer /server.html.
2. Hiroaki Kitano. Robocup. Available at http: //www.robocup.org /RoboCup /New /index.html.
3. A. K. Mackworth. On seeing robots. In A. Basu and X. Li, editors, *Computer Vision: Systems, Theory, and Applications*, pages 1–13. World Scientific Press, Singapore, 1993.
4. Sun Microsystems. Java. Available at http: //java.sun.com/.
5. Sun Microsystems. Java Beans. Available at http: //java.sun.com/ beans/ index.html.
6. M. Sahota and A. K. Mackworth. Can situated robots play soccer? In *Proc. Artificial Intelligence 94*, pages 249 – 254, Banff, Alberta, May 1994.
7. Ying Zhang and A. K. Mackworth. Synthesis of hybrid constraint-based controllers. In P. Antsaklis, W. Kohn, A. Nerode, and S. Sastry, editors, *Hybrid Systems II*, Lecture Notes in Computer Science 999, pages 552 – 567. Springer Verlag, 1995.
8. Ying Zhang and A. K. Mackworth. Constraint Programming in Constraint Nets. Principles and Practice of Constraint Programming, MIT Press, 1995, p.49–68.
9. Ying Zhang and A. K. Mackworth. Constraint Nets: A Semantic Model for Hybrid Dynamic Systems. Journal of Theoretical Computer Science, Vol. 138, No. 1, 1995, p.211–239, Special Issue on Hybrid Systems.
10. Ying Zhang. A foundation for the design and analysis of robotic systems and behaviors. Technical Report 94-26, Department of Computer Science, University of British Columbia, 1994. Ph.D. thesis.

The Small League RoboCup Team of the VUB AI-Lab

Andreas Birk, Thomas Walle, Tony Belpaeme, Johan Parent, Tom De Vlaminck and Holger Kenn

Vrije Universiteit Brussel, Artificial Intelligence Laboratory,
Pleinlaan 2, 1050 Brussels, Belgium,
cyrano@arti.vub.ac.be, http://arti.vub.ac.be/~cyrano

Abstract. The paper describes the VUB AI-Lab team competing in the small robots league of RoboCup '98 in Paris. The approach of this team targets for a longterm evolution of different robots, team-structures, and concepts. Therefore, the efforts for the '98 participation focus on the provision of a flexible architecture, which forms a basis for this goal. In doing so, the development of the so-called RoboCube constitutes a milestone on this road. The RoboCube is an extremely compact robot controller providing quite some computation power, memory, various I/O interfaces, and radio communication. It facilitates the use of many sensors and effectors, including their on-board processing, allowing to explore a large space of different robots and team set-ups. Accordingly, the '98 VUB AI-Lab team consists of heterogeneous robots. The paper describes their important electromechanical features like drive units and shooting mechanisms. In addition, the radio communication, the basic control algorithms and the global vision system are described. Last but not least, first steps towards a general coordination scheme are explained, which is meant to cope with changing compositions of heterogeneous groups.

1 Introduction

Research at the VUB AI-Lab is strongly influenced by the ideas of behavior-oriented AI [Ste94a]. In contrast to "classic" AI, which relies on symbolic approaches and extensive top-down modeling, techniques and experimental set-ups are oriented in a bottom-up manner [Bro91]. In doing so, there are some strong relations with the field of Artificial Life, or short Alife [Lan89], which tries to investigate the basic properties of Life in a constructive manner. The animat (= animal + robot) approach to AI [Wil91] reflects this mutual influence.

The robotics research at the VUB AI-Lab so far centered around the so-called ecosystem [Ste94b, McF94], a set-up where several heterogeneous robots [BB98] engage in complex interactions. In this ecosystem several different subjects are investigated like e.g. autonomy and self-sufficiency [Ste94b, McF94, Bir97a], co-operation, and learning [Ste96b, Ste96a, Bir97b]. RoboCup is for us a supplementary set-up to investigate these subjects, making a direct comparison with other scientific approaches possible. In doing so, we do not see RoboCup as a single event, but as a long-term guideline for concrete experimental set-ups where different concepts and implementations are evaluated on an annual basis. This long-term process will hopefully lead to a co-evolution of robots, concepts, and teams; fostering new scientific insights.

The body-aspects of the actual robots play a significant role within this process. Much like in real soccer, speed, strength, and technical skills are the most basic building-blocks for any kind of successful overall performance. Therefore, we decided that a first step in our RoboCup engagement should focus on the development of a flexible robot architecture, which allows for an easy exploration of different forms of physical interaction between the robot, the ball and its environment, without the need of major re-engineering. In doing so, we chose for the small robots league for following reasons. First, we believe that important scientific issues can be investigated within the small robots league in the following years. Especially hardware for vision as well as substantial computing power and memory can be provided on-board within the size constraints of this league. Second, the needs in respect to infrastructure are the least in the small robots league. The playfield is a simple ping-pong table requiring only few space; this is a major factor especially in academic environments where space is often hard to allocate. In addition, the expenses for small league robots are less as costs increase substantially with increasing robot size due to additional mechanical and electromechanical needs.

The so-called RoboCube is cornerstone in our approach towards a flexible robot architecture, which allows for investigation of different scientific concepts without major re-engineering of the underlying platforms. The RoboCube is an extremely compact robot control computer supporting a substantial amount of various sensors and motors, and providing quite some computation power and memory.

2 The RoboCube

The VUB AI-lab has quite some tradition in developing flexible robot control hardware. Various experimental platforms have been build in the lab starting from the mid-eighties up until now. They were used for different purposes, but the common need for flexibility lead to the *Sensor-Motor-Brick II (SMBII)*[Ver96], based on a commercial board providing the computational core with a Motorola MC68332.

RoboCube is an enhanced successor of the SMBII. In RoboCube the commercial computational core is replaced by an own design, also based on the MC68332, which saves significant costs. Furthermore, the architecture is simplified and additional interfaces are provided. In addition, the physical shape of RoboCube is quite different from the one of the SMBII. First, board-area is minimized by using SMD-components in RoboCube. Second, three boards are stacked on each other leading to a more cubic design compared to the flat but long shape of the SMBII, hence its name Robo*Cube*.

RoboCube has a open bus architecture which allows to add “infinitely” many sensor/motor-interfaces (at the price of bandwidth). But for most applications,

Fig. 1. A picture of the RoboCube

including playing soccer, the standard set of interfaces should be more than enough. RoboCube's basic set of ports consists of 24 analog/digital (A/D) converters, 6 digital/analog (D/A) converters, 16 binary Input/Output (binI/O), 5 binary Inputs, 10 timer channels (TPC), 3 DC-motor controller with pulse-accumulation (PAC), and the option of an intelligent active InfraRed (aIR) subsystem.

An important aspect is that RoboCube's on-board software core supports an easy handling of these ports in combination with common sensors and motors. This means that it is possible to plug-and-play various components, which can be accessed in a high-level manner in software. RoboCube runs the *Process Description Language (PDL)* which combines C with special constructs facilitating behavioral control through networks of dynamical processes.

RoboCube interfaces to an UHF transceiver for medium bandwidth data transfer. The transceiver allows for 40 kBit/s and is extremely small (4cm $\times$ 3cm). During the Paris '98 competition, the radio-communication was a major problem for us. In addition to the normal technical difficulties, we had to face malicious jamming. To resist against simple attack schemes like record-and-playback of data-packets, it is recommended to use suited protocols and encoding.

A detailed description of the RoboCube can be found in [BKW98].

3 The Robot Bases

The drive units of our robots are based on high-quality MAXON^{TM} motor-units, which provide a maximum of power, efficiency, and precision with a minimum of space requirements. The motor-units consist of a 3 Watt DC motor, a 16.58:1 gear, and a 16ppr shaft encoder. The encapsulated units have a diameter of 13mm and a length of 62.15 mm.

For one type of basic players, the drive units are mounted on round plexiglas disks with 150mm diameter and a shaped front for ball manipulation. An additional gear stage is added between the motor-units and the wheels. This has two advantages. First, it allows to explore different final gear ratios in an easy way. Second, mounting the wheels immediately on the motor units would require shorter and hence less powerful motors.

This type of basic players is equipped with a simple shooting mechanism. The mechanism is a linear accelerator based on a rack with pinion. The shooting behavior is controlled in two phases. First, there is an activation of the complete mechanisms, indicating the intention to shoot the ball. Second, there is an actual trigger by a reflective IR-switch indicating the presence of the ball.

In addition to the extremely solid set-ups for one type of basic players, the drive motor-units can be interfaced to LEGO^{TM} parts. This allows fast prototyping and is especially useful to explore different robot shapes and feature combination within the size regulation of the competition. This approach was used to build block-like players without a shooting option.

4 Overhead Vision and Strategies

The position, direction and speed of all actors on the field are tracked by an overhead vision system. A color camera, mounted over the playing field, connects to an off-board PC with a frame grabber. The visual processing algorithms rely on color segmentation to track all objects on the field. The segmentation consists of a global search on hue and saturation of orange, yellow and blue. To speed the algorithm up, a look-up table for the RGB to HSI-values is used. Next, all consistent pixels are grouped into blobs, and a local search is used to locate all actors on the field, using the previous position of all actors. Additionally, some ad-hoc knowledge is used to make the tracking more reliable. The resulting data is passed on to the strategy algorithms at a rate of approximately 7 fps.

The strategies used by the players reside at different levels, with a large amount being completely computed on the robots. On-board the robots, several processes run in simulated parallel. A first class of processes takes care of reactive, task-oriented behaviors, such as shooting and obstacle avoidance. A second, non-reactive class of processes relies on the robots shaft encoders to carry out strategic commands, given by a supervisor in form of a coaching agent residing on an off-field PC. In order to get feedback on its position on the field, each player relies not only on its shaft encoders, but it also gets continuous feedback from the overhead vision. In doing so, the overhead vision is used for large scale positioning, while the shaft encoders are used for quick and local position corrections, e.g. during positioning for the ball.

The complete set of robots for the VUB team consists of ten robots. Four are of the brick-like, defensive type ($T1$) and six are of the agile, offensive type ($T2$). Therefore, there are quite some possible combinations to form different heterogeneous teams with five players.

Coordination of group behavior is especially hard in this case as a "hard-wiring" of reactions to possible situations would require an enormous work due to the combinatorial explosion. Therefore, we introduced the concept of *dynamic formations*, a mixture of distributed and central control. The coaching agent is not seen as an absolute master, who is commanding the players. Instead, the coach gives recommendations, which the players try to execute as well as possible. In doing so, players have roles, which are determined by their capabilities and their current position in a strategic space, here simply their positions on the field. If a certain player B happens to fit a role R better than a player A, which previously was assigned role R, then A and B can switch their roles.

The roles for the '98 team are *goalie (G)*, *defender (D)*, *mid-fielder (M)*, and *forward (F)* with role G, D restricted to robot type $T1$ and M, F restricted to $T2$. Possible team structures are for example $G/D/M/M/F$, $G/D/M/F/F$, $G/D/D/M/F$, and $G/D/D/F/F$. At RoboCup'98 in Paris, we were not able to demonstrate the capability to qualitatively change the team-structure as we only played with the $G/D/D/M/F$ set-up. In doing so, only role switches between *mid-fielder* and *forward* took place.

5 Conclusion

The paper gives a short overview of the VUB AI-Lab team competing in the small robots league of RoboCup '98 in Paris. In doing so, the technical aspects of the flexible robot architecture, which is the basis for the team's longterm goals, is described in some detail. The RoboCube is presented, which is an extremely compact robot controller providing quite some computation power, memory, various I/O interfaces, and radio communication. It facilitates the use of many sensors and effectors, including their on-board processing, which allows to explore a large space of different robots and team set-ups. Substantial amounts of the computation was done on-board our robots in the Paris '98 competition, demonstrating RoboCube's capabilities.

Acknowledgments

The VUB AI-Lab team thanks Sanders Birnie BV as supplier and Maxon Motors as manufacturer for sponsoring our motor-units. Also many thanks to Dany Vereertbrugghen who developed the SMBII. Research on the RoboCube is partially funded with the TMR-grant "Development of a universal architecture for mobile robots" (ERB4001GT965154). Parts of the robotics-research in the VUB

AI-lab are used for grounding within the GOA2-project "Origins of Language and Meaning" (OZR/96/2156). Tony Belpaeme is a Research Assistant of the Fund for Scientific Research - Flanders (Belgium) (F.W.O.).

References

[BB98] Andreas Birk and Tony Belpaeme. A multi-agent-system based on heterogeneous robots. In *Collective Robotics Workshop 98*. Springer LNAI, 1998.

[Bir97a] Andreas Birk. Autonomous recharging of mobile robots. In *Proceedings of the 30th International Symposium on Automative Technology and Automation*, 1997.

[Bir97b] Andreas Birk. Robot learning and self-sufficiency: What the energy-level can tell us about a robot's performance. In *to appear: Proceedings of the Sixth European Workshop on Learning Robots*, 1997.

[BKW98] Andreas Birk, Holger Kenn, and Thomas Walle. Robocube: an "universal" "special-purpose" hardware for the robocup small robots league. In *4th International Symposium on Distributed Autonomous Robotic Systems*. Springer, 1998.

[Bro91] Rodney Brooks. Intelligence without reason. In *Proc. of IJCAI-91*. Morgan Kaufmann, San Mateo, 1991.

[Lan89] Chris G. Langton, editor. *Artificial Life*. Addison-Wesley, Redwood City, 1989.

[McF94] David McFarland. Towards robot cooperation. In Dave Cliff, Philip Husbands, Jean-Arcady Meyer, and Stewart W. Wilson, editors, *From Animals to Animats 3. Proc. of the Third International Conference on Simulation of Adaptive Behavior*. The MIT Press/Bradford Books, Cambridge, 1994.

[Ste94a] Luc Steels. The artificial life roots of artificial intelligence. *Artificial Life Journal, Vol 1,1*, 1994.

[Ste94b] Luc Steels. A case study in the behavior-oriented design of autonomous agents. In Dave Cliff, Philip Husbands, Jean-Arcady Meyer, and Stewart W. Wilson, editors, *From Animals to Animats 3. Proc. of the Third International Conference on Simulation of Adaptive Behavior*. The MIT Press/Bradford Books, Cambridge, 1994.

[Ste96a] Luc Steels. Discovering the competitors. *Journal of Adaptive Behavior 4(2)*, 1996.

[Ste96b] Luc Steels. A selectionist mechanism for autonomous behavior acquisition. *Journal of Robotics and Autonomous Systems 16*, 1996.

[Ver96] Dany Vereertbrugghen. Design and implementation of a second generation sensor-motor control unit for mobile robots. Technical Report Thesis, Tweede Licentie Toegepaste Informatica, Vrije Universiteit Brussel, AI-lab, 1996.

[Wil91] Stewart W. Wilson. The animat path to ai. In *From Animals to Animats. Proc. of the First International Conference on Simulation of Adaptive Behavior*. The MIT Press/Bradford Books, Cambridge, 1991.

CIIPS Glory
Soccer Robots with Local Intelligence

Thomas Bräunl
Dept. of Electrical and Electronic Engineering
The University of Western Australia
Nedlands, Perth, WA 6907
www.ee.uwa.edu.au/~braunl

Abstract. Our team was named after the new and successful *Perth Glory* soccer team. The heart of our robots are the *EyeBot* controllers, which we developed form scratch. We use a Motorola 68332 32-bit controller, which offers a variety of digital/analog I/O facilities. We developed our own operating system *RoBIOS* for these systems, which allows a great deal of flexibility. All image processing and planning is done locally on-board the *EyeBot*. We do not use any global sensor systems. The same *EyeBot* controller is also used for 6-legged and biped walking machines, and – as a boxed version – for undergraduate courses on assembly language programming.

1 Introduction

Our system architecture uses local intelligence without any global sensors. Although our approach is clearly disadvantaged with respect to performing well at a *RoboCup* competition, we do believe that our approach of truly autonomous and locally intelligent systems does make more sense for less restrictive applications. We are more interested in research on general purpose intelligent agents, as opposed to building a system which can serve only in a certain competition, has to rely on global sensors, and reduces mobile robots to remote controlled toy cars. We incorporated a digital color camera and a graphics display to our microcontroller system. All image processing is performed on-board.

Fig 1. *EyeBot* soccer robot with local intelligence

After our experience in the AAAI *Mobile Robot Competition* [2] with large industrial robots and equipment failure, we decided to develop completely new mini-robots to enter the small robot league of RoboCup. Each robot is driven by two DC motors in differential steering and is equipped with two shaft encoders, five infrared range sensors, and a digital color camera. Two PWM servos enable tilting of the camera and activation of the ball kicker bar. A wireless communication unit is currently being developed to allow the robots to talk to each other. We are able to operate without communication between the robots. Each robot is told its starting position and uses its shaft encoders plus infrared sensors to keep track of its current position. This limits our soccer strategies to finding the ball and heading for the goal. Although theoretically possible without robot communication, we will wait with more sophisticated behaviours like passing a ball to another robot, until wireless communication is in place. Since our processing unit is a relatively simple on-board microcontroller, there are obvious limitations on what can be done in terms of image processing in real time.

We are currently operating on images of size 60×80, which we found have sufficient resolution for detecting objects (ball, goals, players, walls) and navigating towards/away from them. The controller is powerful enough to perform real-time on-line image processing, depending on the complexity of the operation.

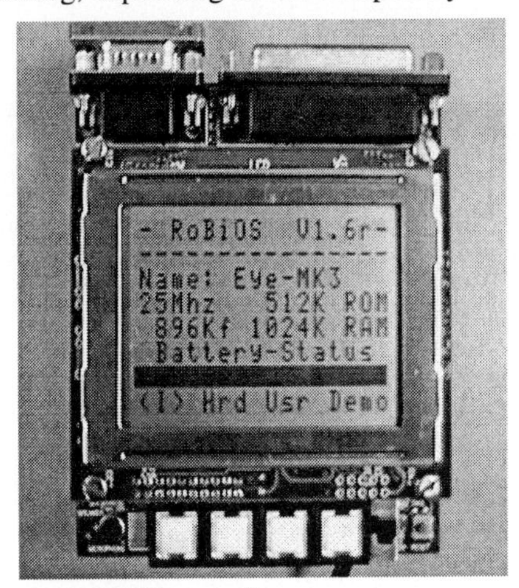

Fig 2. *EyeBot MK3* controller

2 Hardware

Our extendable *EyeBot* base platform is a very compact board (about 9cm × 10 cm), which was developed around the key requirements of image processing. It therefore features a digital camera and an LCD graphics display. It is based on a Motorola 68332 microcontroller [4] and therefore also provides a sufficient number of I/O ports for the connection of various sensors and actuators or any future extensions. While most robot vision systems are either tethered or remote-controlled [1], on-board real-time vision is feasible for large and expensive mobile platforms. Although it seemed very difficult to implement real-time vision on a small and inexpensive system, *EyeBot* accomplishes this goal. *EyeBot* has been successfully used in the construction of a wheel-driven vehicle, a 6-legged walking machine, and two biped walker robots. It is currently considered for the project of a flying robot.

The controller runs at a moderate speed (25 MHz), but it is fast enough to compute basic image operations on a low resolution image in real time. E.g. gray image acquisition, Sobel edge detection and display on the LCD for a 80×60 image can be performed at a rate of more than 10 frames per second. *EyeBot*'s graphics LCD is essential for interaction between the robot and the programmer. One needs to see the robot's view in order to set camera parameters and orientation. Although the camera provides gray scale or color images at medium resolution, the display can only show low resolution black/white images. This is sufficient as a feedback to the programmer when running the robot, but not for program development, which is done on a workstation using the *Improv* tool.

While the hardware was started at Univ. Stuttgart, software design is an ongoing international joint research project between Univ. Stuttgart, Univ. Kaiserslautern (Germany), Rochester Institute of Technology (USA), and The Univ. of Western Australia.

3 Software

The operating system *RoBIOS* (**ro**bot **b**asic **i**nput **o**utput **s**ystem) has been implemented in C plus *m68k* assembly language, after adapting a version of the *gnu* C compiler and assembler tools for the *EyeBot*. This allows program development in a high level language, using assembly routines for time-critical passages [3]. *RoBIOS* comprises a small real time system with multi-tasking scheduler (essential for all robotics applications), libraries for various I/O functions (Table 1), and a number of demonstration applications. The microcontroller's timing processor unit (TPU) is being used for servo control with pulse width modulation (PWM), for sound synthesis and playback, as well as the control of infrared distance sensors.

KEY	key input, e.g. C routine getchar()
LCD	screen output, text and graphics
CAM	camera routines for grayscale and color
OS	operation system specific functions
MT	multi-tasking functions
SEM	semaphore operations
TIME	timer functions
RS232	serial line functions, e.g. prog. download
AUDIO	recording and playing back sounds
PSD	distance sensor routines
IR	infrared sensor readings
SONAR	sonar distance measurement
BUMP	acoustic collision detection
SERVO	servo positioning routines
MOTOR	servo motor control routines
LATCH	digital input/output and analog input
IMG	basic image processing routines

Table 1. *RoBIOS* function groups

The C low level text input and output routines have been adapted for *EyeBot*. This enables us to use the standard C I/O library *clib* together with the *EyeBot* system for user application programs. E.g., a user can call getchar(), in order to read a key input and use printf(..), in order to write text on the screen.

Special care has been taken to keep the *RoBIOS* operating system flexible among several different hardware configurations, because the same system is to be used for wheeled robots *and* for legged robots. Therefore, a hardware description table has been included into the system design, as described in the following. The *EyeBot* operating system *RoBIOS* relies on the **h**ardware **d**escription **t**able HDT, in order to find out which hardware components are currently connected to the system. These hardware components can be sensors or actuators (motors or servos), whose control routines are already available in the general *RoBIOS* system. HDT allows easy detection, initialization, and use of hardware components.

4 Tools

We developed a number of tools to facilitate robot programming. This is especially essential in the case of our mini-robots for which programs have to be cross-compiled and downloaded from a workstation.

4.1 Improv

Improv as a tool for designing the image processing part of a robot control program. It is an application running under Linux/X windows, using *EyeBot*'s digital camera. *Improv* displays the camera image at a higher resolution in a real time, together with five additional windows, representing user defined image processing stages. The *Improv* library comprises a number of low level image processing routines, while additional user defined operators can be added easily. The idea is to use *Improv* to design, test, and debug the robot's image processing component on a PC, until it has reached a stage where it can be tested on the vehicle. Then, the code needs to be recompiled for the microcontroller and downloaded to the robot.

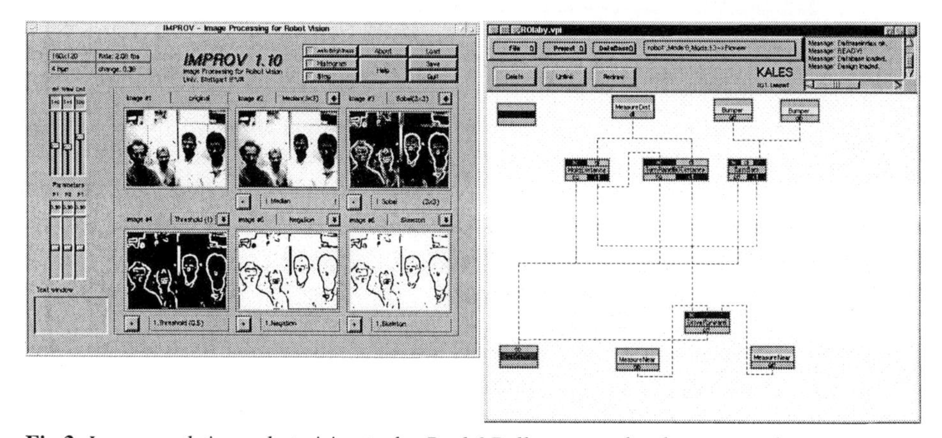

Fig 3. *Improv* real-time robot vision tool – *Rock&Roll* program development tool

4.2 Rock&Roll

On top of the operating system, we developed the integrated tool *Rock&Roll* (**ro**bot **c**onstruction **k**it and **ro**bot **l**ocomotion **l**ink) [5]. This system allows a "click-and-connect" construction of robot control structures. In its data flow model, sensors are sources and actuators are sinks, both representing system-defined module boxes. User-defined control boxes can be added, together with interconnection links between all modules, representing data flow.

4.3 EyeSim

The ***EyeBot*** **sim**ulator *EyeSim* is a valuable tool for the development of robot control programs. The simulator is actually implemented as a library with identical interface to the *RoBIOS* functions. That way a program can be compiled either way, for simulation or the real robot, without a change. Complex robot routines can be debugged and tested much more efficiently. Input/output routines can selected identical to the *EyeBot* LCD in a separate window or via Unix streams. *EyeSim* allows the concurrent simulation of multiple robots, several environment data formats, as well as the inclusion of moveable objects (here: soccer ball).

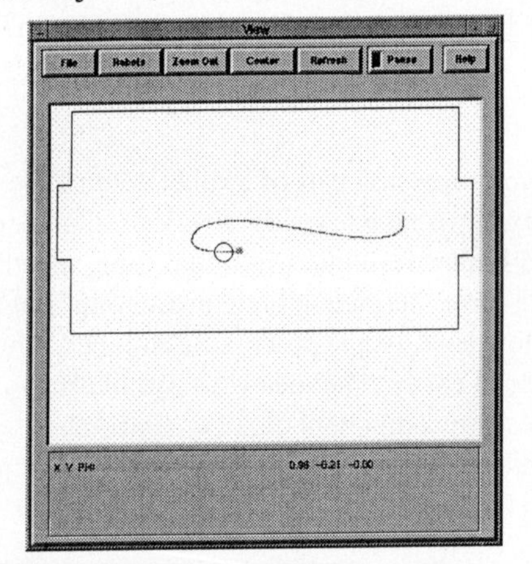

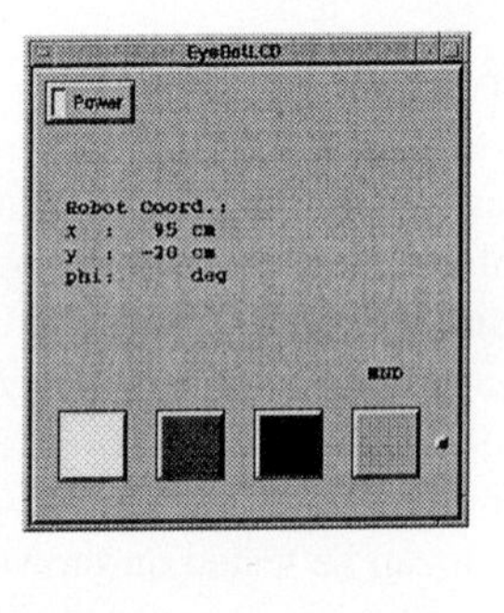

Fig 4. *EyeSim* mobile robot simulator

In Figure 4, the robot drives a spline curve towards the ball. The trajectory is generated by Hermite splines, involving the robot's start position and orientation, together with the perceived ball position and direction towards the goal. Intermediate control points are inserted in certain cases, e.g. when the robot is positioned between goal and ball.

5 Summary and Future Research

EyeBot robot developments include not only the robot soccer vehicles, but also several different 6-legged and biped walking machines. These are all using the same controllers and the same operating system, individually adapted the each robot's sensor/actuator configuration by the concept of a hardware description table.

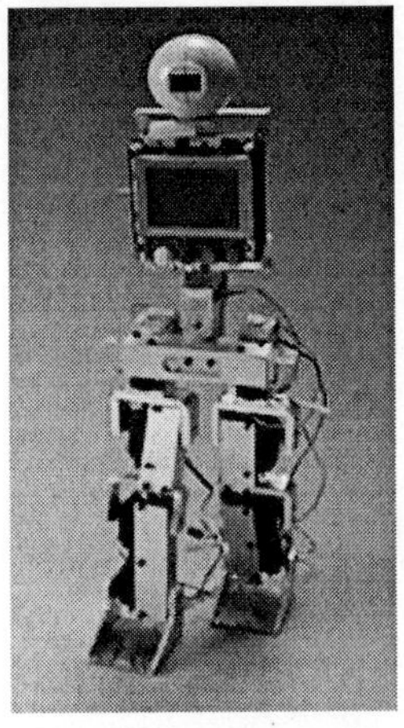

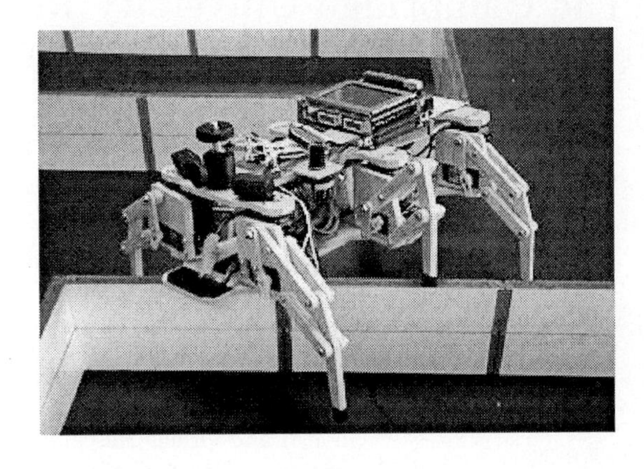

Fig 5. *EyeBot* walking robots are also members of the same family

We have discussed *EyeBot*, a platform used for mobile robots with local intelligence, allowing autonomous real-time vision control. Future research will concentrate on behaviour-based control models for groups of robot. More information is available on the Internet:

http://www.ee.uwa.edu.au/~braunl/eyebot/robots.html

Acknowledgments

The author acknowledges the work Jörg Henne (mechanics), Frank Sautter (electronics), Klaus Schmitt, Barbara Linn, Gerrit Heitsch, Michael Kasper (system software), Thomas Lampart (system software and programming tool), Nicholas Tay, Elliot Nicholls (simulator), and Birgit Graf (soccer software).

References

1. H. Bayer, Th. Bräunl, A. Rausch, M. Sommerau, P. Levi, *Autonomous Vehicle Control by Remote Computer Systems*, Proceedings of the 4th International Conference on Intelligent Autonomous Systems, IAS–4, Karlsruhe, March 1995, pp. 158–165 (8)
2. Th. Bräunl, M. Kalbacher, P. Levi, G. Mamier, *CoMRoS: Cooperative Mobile Robots* Stuttgart, Proc. 13. Nat. Conf. on Artificial Intelligence, AAAI Press, Portland OR, August 1996
3. The GNU Project, *GNU Documentation*, online, Delorie Software, www.delorie.com/gnu/docs/
4. Th. Harman, *The Motorola MC68332 Microcontroller*, Prentice Hall, 1991
5. P. Levi, M. Muscholl, Th. Bräunl, *Cooperative Mobile Robots Stuttgart: Architecture and Tasks*, Proceedings of the 4th International Conference on Intelligent Autonomous Systems, IAS–4, Karlsruhe, March 1995, pp. 310–317 (8)

The Cambridge University Robot Football Team Description

A. Rowstron[2]*, B. Bradshaw[2], D. Crosby[2], T. Edmonds[2], S. Hodges[3], A. Hopper[2], S. Lloyd[2], J. Wang[2] and S. Wray[1]

[1] Cambridge University Computer Laboratory, Pembroke Street, Cambridge, CB2 3QG, UK.

[2] Engineering Department, Cambridge University , Trumpington Street, Cambridge, CB2 1PZ, UK.

[3] Olivetti and Oracle Research Laboratory, 24a Trumpington Street, Cambridge, CB2 1AQ, UK.

Abstract. This paper describes the Cambridge University Robot Football entry and our experiences in the RoboCup'98 Small Robot League of the held in Paris competition.
In the competition we came top of our group, and fourth overall. We had the strongest group, with the team coming second in the group coming second overall in the end. This meant that we were able to play a number of good games, and we were able to evaluate our approach compared to others.
This paper presents an overview of our team, compares it to some of the other approaches used, and highlights the research issues of interest to us.

1 Introduction

The Cambridge University Robot Football Team is a system autonomous five player robot football team. The system uses global vision approach with multiple pitch side servers with multiple platforms linked to the servers by a robust radio communications system[1]. The camera is mounted approximately 4 meters above the centre of the table tennis table.

The pitch side servers perform image processing of the global vision feed and decide where the robots should move, using a novel extension of Potential Fields, called *Time Encoded Terrain Maps.* The image processing software is capable of processing the video stream at approximately 50 fields per second. The software which performs the cooperation planning between the physical agents is written in Java.

The team has a number of unique features; the ability to plan and perform passes between *any* players, and a goal keeper which is capable of capturing and kicking the ball. The cooperation planning software is able to , move another robot into position to receive a pass in advance of the first robot hitting the ball.

* Contact: aitr2@cl.cam.ac.uk

In the RoboCup'98 the team came top of our group and 4th overall. Our overall system was one of the most robust present. None of our robots failed either before or during matches, the vision system was able to be reconfigured to work successfully from the opening match with all the pitches provided, and we were able to change communication frequencies as required, and our communications system was robust enough to be unaffected by local interference.

The next sections provide an overview of our system and contrast it to other approaches.

2 Vision System

The vision system analyses each frame to determine the location and speed of the ball and other players, and the location, orientation and speed of our players. The frame grabber produces 768x576 24-bit RGB frames at a rate of 50 fields per second. Once the initial locations of the objects being tracked have been found the vision system software is capable of processing at full speed (in other words 20 milliseconds per field). Only one other team appeared to be able to process as many fields.

Each frame is translated from RGB space to Hue-Saturation-Intensity (HSV) space, where classifying the marker colours on the mobile platforms and the ball is simpler. The ball and other marker colours are thresholded based on their location in the HSV space, colour being described by hue angle.

Due to the lighting conditions of the area where our test table was placed and the height of the camera above the table we designed our vision system to be able to compensate for any curvature of the table caused by the camera lens. During the calibration stage we allow for the light levels across the table to be calibrated for. Often at some points there were 'hot spots' where light is reflected. Our vision system was robust enough to cope with these whilst many teams were not able to. Also, if necessary due to lighting problems or colour confusion with the opposition robots we could quickly choose other marker colours and re-calibrate the vision system to use them.

Our team uses a team colour plus another colour for each player. This has the advantage that the vision system can uniquely identify each player. However, it requires the vision system to be calibrated to detect eight colours, which sometimes proved to be difficult in the lighting conditions provided in Paris. In the next generation system will use an alternative way to identify the robots and then use the vision system to just work out the exact position and orientation.

More details of the vision system can be found in the team description[2].

3 Cooperation Planning

The Cooperation Planner is responsible for the control and coordination of a team of the physical agents on the pitch. Its input is a stream of location data produced by the vision system and generates command sets for each physical agent. This involves analysis of the situation and prediction of the future game state, from which a team strategy is derived. The team strategy is then broken into individual roles which are assigned to the physical agents and the appropriate commands issued.

The information received from the vision system is a timestamped set of position and orientation coordinates for each object on the field for each frame processed. Relevant velocity information derived from subsequent frames is incorporated using a basic physical model to enable predictions of the play state in the short term (up to a second). The prediction accuracy beyond this is limited by the highly dynamic nature of the environment and its inherent unpredictability.

For the same reason, the cooperation planner is implemented using a stateless design. In this way, the state of the system is reflected only in the state of the physical agents and their environment. This improves the reactiveness of the system to any unexpected changes while a sophisticated system of tolerances promotes stability in the system.

3.1 Strategy Planning

The core of the Cooperation Planner is the Strategy Planner which uses the novel approach of encoding time in a terrain map, these *time encoded terrain maps* are used to represent the current state of play and from it derive a plan of action. In this case, the strategy is basically just path planning for the route of the ball.

The time encoded terrain map is formed from the environment state using information from the physical modeller for the time frame under consideration[2]. It provides a method for combining all of the known information about the given situation including agent positions and velocities into a single map. The map is distorted to represent the different velocities and the time distance travelled between various points. In this way, the map can be said to encode time.

The map in this case, is formed by representing each opposing agent as a hill and each of our own agents as a depression. This leads to a terrain map of a landscape where high ground may be considered as bad and low land may be considered as good. Superposition applies, so if two agents are nearby, their effects will combine. One alteration of the method is made at this point – once the map has been formed, points of low altitude (goodness) are brought back up to the sea-level norm (neutral) as when moving the ball, it is of little difference if it passes near to a player of the same team or through a clear area. The important point is to avoid the opposition players while also allowing for the fact that one of our players may negate the effects of an opposition player in the same area.

Tracing a path through this terrain which attempts to remain at the lowest cumulative altitude will produce a path that avoids opposition and favours locations near our own players. Note, that often with terrain map navigation, the

[2] This is usually advanced by a fraction of a second to allow for the latency through the system from data acquisition to command execution.

path is evaluated on the amount of ascent and descent required; however, in this case, the path is evaluated by the absolute altitude of the path. In this manner of operation, navigation through this terrain is very similar to artificial potential field navigation.

When the terrain map is being created the alterations made to it by a robot's presence is termed that robot's *influence*. The influence of a robot over a particular area may be defined as the likelihood that an area may be reached by the robot in a given time. Hence, a robot's influence is a measure of the possibility of a the robot being at some point. As the possibility of a robot being at a certain point depends on the amount of time it is given to get there, the map effectively encodes the foreseen dynamic variations of the field over time. In robot football, the key interest is the ball, and hence an agent's influence over an area can be interpreted as *the physical agent's influence on the ball should it get to that area.*

Once the map has been formed, finding the optimal path through it to the destination is a simple application of Dijkstra's least distance path finding algorithm for a connected network[3] (The A* pathfinding algorithm[4] is actually used for speed).

3.2 Strategy Action

Once the optimal path for the ball has been determined, the physical agents must be set in action to guide the ball along that path. This involves the assigning of different agents to different roles in the plan. The pathfinder is such that it tends to promote passing the ball to clear areas as well as passing directly to the agents. By doing so, it allows more flexible partitioning of roles and a plan more adept at finding gaps in defences.

Hence, the Cooperation Planner can use a general algorithm to both set up passes between team members and also provide effective blocking and interception behaviour. Additional specialised behaviours are invoked for certain boundary conditions such as when the ball is wedged against the side wall. In this case, the physical agent is directed to flick the ball back out into play using its tail. Additional specialised behaviours are incorporated to control the special abilities of the goalkeeper.

The large number of parameters which govern the operation of the pathfinder in terms of aggressiveness and timeliness allow for in depth tuning of the strategy which could in the future be adjusted automatically on the fly based on analysis of the game to date. Careful tuning of these parameters allowed the Cooperation Planner to achieve a high iteration rate while still maintaining effective performance.

The Cooperation Planner was able to plan passes between any players, and have a physical agent move to receive the ball *before* the first physical agent had hit the ball. Only one other team could perform passes, and this was limited to their attack robots. Although, it was difficult to evaluate the performance of the passing under playing conditions, on several occasions successful passes were observed.

It should also be noted that the ability to alter the characteristics of the system was important. When we arrived we had a system that avoided the oppositions physical agents. However, less than half the teams had this capability. This meant that those teams which avoided the opposition were at a disadvantage. Tuning the system to be more aggressive towards such teams meant that we were not disadvantaged by such behaviour.

4 Physical Agent

We used two different physical agents, a standard player[2] and a goal keeper. Our goal keeper was unique in that it could capture the ball and then kick it at high speed. This feature of the team led to extended periods of play during matches (over two minutes of action). More details of the goalkeeper and its impact can be found in Hodges[5].

The standard player was designed to be 'thin'; this meant that the physical agents should have little computing power, and most processing should be done remotely on the servers at the side, in a similar strategy to the network computer. The platform received commands such as; move forwards x cms, or turn left x degrees.

We were surprised therefore to see that some of the teams had 'thinner' platforms, and purely sent commands of the form, left/right motor on/off. This is something we intend to investigate this coming year. However, these systems appeared to have to use the full bandwidth available to do this, and had no error detection or robustness in the communications systems. This meant they were often unable to change frequency channels, or cope with small amounts of background noise.

5 Communication System

To provide communications between the central server and the mobile platforms we use a wireless network, called Piconet[1]. Piconet is a low-rate, low-range, low-power system intended for use in the investigation of embedded networking. We created a system that allowed the available bandwidth to be focused on certain physical agents. The Cooperation planner told the communication systems which physical agents were important, and the commands to these robots were given much higher priority. Therefore, players not involved in the current action would sometimes appear to pause whilst the ones involved in the action would not.

Our communications systems was very robust, able to survive noise on the frequency we were using (the noise was eventually tracked down to the security guards mobile radio communication system). We were also able to change frequency with about 5 minutes warning.

6 Conclusions

Our system had several novel features, which allowed us to do well. Because of the nature of RoboCup we often disabled parts of our system to give other teams a fair advantage (we often disabled the goal keeper so other teams could track the ball). Our semi-final game against CMU'98 really showed how robust our system was. Firstly we had to change pitches and then recalibrate the system. We then had to change frequencies because CMU could not get their second frequency working, and then we played with only four physical agents because they could only get four platforms working. We then lost the game, however, we were able to show how a robust system could be made.

We will be entering the 1999 games, with a new system building on our experiences at Paris. We intend to push further the limits of sensors and sensing and to create faster, thinner platforms.

Acknowledgements

We were funded and general supported by Prof. Andy Hopper and the Olivetti and Oracle Research Laboratory. Thanks to Stuart Wray, Frank Stajano and others, who helped and advised us during the development of this Robot Team. We also thank Chris at Engineering for his help with platform manufacture.

References

1. F. Bennett, D. Clarke, A. Hopper, A. Jones, and D. Leask. Piconet - embedded mobile networking. *IEEE Personal Communications Magazine*, 4:8–15, October 1997.
2. A. Rowstron, B. Bradshaw, D. Crosby, T. Edmonds, S.Hodges, A. Hopper, S. Lloyd, J. Wang, and S. Wray. CURF: Cambridge University Robot Football Team. In Minoru Asada, editor, *Proceedings of the second RoboCup Competition*, pages 503–510, July 1998.
3. Robert Sedgewick. *Algorithms*, volume 2nd edition. Addison-Wesley, 1988.
4. Stuart C. Shapiro. *Encyclopedia of Artificial Intelligence*, volume 2nd edition. Wiley, New York, 1992.
5. S. Hodges, D. Crosby, A. Rowstron, B. Bradshaw, T.Edmonds, A. Hopper, S. Lloyd, J. Wang, and S. Wray. The challenge of building and integrating a sophisticated goalkeeper robot for the small-size RoboCup competition. Submitted to ??, 1998.

The UQ RoboRoos Small-Size League Team Description for RoboCup'98

Gordon Wyeth, Brett Browning and Ashley Tews

Department of Computer Science and Electrical Engineering
University of Queensland, Australia
{wyeth, browning, tews}@csee.uq.edu.au

Abstract. The UQ RoboRoos have been developed to participate in the RoboCup '98 robot soccer small size league. This paper describes the current level of implementation of the robots, including aspects of hardware design, as well as the software running on the robots and the controlling computer. Key features of the RoboRoos design include the agile and powerful mechanical frame, the robots' navigational techniques and a coordinating planner system based on potential field methods.

Overview

The UQ RoboRoos are a team of five field robots and one specialist goal keeper robot custom designed to play robot soccer in the small-size league of RoboCup '98. The project was initiated at the University of Queensland to provide a highly motivating environment for the development of techniques for robot navigation and multi-agent cooperation. The project also provides a unique educational opportunity for undergraduate engineers to work in a team on a complex mechatronic project.

The team of eight undergraduates, two postgraduates and one academic has been developing the robots for twelve weeks. In that time, the team has made remarkable progress. A high performance mechanical chassis for the field robots has been designed, and five copies constructed. A novel goalkeeper design has also been developed, and partial implementation completed for the contest. Each robot is equipped with a custom Motorola 68332 based controller board. Preliminary communication boards have been developed, with plans for higher bandwidth boards well in hand. Local sensor boards have been designed, but haven't been integrated with the software at this point in time. The playing field has been constructed to RoboCup specifications, with appropriate lighting. The vision system for the project consists of an overhead camera that has been mounted and interfaced to a high-speed frame grabber in a PC.

On the software side, all low-level routines for the robots have been successfully tested with a solid servo loop and reliable communications. The interfaces to this low-level code are duplicated in a kinematically realistic simulator. The simulator supports full game play of the robot team using identical routines to those on the real robot. The virtual physics of the simulator realistically represents the physics of the real field, including such issues as communications bandwidth and delay. Code developed on the simulator can be directly ported from the simulator to the robots. Code for chasing and kicking the ball has been developed in this manner, incorporating collision avoidance and directed passing. A game coordinator has also been developed to coordinate robots in attack and defence. The other major software achievement has been the development of the robot and ball identification and tracking system that currently tracks all objects at 5 fps. Plans to significantly improve the frame rate are well in hand.

The RoboRoos are a work in progress. Their performance at RoboCup '98 is limited by the current communications system and the delay between frames in the tracking system. The current implementation provides a sound basis for the development of a world class team.

Mechanical Design

The mechanical design of the field robots has been optimised for speed, acceleration and cornering ability, while maintaining as broad as possible a profile for contacting the ball. An isometric view of the chassis with electronics removed is given in Figure 1. The wheels are arranged in a wheelchair arrangement, with Teflon skids at the front and rear of the robots providing the third point of contact. The soft rubber tyres provide a high coefficient of friction, minimising slip at the tyre contact points. Table 1 outlines the mechanical abilities of the robot.

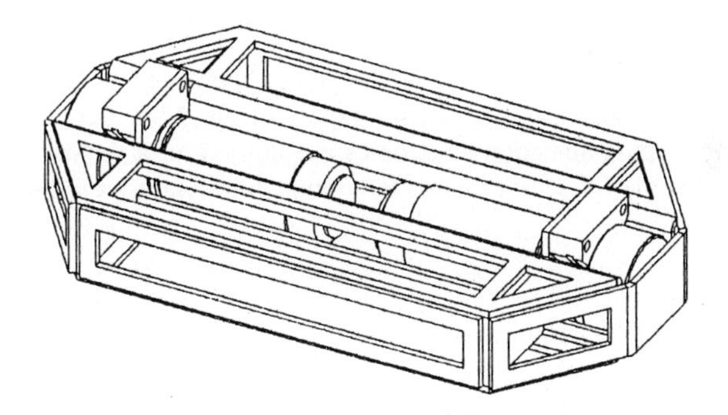

Figure. 1. Mechanical design of the RoboRoos field robots.

	Continuous	Peak
Acceleration (ms^{-2})	1.5	3.6
Speed (ms^{-1})	2	3.4
Time over half field (s)	1.33	0.86
Time for about face (s)	0.6	0.4
Current per motor (A)	0.4	1

Table 1. Mechanical capabilities of the RoboRoos field robots.

The goal-keeping robot has sacrificed cornering ability for raw acceleration capabilities. The four-wheel drive design ensures that all available normal force is used to achieve traction enabling the robot to reach a theoretical acceleration of $8ms^{-1}$. It is hoped that this high-speed manoeuvrability will enable the goalkeeper to block most attempts on goal. In addition, this robot will have a mechanism for trapping and kicking the ball based on a rotating arm. The robot, which is illustrated in Figure 2, is yet to be completed, but is still an effective goalkeeper in its current partial form.

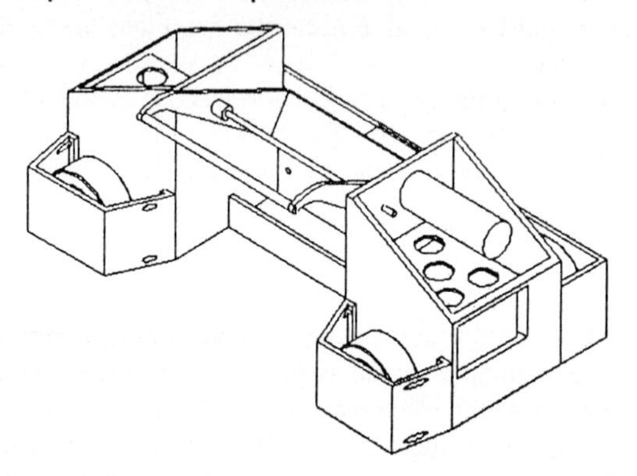

Figure. 2. Chassis design for the goalkeeper robot.

Electronic Design

The electronics are housed on two boards. The main board has the CPU and associated memory, as well as the power electronics for supply of the subsystems and the control of the motors. The RF communication electronics are housed on a separate shielded board. Provision has been made for sensor boards to be mounted to face the front and back of the robot.

Power

Each robot is powered by custom rechargeable NiCad packs that have the capacity to provide 20 minutes of match time under the most strenuous playing conditions. The batteries are regulated to provide isolated supplies for key electronic sub-systems. The motor drives are H-bridges driven by signals from the CPU. Low on-resistance MOSFETs act as switches for the H-bridge, with level shifting provided by gate drivers. The system is rated for continuous operation under motor stall conditions, ensuring reliable motor control in all situations.

CPU

The robot is controlled by a Motorola 68332 microcontroller that carries a great number of useful peripherals and features for the robot. The microcontroller has 16 x 16 bit timer channels supporting PWM and quadrature capture for the motor drive and other timing functions for sensor capture. The CPU has access to 256 kb of SRAM and 256 kb of Flash memory on its 16 bit bus, providing adequate resources for programming. The microcontroller has a built-in asynchronous serial port that is used for radio communications.

Communications

The robots receive a broadcast signal from the external PC for each frame processed (presently 5 fps). The PC broadcasts all player positions and heading vectors, as well as the ball position. In addition the PC sends strategic commands to the players indicating the best activity for each robot. The nature of these commands will be discussed in the software section. Suffice to say, the information transmitted is constrained by the 4800 baud bandwidth of the single broadcast channel. The arrival of new RF modules will soon improve the bandwidth, but supplier delays have prevented the modules being available for RoboCup '98.

On-Board Software

Software on the robots runs in two threads – a cognitive level thread and a schema level thread. The schema level thread is triggered every millisecond and deals with moment to moment navigation problems, such as achieving smooth acceleration or tracking a fast moving ball. The processes that run at this level might be called *behaviours* or *schemas*. The cognitive level thread selects the schemas that are appropriate at a given time, and enforces a hierarchy for arbitration. In addition, the cognitive level maintains planning resources such as paths to goal locations.

Schemas

The primary schemas are *traverse* and *align*, which provide translational and rotational motion respectively. Each schema has a length parameter associated with it that allows the schema to terminate safely in the event of loss of communications. For example, if the robot is facing towards the top of the field and wishes to face the goal, the *align* schema will be instantiated with -90° length parameter. The robot will then turn, accelerating and decelerating in a fashion suitable for a 90° turn. The length of the turn is determined by encoder feedback. Should new information come through during the turn, the length parameter may be modified or the schema switched out in a smooth transition. In the event of no new information the turn will terminate smoothly. The schemas also have variable acceleration and velocity parameters. This means, for example, that the traverse operation can be easily modified for an aggressive kick operation by increasing the acceleration and allowed velocity.

In addition, the schema level keeps track of the motion of the robot to provide a current estimate of position. This process of path integration provides the robot with an accurate position and heading. A delay between event and receipt of that event on the robot is inherent in the use of the overhead vision system. The effects of this delay are minimised by updating beyond the delay using the information from the path integrator.

Cognitive Level

The cognitive level spends most of its time working out how to get to goal locations that are specified in communications received from the planning system. The decisions made by the cognitive level are tactical commands based on a grid representation of the field. The field is divided into 90mm grid units, which roughly fits the field in a 30 x 16 grid rectangle. The choice of grid size corresponds with one-half of the maximum robot dimensions. As data arrives from the communications system the obstacles are added to the grid to represent the current playing conditions. Having established the obstacles, a flood fill operation is performed to determine the path the robot will take to move to its goal. This operation is carried out by flooding the grid with incrementing values from the current goal.

The robot then uses these numbers, termed *weights*, to decide the best motion from the current position. Lower values of weight indicate a point closer to the goal location.

To evaluate the best current action for the robot, the cognitive level observes the weights currently surrounding the robot and the current direction of the robot. If the direction that the robot is facing is favourable, the *traverse* schema is made dominant. If a directional change is required the *align* schema provides the necessary directional correction. This method provides goal-oriented activity with reliable obstacle avoidance. In addition, the robot is able to get to intermediate locations in the case of a blocked path to the goal, and is able to align with target locations for kick operations.

Vision System

The vision system receives input from an overhead colour CCD camera with a variable length lens. The camera interfaces to a high-speed machine vision frame grabber with a PCI interface. The grabber resides in a 233 MHz Pentium II PC that performs all vision processing as well as generating the multi-agent strategy for the team. The primary task of the PC is to accurately locate, identify and track each object on the field. This software must be robust to lighting and positional changes. Software reports a position for each object as well as heading information for our robots at a minimum of 5 Hz.

The image is segmented by colour. As the ball and the robots all have distinct markings this is the obvious way to perform initial segmentation. Colour classification is performed by observing the distance between the RGB vector of each pixel from a specified prototype vector. If the distance is less than a specified threshold then the pixel is classified as belonging to a particular segment. The system is readily adapted by observing RGB values of the desired segmentations for given lighting conditions and adjusting the prototype vectors and thresholds. The ease with which the system can be modified allows the system to be easily adapted to new teams, fields or lighting conditions.

A template match performs the identification of the objects. As the image is segmented, the segmentation routine will call the template matching code when an interesting pixel is found. The template is passed over the region surrounding the seed pixel, searching for the location that maximises the template match. If the match level exceeds a predefined threshold the object is identified and its location reported. Once the objects are identified, the coloured balls on top of the robots are paired to ensure the presence of the robot. This information is also used to determine the heading of the robot.

Planner

The planner provides the strategic commands to each of the robots based on a current assessment of the situation. There are two commands that the planner sends to the robots: GOTO and KICK. Each of these commands carries position parameters that indicate where the robots should "go to" or "kick to". The coordinates used in the planner are the same 30 x 16 grid used by the cognitive level, but the manner in which this grid is used is somewhat different. This highest level of planning is based on a combination of potential fields generated from field position and the position of the robots (shown in Figure 3) and the algorithm described below.

The first aim of the system is to find a good location for the ball that might reasonably be achieved with a single kick from its current location. The evaluation begins with the base potential field that contains values that indicate the desirability of the ball at each location for an empty field. A lower number indicates a more desirable location, forming a potential well to which a ball might roll. The robot potential fields are then superimposed on the base field to create a useful potential map, *P(i,j)*. Friendly robots form a potential well, while opposing robots form a potential lump. For the robot with the ball at *(x,y)*, each point *B(i,j)* is then evaluated by Equation 1.

$$B(i,j) = P(i,j) + \max(i,j...x,y) + \text{length}(i,j...x,y) \quad (1)$$

P(i,j) represents the potential field calculated as above. max() returns the maximum potential field value between *(i,j)* and *(x,y)*. length() gives the Manhattan distance from *(i,j)* and *(x,y)*. If a robot is in a good position to kick the ball, it is issued with a KICK command with the parameters *(i,j)* based on the minimum *B(i,j)*. If no robot is in a suitable position, the planner selects the nearest to ball and directs it to go to a suitable position using the GOTO command.

6	6	6	5	5	4	3	3
9	8	7	5	4	4	2	0
9	8	7	5	4	2	2	0
9	8	7	5	3	1	0	-1
9	8	7	5	3	1	-1	-1
6	6	6	5	4	2	1	2

Figure. 3. Potential field formed by the base field with a robot from each team superimposed.

If the planner deems that the opposition has the ball or is likely to assume control of it, the planner will send defensive GOTO commands to all the other robots that are not going for the ball. The parameters for the GOTO command are designed to occupy the space between each opposition robot and the goal. This corresponds to maximum values in a potential field calculated in similar fashion to Equation 1.

If the planner deems that we have control of the ball, it will issue GOTO commands to each of our field players not involved with manipulating the ball. The coordinates to which each robot goes are determined in similar fashion to Equation 1. As in human soccer, each robot is assigned its position on the field such as left wing or centre. Playing positions are implemented with a field position potential field that encourages the robot to move within its specific area of responsibility. The RoboRoos currently play with a defender, a centre and two wings.

Summary

The UQ RoboRoos robots provide a basis for ongoing research into navigation in dynamic environments and emergent multi-agent cooperation. Further investigation into these two areas combined with improvements in communication bandwidth and vision processing can only lead to improved robot performance, nearing the mechanical capabilities of the robots.

Designing and building the robots has been a highly motivating task for the many undergraduate students involved. The multi-disciplinary nature of the project creates unique project based experiences for the students. Robot soccer will continue to play a role in undergraduate education as well as furthering robotic research.

ISocRob — Team Description*

Pedro Aparício, Rodrigo Ventura, Pedro Lima, and Carlos Pinto-Ferreira

Instituto Superior Técnico/Instituto de Sistemas e Robótica (ISR/IST),
Av. Rovisco Pais - 1, 1096 Lisboa Codex
PORTUGAL
{aparicio,yoda,pal,cpf}@isr.ist.utl.pt

Abstract. The SocRob project was born as a challenge for multidisciplinary research on broad and generic approaches for the design of cooperating society of robots, involving Control, Robotics and Artificial Intelligence researchers. A case study on Robotic Soccer played by a team of 3 robots is currently underway. The team competed at the World Cup of Robotic Soccer, *RoboCup98*, held in Paris, France, and qualified for the quarter-finals of the midlle-size league. In this paper we present the team description, the implemented behaviors and the results obtained.

1 Introduction

Multi-agent systems have become very popular in recent years, especially as a research area in Distributed Artificial Intelligence (DAI) [6]. Simultaneously, several robotic systems based on a fleet of robots have been developed, as an alternative to single-robot systems common in the past [9].

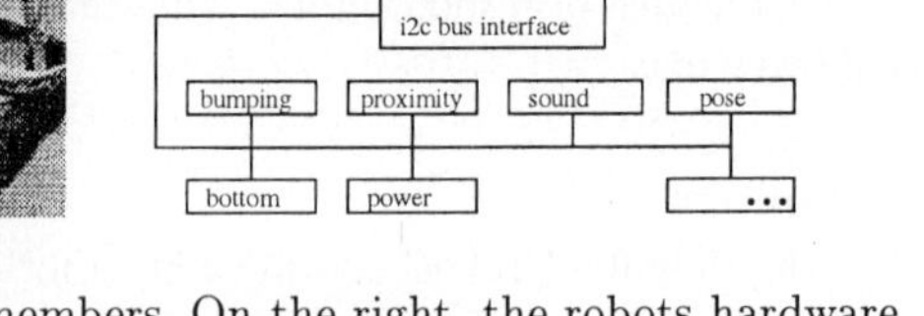

Fig. 1. On the left, three ISocRob team members. On the right, the robots hardware architecture.

The Artificial Intelligence and the Intelligent Control groups of the ISR/IST have started almost one year ago a joint project on Cooperative Robotics, to

* This work has been supported by grants from the following Portuguese institutions: Fundação Calouste Gulbenkian, Fundação para a Ciência e a Tecnologia, PRAXIS XXI/BM/12937/97.

foster research on methodologies for the definition of functional, hardware and software architectures to support intelligent autonomous behavior and evaluate performance of a group of *real* cooperative robots, either as a society and as individuals. The robots were developed from scratch, so that both conceptual and implementation issues are considered. Special attention is taken to cooperation-oriented communication issues, such as the type of information that must be shared and how to distribute that information. This work relies on past experience of both groups regarding topics relevant to robot development [3] and DAI [7]. A case study on Robotic Soccer involving a team of 3 robots, shown in Figure 1, (the **ISocRob** Team) is currently underway.

This paper is organized as follows: Section 2 and 3 describe the details of each robot Hardware and Software Architectures, respectively. The Functional Architecture, presented in Section 4, wraps up the whole picture, relating conceptual issues to the two physical architectures explained in the previous sections. Section 5 presents the details of the implemented behaviors for the soccer competition. Conclusions and foreseen future work are presented in Section 6.

2 Hardware Architecture

Each robot hardware is divided in four main blocks: sensors, main processing unit, actuators and communications. Currently, from the hardware architecture standpoint, the population is composed of homogeneous mobile robots. Figure 1 depicts a block diagram of the hardware architecture of each robot.

2.1 The Processing Unit

Each robot has on-board a PC motherboard with a network adaptor, a video adaptor, a motor control board and interface boards for the sensors. The main processor is an AMD K6, running at 200MHz. The system has 16Mb of RAM and a 1.2Gb hard drive. The PC motherboard was chosen because of its performance/price ratio.

2.2 Sensorial Systems

The sensors of each robot are divided in two main groups:

- **vision sensors**: virtual sensors which extract information from the images acquired by a video camera and its interface board. One physical transducer (the video camera) leads to many (virtual) sensors (e.g. ball location, ball mass).
- **pose, bumping and proximity sensors**, each of them physically associated to one transducer.

The conceptual details of the different sensors/ transducers are described in the sequel.

Video Camera — The video camera is a Phillips XC731/340 interfacing the motherboard through a PCI Captivator board. This combination allows the acquisition of 640×480 images at a frame-rate of 50 interlaced frames per second. Image is used for several purposes, namely, to identify and/or follow/catch the team mates, the opponents, the ball and the goals.

Pose Sensor — Depending on the type of application involved, each robot of the society may need to regularly update its current pose with respect to a reference frame (e.g., located in the field center). This may be accomplished based on the *triangulation principle*: from the measurement of the angles between the robot longitudinal axis and the direction of maximum signal reception from infra-red (IR) beacons whose location in the reference frame is known, the robot is able to compute its pose relative to that frame.

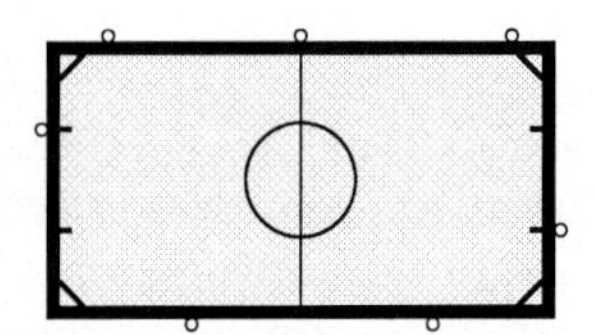

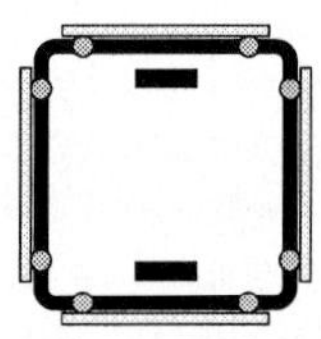

Fig. 2. On the left, some possible beacon locations, and on the right, the micro-switch sensor (bumpers) placement.

Bumping Sensors — These sensors detect collisions. They are made of micro-switches, arranged in a serial connection, divided in 4 sets of 2 micro-switches each. Figure 2 presents a possible location of the bumping sensors around the vehicle (horizontal plane).

Proximity Sensor — Proximity sensors are also based on IR technology. The typical range of those sensors goes from 20 cm to 1.5 m.

2.3 The Actuators

Each robot has a differential drive kinematic configuration. This implies that it has two independent (DC) motors, one for each wheel. The robot speed and heading are set by independently controlling the wheels speed. The motor controller output is a PWM signal and a bit indicating the direction of rotation for each wheel. Those signals are fed to the power amplifiers.

2.4 Communications

A wireless RF Ethernet link (WaveCell from Aaron Tech.) was chosen to support communications between the robots. The devices work on two possible switch-selectable frequencies: 2.4GHz and 2.4835GHz. The bandwidth is 2Mbps, and a range of 150m is covered, inside an office environment.

3 Software Architecture

Each robot's software runs under the Linux [1] operating system. The reasons for this choice were: robustness, lightweight multitasking, scalability, networking facilities, and availability of programming languages compilers, as well as easy integration of programming languages (e.g., Lisp and C).

The top-level software, which is responsible for each robots' behavior is implemented in an agent programming language — RUBA — developed in previous work [7]. Briefly, RUBA is a language that implements a society of agents, tat communicate between them and with the exterior, by the means of a blackboard structure. The software underlying RUBA was re-implemented in the Scheme programming language [2]. The whole team is viewed as a single agent society with a common blackboard (distributed among them, but considered as being unique). Additionally, each robot has an individual blackboard (as RUBA supports multiple blackboards) to handle issues related to it as an individual. The RUBA language is based on production rules interleaved with state machine (multiple machine states are supported) statements, *i.e.*, the rules can be grouped together in a specific state of a state machine. The expressions that fill the `IF`, `THEN` and `ELSE` fields are essentially Scheme expressions which manipulate internal symbols and primitive tasks (i.e., functions executed using the robot hardware). This mechanism bridges the gap between high-level agent programming and specific robot actions/sensing.

4 Functional Architecture

From a functional standpoint, the whole robot society is composed of functionally heterogeneous robots. In the particular case of soccer robots, the functionalities are *Goal Keeper*, *Attacker*, etc. The functional architecture is *scalable* regarding the number of robots (or *agents*) involved. This means that, when a new robot joins the society, no changes have to be made to the overall system. The functional architecture establishes three levels, inspired in [4, 5]. Their description is detailed in the sequel.

Organizational level — This level deals with the issues unconditionally common to the whole society. In the soccer team context, these are:

- the state of the game according to the rules (before kick-off? in-game? off-side?) and the way the team has to behave in order to follow them;
- the global strategy of the team (time to re-positioning of the team? time to attack? time to defend?).

These issues involve the whole team, and thus have to be established at this global level.

Relational level — In a cooperative robotics context, in order to accomplish useful cooperation, relationships between robots have to be accomplished. This involves an important characteristic of the agent concept: social ability [8]. This means that one given agent has to be aware of the existence of other agents like him, with whom it has to negotiate. At this level, groups of agents negotiate and eventually come to an agreement about some objective (common or not). The issues involving the formation of groups and its disbanding are handled at this level. The key idea of this process is negotiation among agents. The blackboard structure provides the common medium through which the necessary communication circulates.

For instance, if one robot wants to pass a ball to another player (of the same team, hopefully), it has to find someone available and sufficiently well positioned. The ball pass is arranged via a negotiation process, and after an agreement is reached, the pass is performed.

Individual level — The individual level encapsulates each robot as an entity, comprising all aspects of a robot as an individual. This includes the individual primitive tasks, invoked by the relational level (e.g., path planning, motion with collision avoidance), robot performance and monitoring.

5 Individual Behaviors

The individual level is responsible for accomplishing each behavior, running the *sense-think-act* loop of its primitive tasks. This involves processing vision (as well as other sensors) data and driving the motors according to the desired behavior. Behaviors, comprising individual primitives, are generated at the relational level.

The player behaviors so far implemented are:

Goal Keeper – A good goal keeper is essential in a team that wants to win. Being so, a significant part of our effort has been devoted to the development of a efficient goal keeper. A state machine that illustrates the goal keeper behavior is presented in Figure 3.

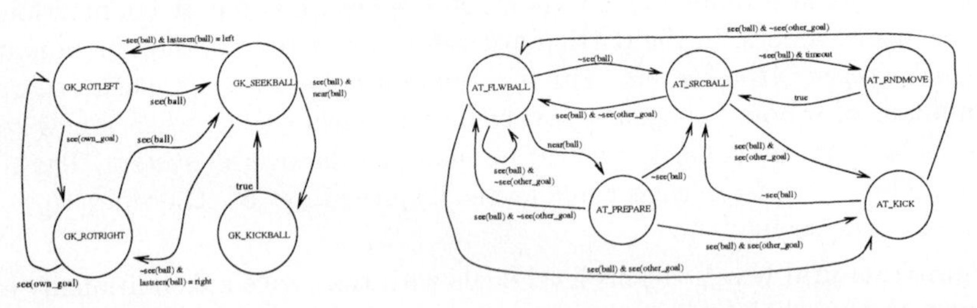

Fig. 3. Goal Keeper (left) and Attacker (right) State Machine. Arc labels represent predicates which allow transitions between states when true. Their values are obtained by vision sensors. States represent primitive tasks.

In short, the implemented behavior keeps the robot continuously seeking for the ball if it is visible (GK_SEEKBALL). If the ball is not visible, it starts rotating in the direction it was last seen (GK_ROTLEFT or GK_ROTRIGHT). If its own goal is visible and the ball is not, it starts rotating in the opposite direction and the process repeats. If the ball is found, and its estimated distance to goal is less than a given threshold, the player *kicks* (GK_KICKBALL) the ball and restarts its seeking movement.

Attacker – The attacker behavior (depicted in Figure 3) starts by looking for the ball, rotating on spot (AT_SRCBALL). If the ball is not visible after a given time, it makes a random move (AT_RNDMOVE) and resumes the search procedure. If the ball is found and the opponent goal is also present in the image, it starts the *kicking* process (AT_KICK). This state directs the ball towards the opponent goal. If, when the ball is seen, the opponent goal is not visible, the player starts the ball following movement (AT_FLWBALL). This state moves the player towards the ball. When the ball is close enough, it starts moving around it, in order to have the ball and the opponent goal visible (state AT_PREPARE).

The player keeps a record of the last side it has seen the ball, its own goal and the opponents goal. If, during any movement, the ball is lost, this mechanism gives the direction for the search direction.

6 Preliminary Results and Conclusions

Currently, our robots are capable of relatively simple behaviors, composed of primitive tasks, such as following a ball, kicking a ball, scoring goals and defending the goal, using vision-based sensors. We are now moving towards the development of more primitives, including those which require information from sensors other than vision, as well as establishing the link between rules and primitives, so that more behaviors can be specified and implemented.

The implementation of RUBA *blackboard* is one of the current open issues. A centralized version based on an external computer is one of the alternatives. A more autonomous solution would consist of distributing the blackboard by the team members, by repeting initial information and then update it simultaneously for all robots. Another open issue is cooperation, i.e., what info should be exchanged between team members.

References

1. Linux online. URL: `http://www.linux.org`, 1998.
2. H. Abelson, G. J. Sussman, and J. Sussman. *Structure and Interpretation of Computer Programs.* MIT Press and McGraw-Hill, 1985.
3. Pedro Aparício, João Ferreira, Pedro Raposo, and Pedro Lima. Barbaneta - A Modular Autonomous Vehicle. In *Preprints of the 3rd IFAC Symposium on Intelligent Autonomous Vehicles*, March 1998.
4. Alex Drogoul and C. Dubreuil. A distributed approach to n-puzzle solving. In *Proceedings of the Distributed Artificial Intelligence Workshop*, 1993.
5. Alex Drogoul and J. Ferber. Multi-agent simulation as a tool for modeling societies: Application to social differentiation in ant colonies. In *Actes du Workshop MAAMAW'92*, 1992.
6. Peter Stone and Manuela Veloso. Multiagent Systems: A Survey from a Machine Learning Perspective. Technical Report CMU-CS-97-193, CMU, School of Computer Science, Carnegie Mellon University, May 1997.
7. Rodrigo M. M. Ventura and Carlos A. Pinto-Ferreira. Problem solving without search. In Robert Trappl, editor, *Cybernetics and Systems '98*, pages 743–748. Austrian Society for Cybernetic Studies, 1998. Proceedings of EMCSR-98, Vienna, Austria.
8. Mike Wooldridge and Nick Jennings. Intelligent agents: Theory and practice. *Knowledge Engineering Review*, 10(2), 1995.
9. Alex S. Fukunaga Y. Uny Cao, Andrew B. Kahng, and Frank Meng. Cooperative Mobile Robotics: Antecedents and Directions. In *http://www.cs.ucla.edu:8001/-Dienst/UI/2.0/Describe/ncstrl.ucla_cs%2f950049?abstract=Cooperation*, December 1995.

Real MagiCol 98: Team Description and Results

E. González [2], H. Loaiza [1,3], A. Suárez [2], C. Moreno [2]

[1] Universidad del Valle - PAyRA. Escuela de Ingeniería Eléctrica y Electrónica, Ciudad Universitaria, Cali, Colombia
Université d'Evry Val d'Essonne, LAMI[2] – CEMIF[3],
Cours Monseigneur Roméro, 91025 Evry Courcouronnes, France.
{suarez, gonzalez, moreno}@lami.univ-evry.fr, hloaiza@eeie.univalle.edu.co,

Abstract. The hardware and software architectures of the Real MagiCol robots are presented. The robots were built for the RoboCup98 competition, but future research in other subjects has been also considered. Our programming methodology is called *Behavior Oriented Commands (BOCs).* Relevant aspects of the vision system, the BOC model of our goalie and the team strategy are presented.

1. Introduction

The middle size team "Real MagiCol" (Realismo Mágico [1] Colombiano) is a joint effort of institutions in Colombia and France. In addition to participating in RoboCup98, the robots will be used in the future for research and educational activities in colombian universities and will be employed in a permanent exhibit of the interactif science museum "Maloka". In fact, we are the first latinamerican middle size team in RoboCup.

We decided to build our own robots. This allows a greater insight and a complete mastering of the robot's technology. We designed an open, easily reconfigurable PC based architecture in order to allow for future evolutions. This decision implied a greater amount of work in hardware aspects than expected but the time left for debugging and refinig the strategies was short. In the short term, the use of commercial robots allows to concentrate all efforts into the individual and team strategies. In fact, the teams that played the RoboCup98 final chosed this approach. Next year should allow teams like ours to work more on higher level tasks and to get better results.

The Real MagiCol team features our hybrid software architecture called Behavior Oriented Commands (BOC) [2] [3]. It allows a soccer robot to plan complex deliberative actions while offering good reactivity in a very dynamical environment. BOCs provide a high level distributed intelligence model which is directly translated into a real time application.

This article presents the main aspects of the robots hardware and vision system, as well as their control architecture and the team strategy.

2. Robot Description

The Real MagiCol team architecture is local vision, multi CPUs. Explicit communication between the players has not yet been implemented. It consists of five robots sharing the same hardware design. Each robot has an external diameter of 44 cm and a height of 18 cm (37 with the optional turret and vision system). Lineal speeds of almost 2m/s with accelerations of 1 m/s^2 are possible. The hardware architecture of the robots is shown in Figure 1. Figure 2 shows our goalkeeper in action.

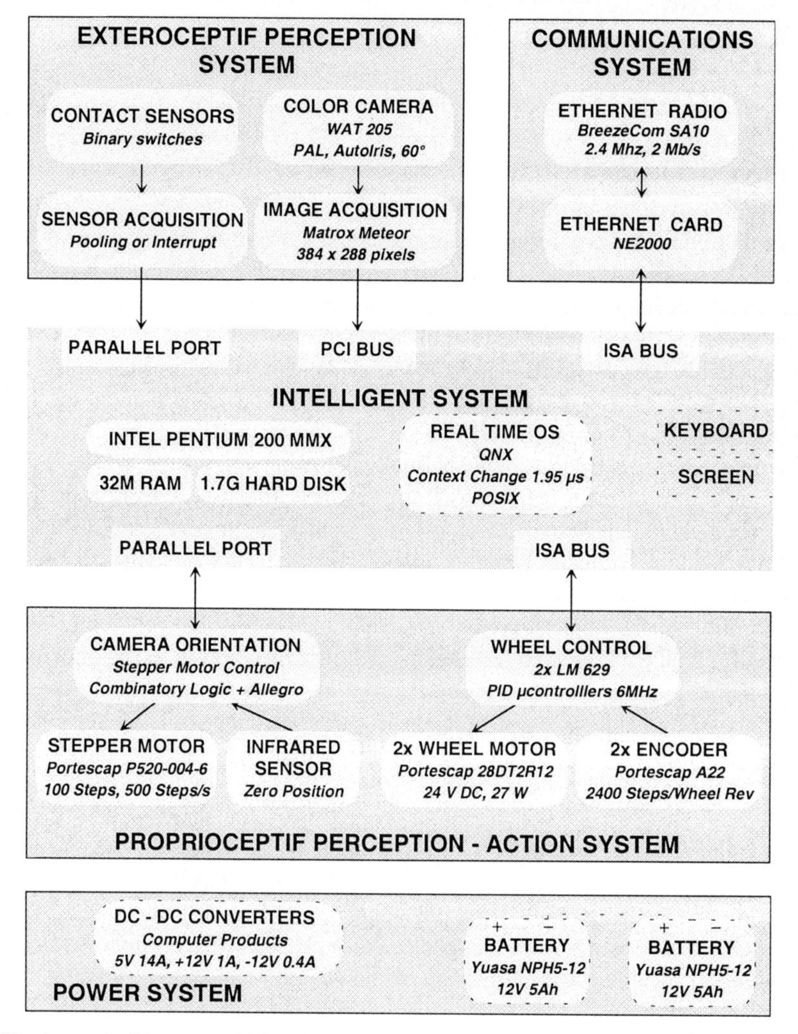

Fig. 1. Hardware Architecture of MagiCol Robots.

Fig. 2. General View of Robot and Ball

3. Local Vision System

Each robot possesses a color mini-camera with a 3.8 mm focal distance and a 51°(h) x 40°(v) vision field. The images are acquired in the RGB color model [4]. The detection algorithm of the elements (color-objects) of the terrain (ball, robot markers, goals, etc.) uses thresholded LUTs applied to the three color fields. A logical function between these images results in eight binary images, one for each color-object. A composed 8-bit image is obtained after bit-shifts and combination of the binary images. While performing the last step a LUT is applied to discriminate and label ambiguous pixels. A 9-level image is finally constructed.

The attributes of each color-object are obtained after segmentation of the 9-level image. For each detected color-object, the center of gravity, the surface and the enclosing window are calculated in image coordinates. Color-objects presenting a small size are rejected. The vision system was calibrated to carry out a 3D position reconstitution of the objects, taking advantage of the fact that the height of the ball, goals and robots are known to reduce the unknown variables in the camera model.

The *local map* is the set of objects recently seen by the robot. An object is characterized by its relative position, speed and uncertainty in local coordinates. The color-object information provided by the image treatment module and the odometer allows to estimate the position and speed of objects. Newly detected objects are incorporated into the map with an initial uncertainty value based on their distance and image surface. Objects re-detected in new images are updated. The uncertainty of undetected objects is increased until they reach the forget threshold.

The robot localization is carried out using objets known to be static. We plan to integrate the information from the local modules in the different robots into a *global map*, in order to improve the precision of the position and speed estimates of objects of interest.

We expect to improve the robustness by implementing an HSI (Hue, Saturation, Intensity) model based algorithm for the color-objets detection. These model could allow the color camera calibration under unstable illumination conditions.

4. Roles and Formations

In a soccer game robots need to be organised to play coherently [9]. Our robots incorporate the rules to execute individual actions depending on its role, team formation and game context.

A role assigns responsibilities and actions to a robot. The generic roles of goalie, defender and attacker are defined. The goalie stays parallel and close to the goal, trying always to be directly between its goal and the ball. The attacker moves to the ball trying to kick towards the opponent's goal; when the ball is in the attacking side it attempts to have a good non interfering attacking position. The defender maintains a good defending position between its goal and the ball/opponent, to move near the ball and pass it to the attackers; it also tries to place itself between the ball and its own goal when the ball is far [5][6]. We also define a new role, the *coach*. It performs global localisation, role and formation distribution, supervision activities, and manages external information.

The generic roles are specialised in sub-roles according to the robot playing region in the field and its attitude towards the game. The field is divided into three regions: left, right and central allowing to decline the roles as left-handed, right-handed and central. Robots can also play a role having different attitude towards the game, for example, a defender may be *prudent* (always staying in a defensive position) or *aggressive* (always trying to kick the ball). This specialisation by attitude allows to easily built teams playing different tactics without modifying its formation.

A formation is a team structure that defines a set of roles in a particular game [6]. A formation assigns a specialised role for each robot. The selection of the team formation depends on the game situation, particularly on the score and opponent's strategy. At start information concerning global team formation affects the way individual sub-role rules are interpreted which allows to have collective conscience.

5. BOC Implementation

A real time control architecture should be used to implement a soccer mobile robot that deals with a dynamic environment. Our robot control system is implemented using the hybrid architecture BOC, which combines reactivity and deliberative reasoning by the distribution of the knowledge system into modules called *behaviours*. A *BOC* is a service carried out by a set of cooperative associated behaviours (*ABs*) executing in parallel. The co-ordination of the *AB*s is performed by a control unit (CU) using synchronisation signals.

A general description of MagiCol robots using the *BOC* architecture was presented in our previous paper [3]. Figure 3 presents relevant aspects of the actual BOC model implementation of our goalie; ellipses and rectangles represent *behaviours* and *control units* respectively.

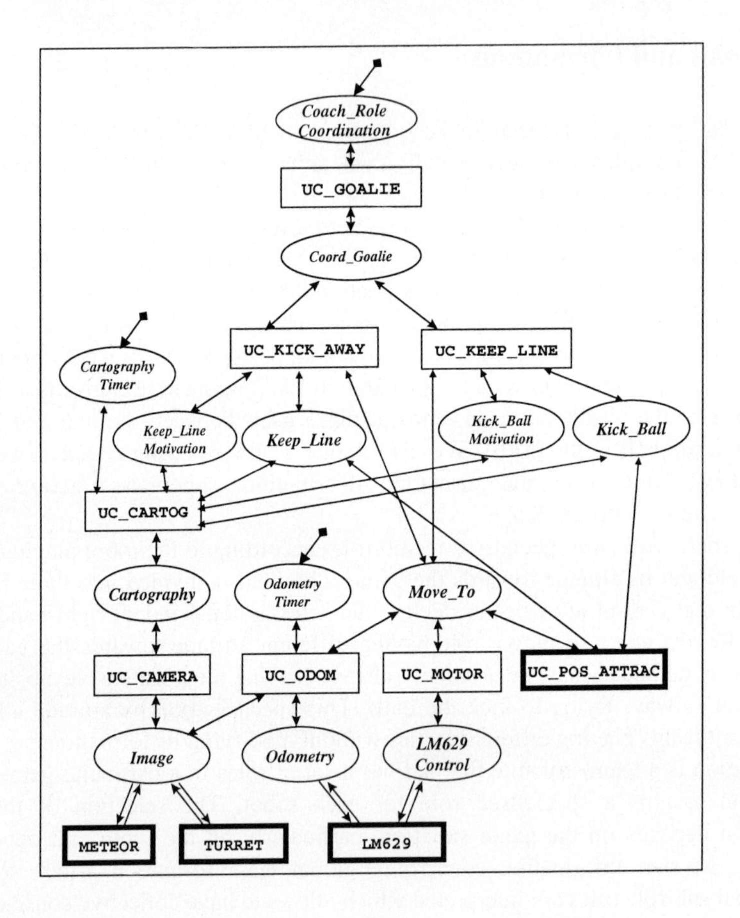

Fig. 3. BOC Model of our Goalie for RoboCup98.

The coach assigns roles by requesting a high level *BOC* to each robot. The command parameters include zone and attitude specialization attributes of the role. In our goalie, UC_GOALIE activates the *behavior Coord_Goalie*, which decides the next action to execute. Two possible high level goalie actions are defined: *Keep_Line* and *Kick_Away*. A request to the concerned *UC* is performed. When the action is finished an acknowledge is received, the state is modified and the next action to perform is selected using the control rules embedded in the *behavior*.

The execution control of the command `Keep_Line` is performed by UC_KEEP_LINE that activates three *ABs*. The behaviour *Position_Line* modifies the attraction point to make robot move to the best defensive position over its line (parallel to the goal entry). The *behavior Keep_Line_Motivation* monitors if the motivations for doing this action are still valid, when motivation falls under a threshold (specified by *Coord_Goalie*) an end signal is generated thus stopping all *ABs* and finishing the *BOC*. The third *AB Move_To* drive the robot to arrive to the attraction point specified by other cooperative *behaviors*. High level behaviors have access to sensors and actuators through low level commands.

6. Conclusions

The hardware and software architectures of the Real Magicol Robots were presented. These robots were built for the RoboCup competition, but future research in other subjects has also been considered. Teams using commercial robots performed generally better in Robocup98. In our case, building our own robots demanded a greater effort than expected, leaving little time for software debugging. In fact, the team was developed in only six months. We still believe that a greater insight and a more prospecting architecture has been obtained. We currently work on increasing the overall robustness of the robots and study the addition of new sensors and fusion algorithms.

The BOC architecture was used with few modifications and proofs to be well adapted to this kind of challenge, allowing a straightforward well structured real-time implementation of the proposed concepts.

In our current implementation, collective behaviours emerge as a result of role attribution and team formation. We plan to extend the task parallelism to the team as a whole by adding explicit communication between the players. The increased information exchange, should also allow our coach to detect specific strategy patterns from the opponent team in order to adapt our own strategy.

Acknowledges. We wish to express our gratitude to A. Barandica, E. Caicedo, L. Ebrard, A. Ramonaite, L. Boutté, S. Lelandais. This work was financed by the Fundación de la Universidad del Valle and Colciencias-Colombia. We also wish to thank the Maloka Museum, Sinfor, Renault Automation, Allegro Microsystems, QNX, Societé M.A.F.

7. References

1. Realismo Magico. http://artcon.rutgers.edu/artists/magicrealism/magic.html
2. Cuervo J., González E.,Suárez A., Moreno , «Behavior-Oriented Commands: From Distributed Kwoledge Representation to Real Time Implementation », *Proc.Euromicro Workshop on Real-Time Systems,* Jun. 1996. pp. 151-156
3. Loaiza H., Suarez A., González E., Lelandais S., Moreno C., «Real MagiCol: Complex Behavior through simpler Bahavior Oriented Commands », *Proc.of the second RoboCup Workshop,* July. 1998. pp. 475-482
4. Marszalec E., Pietikäinen M., «Some Aspects of RGB Vision and its Applications in Industry», *Machine Vision for Advanced Production,* Series in Machine Perception and Artificial Intelligence, vol. 22, 1996, pp 55-72.
5. Shen Wei-Min, Adibi Jafar, Adobbati Rogelio, Cho Bonghan « Building Integrated Mobile Robots for Soccer Competition », *http://www.isi.edu*
6. Veloso Manuela, Stone Peter, Han Kwun « The CMUnited-97 Robotic Soccer Team : Perception and Multiagent Control, *http://www.cs.cmu.edu/{~mmv,~pstone,~kwunh}*

Agilo RoboCuppers: RoboCup Team Description

Michael Klupsch, Maximilian Lückenhaus, Christoph Zierl, Ivan Laptev, Thorsten Bandlow, Marc Grimme, Ignaz Kellerer, Fabian Schwarzer

Forschungsgruppe Bildverstehen (FG BV) – Informatik IX
Technische Universität München, Germany
{klupsch,lueckenh,bandlow,schwarzf,zierl}@in.tum.de
http://www9.in.tum.de/research/mobile_robots/robocup/

Abstract. This paper describes the *Agilo RoboCuppers* [1] – the RoboCup team of the image understanding group (FG BV) at the Technische Universität München. With a team of five Pioneer 1 robots, equipped with a CCD camera and single board computer each and coordinated by a master PC outside the field we participated in the medium size RoboCup league in Paris 1998. We use a multi-agent based approach to represent different robots and to encapsulate concurrent tasks within the robots. A fast feature extraction based on the image processing library HALCON provides the necessary data for the onboard scene interpretation. These features as well as the odometric data are checked on the master PC with regard to consistency and plausibility. The results are distributed to all robots as base for their local planning modules and also used by a coordinating global planning module.

1 Introduction

Our research group started working on robot soccer at the beginning of 1998 considering it a very challenging and interesting research domain for several reasons. The main challenge is to combine several complex computer domains, like vision, robotics, and artificial intelligence to one real system of several autonomous hard- and software components which perform together one common task. For this, multiple agents need to collaborate. They should be able to organize themselves, to learn how to act in specific situations, and to handle with uncertain data. The basic conditions are quite harsh: a dynamically changing environment is to be observed in real time and fast moving objects like ball and opponent must be recognized, tracked, and considered within planning methods for controlling the movement of the own robots.

The aim of our activities on robot soccer is to develop software components, frameworks, and tools which can be used flexibly for several tasks within different scenarios under basic conditions, similar to robot soccer. This can be used for teaching students in vision, artificial intelligence, robotics, and, last but not least, in developing large dynamic software systems. For this reason, our basic

[1] The name is derived from the Agilolfinger, which were the first Bavarian ruling dynasty in the 8th century, with Tassilo as its most famous representative.

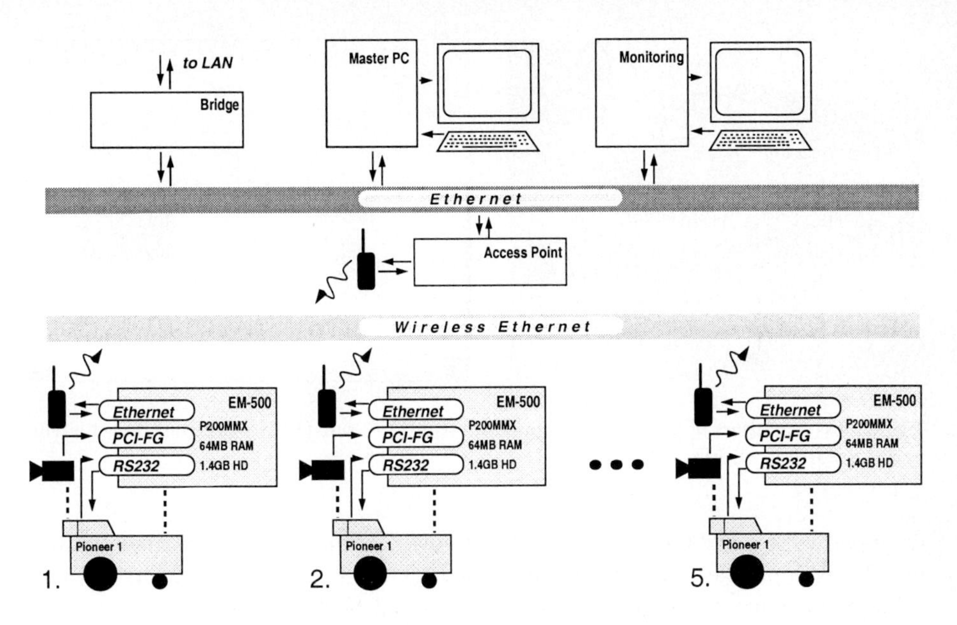

Fig. 1. Hardware architecture.

development criterion is to use unexpensive, easy extendible standard components and a standard software environment.

This paper is organized as follows. Section 2 gives an overview about the employed hardware, namely robots and computers. It follows in section 3 the description of some fundamental concepts concerning the overall system structure. Details on the design of the most important components are given in section 4.

2 Hardware Architecture

Our RoboCup team consists mainly of five Pioneer 1 robots, five single board computers and one master PC. The single board computers are mounted on the topside of the robots, firmly fixed – mechanically and electrically. All robot computers are linked together via a radio ethernet network[4]. The master computer is linked to the radio ethernet, too, and is located outside the soccer field. For debugging during software development and monitoring the robots' behaviors and features extracted from the sensor data, we use an additional monitoring computer. The operating system for all computers is Linux. Figure 1 gives an overview about the hardware architecture.

Figure 2 (a) shows one of our Pioneer 1 robots [1]. Each of them measures 45 cm × 36 cm × 30 cm in length, width, and height and weighs about 12 kg. Inside the robot a Motorola microprocessor is in charge for controlling the drive motors, reading the position encoders, for the seven ultrasonic sonars, and for communicating with the client. In our case this is a single board computer (EM-500 from [2]) which is mounted within a box on the topside of the robot. It is

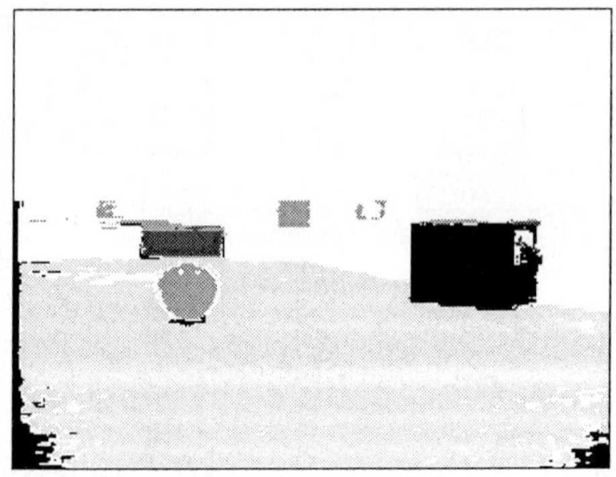

Fig. 2. (a) Hugibert – one of our Pioneer 1 robots – and (b) a example view for the world according to him.

equipped with a P 200 processor, 64 MB RAM, 2.5" hard disk, onboard ethernet and VGA controller, and an inexpensive PCI video capture card [3]. PC and robot are connected via a standard RS232 serial port. A color CCD camera is mounted on top of the robot console. For better ball guidance we mounted a simple concave-shaped bar in front of each robot.

3 Fundamental Software Concepts

The software architecture of our system is based on several independent modules which perform each a specific task. Software agents control the modules, they decide what to do next and are able to adapt the behavior of the modules they are in charge for according to their current goal. For this, several threads run in parallel. In Figure 3 an overview about the software architecture of our system is shown.

The modules are organized hierarchically, within the main modules basic or intermediate ones can be used. The main modules are image (sensor) analysis, robot control, local planning, information fusion, and global planning. The latter two run on the master PC outside the field, the others on the single board computers on the robots.

Beside the main modules there are some auxiliary modules, one for monitoring the robots, extracted sensor data and planning decisions, one for interacting with the system or with particular robots, and one for supervising the running processes. A large number of basic functions define fundamental robot behaviors, provide robot data, and realize different methods for extracting particular sets of vision data.

As for the communication between different modules, we strictly distinguish between controlling and data flow. One module can control another by sending messages to the appropriate agent. Data accessed by various modules is handled in a different manner. For this, a special sequence object class was defined. This

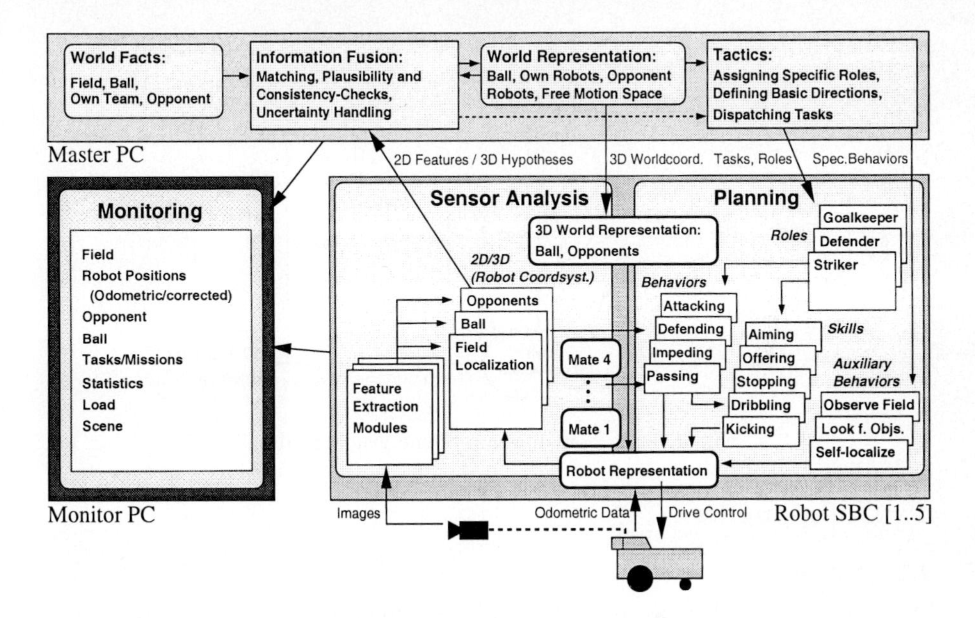

Fig. 3. Software architecture.

offers a consistent concept for exchanging data between arbitrary components [6]. The most important features of these objects are:

- A list of previous sequence values is held allowing to access values from the past. Values to old are "forgotten".
- A uniform time stamp is attached to all sequence values.
- The sequence objects "know" their functions for updating the sequence.
- Side effect functions can be defined which are performed after updating a sequence, and others, if the update failed.
- An update is automatically triggered when querying for a new sequence value.
- Sequence values can be interpolated or predicted in the case that a value is requested for a time between two available values or for a future time, respectively.
- Sequence objects are global and thread-save, i.e. one module can be in charge for setting the values appropriately and other modules can use their values at the same time.
- Sequence data can easily be made transparently global over a network. So all robots as well as the master PC can access the data of the other robots.
- They can be (re)configured online dynamically during the running process.

The agents are responsible for triggering the sequences, which are needed in the current situation and to configure them according to the actual robot task.

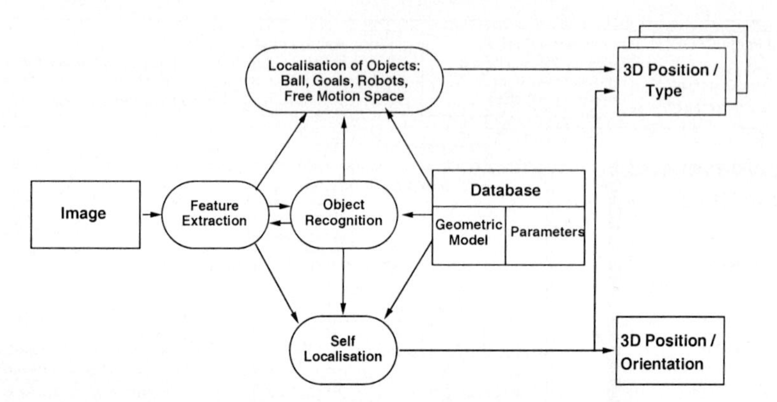

Fig. 4. Data flow diagram of the vision module.

4 Components

4.1 Vision

The vision module is a key part of the whole system. Given a raw video stream, the module has to recognize relevant objects in the surrounding world and provide their positions on the field to other modules. This is done with the help of the image analyzing tool HALCON (formerly known as HORUS [5]). The extracted data is provided by sequence objects as described in section 3.

In general, the task of scene interpretation is a very difficult one. However, its complexity strongly depends on the context of a scene which has to be interpreted. In RoboCup, as it is defined in the present, the appearence of relevant objects is well known. For their recognition, the strictly defined constraints of color and shape are saved in the model database and can be used. These constraints are matched with the extracted image features such as color regions and line segments (see Fig. 2 (b) and 4).

Besides recognizing relevant objects, a second task of the image interpretation module is to localize the recognized objects and to perform a self-localization on the field if needed. To localize objects we use their estimated size to determine their position relative to the robot. Self-localization is performed by matching the 3D geometric field model to the extracted line segments of the border lines and the visible goals.

4.2 Information Fusion

The information fusion module has to combine the fragments of information, which are provided by our robots, and form a consistent view of the world. This component is supposed to have two representations of the world: The geometrical one consists of the position, orientation, and velocity vector of all robots and the ball. Another representation is based on a grid covering the soccerfield and its contents. This will mainly be used for consistency checks and for global planning.

Information concerning uncertainties of the robot data is provided by the robots in conjunction with the odometric data. It is represented by ellipsoids where the volume of the ellipsoid corresponds with its uncertainty.

4.3 Planning

Planning is done on two levels: A *central, global* planner coordinates the different team members and a *local* planner, that resides on each robot, controls the behavior of a single robot. Each robot has its own view of the current situation and incorporates global data from the information fusion module to choose an appropriate action (behavior or skill). This is done by evaluating the actual values of the sequence objects. The decision is also influenced by the current *role* of the robot that may change during the game (except for the goalkeeper).

The task of the *global planner* is to coordinate the robots. E.g., if several robots try to reach the ball, the global planner chooses the most promising of them as forward and asks the others to change their role so that they do not impede the selected forward. Decisions of the global planner are driven by the world view provided by the information fusion module. Therefore, the global planner depends on a stable connection to all robots. In case of an instable or disturbed interconnection the robots may also work properly controlled only by their local planner.

5 Experiences during RoboCup'98 in Paris

As a result of our participation, we summarize the following:

- We have made good experiences with concentrating on our main scientific goals – robot vision and intelligent control of distributed systems – by combining robust standard components for the robot hardware. However, one should not underestimate the effort to establish and maintain a running robot system.
- A good monitoring system is inevitable for debugging and analyzing within a distributed robot system. This is particularly true with a system sharing global data for coordination.
- The lack of a kicking device is a main disadvantage in terms of competition.
- The different interpretation of the rules (e.g. size restrictions, penalty shoot-out) by teams and referees made it difficult to enable a fair competition.

Building up and maintaining a RoboCup team is a great challenge and needs huge personal efforts and a lot of time. Thus we hope that we will still have enough resources in future to continue our promising work.

References

1. ActivMedia Robotics, *http://www.activmedia.com/robots/*
2. Lanner Electronics Inc., *http://www.lannerinc.com/*
3. Videologic Inc., *http://www.videologic.com/ProductInfo/capt_pci.htm*
4. AAEON Technology Inc. *http://www.aaeon.com.tw/html/prod.htm*
5. Eckstein W., Steger C., *Architecture for Computer Vision Application Development within the HORUS System*, Electronic Imaging: 6(2), pp. 244–261, April 1997.
6. Klupsch, M., *Object-Oriented Representation of Time-Varying Data Sequences in Multiagent Systems*, 4th International Conference on Information Systems and Synthesis – ISAS '98, pp. 33–40, 1998.

The Ulm Sparrows: Research into Sensorimotor Integration, Agency, Learning, and Multiagent Cooperation

Gerhard K. Kraetzschmar[1], Stefan Enderle[1], Stefan Sablatnög[1], Thomas Boß[1], Mark Dettinger[2], Hans Braxmayer[1], Heiko Folkerts[1], Markus Klingler[3], Dominik Maschke[1], Gerd Mayer[1], Markus Müller[1], Alexander Neubeck[1], Marcus Ritter[1], Heiner Seidl[3], Robert Wörz[1], and Günther Palm[1]

[1] Dept. Neural Information Processing,
[2] Dept. Software Engineering and Compiler Construction,
[3] Dept. Mathematics and Economics,
University of Ulm, Oberer Eselsberg, 89069 Ulm, Germany
`sparrows@neuro.informatik.uni-ulm.de`,
WWW: `http://smart.informatik.uni-ulm.de/SPARROWS/`

Abstract. We describe the motivations, research issues, current results, and future directions of THE ULM SPARROWS, a project that aims at the design and implementation of a team of robotic soccer players.

1 Motivation and Research Goals

Engineering a team of robotic soccer players that is able to successfully compete in championship games requires truely interdisciplinary research and the effective cooperation of many disciplines. This is our main scientific reason for setting up a local research effort related to ROBOCUP and to implement THE ULM SPARROWS, a team of soccer agents that is currently projected to participate both in the mid-size real robot league as well as the simulation league. A secondary reason is the coincidence of ROBOCUP research issues[1] with midterm research goals of SMART[2][1], a local, large, interdisciplinary joint research project that aims at the development of a complete cognitive architecture for an adaptive robotic agent performing complex tasks. Not the least reasons for starting ROBOCUP activities were their attractiveness to students and its potential for appealing educational activities.

Considering the wide variety of fields involved with ROBOCUP, no single group or team can equally well address all research issues of interest. In order to ensure high-quality research results, both a long-term commitment to ROBOCUP activities and a well-focused research agenda are essential. For our team, the following open research issues are of particular interest: *Learning* elementary football skills from demonstration, by reinforcement or teaching, *Spatial Modeling* of highly dynamic environments including their rapid update by sensor

[1] See `www.uni-ulm.de/SMART/` for more information.

interpretation and fusion, *Emergent multiagent cooperation* for achieving team play without explicit communication, *Robot control architectures*, *Neurosymbolic integration* in robot control architectures. Furthermore, we also favor the simultaneous use of *simulated and real-world* testbeds for robotics research, in particular the development of good simulation models of sensors and actuators of (increasingly complex) robotic agents.

Solving these problems, even making good progress, will undoubtedly take many years, and our initial focus is to lay the groundwork on which these problems can be studied.

2 Architectural Overview

Developing both a simulation *and* a real robot team at the same time is a serious challenge. In order to permit as much coherence as possible between both development efforts, we try to develop a *common* control architecture to be used by agents in both teams.

The architecture of our agents is structured into a reactive and a deliberative layer (see Figure 1). In the reactive layer, a sensor interpretation module, a dynamically determined set of behavior modules, and an arbiter for prioritizing conflicting primitive actions are concurrently executed. Together, they implement a fast sensorimotor feedback loop and ensure timely response to changes in a highly dynamic environment. The deliberative layer itself consists of three levels: *i)* elementary tasks (behaviors), *ii)* coherent action patterns (combinations of behaviors to form simple playing patterns like passing the ball, shooting a goal, etc.), and *iii)* cooperative action patterns (patterns, where several team players take different roles and interact). On each level, a situation/play pattern associator evaluates the currently perceived situation and matches it against a database of situation/behavior patterns (aka plan library). If a matching pattern is found, it dynamically modifies the set of concurrently executed behaviors in a controlled way. The *elementary task behaviors* implement a set of soccer-oriented reactive action patterns. These patterns are abstractions of short sequences of low-level commands which the agent can execute directly. At the same time, the behaviors provide a certain degree of isolation from low-level details, thereby reducing dependency on particular details of the soccer server protocol or the hardware.

Building upon the behavior level, *coherent action patterns* focus the selection of behaviors to locally coordinated action sequences, which exhibit coherence over time. Thus, action patterns provide situation-oriented partitioning of the space of possible behavior sequences and focus of attention for a single agent.

Coordinated team play is dealt with on the group level. Albeit we assume explicit mechanisms to achieve cooperative behavior of soccer agents, we do neither assume nor rely on explicit communication. Cooperative behavior among multiple agents in games like soccer is an emergent phenomenon. Agents have a database of stored *cooperative play patterns*. Each agent continuosly tries to match the currently perceived situation against this database. If it can match

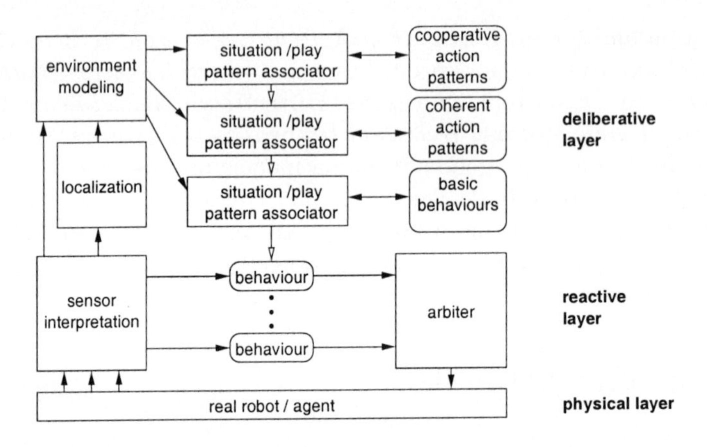

Fig. 1. Architectural Overview

itself *and* at least another teammate with roles in a tactical move, this move is executable. Currently, such a move is executed directly, but situated valuation and selection of several executable moves will be necessary in the future. Whenever no cooperative play patterns are executable or in the case of failures, the agent falls back to using single agent coherent play actions. Currently, all players except for the goalkeeper use the same database of cooperative play patterns.

3 The Robot Sparrows: The Middle Size Robot Team

The real robot team of The Ulm Sparrows consists of four robots: three Pioneer-1 and a tracked vehicle as goalie (see Figure 2). Our robots are com-

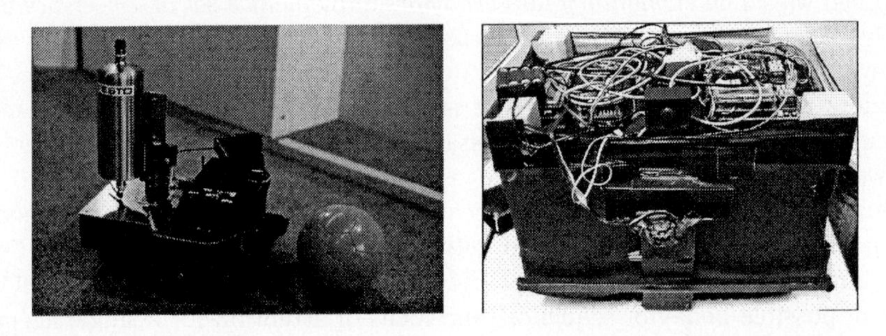

Fig. 2. The Ulm Sparrows: a field player (left) and the goalkeeper (right).

pletely autonomous, i.e. all computation is performed onboard; no radio links to offline workstations are not needed for play action, but may be added for debugging purposes. The robots are equipped with a Pentium CPU board in PC 104 technology, including various I/O interfaces, harddisk, and a framegrabber. A wide angle color camera is mounted on a servo providing 180 degrees of pan angle. Thus, our field players do have a neck, which permits (largely) independent control of view direction and movement heading. Together with the wide angle camera, the robot has almost complete view of its surroundings without having to turn its body. This proved to be very usefull as our robots proved to be able to detect the ball without long exploratory movements on the field.

The vision system implements a real-time color blob tracking algorithm. Research to invesitigate the combined use of color and other features like shape or texture is underway.

A real problem for Pioneer robots is kicking the ball. Their acceleration and top speed is too slow to allow for competitive kicks by just pushing the ball. Given the size limitations for the mid-size league robots, designing kickers for the Pioneers is not easy. We opted for a solution using pneumatics. A high pressure air tank was added. An electrically triggerable valve gives air into a pneumatic cylinder, which carries a plastic paddle at its end that actually exerts force onto the ball.

Movement Control: The movement of the robots is controlled by a behavior-based reactive control system as sketched in Figure 1. All lower-level action is determined by *behaviors.* The actions a behavior can control are setting the speed or direction of the robot, setting the camera angle, or changing state variables, like *ball-is-visible.*

A pool of mutually exclusive basic behaviors (see bottom of the deliberative layer) are waiting to be activated. Each one of these behaviors implements a small, simple task, like searching for a goal, moving to the ball, or kicking. They all have in common that they cannot be performed simultaneously and must therefore be activated sequentially. An example sequence for shooting a goal could be: *search for goal, search for ball, move to ball, kick ball into goal.* Behavior sequences are invoked by deliberative level, which corresponds to the levels coherent action patterns and group patterns in the common architecture and at the moment holds a number of hand-coded plans.

Aside of the mutually exclusive sequential behaviors, the behaviors in the reactive layer can be activated in parallel. For example, an *avoid* behavior for avoiding collisions with walls or other robots or a *stall* behavior for detecting stalled wheels may be executed simultaneously. However, when the robot approaches the ball, the *avoid* behavior must be temporarily suppressed. The concurrent execution of parallel behaviors is also represented in higher-level plans. We use a very simple fixed priority scheme to arbitrate the proposed actions given by the active behaviors.

Orientation Estimation: The absolute position is of the robot is not tracked. Only information about the orientation is collected. As soon as the vision system identifies a goal, the orientation is updated accordingly. While none of the goals is visible the odometry gives us a good enough estimate of the orientation. Competition experience proved that this concept suffices for achieving coherent action patterns. But as CS-Freiburg demonstrated, exact information about the own (and the team members) position is essential for cooperative group play. Our future research will therefore try to extend our capabilities in position estimation.

4 Simulation Team

THE ULM SPARROWS simulation team consists of two different variants of agents: a generic field player agent and a goalie agent. The software architecture of both types of agents consists of two modules: *i)* the world model, which collects and integrates the information sent by the soccer server, and *ii)* the action selection module, which itself is divided in three levels as described in Section 2. World modeling and action selection are run in separate threads. This proved problematic as our development platform (Linux) was different from the platform used in Paris (Solaris).

World Model: The world model parses the messages from the server and tries to keep a consistent model of the situation of the game by accumulating the information provided by the soccer server into an allocentric representation. By fusion of extrapolated information with the actual snapshots a more complete and stable representation of the current situation is achieved. Future versions of the world model will try to integrate more sophisticated approaches for modeling the uncertainty which is involved in the perception. The competition gave us the important possibility to collect data sets with a variety of opponents, which exhibited a lot of different strategies.

Action Selection: The action selection module largely follows the common agent architecture described in Section 2.

Behaviors: At the elementary task level a set of robust behaviors is provided which serve as basic building blocks for higher-level coherent single agent action and coordinated group behavior. Behaviors implement a soccer-oriented abstraction of primitive actions by sequencing multiple atomic actions over relatively short periods of time. In order to build a basic soccer player, an agent should at least be able to perfom the following list of behaviors: moving with and without the ball while avoiding collisions, dribbling, shooting (to score a goal), intercept, pass to stationary and moving team members. At any given time only one of these behaviors is active. It remains active until the task is believed to be completed. Deciding the termination condition is usually performed by the upper levels of our architecture, but behaviors can themselves provide hints whether they consider their task finished or not, or whether they cannot fulfill the desired action. For example the passing behavior would signal an error if the player does not have the ball.

Coherent Action Patterns: The space of all possible situations is partitioned into six prototypical situations, each of which induces a different action pattern. Partitioning the situation space is performed according to the following two criteria: *i)* the distance of the individual player to the ball (near/far), and *ii)* the team currently in possession of the ball (own/opponent/unknown). According to the result of the match of the actual situation with the six prototypical situations, the player follows a predefined strategy invoking a reasonable behavior of the agent.

Group Play Patterns: For emergent multiagent cooperation in soccer, an agent has to recognize situations where cooperative play is possible. For that, the agent must be able to match itself and other team member with different roles required by group play patterns. The group play pattern module examines the overall situation as given by the world model. As soon as it is able to match one of the prestored situation patterns to current perception, the agent identifies its role in this pattern and starts executing the predefined plan, which is bound to this role. Cooperative group play will only emerge, however, if several agents dynamically activate the same group play patterns with the correct role instantiations. While executing a group play pattern, an agent monitors its progress.

5 Conclusions

In this paper, we presented an overview on the research done in the THE ULM SPARROWS team as well as the current state of our soccer teams. Most work on teams has been performed within the last 2–3 months, after a sufficient number of students could be interested and engaged itself into team implementation. As all people involved were new to ROBOCUP, getting working teams of robot soccer players (and a collection of software agents) was our main focus for ROBOCUP-98. With all the basic groundwork done, the foundation for doing serious research on the issues discussed earlier is laid. Modelling the environment, the overall agent architecture, and learning schemes for basic skills will be in the center of research within the next year.

The results of Paris showed that our architecture is sufficient to be used for our further research. Minor hardware and software problems that harmed performance in Paris could be identified and resolved with the experiences gained in a real competition.

References

1. Hiroaki Kitano, Minoru Asada, Yasuo Kuniyoshi, Itsuki Noda, Eiichi Osawa, and Hitoshi Matsubara. RoboCup — a challenge problem for ai. *AI magazine*, pages 73–85, Spring 1997.
2. Günther Palm and Gerhard Kraetzschmar. Integration symbolischer und subsymbolischer Informationsverarbeitung in adaptiven sensomotorischen Systemen. In Matthias Jarke, Klaus Pasedach, and Klaus Pohl, editors, *Informatik 97: Informatik als Innovationsmotor*, Informatik aktuell, pages 111–120, Heidelberg, Germany, 1997. GI, Springer.

ART
Azzurra Robot Team

Daniele Nardi, Giorgio Clemente, and Enrico Pagello

Progetto RoboCup-Italia
Consorzio Padova Ricerche, Corso Spagna 12, I-35127 Padova, Italy
nardi@dis.uniroma1.it
http://www.dis.uniroma1.it/~nardi/ART.html

1 The ART team

Azzurra Robot Team is the result of a joint effort of seven Italian research groups from Univ. of Brescia, Univ. of Genoa, Politecnico of Milano, Univ. of Padua, Univ. of Palermo, Univ. of Parma, Univ. of Roma "La Sapienza", and the Consorzio Padova Ricerche which has provided resources and a set up of the soccer field in its Center in Padua. Our goal at Robocup 1998 has been to provide a flexible and low-cost experimental team to make experience before undertaking a larger project. Our long term goal is to foster the development of research and education projects on autonomous mobile robots by exploiting the RoboCup challenge.

In order to exploit everyone's experience we need to accomodate different kinds of players in terms of both hardware and software architecture. We thus have a first set of players that have been designed with a common basis and constitute the main skeleton of the team, but we also have another type of player, named Mo2Ro, which has been designed by the group at Politecnico di Milano. The ART team in action is shown in Figure 1, where one can see the player Mo2Ro close to the ball, three other middle field players in the center, and the goal-keeper defending the goal. It is obvious that such a diversity raises several problems from both the organizational and development viewpoint, but we have considered to be an interesting challenge of the overall project the ability to design players with different features, yet capable to interact with the team mates. In the following section, we first focus the presentation on the first group of players and then briefly address the second type. In the last section we describe our research perspectives within the Robocup framework.

2 The ART Players

The design of the basic ART player has been mainly guided by the goal of achieving an open architecture where new hardware and software components could be easily added. In addition, we aimed at realizing a good development environment to carry on experiments in the field, thus allowing many different technical solutions to be tested.

Fig. 1. The ART team in action

The hardware architecture is illustrated in Figure 2. The first building block is constituted by the Pioneer mobile basis and the second is constituted by a conventional PC for onboard computing. We have reached a compromise between weight and power consumption, where the player has enough autonomy to play games.

The third building block is constituted by a wireless high bandwidth connection that we consider necessary to have a development environment that allows the programmer operating on a standard platform (connected to the robot) to obtain accurate information about the situation onboard. The wireless connection supports also the exchange of information among the players, but it is not used to transfer raw data among the players.

The fourth component is the vision system which is constituted by a low-cost frame grabber based on the BT848. At Robocup 98 we have used a very cheap color CCD camera with a resolution 380 TV lines. On the middle field player the camera is mounted in front part of the robot, while on the goal-keeper the camera points at a convex mirror provinding a 360^0 view of the field.

The remaining components are devices that we are able to connect (as indicated in Figure 2) through the Pioneer input/output ports or through an ad hoc made board on the ISA bus. Aa for additional devices, we have used a compass, infrared sensors, a bumper for the goal keeper and a kicker for the middle field player. The kicker, operating with air pressure, allows the player to choose the direction to give to the ball. In fact, it is constituted by two independent devices that allow for a kick left, a kick right and a kick straight action, when they are used simoultaneously. We consider the capability of performing kicking actions

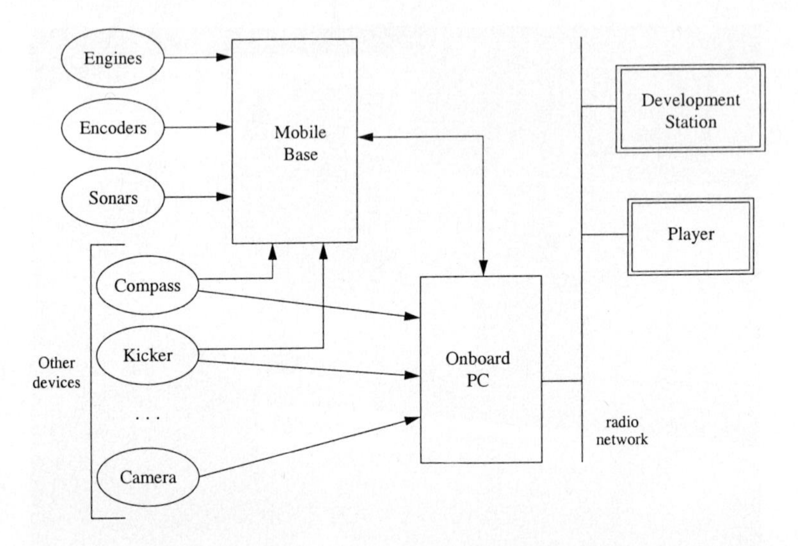

Fig. 2. The basic hardware components

essential not just to pursue the football challenge, but to provide a setting where the knowledge about the environment and the game strategy play a significant role in the robot performance.

The software architecture is centered around two main components: LINUX and SAPHIRA. SAPHIRA is an environment that has been designed to develop fuzzy controllers and is provided with the Pioneer mobile base. The features of SAPHIRA that we have considered critical for software design and engineering are Client Server Architecture and the development environment suited for rapid prototyping. SAPHIRA provides a language for specifying concurrent activities, where the primitive activities can be either programmed or implemented as fuzzy behaviours and composed as described in [6]. We have experimented different strategies for designing the control systems, either by straight activity programming and by designing fuzzy behaviours.

Figure 3 shows the main components of the software architecture. The Local Perceptual Space (LPS) is the structure where all the information about the status of the system is stored. Specifically, the LPS includes a description of static objects: goals, walls, lines and field, as well as information about dynamic objects: ball and other players. The information stored in the LPS is used by the controller to drive the actions of the agent. The other modules shown in the figure provide some details on the process of acquiring information and putting it in the LPS. The sources for new information can be either sensing devices or other agents.

The module labelled Sensor Data Processing is dedicated to the processing of data gathered from the sensors and gives as output information about the objects in the LPS, such as position, speed etc. In acquiring information from sensors a central role is played by the vision system. We have tried various approaches to extract from the images information about the objects that are in the field and are represented in the LPS. In particular, we have developed a system based

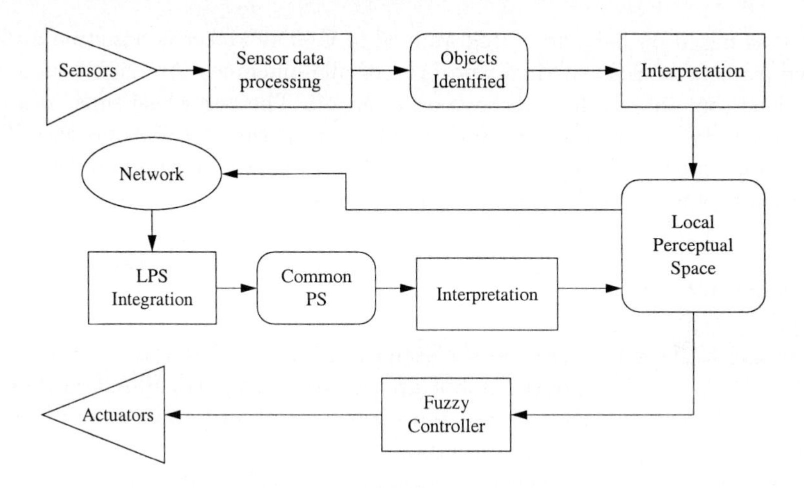

Fig. 3. The basic software components

on HSV color filtering to identify the ball and evaluate its distance and angle relative to the robot's position at a rate of 10 frames per second.

The module labelled Interpretation has the goal of turning the object descriptions obtained by the sources into the proper representation in the LPS. This process is accomplished by taking into account the reliability of the various input devices, the frequency of acquisition and specific features of the objects. Information about static objects may lead to adjustments of the position of the robot, while information about dynamic objects changes their position relatively to the player's position.

Information about the status of the system can also be acquired from other agents, which broadcast the information gathered locally to other agents. To this end, we have designed a communication protocol suitable and efficient for the selected domain, and a high-level inter-robot interface dealing with the exchange of information. However, communication failures do not block the functioning of the robot, that is able of acting on the basis of the information locally available.

2.1 Mo2Ro, a modular, mobile robot

The second type of player, called, Mo2Ro, is a mobile robot consisting of two main modules: the electromechanical base, including the electronic devices providing power supply and low-level sensor management, and the PC. It has two wheels (diam. 12.5 cm) each moved by an independent engine, and a free wheel. Its maximum speed is 100 cm/sec, and it can turn on place. The payload is about 60Kg. It has bumpers, encoders and battery level sensors. It is possible to add to the base any kind of modules, including sensorial modules (sonars, cameras) and actuator modules (a kicker and a robotic arm). The add-on modules can also extend the size of the base, relying on an additional wheel and hinging on the base. A basic fuzzy control system ensures the accomplishment of movement commands given as position and/or speed setpoints. The software architecture runs on the on-board PC (a PENTIUM 233MMX) under Linux operating system. It is based on behaviors programmed in C. The kicker is based on an idea drawn from Leonardo da Vinci's projects of pile machines. An engine mounts a cog-wheel some of which teeth have been cut off. The cog-wheel runs on a rack charging a spring-piston until a sensor detects that the teeth are present. Then the engine stops, and can be turn on again by a single bit command from the PC, thus leaving the spring piston to kick the ball.

3 Research goals

The research issues that arise in the scenario of the football game proposed for the "Middle-size" league of the RoboCup are discussed in [1]. Based on the kind of capabilities of the players we are addressing the following two issues: Sensing, Action and Control; Agent Coordination.

From the viewpoint of Sensing Action and Control we have tried to keep the player architecture as open as possible to make it easy to adapt different kinds of sensing devices and actuators. The underlying idea is to introduce many sources of information as well as different action capabilities. The approach proposed in [6] has been evolving in response to the need of controlling a much more complex robot, with different sensing devices, action capabilities and in a multi-agent environment. Many such aspects are not quite settled and we are currently proposing a possible extension through the software architecture outlined above. In particular, we emphasize the separation between the processing for acquiring new data and their interpretation according to the internal state of the robot and the knowledge about different sources of information.

However, we believe that in order to handle properly a variety of sensing and action devices, possibly trying to apply specific game strategies, a deliberative layer on top of a fuzzy control system is required. In previous work we have proposed such an approach, by relying on a declarative language for representing and reasoning about actions[3]. In the proposed setting both static and dynamic knowledge about the system is used to generate a plan, possibly including sensing actions [4], to achieve a given goal. The main issue we are investigating is to establish a proper and effective connection between the deliberative and the control level. Specifically, there is the need both to find a framework where there is a precise relationship between logical specification of actions and the underlying control system and methodologies for deciding which aspects of the control are better handled at the deliberative level. In addition, we are developing tools to support the development and debugging of the control system.

The second issue we have addressed is coordination. We have developed algorithms allowing the fusion of the multi-robot sensor data (in particular of the perception of the robots' and ball position) to reduce the overall error. These components have been integrated in SAPHIRA.

In a future perspective, we aim at the design and development of a novel distributed software architecture, not only to meet the robocup challenges, but also for more general tasks. This activity will build on the previous experiences in the field of distributed software architectures [8] and of robot cognitive modelling. Moreover, part of the research activity will also focus on the solution of navigation and planning problems throughout forms of analogical representation and reasoning [5].

Finally, we are transferring the experience gained in participating to the Simulation League into the middle size-league on the particular aspect of getting an emergent collective behavior [7]. The limitations in the ability of performing behaviors in the case of middle-size robots, makes the overall performance of the real multirobot system weaker than the one obtained by simulated agents. However, some basic cooperative patterns have been identified and complex support actions have been experimented.

Acknowledgments The project has been supported by Facoltà di Ingegneria, Università di Roma "La Sapienza", by Politecnico di Milano and by Consorzio Padova Ricerche, for the equipment. All the other institutions of the team members have provided various forms of support to the individual members of the project. Daniele Nardi is the team coordinator. Giovanni Adorni, Andrea Bonarini, Riccardo Cassinis, Giorgio Clemente, Enrico Pagello, Maurizio Piaggio, and Pietro Storniolo have coordinated the activity at their sites. The list of people that we want to acknowledge is very long and it includes colleagues and students who have provided most of the motivation and enthusiasm that was required for setting up the team. We thank Luigia Carlucci Aiello for providing stimuli and various forms of support to the initiative; Kurt Konolige for the technical support; Adriano Maurizio and Gianni Zampieri for their work, especially in setting up the environment for experimentation.

References

1. Asada. M., et alii. The RoboCup Physical Agent Challenge: Goals and Protocols for Phase-I. In *Robocup-97: Robot Soccer World Cup I*, Nagoya H. Kitano Ed., L.N. on A.I., Springer Verlag 1998, pp. 42-61.
2. A.Chella, M. Frixione, S.Gaglio: A Cognitive Architecture for Artificial Vision, *Artificial Intelligence*, 89, No. 1-2, pp.73-111, 1997.
3. G. De Giacomo, L. Iocchi, D. Nardi, and R. Rosati. Moving a robot: the KR&R approach at work. In *Proceedings of the Fifth International Conference on the Principles of Knowledge Representation and Reasoning (KR-96)*, 1996.
4. G. De Giacomo, L. Iocchi, D. Nardi, and R. Rosati. Planning with sensing for a mobile robot. In *Proceedings of the European Conf. on Planning (ECP-97)*, 1997.
5. M. Frixione, G. Vercelli, R. Zaccaria, Dynamic Diagrammatic Representations for Reasoning and Motion Control- *Proc. ISIC/CIRA/ISAS '98*, U.S.A., 1998.
6. K. Konolige, K.L. Myers, E.H. Ruspini, and A. Saffiotti. The Saphira architecture: A design for autonomy. *Journal of Experimental and Theoretical Artificial Intelligence*, 1997.
7. E. Pagello, F. Montesello, A. D'Angelo, and C. Ferrari. Emergent Cooperative Behavior for Multirobot Systems. *Proc. of 5th Int. Conf. on Intelligent Autonomous Systems (IAS-5)*, Sapporo, Japan, June 1-4, 1998
8. M. Piaggio and R. Zaccaria, Distributing a Robotic System on a Network: the ETHNOS Approach - *Advanced Robotics Journal*, Vol. 12, N.8, VSP, 1998.

Design and Evaluation of the T-Team of the University or Tuebingen for RoboCup'98

Michael Plagge, Boris Diebold, Richard Günther, Jörn Ihlenburg, Dirk Jung, Keyan Zahedi, and Andreas Zell

W.-Schickard-Institute for Computer Science, Dept. of Computer Architecture
Köstlinstr. 6, D-72074 Tübingen, Germany
{plagge, diebold, guenther, ihlenburg,
jung, zahedi, zell }@informatik.uni-tuebingen.de
http://www-ra.informatik.uni-tuebingen.de/aktuelles/robocup.html

Abstract. In this paper we present the hard- and software architecture of the robots of the T-Team Tuebingen, which participated in the RoboCup'98. This paper describes how we try to accomplish the basic skills of our robot team capable of successfully playing robot soccer by designing our hard- and software and the experiences we made with our team at the RoboCup'98 in Paris.

1 Introduction

The RoboCup contest, which first took place in 1997 in Nagoya, Japan, is an interesting environment for teams of autonomous, mobile robots [1]. To accomplish the demands of this contest each individual robot player and the team as a whole must possess a set of basic skills. These skills can be separated in three different categories: sensorial skills, control skills and cooperative skills.
Detecting the ball and the goals, detecting the other robots and determining the own position are the basic sensorial skills. The RoboCup'98 showed that there are several different ways to fulfil these tasks [3], [4], [5]. Our approach to these three tasks relies mainly on vision processing. Only for the detection of the other robots and the walls we use additionally the available sonars. The skills of the robot control unit should be obstacle avoidance, the ability to drive to a singular position with respect to the ball position and the ability to move the ball along a specific trajectory. The assignment of the tactical role of each robot, e.g. defender or attacker and the cooperation of the robots in specific situations belong to the last category. For this last skill either direct or indirect communication is necessary.
Especially direct communication simplifies the process of team-play a lot [6]. It is possible, for example, to assign the tactic role of a robot in a dynamic manner with respect to the current situation. But the problem with direct, explicit communication is the reliability of the wireless connection. In RoboCup'98 several teams had problems with this reliability.

The remainder of the paper is structured as follows: section 2 gives a short introduction to the RoboCup setup. Section 3 describes the hardware architecture of the T-Team. Chapter 4 gives an overview of the software structure. The next section 5 presents the "tactical" setup of the team, before we summarise our experience of the RoboCup'98 in section 6.

2 The RoboCup Setup

The rules for RoboCup [2], which were applied at Paris, define the following properties of the field and the involved objects in the Mid-Size-League: The field is 8,22 m long and 4,58 m wide. The surrounding wall is painted white and the height is 50 cm. A flat panel is placed on every corner of the field. It is located 20 cm from the corner for each axis. Green strips of width 30 mm are painted to identify the edge of the panel. The surface of the field is also painted green. One of the goals, whose width and height are 1,50 m and 0,50 m, is painted yellow and the other is painted light blue. The ball is a usual leather football with a diameter of 20 cm and is painted orange-red. The robots should be mainly black. Between 30 cm and 60 cm there should be a colour-marking of at least 10 cm in any dimension. These markings were not used in Paris, because nearly no team did rely on them.

3 Hardware

To successfully operate in the RoboCup a robot has to sense players, ball, goals, and field borders, must have the ability to move around and manipulate the ball and must bridge the gap between sensing and acting. For these abilities the hardware of the robot can be separated in three different building blocks: sensors, control unit and actuators.

3.1 Actuators

Drive: The basis of our robot team are the commercially available robots Pioneer1 and Pioneer AT from Activmedia. These robots already existed at our robotics laboratory for the purpose of student education and research in the field of autonomous mobile robot systems. The four three-wheeled Pioneer1 robots are equipped with a differential drive and a free caster wheel at the back of the robot. On the Pioneer AT, which serves as goalkeeper in the team, each of the four wheels is driven by its own motor. The wheels on one side are coupled with a belt.
Kicker: As the maximum speed of the robots is about 60 cm/s, this is hardly sufficient for passing the ball or scoring a goal against a working goalkeeper without a kicking device. So we decided to develop a kicker for achieving higher ball speeds. In fact we developed two different kinds of devices. One consists of

a)

b)

Fig. 1. a) goalkeeper robot with mounted SICK laser scanner. b) one of the field players. In the black box on the back of the robot is the PC. On top of the robot one can see the Sony camera.

a pneumatic cylinder, controlled via a solenoid valve. The compressed air tank with a volume of 500 cm^3 is located in the back of the robot. This volume of air is sufficient for nearly 30 kicks. The other device consists of a spring mechanism which is wound up by a strong motor. In Paris we used only the device that works with the compressed air, as the second device could not be tested well enough in time.
To prevent timing problems with releasing the kicking device and to use the kicker in an effective way, we developed a hard-wired release, which actuates the kicker in the moment the ball touches a microswitch at the front of the robot. To prevent the kicker from being actuated every time the ball touches the microswitch, the hard-wired release can be switched on and off by the robot control software.

3.2 Sensors

Cameras: Sony ECI 21 cameras, which produce a colour PAL signal, are mounted on the top of each field player. These cameras are equipped with a pan-tilt-unit and several other adjustable features like zoom and colour temperature. The view angle of the cameras is about 45^0. They are mounted on the robots with an angle of 15^0 downwards in such a way, that the upper edge of the surrounding wall still can be detected. Our on-board vision PC was designed large enough to hold a Matrox Meteor PCI bus frame grabber, which delivers 25 fps PAL images to main memory with a resolution up to 768*576. The goalkeeper uses a simple, colour based, commercial image processing system from NewtonLabs with a NTSC CCD camera. This system is capable of tracking coloured objects

with 60 Hz. It sends information, e.g. the size and centre of gravity, about these objects over a serial device.
Laser scanner: The employed laser scanner is a LMS200 from SICK AG. This device has a 180^0 field of view and a angular resolution of $0{,}5^0$. It can measure distances up to 15 m with an accuracy of 10 mm.
Ultrasonic Transducers: The Pioneer robots are equipped with seven Polaroid 6500 Ultrasonic transducers. These transducers are used for local obstacle avoidance and for self localisation of the goalkeeper.

3.3 Control Unit

Field players: A PC from standard components with Intel Pentium 166 MMX processor is mounted on each of the four Pioneer1 robots. At the time the system was designed the decision for this processor type was made because of the favourable relation between computing power and power consumption. Each PC has 64MB RAM, a 1,2 GB Hard Disk Drive and the before mentioned framegrabber. Additionally each PC has a wireless Ethernet card from ARtem Datenfunksysteme GmbH, Ulm, an ISA adaptor to a PCMCIA card, and is connected via this card and an AccessPoint with an external Computer. At Paris, we used external universal adaptors (small boxes placed on top of the robots), as we obtained the PCMCIA cards late for finishing the Linux device driver development. The connection to the microcontroller board, which controls the robot, is realized over a serial device. The standard Pioneer1 controller board is based on a Motorola MC68HC11. This microprocessor controls the motors of the robot and calculates the x,y position from the data it obtains from the wheel encoders, among other things.
Goalkeeper: The data of the onboard vision processing, based on a MC68332 CPU is evaluated by a microcontroller board based on a second Motorola MC68332 CPU, which was developed by us. This board is capable of realising a closed loop controller for the robot position in respect to the ball position, which works at 60 Hz.

4 Software Architecture

Our software architecture (Fig. 2) is build up from a set of independent modules. As the underlying operating system we use Linux, because of its simple hardware access mechanisms and the interprocess communication capabilities. The modules can be classified by the basic skills they are designed for to accomplish. The image processing, the laser data processing and the receiver part of the base server fulfil the basic sensorial skills described in section 1. The robot control and the sender part of the base server realise the control skills, and the global data processing is responsible for the team cooperation. Besides the software, which is involved in direct robot control, we developed two tools to handle the training of colours for the image processing system (Fig. 3) and to visualise the

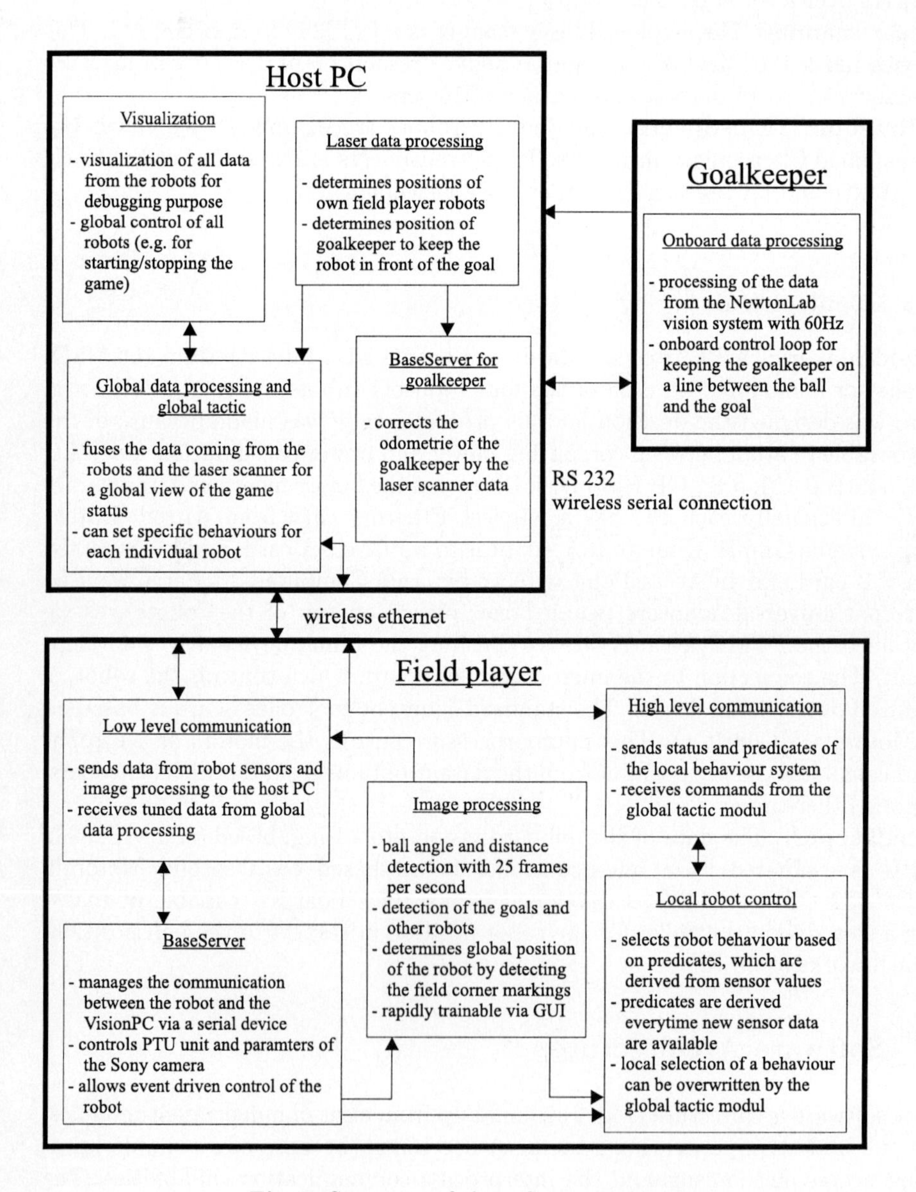

Fig. 2. Structure of the software architecture

robot sensor values and status for debugging purposes. The later tool also has the possibility to manage start and stop events during a RoboCup game in a comfortable way.

Image processing: The image processing system should handle the following three tasks in real time: Detection of the dynamic objects, detection of the statical environment (especially the goals), and estimating the position of the robot with the use of the markers in the corners of the field [7]. The image processing works with a resolution of 256*192 pixels in the YUV image format. With this resolution it is possible to detect the ball over a distance of 8 m and estimate the ball size and therefore the ball distance with an accuracy of 5 percent. The accuracy for the distance estimation of the other objects is worse because of the unknown size of these objects (only the maximum allowed size is specified and can be used for distance estimation). The error in the angular position estimate of all objects is about 1 degree. For the reason of saving time, the image processing does not search the whole image for an object, but uses a history of object positions in old images to predict the position of the object in the next new image. Only if the object is not at the predicted position, the whole image is searched for the object, starting the search at the predicted position.

The statical environment is detected by an algorithm, which predicts the lines which should be seen with respect to the current robot position. These lines are compared with the lines extracted from image processing. Even lines, which are partly hidden by an object, can be detected, if there are at least three scan points.

Experiments show, that the image processing needs 3 ms per frame in the worst case (no object at the predicted position). The average processing time for one frame is less than 1 ms. Therefore the image processing is capable of handling the 25 frames the framegrabber writes to main memory.

Base server: The base server is responsible for the communication with the microcontroller board on the robot. By using the signal handling scheme from Linux the module allows an asynchronous control of the clients. Data from the robots is put in a shared memory segment and can be accessed by the client via this segment.

Laser data processing: The data from the laser scanner, which is mounted on the goalkeeper, serves two different purposes. First, the global position of the goalkeeper is calculated and, if necessary, the position on the goalkeeper's microcontroller board is corrected. Second the laser data processing tries to localise all the other robots belonging to the team to provide data for them for an exact localisation.

Communication: In our architecture we use two different levels of communication, the so called "high level"- and "low level"-communication. The high level communication submits commands from the global data processing to each of the robot controls. The robot controls return their status back to the global data processing. The low level communication submits the fused data from the global data processing, for example the position of a robot evaluated by the laser data module. The robot itself sends back his sensor values and status data via the low level communication.

Robot Control: The architecture of our robot control is behaviour based [8]. But in contrast to the usual behaviour based control algorithms, which either use hard wired priorities or in which an arbiter controls the reaction of the system by merging the output of the behaviours, our system first evaluates a set of predicates (which can be observed as the output from a kind of virtual sensor) on which an appropriate behaviour is selected. That means that only one behaviour has to be computed at a time. The predicates are computed every time when there is new data from the vision system or the sensors of the robot. After a behaviour is selected it is accomplished until it gets a programmable timeout or a set of predicates for aborting the behaviour becomes true. Some of the behaviours can be selected directly by the global robot control.
Global data processing: The first task for the global data processing is to fuse the data it gets from each of the robots and from the laser data processing. It sends back this fused data to the robots. Additionally it provides the robots, which cannot see the ball, with information about the ball position obtained by robots, which see the ball. The second task is to select special behaviours on the robots for cooperation.

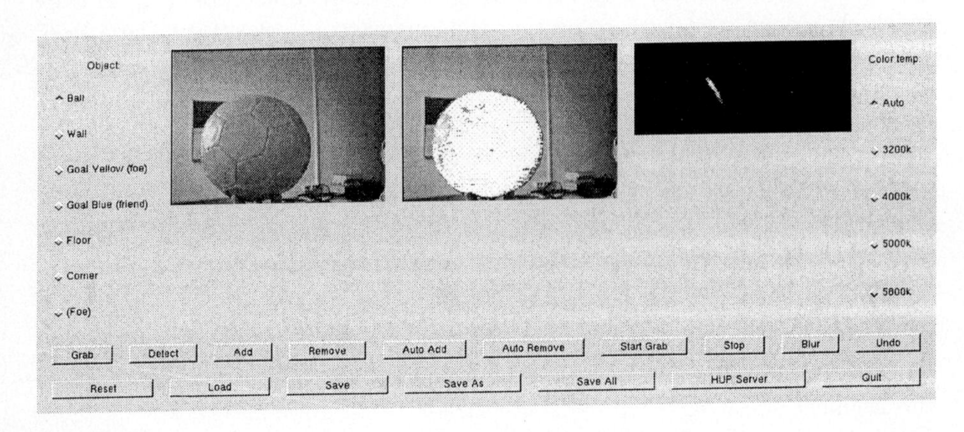

Fig. 3. GUI of our colour training centre. The leftmost picture shows the image, which was grabbed last. One can add a colour area to an object colour area by selecting the appropriate area in the grabbed picture with the mouse.

5 Team tactics

Because a team should be capable of playing soccer even in a situation where the direct communication does not work, it is safer to assign the tactical role to each team member statically, as long as there is no reliable indirect communication mechanism. So the following tactical roles are assigned to the robots:

Goalkeeper: Because of its limited field of view the CCD-Camera from the NewtonLab system is not able to sense the whole area around the goalkeeper. Therefore the camera is mounted in such a way, that the whole right part of the field with respect to the mounting point of the camera can be observed. Now the goalkeeper tries to stay on a line between the ball and the goal. Only in times, when the goalkeeper does not see the ball, it returns to its default position in front of the left side of the goal.
Defender: The two defenders are located at certain points right and left in front of the goal. To prevent the defenders from disturbing each other, each defender is responsible for its own half of the field. If the ball is farer away than 1,50 m the only task for the defenders is to track the ball. At the moment the ball comes nearer than this 1,50 m the defender should move to the ball and kick the ball into the opponent's half of the field. Afterwards it should return to its default position.
Attacker: The attacker's task is to push or kick the ball into the opponent's goal. After it detects the ball the attacker drives to the ball and turns with the ball into the direction of the opponent's goal. Because the attackers are not assigned to a specific area of the field they use a different mechanism to prevent disturbing each other. Only the attacker, which is nearest to the ball, goes for the ball. The other attacker just tracks the ball and does not move further. If the robot, which possesses the ball, detects the opponent goal, it enables the kicker release. Then the next time the ball touches the microswitch in front of the robot, it will be kicked in the direction of the opponent's goal.

6 Experience from RoboCup'98

The RoboCup'98 in Paris showed that there were a lot of different views about the way the task of playing soccer with robots could be accomplished. But it also showed that sometimes the realisation of these ideas failed due to problems on very low levels, e.g. mechanical failures. Especially problems, which arise from facts that can not easily be simulated or tested at the home laboratory were quite hard to work with. Our team in particular had problems with the reliability of the wireless communication between the host PC and the laser scanner. Therefore we could not use the laser scanner data for self localization of the robots. Additionally our cameras sensed the very dark green from the edge markings as black so that the image processing was not able to detect them. Due to these problems the robots were not able to relocalize when the odometry data deteriorated after a certain time. That led to several situations where the robots almost scored an own goal. But the rather successfull result of our team at the RoboCup showed that our architecture was robust enough to cope with problems like these. Besides to the robustness of our design there were some more points which were responsible for the successfull outcome. The first was the usage of a commercially available robot platform, which saves time to focus on more important aspects. The next was the ball handling and ball kicking mechanism we developed. Our PC based vision system with camera and framegrabber was more suitable and more flexible than commercial products, especially with respect to a faster and easier on-site colour retraining. The development of the microcontroller board for the goalkeeper allowed to realize a closed loop control of the goalkeeper's position with respect to the ball, working with 60 Hz. This was the key to success in penalty shooting (despite software problems in the preliminary round). In the future we will focus on a better cooperation between the robots. It should be self organising without the need for a centralised control. A second point of interest will be a probabilistic representation of the environment.

References

1. Kitano H., Asada M., Kuniyoshi Y., Noda I., Osawa E. RoboCup: The Robot World Cup Initiative, Proc. of the first Int. Conf. on Autonomous Agents, Marina del Rey, CA, 1997, 340-347.
2. Rules http://www.er.ams.eng.osaka-u.ac.jp/robocup/middle98/
3. Gutmann J-S., Hatzack W., Herrmann I., Nebel B., Rittinger F., Topor A., Weigel T., and Welsch B. The CS Freiburg Team, Proc. of the second RoboCup Workshop, France, 1998, 451-459.
4. Kraetzschmar G. K., Enderle S., Sablatnög S., Bos T., Ritter M., Braxmayer H., Folkerts H., Mayer G., Müller M., Seidl H., Klinger M., Dettinger M., Wörz R. and Palm G. The Ulm Sparrows: Research into Sensorimotor Integration, Agency, Learning, and Multiagent Cooperation, Proc. of the second RoboCup Workshop, France, 1998, 459-464.
5. Kuth A., Bredenfeld A., Guenther H., Kobialka H.U., Klaassen B., Licht U., Paap K.L., Ploeger P.G., Streich H., Vollmer J., Wilberg J., Worst R. and Christaller T. Team Description of the GMD RoboCup-Team, Proc. of the second RoboCup Workshop, France, 1998, 439-449.
6. Yokota K., Ozaki K., Watanabe N., Matsumoto A., Koyama D., Ishikawa T., Kawabata K., Kaetsu H. and Asama H. Cooperative Team Play Based on Communication, Proc. of the second RoboCup Workshop, France, 1998, 491-496.
7. Shen W., Adibi J., Adobbati R., Lanksham S., Moradi H., Salemi B. and Tejada S. Integrated Reactive Soccer Agents, Proc. of the second RoboCup Workshop, France, 1998, 251-263.
8. Arkin R.C. Behavior-based robotics, MIT Press, 1998

Team Description of the GMD RoboCup-Team

A. Siegberg, A. Bredenfeld, H. Guenther, H.U. Kobialka, B. Klaassen, U. Licht, K.L. Paap, P.G. Ploeger, H. Streich, J. Vollmer, J. Wilberg, R. Worst, and T. Christaller

GMD,
Institute for System Design Technology,
D-53754 Sankt Augustin, Germany,
siegberg, bredenfeld, guenther, kobialka, klaassen, licht, paap, ploeger, streich, vollmer, wilberg, worst, christaller@gmd.de

Abstract. The article describes the structure of the GMD robots developed for the RoboCup '98. The hardware of these robots consists of an aluminum chassis with differential drives. They have low level sensors (odometry, distance sensors etc.) and a NewtonLab vision system. The software is organized in a layered structure using a uniform design pattern on each layer. A synchronous communication paradigm is adopted for the information exchange between the different layers.

1 Introduction

In order for a robot team to actually perform a soccer game [1] [3] various technologies must be incorporate: design principles of autonomous robots (hardware and software design), networking, strategy acquisition, sensor fusion and actor controlling. We present the design methodology for a team of middle sized robots.

The next section contains our design guidelines for the RoboCup scenario. Section 3 gives a detailed description of the GMD-robot. The system software is structured in four layers. Each layer uses a similar design pattern. Section 4 and 5 present the details of this structure. In section 6 we introduce our first strategy planning concepts. Concluding the paper we present results of our robotic activities.

2 Design Guidelines

For the RoboCup project we define some design guidelines to develop skillful soccer playing robots. The robots should cope with a dynamically changing environment and a quickly moving ball.

Autonomous systems, such as these robots, need to be physically strong, computational fast and behaviorally accurate to behave sensible in the rapidly changing environment of a soccer field. In our general architecture we implement therefore strong motors, simple sensors and simple stereotype movements to guarantee the basic qualifications.

Since teamwork and individual skills are fundamental factors on the soccer playground the complexity of the design increases. The soccer robots must have the basic soccer skills, such as shooting, passing and recovering the ball from an opponent. The robot must be able to evaluate its position with respect to its teammates and opponents. The robot has to decide its next action while at the same time following the rules of the game [4]. Finally a communication structure is needed because this offers the robot the possibility to communicate with its teammates and an external PC.

In the next section we describe the first prototype of the GMD robot.

3 GMD Robots

The GMD-robots consist of the following components:

PC and Wavelan The PC on each robot is a small laptop and a wavelan ethernet card is used for the wireless communication.

Vision System Our vision system is a camera together with a board from NewtonLab, which is used to detect and track the ball. A servo motor is used to change the viewing direction of the camera. This allows an angular ball tracking range of approximately 270 degree. The soccer robot depends almost entirely on its visual input to perform its task.

Sensors 4 color detectors determine the color of the surrounding objects. For RoboCup each robot must distinguish the colors of the ball, goals, wall and other objects in the playground. After learning a very fast object recognition is possible. 16 gray scale sensors are placed around the robot's body. They are able to measure the light intensity in the immediate vicinity of the robot.

Touch sensors Bumper, constructed at the GMD, detects a collision of the robot e.g. with opponents or the wall. Furthermore the robot uses infra-red distance sensors.

Motor The motor speed is controlled by PWM (pulse width modulation). A special brake function is included.

Micro-Controller The robot is equipped with three 16-Bit Micro-Controllers [5] which are connected via a CAN-bus.

The total implementation of the robot was performed in different layers. The four major layers are described in the next section.

4 System structure

For a robot which has to play soccer a wide range of different components need to be integrated. Because of the complexity of the whole structure we divide it in four layers. On each layer we uses a uniform software design pattern (Fig. 1). A synchronous communication scheme is adopted for the communication among the layers (Fig. 2).

The server PC is an external PC which is able to communicate via a non-real-time LAN (local area network) with the Robot PC. The external robot

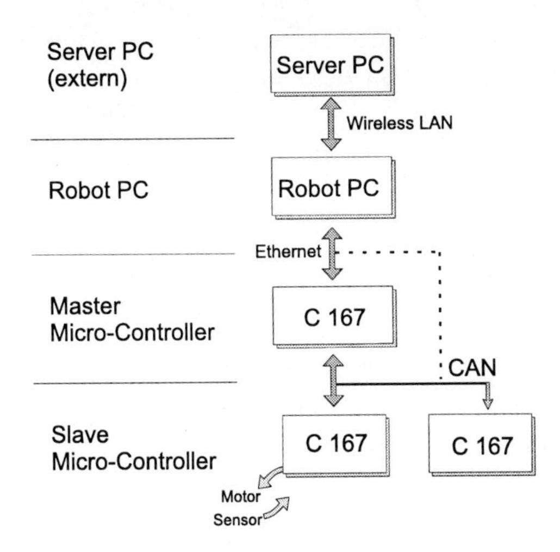

Fig. 1. The system structure with different layers of implementation

is used to exchange the position information among the robots. A star-type network topology is used. Each robot sends its local vision data (own position, ball position, and distance to other robots) to the server. The server merges the data to for a coherent data set which is broadcasted to the different Robot PCs.

The Robot PC is installed on top of each robot. On this PC a subset of local strategies is implemented because this makes each robot able to act in a useful autonomous way even when the network is down and the connection to the server PC is not active.

The master-controller and the slave-controllers define the real-time layers. There are three 16-Bit Micro-Controllers (C167). These controllers are connected via a CAN-bus. The master-controller performs the low level decision making. The first C167 slave-controller processes the motor-, odometry, color cell-, bumper- and line detector-data and it is responsible for the obstacle avoidance. The second C167 slave-controller processes the data of the distance- and gray-scale-sensors. The three micro-controllers perform the real-time processing of the robot system and communicate via the CAN bus whereas the server-PC and the robot-PC communicate via ethernet. If real-time processes are necessary in the whole system it is also possible to connect the controllers via the CAN bus directly to the robot-PC (this involves a changing of the robot-PC operating-system).

The whole communication and data transmission between the four layers is organized in a synchronization scheme with slot frames (Fig. 2).

The master-controller acts like a synchronous time clock and is responsible for the synchronization task between the different layers. The schedulers of each layer operate at different synchronization rates. The master controller runs at the highest synchronization rate. The rates of all other layers are derived from this rate. Schedulers on each layer are synchronized by the master controller.

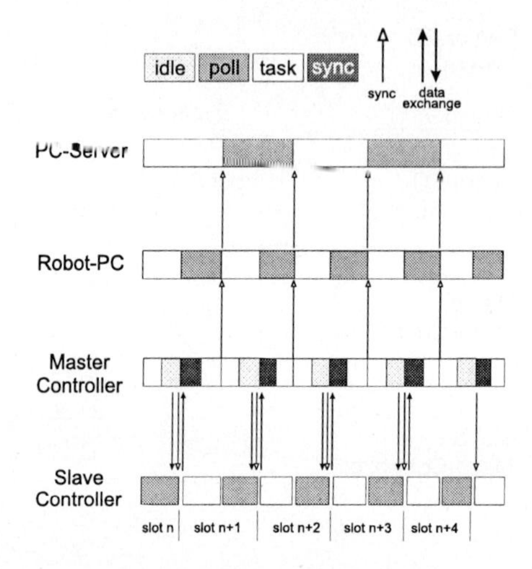

Fig. 2. Synchronization of the different layers with slots.

The conditions of the server-PC, the robot-PC and the slave-controller change between polling and task processing.

5 Software Design

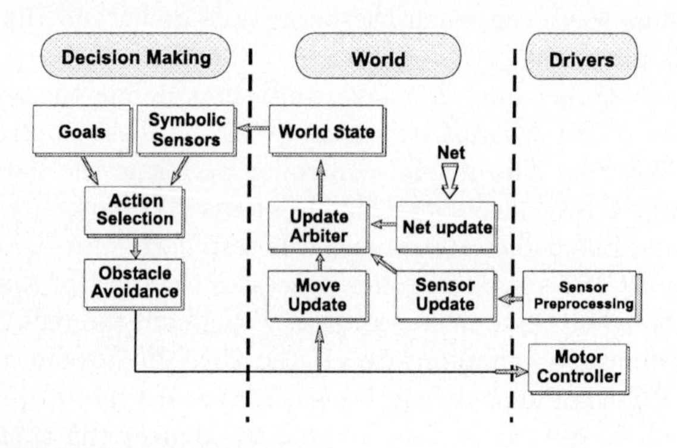

Fig. 3. The three most important parts of the software design.

A uniform software design pattern is used on each of the system layers. The pattern is depicted in Fig. 3. The software pattern is divided into three parts.

Drivers A sensor preprocessing is performed and the motor controller is implemented. The sensor preprocessing maps real world events to an internal representation.

World The world models the actual playground and the robot state depending on the sensor values and the world state received via the network.

Decision Making For the decision making part the values of the world state are reduced to symbolic sensors. The symbolic sensors and the robot goal are the input for the decision making process. After an action is selected a check for a collision is made. If there is no collision, the action command is transmitted to the motor controller.

A detailed description of our first strategies are described in the next section.

6 Strategy Planning

Each robot must react appropriately to different situations on the soccer field. The robot needs information about the environment and itself. This information delivers the world. The world presents the actual states of the sensors.

For the RoboCup competition in 1998 we implemented a set of strategies. These strategies are the basic skills for the robot to act on the playground.

We divided the strategy in two parts (1) actions in the opponent half and (2) actions in the own half. Important values for the robot are the X- and Y-coordinates of itself, his team members, the opponents and the ball. These values are transmitted to each robot via the world interface.

The first priority for our robot in the opponent half is the ball-goal line (Fig. 4). When the robot detects the ball, it drives behind the robot on the ball-goal line. Next, it turns into the direction of the goal and shoots the ball to the goal.

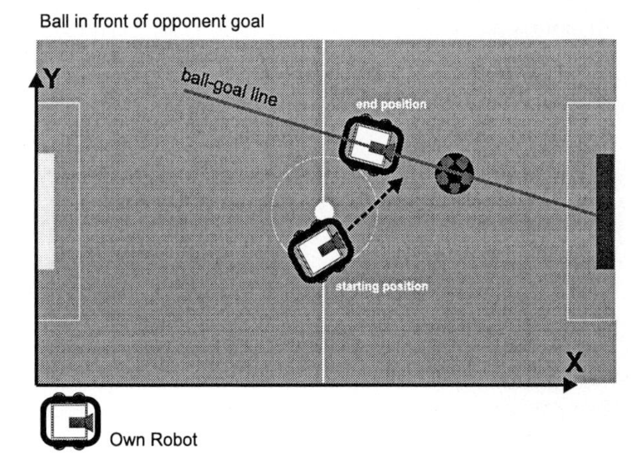

Fig. 4. Robot on the soccer field.

The first priority for the robot in its own half is the protection of the own goal. Therefore it always drives behind the ball. In this situation the robot ignores the ball-goal line. If the robot detects the opponent *and* the ball, it has the possibility to check the value of delta x. Delta x delivers the robot the distance between ball and itself. When the value is large it can directly drive on the ball-goal line. When the value of delta x is small the robot has to perform an obstacle avoidance.

7 Results and Future Work

Building a robot to play soccer is a big challenge in different fields. The range of technologies spans AI, robotic research and embedded system design. What we have learned after the RoboCup 1998 in Paris is the great importance of the sensors. In Paris we used mainly the vision sensor. But especially the vision sensor is a very complex one. We must take into account the movement of the sensor itself (because it is installed on a servo), the movement of our robots, and the movement of the ball.

In our system of the Paris RoboCup we tried to derive the movement operators directly from the vision data, i.e., estimate the ball distance (e.g., 20cm) and generate a movement command (e.g., `move 20`). This leads to oscillations and unstable behavior (in particular, when using fast turns). Thus only slow robot movements can be realized in this way. The situation will be remedied by using a behavior-oriented robot control [2].

The GMD-robots are developed for a spectrum of different missions. Of course for the RoboCup they will act as soccer players but our research goal is the implementation of a flexible and changeable platform. Researchers from different fields can use the robot platform to tackle problems like real-time sensor fusion, reactive behavior, strategy acquisition, learning, real-time planning, multi-agent systems, context recognition, vision, strategic decision making and intelligent motor control.

References

[1] S. Hedberg. Robots playing soccer? RoboCup poses a new set of AI research challenges. *IEEE Expert Intelligent Systems and Their Applications*, pages 5–9, September/October 1997.

[2] H. Jäger and T. Christaller. Dual dynamics: Designing behavior systems for autonomous robots. In *Proceedings of International Symposium on Artificial Life and Robotics*, pages 76–79, 1997.

[3] H. Kitano, M. Asada, Y. Kuniyoshi, I. Noda, E. Osawa, and H. Matsubara. RoboCup – a challenge problem for AI. *AI Magazine*, 18(1):73–85, Spring 1997.

[4] RoboCup. Regulations & rules. http:// www.robocup.org /regulations /4.html, 1998.

[5] Siemens AG, Munich, Germany. *C167 Derivatives – 16-Bit CMOS Single-Chip Microcontrollers*, 3.96 edition, 1996.

UTTORI United: Cooperative Team Play Based on Communication

K. Yokota[1], K. Ozaki[1], N. Watanabe[1],
A. Matsumoto[2], D. Koyama[2], T. Ishikawa[2],
K. Kawabata[3], H. Kaetsu[3] and H. Asama[3]

[1] Utsunomiya University,
7–1–2 Yoto, Utsunomiya–shi, Tochigi 321–8585, JAPAN.
[2] Toyo University,
2100 Kujirai, Kawagoe–shi, Saitama 350–8585, JAPAN.
[3] The Institute of Physical and Chemical Research (RIKEN),
2–1 Hirosawa, Wako–shi, Saitama 351–01, JAPAN.

Abstract. In order for multiple robots to accomplish a required mission together, they need to organize themselves, cooperate and share information. We regard such actions as "team play" and believe communication is the essential tool for team plays. This paper discusses communication in the distributed autonomous robotic system and development of cooperative actions for football playing robots. The discussed communication framework and cooperation are implemented in our omni-directional mobile robots which has vision for sensing and a wireless device for communication.

1 Introduction

The distributed autonomous robotic system (DARS) is a flexible and robust robotic system in which multiple robots and other agents act in a cooperative and coordinated manner to achieve tasks which are not easy for a single robot [1]. We believe communication is essential tool for cooperative actions of autonomous robots and have been developing communication technology and demonstrating cooperation based on communication [2, 3].

We regard the football game as a collection of cooperative actions of autonomous robots. In order for a team to score a goal, its players must cooperate in bringing the ball forward, maneuvering between fore players. We assembled a team of omni-directional, autonomous robots which are capable of communicating and sharing information between them[4, 5].

2 Communication-Based Cooperation

We regard explicit communication is the essential tool to realize team plays in the multi agent robotic system [2]. It is sometimes argued that explicit communication is too complicated to implement and implicit communication performs better in the environment where agents must act in real-time [6, 7]. However, explicit communication is more direct means to pass information and intention of a robot to other robots. The explicit communication is also expected to be reliable in that ambiguity can be avoided. The implicit communication, or communication based on observation, always is susceptible to errors arising from misunderstanding and misinterpretation of the situation.

3 Communication among Robots

We classify communication patterns as shown in Fig. 1. Of these patterns, "announcement", "order", "synchronization" are the communication which does not expect answers. Answers and interaction are expected in the other communication patterns.

Announcement This is broadcast communication. It is used to pass piece of information to the other agents.

Order When an agent asks the others to take certain action, this communication is used. Unlike "request" and "contract" described below, this communication is of broadcast type and does not expect answer, it is not guaranteed that the recipient agent will take requested action.

Request This is point-to-point communication and is used to retrieve particular information from a particular agent.

Contract When an agent wants to request action or service from one or more agents, this protocol is followed. It is based on the Contract Net Protocol [8]. Both parties of the contract are fully in agreement as to what is going to be done.

Collective Decision This communication provide "voting" mechanism. A robot initiates the voting procedure, and the other robots will give their votes. Result is summed up by the first robot and transmitted to the rest.

Synchronization This communication is used when synchronized action is required by more than one robot. All robots start intended action upon receiving this message.

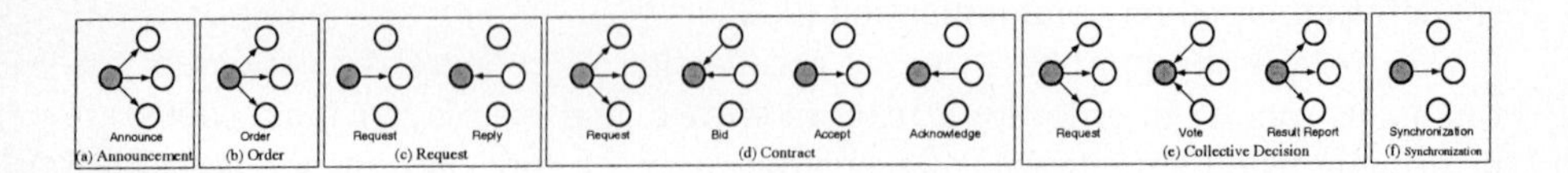

Fig. 1. Communication for "team play"

4 Autonomous Robots for Team Play

Fig. 2 and fig. 3 show our football playing robot and its system diagram. The robot is based on the omni-directional mobile robot. All the robots share the same basic design, yet their control system, sensors and kicking mechanisms differ.

"ZEN" has an omni-directional driving mechanism which is driven by three actuators [9]. The primary sensing device of the robot is vision; a CCD camera is connected to a frame buffer whose resolution is 512 dots by 512 dots. A simple kicking mechanism (fan or repulsing plate) and tactile sensors are connected via parallel I/O.

A single microcomputer (i486/DX4 100MHz or Pentium MMX 200MHz) controls all actuators, manages sensory inputs, and processes image input from the CCD camera. A wireless LAN adapter is provided so that the robot program can exchange data with the other robots. The current communication program module uses UDP/IP as communication protocol and implement message exchange protocol on top of it.

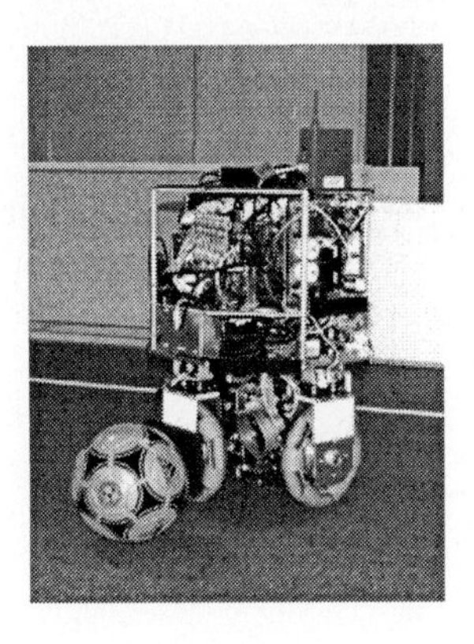

Fig. 2. Omni-directional mobile robot "Zen"

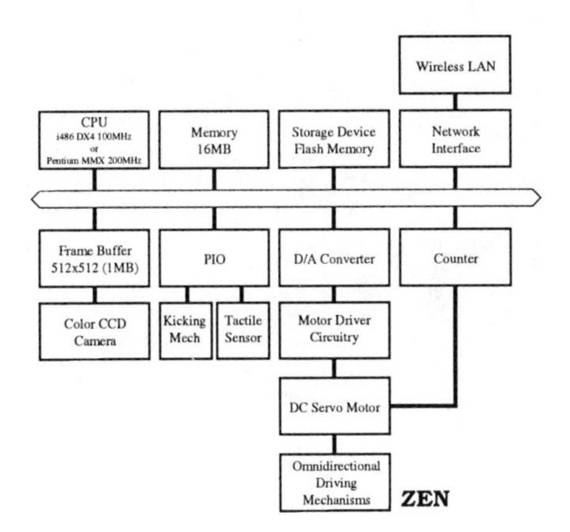

Fig. 3. System diagram of the football playing robot

5 Cooperative Team Play

Fig. 4 lists the cooperative actions implemented by the football playing robot.

Coordinated Offense When a robot thinks that it is in control of the ball, it will let the other robots know that the ball is owned by the first robot. The other robots will not approach the ball and will prepare for the next attacking action. These actions are coordinated by the broadcast communication "announcement" and will give the team the ability to infiltrate into the enemy half of the field quickly towards the goal.

Defense When the goal keeper finds the ball near the goal, it will assume that the opponent is attacking the goal and will order the other robots to backup and defend the goal. This is implemented by the broadcast communication "order".

Pass Explicit agreement between the two robot is sought before the ball is passed. The action must be negotiated by the robots as described in Section 3.

Distributed Sensing It so often happens that a robot may loose sight of the ball. Another robot may be able to find the ball and will tell the others of the location of the ball, if possible. This is a form of distributed sensing. The robot which finds the ball will use the broadcast communication "announcement" to tell the robots of the location of the ball.

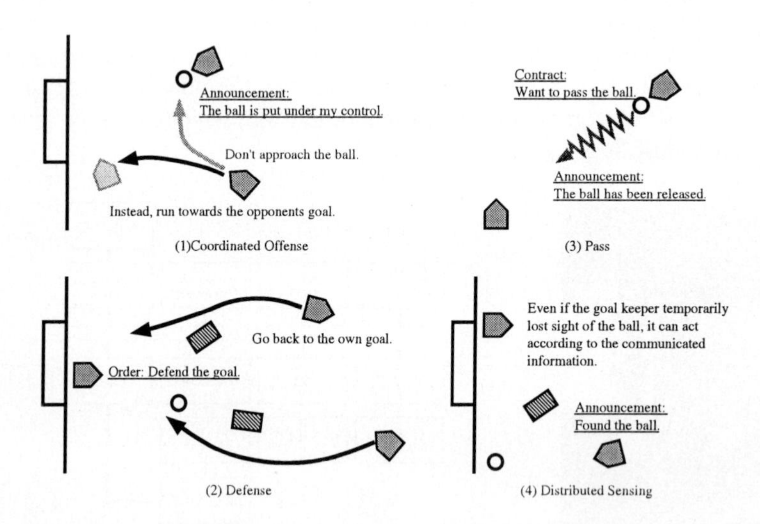

Fig. 4. Cooperation among robots

6 Performance of UTTORI United at RoboCup-98 Paris

The team UTTORI United brought three robots to Paris: two forward robots and a goal keeper. A laptop computer was used to instruct the robots to start and end playing the game. However, the computer was not used in any way to tell the robots what to do during the game. There is no centralized control in the team. The entire system is fully autonomous and distributed; the robots acts, based only on sensory information and communication from the other robots.

Our system worked very well throughout the competition. Table 1 (a) shows communication log in one of the games. Soon after kick off, the two forward robots *stmr2* and *omni4* dashed towards the ball. The robot *omni4* reached the ball first and declared that it was in control of the ball by sending the message "CONTROLLED" to the other robots (it continued to transmit the same message while it is moving the ball). The robot *stmr2* then stopped approaching the ball and turned towards the goal. The two robots moved together, side by side, to the goal. This is one instance of the most powerful performance of UTTORI United; two robot showed coordinated offense action. This is not achieved by the centralized control or planning; it is realized by cooperation by the two robots based on communication.

Table 1 (b) shows negotiation process (Fig. 1 (d)) of the two forward robots. The robot *stmr2* wants to pass the ball to another robot. The argument "5" for the message "PASS" indicates that the robot *stmr2* wanted to kick the ball towards the very front of the goal. The negotiation succeeded and the robot *stmr2* kicked the ball. However, the ball did not successfully reached the robot *omni4* in this particular case.

Table 1. Communication log

(a) Coordinated offense

Time	From	To	Message	Notes
0000				stmr2 and omni4 dashes to the ball
0017	omni4	all	CONTROLLED	omni4 declares it has the ball
0017	omni4	all	CONTROLLED	two robots run side by side
0017	omni4	all	CONTROLLED	
0018	omni4	all	CONTROLLED	

(b) Pass

Time	From	To	Message	Notes
3184	stmr2	all	CONTROLLED	stmr2 is in control of the ball
3184	stmr2	all	CONTROLLED	stmr2 is in control of the ball
3185	stmr2	all	PASS 5	stmr2 wants to pass the ball
3185	omni4	stmr2	RECEIVE 5	omni4 offers to receive the ball
3185	stmr2	omni4	PASS_OK 5	stmr2 accepts the offer
3185	stmr2	all	CONTROLLED	stmr2 is still in control of the ball
3185	omni4	stmr2	PASS_ACK 5	omni4 acknowledges
3185	stmr2	all	CONTROLLED	stmr2 is about to kick the ball
3193	stmr2	all	RELEASED	stmr2 has kicked the ball
3193	stmr2	all	RELEASED	

Other communication patterns, defense and distributed sensing, also appeared during the games. Communication-based cooperation made UTTORI United one of the most powerful teams during preliminary games; the team scored 8 goals in three games, although it has only three robots and they are slower than most robots from other teams.

However, when wireless communication does not work due to various disturbances, performance of the team suffers. For some reasons, wireless communication in our team was disappointingly inferior during the final tournament and the team failed to score any goals.

Fig. 5 shows the proportion of cooperative actions and autonomous actions by robots in terms of time. Figures are derived by examining communication logs and adding duration of valid messages. The figure clearly indicates that the performance of the team is better when the robots cooperate than when they behave independently. These results proves that cooperation based on communication significantly amplified the performance of the team.

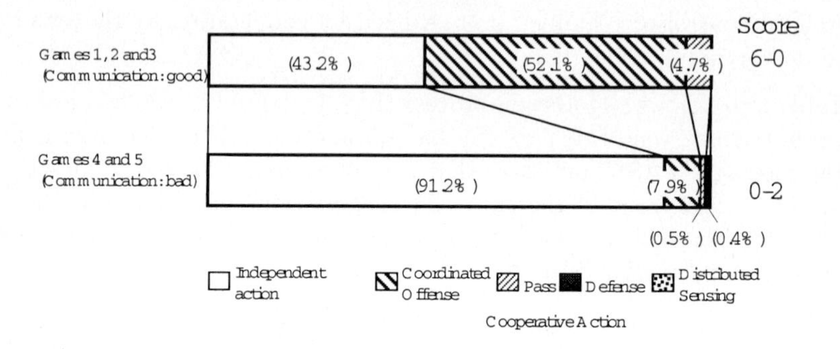

Fig. 5. Cooperative actions and independent actions

7 Conclusion

We believe communication is the essential tool for team plays. The paper discussed communication among autonomous robots and development of cooperative actions by them. The discussed communication framework and cooperation are implemented in our omni-directional mobile robots which has vision for sensing and a wireless device for communication. The games in Paris demonstrated the validity of our approach.

References

1. Asama, H., et al., *Design of an Autonomous and Distributed Robot System: ACTRESS*, IEEE/RSJ Int. Workshop on Intelligent Robots and Systems, (1989) 283.
2. Matsumoto, A., et al., *Communication in the Autonomous and Decentralized Robot System ACTRESS*, IEEE/RSJ Int. Workshop on Intelligent Robots and Systems, (1990) 835.
3. Ozaki, K., et al., *Negotiation Method for Collaborating Team Organization among Multiple Robots*, in *Distributed Autonomous Robotic Systems*, Springer–Verlag, (1994) 199.
4. Yokota, K., et al., *Omni-directional Autonomous Robots Cooperating for Team Play*, in Kitano, H., (ed), *RoboCup-97: Robot Soccer World Cup I*, Springer, (1998) 333.
5. Yokota, K., et al., *Modeling Environment and Tasks for Cooperative Team Play*, in *Distributed Autonomous Robotic Systems 3*, Springer, (1998) 361.
6. Beckers, R., et al., *From Local Actions to Global Tasks: Stigmergy and Collective Robotics*, 4th Int. Workshop on the Synthesis and Simulation of Living Systems, (1994).
7. Montesello, F., et al., *Implicit Coordination in a Multi-Agent System using a Behavior-based Approach*, in *Distributed Autonomous Robotic Systems 3*, Springer, (1998) 351.
8. Smith, R. G., *The contract net protocol; high-level communication and control in a distributed problem solving*, IEEE Transactions on Computers, (1980).
9. Asama, H., et al., *Development of an Omni-Directional Mobile Robot with 3 DoF Decoupling Drive Mechanism*, IEEE Int. Conf. on Robotics and Automation, (1995) 1925.

Quadruped Robot Guided by Enhanced Vision System and Supervision Modules

Vincent Hugel, Patrick Bonnin, Jean Christophe Bouramoué, Didier Solheid, Pierre Blazevic and Dominique Duhaut

Laboratoire de Robotique de Paris, 10/12 avenue de l'Europe, 78140 Vélizy, France
{hugel, bonnin, blazevic,duhaut@robot.uvsq.fr}

Abstract. Legged robots taking part in real multi-agent activities represent a very innovative challenge. This domain of research requires developments in three main areas. First the robot must be able to move efficiently in every direction in its environment. The faster the motion, the better it is. Special care must be taken when designing walking pattern transitions. Then, without any exteroceptive sensor to get information about its surroundings, the robot is blind. Fortunately, the quadruped prototype on which all experiments are carried out is equipped with a enhanced vision system and vision is the best means of getting a representation of the world that can be found in Nature. Finally the machine should be brought a minimum of intelligence since it has to manage vision information and its walking gaits by itself. When involved in cooperation, confrontation or both like in the soccer play, a high level supervision task is welcome. This paper presents detailed developments of these three points and describes how they are implemented on the real robot.

1 Introduction

The LRP legged machines team has been given the opportunity to participate in the Legged robots RoboCup competition which was held in July 1998 in Paris. Sony Corporation and specifically the D21 Laboratory in Tokyo lent very kindly three pairs of "pet robots" to three teams in the world. These prototypes represent a high level development platform on which to put into practice optimized algorithms of locomotion and vision. Behavior strategies can also be included after testing on simulation.

Having in mind that three robots must cooperate to score a goal in the opponent player goal, two kinds of strategies are considered. The first one has to answer the following question : how to incorporate the three functions mentioned above : locomotion, vision and supervision, and how are they going to interact with each other in an efficient manner ? This strategy of implementation is very important since all tasks are to execute on-board in real time. The second strategy is the decision-level one to adopt on the soccer field to win the game.

The first section of this paper is dedicated to the walking patterns used to make the robot move quickly in every direction. In the second section the different steps of the vision recognition and localization procedures are explained. The third section is devoted to the strategy employed for the robots to play football.

2 Legged Locomotion

Legged locomotion has been implemented using the fundamental principles developed by McGhee in [1],[2]. Forward, bacward, turning and rotation motions have been derived from the crawl model. Forward and backward gaits have been improved thanks to a special sideways motion [3]. The two next sections deal with turning and rotation gaits.

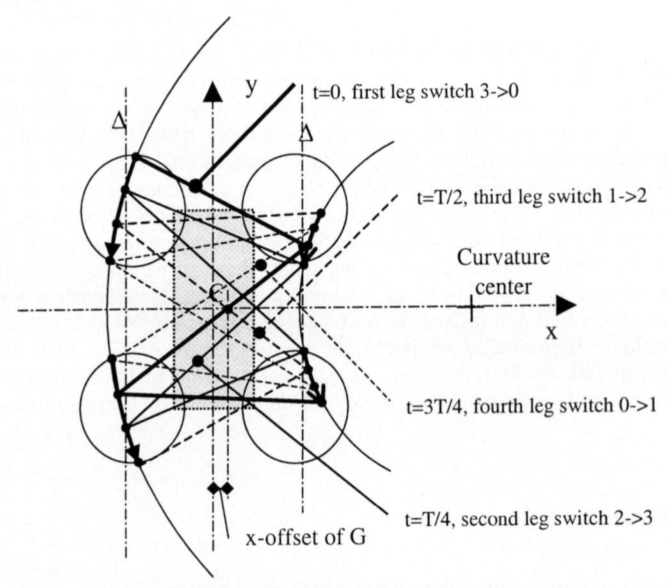

Fig. 1. Successive support polygons for right turning gait at fractions of cycle time of 0,1/4,1/2 and 3/4

2.1 Turning gaits

This section focuses on turning gaits. The strategy adopted here is to define turning motions whose curvature center is located on the transverse axis of the robot. This will enable the machine to turn right or left with a varying turning circle. The idea consists in adapting the forward sideways crawl. Since it has been decided to exclude prediction features in the definition of the gait, such as predicted duty times or predicted footholds for instance, it seems difficult to master transitions at any moment of time. To avoid this problem turning gaits are designed in such a manner that they share a common set of footholds with the forward walking pattern at a particular instant of cycle time. Fig. 1 shows the common set of footholds for right turns on lines Δ and Δ'.

The configuration for right turning refers to the foothold positions of the legs in the forward motion at instant of T/4, and for left turning, it corresponds to positions of legs at 3T/4. The exact distance traveled by the leg in the traction phase with respect to the body reference frame is computed to be approximately equal to the one defined in the forward motion, this is to guarantee continuity of speed. The sequence of legs is the same as in the crawl leg state diagram [1].
However the COG must be shifted so as to be on the diagonal supporting line when transitions 3->0 and 1->2 occur, see fig. 1. Moreover the sharper the turning circle, the larger the amplitude of the sideways motion is. When the turning circle is shorter, the incline of the diagonal supporting line increases, and amplitude must be increased. Besides, lateral stability limits the magnitude of lateral oscillations. Therefore, there is a threshold turning circle under which it is not possible to use the same technique any more. This threshold is approximately equal to the length of the machine, that is about 15 cm. Finally it must be noticed that right and left turns converge towards forward crawl gait when the turning circle tends towards infinity.

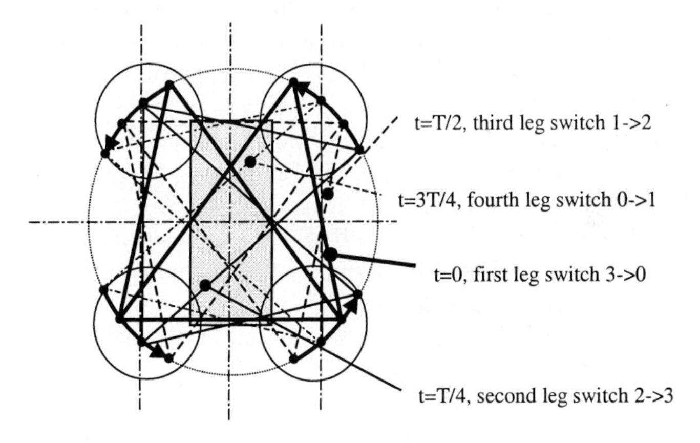

Fig. 2. Successive support polygons for right rotation at fractions of cycle time of 0,1/4,1/2 and 3/4

2.2 Rotations

Since the goal is to design efficient motion of the robot in every direction, it is interesting to define a rotation motion around the COG. The first step is to find a common set of leg footholds for the transitions. The configuration is given by the intersection between the trajectories in straight line motion and the circle spotted in fig. 2.
Transitions can occur at T/4 or 3T/4, depending on the initial and final motions.
One important point in rotation motion is that phase differences between legs are changed. It is not a symmetric gait any more, that is, opposite legs are no more half a cycle period out of phase. The successive support polygons are plotted in fig. 2.
Obviously there is no need for sideways displacement of the COG. Rotations in clockwise and counterclockwise directions are straightforward, transitions between them can trigger at any time without problem.

3 Vision System Module

The goal of the vision system is to detect, to identify and to spot the different elements constituting the scene during the play. "Detecting" means extracting all connected components belonging to the scene elements from the color images . "Identifying" means finding the sole or the several connected components constituting an object of the scene and giving it a symbolic label such as ball, beacon, own or opponent goal, partner or opponent player, edge of the soccer field. "Spotting" beacons or goals means determining the view angle in azimuth with respect to the head direction and a "rough" measurement of the distance between head and target.

3.1 The detection Step

The detection step is composed of the five following algorithms :

1. Color Detection, performed by the Sony specific hardware using threshold values as input controlled parameters and providing output in the form of 8 bit-planes, each corresponding to a color template.
2. Opening performed simultaneously on each of 8 bit-planes, using an isotropic 3 by 3 centered neighborhood. It cleans the "8- binary" image of pixels resulting of color detection from noise.
3. Connected Component Extraction [4] which requires a single scan of the image, detects equivalent labels, and computes the connected component attributes : gravity center, surface, including box.
4. Filtering on the surface attribute of the connected components to remove too small components due to bad lighting conditions.
5. Merging of connected components, to deal with bad lighting conditions responsible for the decomposition of a scene object into several connected components. To be merged, two connected components must have their including box close to each others. An „image object“ resulting from the merging of several connected components is not a connected component. But its parameters such as surface, gravity center, including box are computed directly from parameters of merged connected components.

Steps 1, 2 and 3 are low level image processing, applied to image data. Steps 4 and 5 are intermediate level image processing, applied to feature attributes : they are fast to perform.

3.2 The identification Step

The ball is a small scene object which may be partly or wholly occluded, and generally it does not produce more than a single image object. So the ball identified in the image is the image object corresponding to the orange color template, whose surface is maximum.

Beacons are also small scene objects, but they are localized in such a way that they cannot be occluded. They are composed of two colors : one necessary pink, the other

yellow or blue or green. Taken into account the geometry of the camera and the soccer field, two beacons can be viewed at most. An identified beacon is composed of two image objects, the first is necessary pink, the second is yellow, blue or green, and the two image objects are located one above the other.
Goals are yellow or blue, but they have the same colors as the beacons. The goals are seldom viewed entirely. Either the robot is far from the goal, and the goal is certainly occluded by one or more players, or the robot is close, and the goal comes out of its field of view. So, the identified goals are the image objects (blue or yellow) which are not identified as a beacon.
Players are either dark blue or red. But, depending on their direction, they appear constituted by one or two image objects.
The edges of the soccer field are white. It is the last available template color!

3.3 The localization Step

The angle in azimuth with respect to the head direction is computed from the x component of the gravity center of the identified object, knowing the image resolution on the x axis and the horizontal field of view of the video camera (52°), assuming the pin hole model of camera.
A "rough" measurement of the distance between the head and a scene objet is given by a look up table based on the surface attribute for the ball and the beacons, the height of the including box for the goals. But it seems impossible to give a reliable rough measurement of the distance from another player, because all feature attributes change for a great part when the player rotates on itself at the same distance! But an "alert" is generated if the surface of the image object of a player grows up beyond a threshold value, to stop the robot and consequently to avoid collisions.
A "rough" measurement of the distance from the robot and the edge of the soccer field is geometrically computed.

In conclusion, the cadence of image processing, measured by two different benchmarks is about 14 or 15 frames per second, while the robot is walking. In these conditions, the head can rotate smoothly to track the ball.

4 Behavior and Strategy of Quadruped Robots

Having two teams of three robots allowed to considerate the whole game as a Multi-Agents System. The main difficulty we encountered was the communication between agents (i.e. robots). As a matter of fact, viewing at each over, using an embedded vision system, was the only way for robots to communicate and locate their own position on the field.
The idea was to allow the quadruped to be able to quickly swap between planning and reactivity; that is to say that pet should be able to generate a trajectory, for instance, to kick the ball, or protect is own goal, and give up all is plans at the next decision cycle if the situation has changed.

With regard to strategy, we used a "Buckett Brigade" type algorithm : Each agent has a rule system, and a weight is associated to each rule. Applying a rule means, for an agent, to choose a role, that is to say a specific behavior, for instance a role as a Kicker or a Defender. During the game, weights are dynamically modified : the weight of a rule is increased if applying this rule allow the pet to kick the ball or even to score a goal, otherwise, the weight is decreased.

We build a simulator for Windows 95 which uses the real pet's odometry parameters. Playing several games in this simulator allows the quadrupeds to learn how to select the right role for a given game situation by adjusting the rules' weights.Further work can be explored in the two following domains:• Pattern recognition and Learning : to learn the pet how to build a model of his environment is not a trivial problem. The robot must be reactive enough to avoid unexpected problems such as a chair or a table which has been recently moved in the room. At the same time, it must be able to reach is goal, in an efficient way, that means to plan his trajectories and actions.• Cooperation between pets, to achieve a common task, for instance cleaning a room by pushing objects too heavy for one and only robot. This point is directly linked to soccer strategy. That domain is strongly dependant of vision and walk : to cooperate, robots need to improve their abilities to communicate and move.

5 Conclusion

This paper has described the implementation of turning and rotation gaits derived from the crawl gait. All these gaits were designed under specifications of efficiency and increased velocity. The main axes of the vision system associated with the legged locomotion module have been presented, together with some implementation details of the strategy level. The 1998 Robocup legged robot challenge was the opportunity for our team to face other teams from other foreign countries. Our prototypes showed good abilities in moving on the soccer field tracking the ball.
Efficient implementation of walking is very useful since it makes robots operational to take part in multi-agents cooperation activities. Soccer play appears to be a good testing ground for research in strategy and multi-agent behavior.

References

1. R. B. McGhee, Some Finite State Aspects of Legged Locomotion, Mathematical Biosciences 2, 67-84 (1968)
2. R. B. McGhee and A. A. Frank, On the Stability Properties of Quadruped Creeping Gaits, Mathematical Biosciences 3, 331-351 (1968)
3. S. Hirose, Y. Fukuda and H. Kikuchi, The Gait Control System of a Quadruped Walking Vehicle, Advanced Robotics, Vol. 1, N°4, pp. 289-323, 1986.
4. A. Rosenfeld, JL. Pfalz, Sequential operations in Digital Picture Processing, Journal of ACM, vol. 13, n°4, pp 471-494, 1966.

The CMTrio-98 Sony Legged Robot Team

Manuela Veloso and William Uther

Computer Science Department
Carnegie Mellon University
Pittsburgh, PA 15213
{veloso,will}@cs.cmu.edu

Abstract. Sony has provided a remarkable platform for research and development in robotic agents, namely fully autonomous legged robots. In this paper, we describe our work using Sony's legged robots to participate in the RoboCup'98 legged robot demonstration and competition. The robots are fully autonomous with on-board vision, control, and navigation. The challenges we addressed in this framework include the color calibration of the vision hardware, the landmark-based robot localization on the playing field, and the development of robot behaviors for the actual play of the game. The paper presents our approach and contributions to these issues. We apply machine learning techniques for automated color calibration and we develop effective vision-servoed navigation. We present our Bayesian localization algorithm. Team strategy is achieved through pre-defined behaviors. Our team of the Sony legged robots, CMTrio-98, won all of its games in the RoboCup-98 competition, and was awarded the first place in the championship.

1 Introduction

RoboCup-98 included a new demonstration league: the Sony legged robot league. The three participating teams, namely Osaka University, University of Paris 6, and Carnegie Mellon University, all used identical robots supplied by Sony Corp. These small autonomous legged robots provide a very challenging platform for robotic soccer.

The Sony legged robot as a robotic soccer player is a fully autonomous robotic system without global vision or wireless remote operation. The RoboCup-98 legged robot exhibition match was therefore a software competition between robots with the same hardware platforms.

The Sony legged robots use on-board color-based vision as their only sensing. The vision processor is past of the robots' hardware. It provides a robust eight-color discrimination when calibrated correctly. Robots need to act solely in response to the visual input perceived. The work to build our Carnegie Mellon team, namely CMTrio-98, was decomposed along the following aspects:

- Reliable detection of all of the relevant colors in the game: orange (ball), light blue (goal and marker), yellow (goal and marker), pink (marker), light green (marker), dark blue (teammate/opponent), and dark red (opponent/teammate).
- Active ball chasing: the robot actively interleaves searching for the ball and localization on the field to evaluate both an appropriate path to the ball and final positioning next to the ball.
- Game-playing behaviors: robots play attacking and goal keeping positions.

This chapter is organized as follows. Section 2 presents our machine learning algorithm to automatically learn the YUV thresholds for color discrimination. Section 3 describes our Bayesian probabilistic localization. Section 4 outlines the basic robots playing behaviors. Section 5 summarizes the CMTrio-98 RoboCup-98 demonstrations and games and concludes the paper.

2 Supervised Learning of *YUV* Colors

The Sony legged robot has specialized hardware for the detection of colors. This hardware quickly maps the high-resolution colors of YUV color space into one of eight *symbolic colors*. However, this hardware still requires pre-setting of the appropriate thresholds in YUV color space for the desired mapping. It is well known that color adjustments are highly sensitive to a variety of factors, such as lighting and shading. Given that the legged robots inevitably act under many different conditions, we developed a method to automatically acquire the necessary YUV color thresholds.

The YUV color space describes a color in terms of three numbers. The Y value describes the brightness of the color. The U and V values together describe the color (without brightness information). Initially we developed a tool to manually experiment with different boundaries in the UV plane for each symbolic color, and ignored the Y axis.[1] Figure 1 (a) shows some images taken through the robot's camera, the projection of the colors in those images onto the UV plane along with rectangles specifying which areas of the UV plane map to which symbolic colors and then the images with just the symbolic colors marked.

Based upon the experience using the UV plane tool, we understood that it would be necessary to adjust the color thresholds separately for different brightness levels. Instead of modifying the previous tool to handle the different Y values separately, we developed a classification algorithm to automatically adjust the thresholds to maximize the accuracy of the desired color detections.

Our algorithm relies on supervised classification using a set of training and testing images. By moving the robot to different positions on the field, we accumulate a series of images. For each image, we manually classify the regions of the different symbolic colors using an interface that overlays the original image and the supervised classification (See Figure 1 (b)). The result is a list of labelled pixels. Each has a position in YUV color space representing its actual color, and a label specifying which of the symbolic colors we think this is. The position in the image it came from is ignored.

Once the data has been labelled, the YUV color thresholds are learned separately for each symbolic color using a conjugate gradient descent based algorithm. For each symbolic color and Y value, the edges of a rectangle in UV space are adjusted to minimize sum-squared classification error in the labelled data. We can view each boundary of the rectangle as a function between either the U or the V value and the probability that this pixel is in the correct range. These

[1] This tool was developed by Kwun Han, for which we thank him.

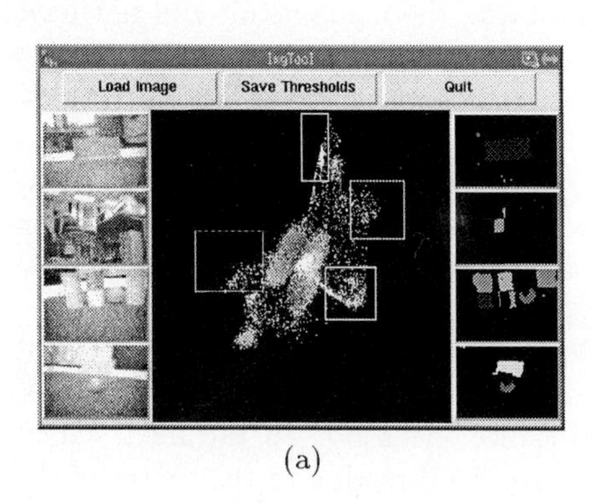

(a)

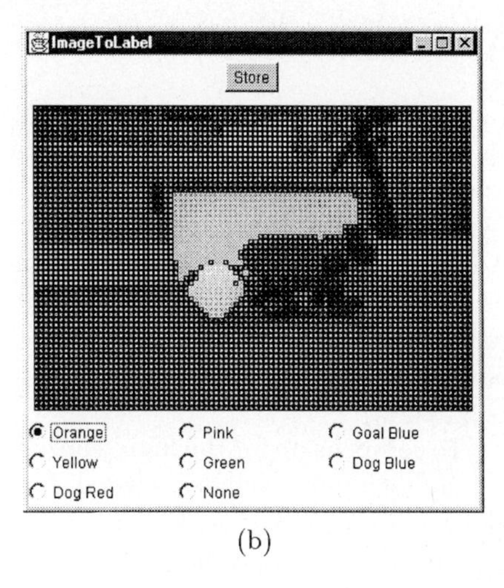

(b)

Fig. 1. The tools for color separation: (a) manual specification; (b) labelling data for automatic separation.

boundaries are each of the following form:

$$C = \begin{cases} 1, \text{if } x > a_i \\ 0, \text{otherwise}, \end{cases} \tag{1}$$

where C is the probability this pixel lies above the threshold, x is the current U or V value, and a_i is the value of this threshold.

To represent a rectangle in UV space, the probability that we are on the correct side of the individual thresholds was multiplied. In order to use gradient descent, we need to calculate derivatives for the model parameters, i.e., the rectangle boundaries, a_i. This is not possible for the step function shown. To solve this, each threshold was replaced by a sigmoid function:

$$C = \frac{1}{1 + e^{t(a-x)}}, \tag{2}$$

where C, x and a are as above, and t is a measure of the 'smoothness' of a threshold. We know which side of the rectangle each pixel should fall given its label, i.e. if the probability of being in the rectangle should be 0 or 1. Gradient descent is performed on the sum-squared error in probability. Initially t is small leading to very smooth thresholds. t is gradually increased over time, hardening the thresholds and making them more like the step functions used by the hardware.

In our experiments, we generally use about twenty images for training. The learning algorithm converges in about three hours achieving a high classification accuracy.

3 Bayesian Probabilistic Localization

In order to kick a ball towards a goal, it is necessary to know the direction to that goal. However, the Sony robots used during the competition did not have any built-in localization mechanism. Relying on dead-reckoning in the legged robot for localization is completely unrealistic because of variability in the gait. Additionally, because of the limited field of view of the robot's camera, it was usually not possible to simultaneously keep the ball and the goal in view at the same time. Our solution to this problem was to have the robot track its position on the field and the direction to the goal using a Bayesian localization procedure (e.g., [2, 1]).

To compensate for the highly unreliable dead reckoning, the field environment for RoboCup-98 includes several fixed colored landmarks. These landmarks, along with the goal itself, were used to localize the robot on the field, calculate its orientation on the field and to estimate the direction to the center of the goal.

3.1 Position Localization

For position/angle localization, the field was divided into a $20 \times 30 \times 8$ (X, Y, θ) state grid. We recorded for each state the probability that the robot was in that state. This probability distribution was updated as observations were made about the world, and as the robot moved in the world.

A table of values was chosen because of some of the distributions we wish to represent do not have a nice parametric form. For instance, given a uniform prior distribution, the observation of the angle between two markers gives a high probability circle through the state space that is not representable by a Gaussian distribution.

Incorporation of observations is based upon Bayes' Rule:

$$P(S_i|O) = \frac{P(S_i)P(O|S_i)}{\sum_j P(S_j)P(O|S_j)}, \tag{3}$$

where $P(S_i)$ is the apriori probability that the robot is in state S_i, $P(S_i|O)$ is the posterior probability that the robot is in state S_i given that it has just seen observation O and $P(O|S_i)$ is the probability of observing O in state S_i.

Incorporation of movement is based upon a transition probability matrix. Given a previous movement M, for each state the algorithm computes the probability that the robot will end up in that state:

$$P(S_i|M) = \sum_j P(S_j)P(S_j \Rightarrow S_i|M), \tag{4}$$

where $P(S_j)$ is the apriori probability of state S_j and $P(S_j \Rightarrow S_i|M)$ is the probability of moving from state S_j to state S_i given the movement M. It is assumed that the transition probabilities, $P(S_j \Rightarrow S_i|M)$, take into account any noise in M.

For example, imagine the robot sees an angle of 90° between two markers, turns to the left and then sees an angle of 90° between two more markers. Initially it does not know where it is - our prior distribution is flat. After its first observation, the projection of the state probability matrix onto the X, Y plane would be as shown in Figure 2(a). During the turn, the projection spreads over the X, Y plane - representing the increased uncertainty introduced by the dead reckoning as the robot turns (see Figure 2(b)). The second observation finally localizes the robot (see Figure 2(c)).

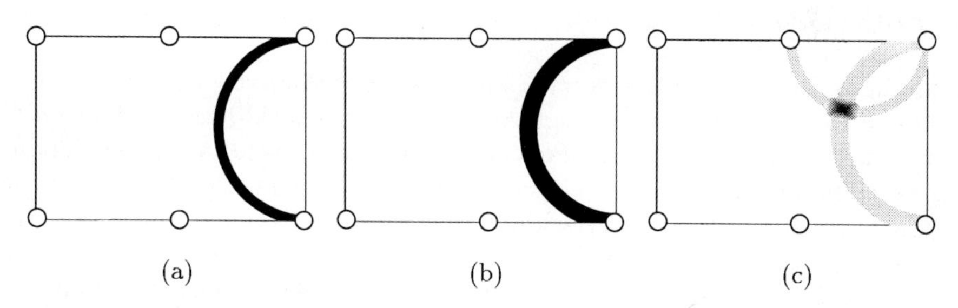

Fig. 2. The positioning probability: (a) after the 1st observation; (b) after a 90° turn; (c) after the 2nd observation.

In fact, there were two different types of observation used in our algorithm. The first was the angle between landmarks. These angles were pre-computed for each state. The second was the size of the landmarks in the robot's camera image. This size is linearly related to the inverse square of the distance to the marker. Once calibrated this was the most informative form of observation.

In order for this algorithm to be computationally tractable, one important approximation is needed. Probabilities that are close to zero are set to zero. Bayes' Rule never changes a zero probability, so these be checked for during updating and ignored. Once the robot is reasonably sure of its location most states have zero probability leading to a dramatic increase in speed.

3.2 Multiple Localization Levels

Having only 8 directions is not enough to accurately aim a ball. However due to both memory and time constraints, it is not possible to increase the angular resolution of the previous discretization. To increase the angular resolution, we introduced two more localization systems that ran in parallel with the one described above. All three used the same basic algorithm, but the number of states, the types of observations and the movement transition matrices are changed.

The second localization system tracked absolute angle on the field using a higher resolution (100 states) under the assumption that the robot was at the maximum likelihood X, Y location given by the first system. The third localization system tracked the angle of the goal. Again this was a high resolution angle and the system used the first two systems as well as vision and movement information.

3.3 When to Localize?

Our localization algorithm is passive in that it updates based on any information seen, without needing to control the robot to gather data. Unfortunately, sometimes this is not enough. When the maximum state probability is lower than a pre-defined threshold the localization becomes active. The robot searches for landmarks to improve its knowledge of its location. In our experiments, the robots localize themselves with high accuracy.

4 Role-Based Behaviors

Following up on our experience with our other RoboCup leagues, namely the small-size wheeled robots [5, 4], and the simulator league [3], we developed different behaviors based on positioning of the robots on the field. As of now, robots play two different roles, namely attacking and goal keeping.

The procedure used by an attacking robot consists of the following steps: (i) find the ball; (ii) localize attacking goal; (iii) position behind the ball, aligned with the goal; (iv) shoot. CMTrio-98 includes the simplest model for collaboration between the two attacking robots, namely the spatial division of the field. The field is divided in two slightly overlapping left and right longitudinal regions and the robots move in their own regions, as detected by their localization.

The procedure used by the goal keeping robot is a simplified version of a combination of the goal keeper and defender of the CMUnited-98 small-size team [4]. Its behavior consists of the following steps: (i) find the ball; (ii) remain close to the goal; (iii) move sideways aligned with ball; (iv) clear the ball when it gets close to it.

5 Discussion about Real World Conditions

All the algorithms described above were developed in our lab. While we had a complete mock up of the competition field, conditions at the RoboCup-98 site, namely La Cité des Sciences in Paris, were still significantly different from our laboratory conditions in a number of ways.

As we had anticipated the lighting conditions were significantly different between the lab and the competition field. With our automated color calibration we were able to effectively adjust for this. However, the program being automated caused one unforeseen difficulty. With some early versions of the vision hardware there are initialization problems that occasionally cause pictures taken by the pet to be shifted in color space until the vision is restarted. This shift, if undetected during the manual classification phase, causes incorrect parameters to be learned.

The second large change between the lab and competition conditions, which we anticipated but underestimated, was the effect of the audience. The robots, when placed on the field, need to look up at the localization markers which are on poles about the edge of the field. The background in these pictures is the audience - wearing whatever colored clothing they chose. This problem was exacerbated because we could not take calibration pictures with the audience - only with the empty stadium.

Finally, there was an unforeseen interaction between the penalty system and our localization routines. When a team member was given a penalty it was picked up by the referee and placed elsewhere on the field. Our Bayesian localization system did not expect to be teleported at any stage and so the robot lost track of its location when this happened. Luckily we had an auto-reset mechanism if our internal model of location was significantly different from our sensor information, but this was not meant to be relied upon as much as it was.

6 Conclusion

In this paper, we reported on our work using the Sony quadruped legged robots to play robotic soccer. We briefly described the components of Sony's legged robots. We then presented our vision-based navigation, Bayesian localization, role-based behaviors.

Our color calibration, vision-based navigation, and localization algorithms showed to be effective. Our robots were the only ones to complete the RoboCup-98 physical challenge, where a single robot had to push the ball into the goal starting from a configuration where the the ball was not lined up with the goal from the robot's view point. Our CMTrio-98 team was also victorious in the competition games, winning 2-1 against both the French and Japanese teams.

References

1. W. Burgard, A.B. Cremers, D. Fox, D. Haehnel, G. Lakemeyer, D. Schulz, W. Steiner, and S. Thrun. The interactive museum tour-guide robot. In *Proceedings of AAAI-98*, Madison, WI, July 1998.
2. Alberto Elfes. *Occupancy Grids: A Probabilistic Framework for Robot Perception and Navigation.* PhD thesis, Department of Electrical and Computer Engineering, Carnegie Mellon University, Pittsburgh, PA, 1989.
3. Peter Stone and Manuela Veloso. Task decomposition and dynamic role assignment for real-time strategic teamwork. In *Proceedings of the ATAL workshop (Agent Theories, Architectures and Languages)*, Paris, July 1998.
4. Manuela Veloso, Michael Bowling, Sorin Achim, Kwun Han, and Peter Stone. The CMUnited-98 champion small robot team. In Minoru Asada and Hiroaki Kitano, editors, *RoboCup-98: Robot Soccer World Cup II.* Springer Verlag, Berlin, 1999.
5. Manuela Veloso, Peter Stone, Kwun Han, and Sorin Achim. CMUnited: A team of robotic soccer agents collaborating in an adversarial environment. In Hiroaki Kitano, editor, *RoboCup-97: The First Robot World Cup Soccer Games and Conferences.* Springer Verlag, Berlin, 1998.

BabyTigers-98: Osaka Legged Robot Team

Noriaki Mitsunaga and Minoru Asada and Chizuko Mishima

Dept. of Adaptive Machine Systems, Osaka University, Suita, Osaka, 565-0871, Japan

Abstract. The Osaka Legged Robot Team, BabyTigers-98, attended the First Sony Legged Robot Competition and Demonstration which was held at La Cite La Villeta, a science and technology museum in Paris in conjunction with the second Robot Soccer World Cup, RoboCup-98, July, 1998. This article describes our approach to the competition. The main feature of our system is to apply a direct teaching method to behavior acquisition for four-legged robots as one of the robot learning issues. Since the robots recognize objects in the field by color information, the color calibration is another issue. To cope with color changes due to lighting conditions, we developed an interactive color calibration tool with graphic user interface. Based on the color calibration and the learning by teaching methods, the robot successfully exhibits basic skills such as ball approaching and kicking.

1 Introduction

The Osaka Legged Robot Team, BabyTigers-98, attended the First Sony Legged Robot Competition and Demonstration which was held at La Cite La Villeta, a science and technology museum in Paris in conjunction with the Second Robot Soccer World Cup Competition and Conference, RoboCup-98, July 2-9, 1998.

The final goal of the Legged Robot Project in our group is to establish the methodology to acquire behaviors for team cooperation in the RoboCup context from the interactions between the legged robots through multi sensor-motor coordinations. The desired behaviors can be categorized into three levels; a basic level, a basic cooperation level, and a higher team cooperation level. In this article, we briefly explain our first step for the first level skill acquisition with preliminary results. That is behavior acquisition by direct teaching.

The most fundamental feature of the legged robot is that it moves by its four legs (12 DOFs), which is quite different from conventional mobile robots (2 or 3 DOFs). From a viewpoint of sensor-motor learning and development, multi modal information and multi DOFs control should be established simultaneously, that is, affecting each other, the sensory information is abstracted and the multi joint motions are well coordinated at the same time [1]. However, it seems very difficult for artificial systems to develop the both together. Our goal is to design such a method.

From our experiences on robot learning, we realized that the number of trials by real robot is limited and a good trade-off between computer simulations and real robot experiences is essential for good performance. However, the computer simulation of the legged robots seems difficult to build, then we decided to adopt a direct teaching method in order to reduce the number of trials by real robot.

Since the robot detects objects in the field using color information such as a orange ball, aqua-blue and yellow goals, color calibration is another issue for the robot to robustly detect objects in the field and then to learn by teaching. Since color calibration during the game seems difficult to realize due to the limitation of on-board computation power, we adopt an off-line color calibration method. To make the calibration process much easier, we developed an interactive color calibration tool with graphic user interface.

This article is structured as follows. First, we show an interactive off-line color calibration system we developed. Next, we describe the method of robot learning by direct teaching. Finally, we show some experimental results and discuss the future issues.

2 Vision System

2.1 Color detection

In the legged league field [3], 7 colors (aqua-blue, yellow, orange, blue, red, pink, green) are used and robots need to detect all of them. The Sony's legged robot has the color detection facility in hardware that can handle up to 8 colors at frame rate (60-80[ms]). To detect colors with this facility, we need to specify each color in terms of subspace in *YUV* color space. *YUV* subspace is expressed in a table called Color Detection Table(CDT). In this table, Y are equally separated into 32 level and in each Y level we specify one rectangle $(u_{\min i}, v_{\min i})$,$(u_{\max i}, v_{\max i})$ $(i = 1, ..., 32)$.

The color observed by the robots dynamically changes due to the robot and/or object motions. The reason is that it depends on various kinds of parameters such as, the color of object itself, the color of light source, the surface material of object, the surface orientation of object, the angle between the view line of the camera and the beam of the light and so on. In order to cope with color changes, *YUV* subspaces should be large enough to involve the changes, but small not to include different colors. Then we developed an interactive tool so that we can make and modify CDTs that can handle expansion and reduction of the CDT easily.

Basic procedures are:

1. To make a CDT, we specify multiple pixels that should be treated as the same color in an image taken by the legged robot. Each pixel has a pair of (y, u, v) value or a point in *YUV* space. According to the value of y, each point is classified into 32 Y level. In each y level, $(u_{\min i}, v_{\min i})$,$(u_{\max i}, v_{\max i})$ $(i = 1, ..., 32)$ are set to indicate the bounding box that includes all points in that y level (see Fig.1(a)).
2. To add a point in *YUV* to the CDT which should be treated as the same color, expand the bounding box of the corresponding y level to include the point (see Fig.1(b)).

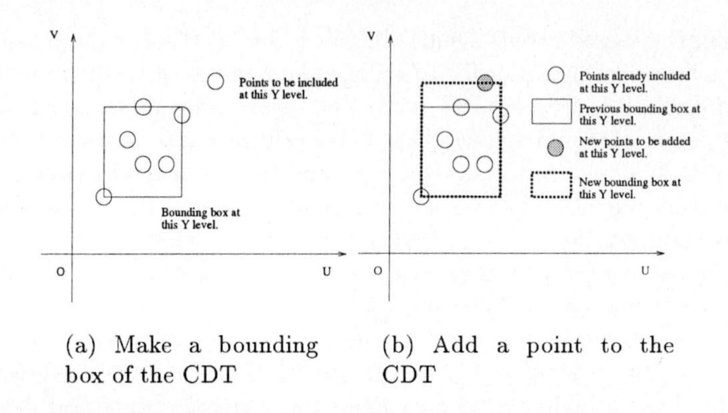

(a) Make a bounding box of the CDT

(b) Add a point to the CDT

Fig. 1. Making a CDT first time and expansion of a CDT

3. To delete a point in YUV from the CDT which should not be treated as the same color, shrink the bounding box of the corresponding y level to exclude the point. To exclude the point, we have four options (see Fig.2) and we take the one which is the least reduction of the area (in this case Fig.2(a)).

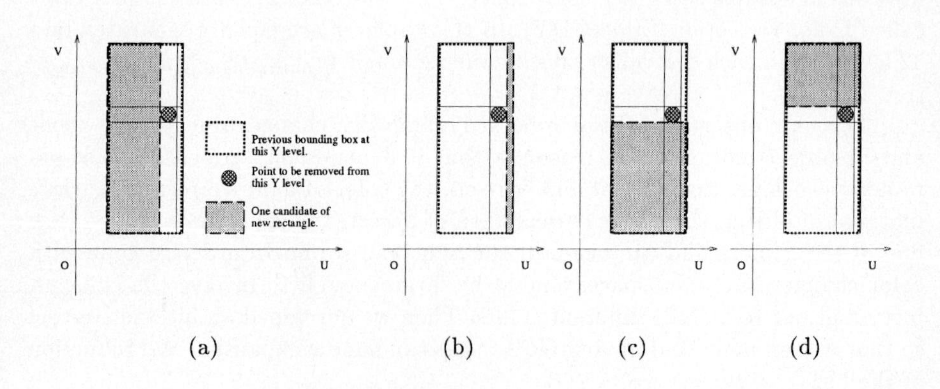

(a) (b) (c) (d)

Fig. 2. Deleting a point from CDT(four options)

We developed a GUI tool written in Tcl/TK to realize the above procedures easily. The appearance of the tool and emulated color detection is shown in Fig.3. We show an example of CDT construction with images shown in Fig.4, which shows ball images observed by the robot from different positions.

First we specify some pixels in Fig.4(a) and make a CDT. The distribution in

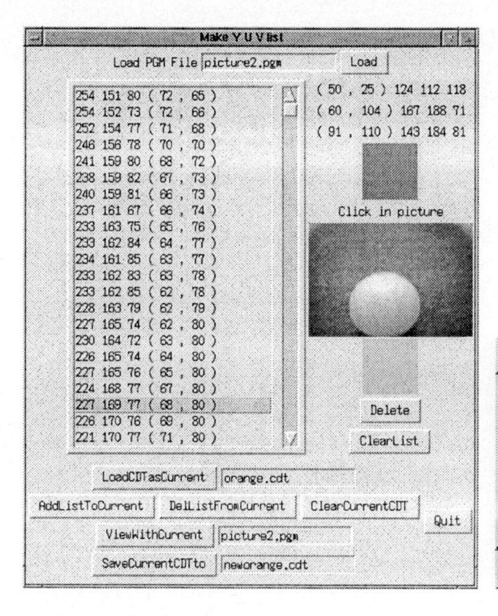

(a) making color detection table with the tool

(b) emulation of color detection

Fig. 3. GUI tool to make CDTs

YUV space projected to *UV* plane (at y level 16) is shown in Fig.5 (the bounding box is also shown). Distributions in *YUV* space of the ball color in the picture Fig.4(a) and (b) are shown in Fig.6. It indicates that the distribution in Fig.4(a) does not cover the one in Fig.4(b). That is, the CDT constructed by Fig.4(a) does not cover the color distribution of the ball in Fig.4(b) (see Figs.8(a) and (b)). Then, we specify some pixels that do not seem to be detected in Fig.4(b) and add to the CDT. Fig.7 shows the expansion of the bounding box in y level 16. After all of addition, the robot can detect Fig.4(b) as Fig.8(c).

2.2 Object detection

Since the legged league field includes multiple objects, some of which have the same color, and observed images are noisy, object detection is carried out with noise reduction based on a priori knowledge of the environment.

First, using the color detected image, pixels in the same color are connected into regions by their 8-neighbors. At the same time, area sizes, centroids and bounding boxes of the regions are calculated. Next, the small regions less than 4 pixels are deleted because the minimum size of the smallest object (ball) in the image plane is 4. Then, the order of object detection is determined, consid-

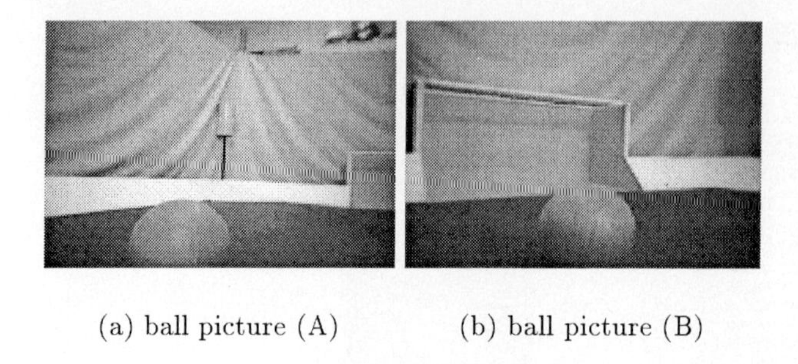

(a) ball picture (A) (b) ball picture (B)

Fig. 4. Pictures of a ball

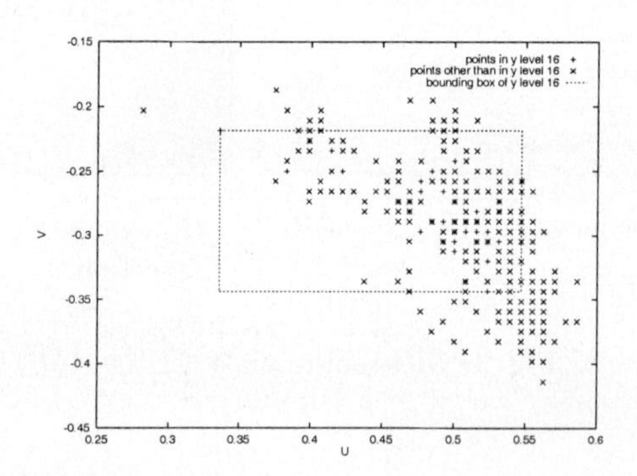

Fig. 5. Distribution of the ball color (A) in UV space

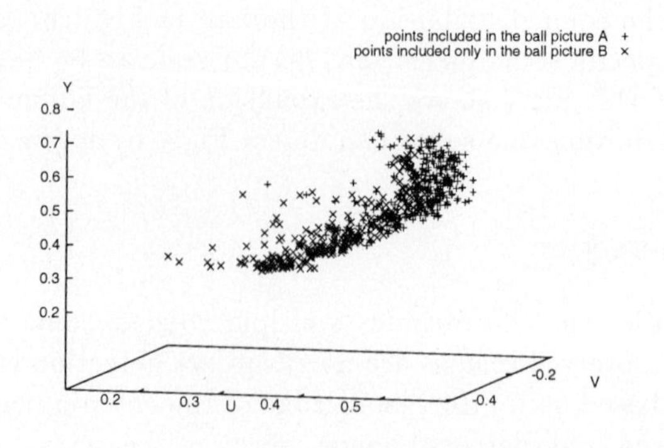

Fig. 6. Distribution of the ball color in YUV space

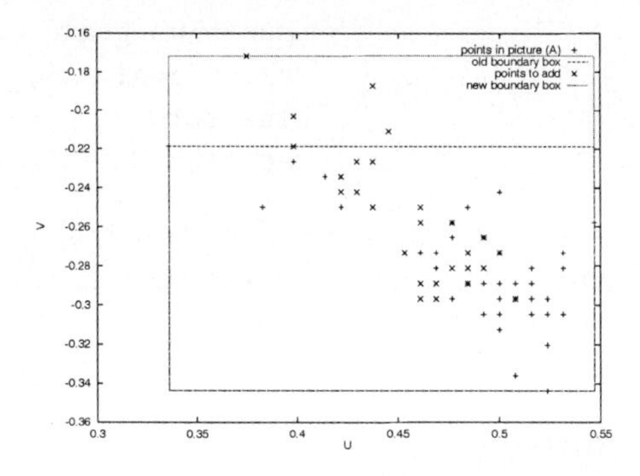

Fig. 7. Distribution of the ball color in UV space y level 16

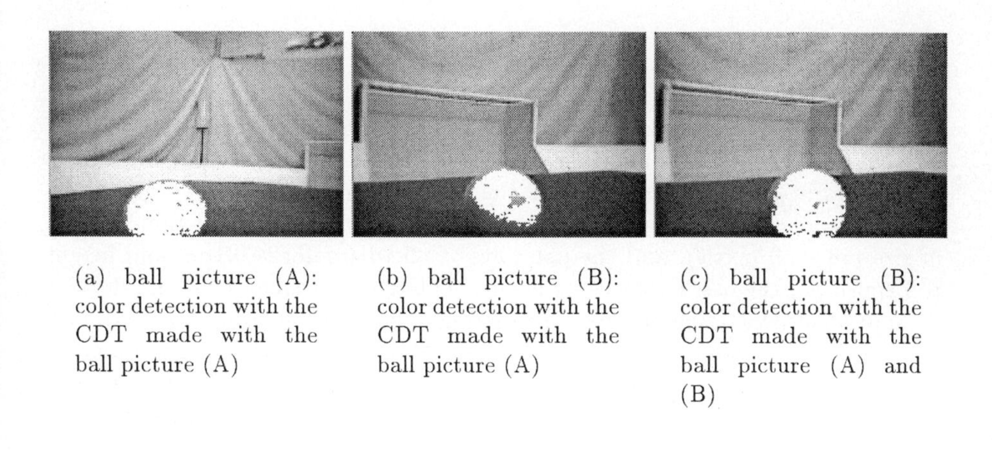

(a) ball picture (A): color detection with the CDT made with the ball picture (A)

(b) ball picture (B): color detection with the CDT made with the ball picture (A)

(c) ball picture (B): color detection with the CDT made with the ball picture (A) and (B)

Fig. 8. Result of color detection with CDTs (emulation by software)

ering the importance of object, easiness of color discrimination and visible constraints of multiple object colors. Constraints means that maximum number of objects that can be observed simultaneously is known, orange(1), aqua-blue(3), yellow(3), green(3), pink(3), blue(3) and red(3). This can also be used to noise reduction. The order we used is as follows.

1. ball
2. poles: (top color)pink (bottom color)aqua-blue

(top color)	(bottom color)
pink	aqua-blue
aqua-blue	pink
pink	yellow
green	pink
pink	green

3. goals: aqua-blue goal
   yellow goal
4. robots: blue robots
   red robots

The color of robot's body and head was blue or red, of which saturation was low and difficult to detect. That is why we put them in low priority in object detection.

3 Behavior Acquisition

In the competition, we assigned the roles of attacker and goal keeper to three robots, that is, two attackers and one goal keeper. In the following, we explain how the behaviors for the goal keeper and the attackers are obtained.

3.1 Goal-keeper

We adopted a strategy for the goal keeper to save the goal by positioning itself in front of the goal. The angles of the visible area is 53 degrees in width and 41 degrees in height. It means that the robot cannot see as many goals and landmark poles to determine its position. The result of color detection is small in size (88×59 pixels), and the poles are small (diameter is 0.1[m] and height is 0.2[m]). So, the robot rotates its head to enlarge the visible field, and decides its movement depending on rough estimation of its position. Since the landmark poles near the opponent goal is difficult for it to observe, the robot compares the angles between its own goal and four poles near own goal in front of its own goal, and determines which direction to move. Further, the robot tracks the ball for a while when it is near the ball.

3.2 Attacker

The behavior of two attackers is obtained by direct teaching. More precisely, the robot learns its behavior by classifying the data taught by a human trainer from a viewpoint of information theory. A human trainer can teach the robot via serial console with PC.

First, the robot collects test data, pairs of an action command given by human trainer and the sensory information during the action execution in every 300ms. Next, C4.5 [2] is applied to the test data to extract rule sets. Then, the validity of the rule sets are checked against test data by applying the rule sets. Specifications of the data for shooting skill acquisition are as follows:

- **action command** forward, backward, left-shift, right-shift, left-rotation, and right rotation (20 degs/sec): these abstracted action commands will be decomposed into more primitive motor commands in future.

- **sensory information** head direction (rad.) and image features of both the ball and the goal in the observed 88x59 image: area (pixels), position (x-y coordinates), bounding rectangle (x-y coordinates of corners), height, and width. See Figure 9.
- **training position and sampling rate:** initial positions of direct teaching by serial line connection are evenly distributed in the field heading the goal, and the sampling rate is 300ms. One trial from the initial position to the goal takes about 10 seconds.

In the experiments, 740 pairs of an action command and sensory information for training data to make rule sets, and 500 pairs for test data to check the validity of the rule sets are given. The both data are obtained in the same manner starting from the similar initial positions, but individual pairs are different from each other trial by trial. The number of rules obtained is about 30, and typical ones for forward(F), left-rotation (LR), and right-shift (RS) are shown below. Figure 9 shows the typical situation of right-shift motion.

```
Forward: BallArea>56   Left-      BallArea>49    Right- BallArea<=40
         HeadDir>-.14 Rotation: HeadDir>.52    Shift: BallYcen>8
         HeadDir<=.37             HeadDir<=1.10         BallXmin>3
         GoalXmin<=11             BallWidth<=11         HeadDir<=-.24
                                                        GoalArea<=665
                                                        GoalXmax<=64
```

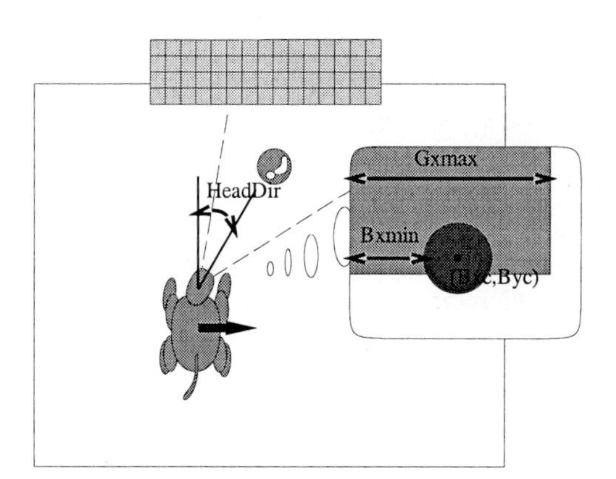

Fig. 9. Typical situation to take Right-Shift motion

The following table indicates a confusion matrix showing where the miss-classification of the training cases occur. Due to the wrong teaching by a human trainer (actually, she has not got used to teaching), this matrix includes many miss-classifications, especially LEFT-TURN does not have any correct classifications. We are planning to skill up teaching and also to use more training data to construct robust classification. Generalization is one of the big issue of direct teaching, and EBL seems one alternative to solve the problem. These are under investigation.

classified as:	LEFT	LEFT-TURN	RIGHT	RIGHT-TURN	FORWARD	BACK
	111	13	3		4	
	14			2	9	
			8	16	9	
			20	151	34	
	1		3	13	91	

4 Conclusion

This paper has described the implementation of vision system and strategies used in BabyTigers-98. Although it was not the best, the resultant behaviors were encouraging. The followings are future issues.

1. The current color calibration system includes an off-line interactive process, therefore, it cannot cope with dynamic changes in color during the match. Full autonomous on-line color calibration system should be developed.
2. Direct teaching seems useful under the condition that the robot sensation can be informed to the human trainer. However, it does not generally true, and often perceptual aliasing happens. As a short range research issue, the robot should reconstruct its state space so that the instructions given by the trainer can be consistent and effective. As a longer one, both the robot and the trainer should learn each other, that is, the robot build the trainers model and vice versa. Through the model developing process, the robot learns how to behave according to the instructions, and the trainer learns how to teach it.
3. Currently, only the basic skills have been obtained and not the cooperative ones yet. Direct teaching should be useful for such behavior acquisition. Mutual learning (co-evolution) process is indispensable to obtain the cooperative and competitive behavior in dynamic multi-agent environment.

References

1. Asada, M., "An Agent and an Environment: A View on "Having Bodies" - A Case Study on Behavior Learning for Vision-Based Mobile Robot," *Proceedings of 1996 IROS Workshop on Towards Real Autonomy*, pp.19-24, 1996.
2. Quinlan, J. R., "C4.5: Programs for Machine Learning," *The Morgan Kaufmann Series in Machine Learning*, 1993.
3. Veloso, M., Uther, W., Fujita, M., Asada, M. and Kitano, H., "Playing Soccer with Legged Robot," *Proceedings of the 1997 IEEE/RSJ International Conference on Intelligent Robots and Systems*, Vol. 1, pp. 437–442, 1998.

Author Index

Lecture Notes in Artificial Intelligence (LNAI)

Lecture Notes in Computer Science

Vol. 1614: D.P. Huijsmans, A.W.M. Smeulders (Eds.), Visual Information and Information Systems. Proceedings, 1999. XVII, 827 pages. 1999.

Vol. 1615: C. Polychronopoulos, K. Joe, A. Fukuda, S. Tomita (Eds.), High Performance Computing. Proceedings, 1999. XIV, 408 pages. 1999.

Vol. 1616: P. Cointe (Ed.), Meta-Level Architectures and Reflection. Proceedings, 1999. XI, 273 pages. 1999.

Vol. 1617: N.V. Murray (Ed.), Automated Reasoning with Analytic Tableaux and Related Methods. Proceedings, 1999. X, 325 pages. 1999. (Subseries LNAI).

Vol. 1618: J. Bézivin, P.-A. Muller (Eds.), The Unified Modeling Language. Proceedings, 1998. IX, 443 pages. 1999.

Vol. 1619: M.T. Goodrich, C.C. McGeoch (Eds.), Algorithm Engineering and Experimentation. Proceedings, 1999. VIII, 349 pages. 1999.

Vol. 1620: W. Horn, Y. Shahar, G. Lindberg, S. Andreassen, J. Wyatt (Eds.), Artificial Intelligence in Medicine. Proceedings, 1999. XIII, 454 pages. 1999. (Subseries LNAI).

Vol. 1621: D. Fensel, R. Studer (Eds.), Knowledge Acquisition Modeling and Management. Proceedings, 1999. XI, 404 pages. 1999. (Subseries LNAI).

Vol. 1622: M. González Harbour, J.A. de la Puente (Eds.), Reliable Software Technologies – Ada-Europe'99. Proceedings, 1999. XIII, 451 pages. 1999.

Vol. 1625: B. Reusch (Ed.), Computational Intelligence. Proceedings, 1999. XIV, 710 pages. 1999.

Vol. 1626: M. Jarke, A. Oberweis (Eds.), Advanced Information Systems Engineering. Proceedings, 1999. XIV, 478 pages. 1999.

Vol. 1627: T. Asano, H. Imai, D.T. Lee, S.-i. Nakano, T. Tokuyama (Eds.), Computing and Combinatorics. Proceedings, 1999. XIV, 494 pages. 1999.

Col. 1628: R. Guerraoui (Ed.), ECOOP'99 - Object-Oriented Programming. Proceedings, 1999. XIII, 529 pages. 1999.

Vol. 1629: H. Leopold, N. García (Eds.), Multimedia Applications, Services and Techniques - ECMAST'99. Proceedings, 1999. XV, 574 pages. 1999.

Vol. 1631: P. Narendran, M. Rusinowitch (Eds.), Rewriting Techniques and Applications. Proceedings, 1999. XI, 397 pages. 1999.

Vol. 1632: H. Ganzinger (Ed.), Automated Deduction – Cade-16. Proceedings, 1999. XIV, 429 pages. 1999. (Subseries LNAI).

Vol. 1633: N. Halbwachs, D. Peled (Eds.), Computer Aided Verification. Proceedings, 1999. XII, 506 pages. 1999.

Vol. 1634: S. Džeroski, P. Flach (Eds.), Inductive Logic Programming. Proceedings, 1999. VIII, 303 pages. 1999. (Subseries LNAI).

Vol. 1636: L. Knudsen (Ed.), Fast Software Encryption. Proceedings, 1999. VIII, 317 pages. 1999.

Vol. 1638: A. Hunter, S. Parsons (Eds.), Symbolic and Quantitative Approaches to Reasoning and Uncertainty. Proceedings, 1999. IX, 397 pages. 1999. (Subseries LNAI).

Vol. 1639: S. Donatelli, J. Kleijn (Eds.), Application and Theory of Petri Nets 1999. Proceedings, 1999. VIII, 425 pages. 1999.

Vol. 1640: W. Tepfenhart, W. Cyre (Eds.), Conceptual Structures: Standards and Practices. Proceedings, 1999. XII, 515 pages. 1999. (Subseries LNAI).

Vol. 1642: D.J. Hand, J.N. Kok, M.R. Berthold (Eds.), Advances in Intelligent Data Analysis. Proceedings, 1999. XII, 538 pages. 1999.

Vol. 1643: J. Nešetřil (Ed.), Algorithms – ESA '99. Proceedings, 1999. XII, 552 pages. 1999.

Vol. 1644: J. Wiedermann, P. van Emde Boas, M. Nielsen (Eds.), Automata, Languages, and Programming. Proceedings, 1999. XIV, 720 pages. 1999.

Vol. 1645: M. Crochemore, M. Paterson (Eds.), Combinatorial Pattern Matching. Proceedings, 1999. VIII, 295 pages. 1999.

Vol. 1647: F.J. Garijo, M. Boman (Eds.), Multi-Agent System Engineering. Proceedings, 1999. X, 233 pages. 1999. (Subseries LNAI).

Vol. 1649: R.Y. Pinter, S. Tsur (Eds.), Next Generation Information Technologies and Systems. Proceedings, 1999. IX, 327 pages. 1999.

Vol. 1650: K.-D. Althoff, R. Bergmann, L.K. Branting (Eds.), Case-Based Reasoning Research and Development. Proceedings, 1999. XII, 598 pages. 1999. (Subseries LNAI).

Vol. 1651: R.H. Güting, D. Papadias, F. Lochovsky (Eds.), Advances in Spatial Databases. Proceedings, 1999. XI, 371 pages. 1999.

Vol. 1652: M. Klusch, O.M. Shehory, G. Weiss (Eds.), Cooperative Information Agents III. Proceedings, 1999. XI, 404 pages. 1999. (Subseries LNAI).

Vol. 1653: S. Covaci (Ed.), Active Networks. Proceedings, 1999. XIII, 346 pages. 1999.

Vol. 1654: E.R. Hancock, M. Pelillo (Eds.), Energy Minimization Methods in Computer Vision and Pattern Recognition. Proceedings, 1999. IX, 331 pages. 1999.

Vol. 1663: F. Dehne, A. Gupta. J.-R. Sack, R. Tamassia (Eds.), Algorithms and Data Structures. Proceedings, 1999. X, 367 pages. 1999.